Handbook of Transmission Electron Microscopy

Handbook of Transmission Electron Microscopy

Edited by **Lisa Page**

NY RESEARCH PRESS
New York

Published by NY Research Press,
23 West, 55th Street, Suite 816,
New York, NY 10019, USA
www.nyresearchpress.com

Handbook of Transmission Electron Microscopy
Edited by Lisa Page

International Standard Book Number: 978-1-63238-283-2 (Hardback)

Printed in the United States of America.

Contents

Preface

The purpose of the book is to provide a glimpse into the dynamics and to present opinions and studies of some of the scientists engaged in the development of new ideas in the field from very different standpoints. This book will prove useful to students and researchers owing to its high content quality.

This book compiles various research papers written by experts and scientists. It is a comprehensive summary of the various features of Transmission Electron Microscopy (TEM), from fundamental mechanisms and diagnosis to the recent developments in this area. This book discusses electron microscopy, new techniques for projection of images, experimental techniques for the study of nanomaterials, among others. The book also elucidates theoretical and practical aspects of modern microscopy techniques and the usage of TEM in material characterization. It will be beneficial for students, scientists, engineers and researchers working in this field.

At the end, I would like to appreciate all the efforts made by the authors in completing their chapters professionally. I express my deepest gratitude to all of them for contributing to this book by sharing their valuable works. A special thanks to my family and friends for their constant support in this journey.

Editor

1

Indexing of Electron Diffraction Patterns of Icosahedral and Decagonal Phases

Rajiv Kumar Mandal
Department of Metallurgical Engineering, Centre of Advanced Study, Institute of Technology, Banaras Hindu University, Varanasi, India

1. Introduction

The atomic arrangements of solids fall into two broad categories. First refers to long range translational order giving rise to sharp diffraction patterns. Second relates to an atomic order that displays diffuse halos. Prior to the discovery of Quasiperiodic translational order in rapidly solidified Al-Mn alloys [1], sharp diffraction peaks were considered to be synonymous to possession of periodic translational orders in solids. The electron diffraction patterns reported in reference [1] were found to be invariant under icosahedral point group ($m\bar{3}\bar{5}$) symmetry. Any elementary text on crystallography begins by showing that 5-fold symmetry is incompatible with periodic lattice translations. Thus, it was proved beyond doubt that Shetchman et al.[1] have discovered a new state of order in solids. Readers may go through the notes given in annexure **A** for the excitement and importance of this discovery. The underlying atomic arrangement [1] was believed to possess "Quasiperiodic" translational order and due to invariance of diffraction patterns under icosahedral point group, such a class of solids was later termed as icosahedral quasicrystals (IQC). If one observes the location of diffracted spots in this class of solids then it is not periodic but all of them are as sharp as any crystalline diffraction patterns. These observations clearly established the fact that IQC displays new type of long range translational order known as Quasiperiodic translational order. We refer the readers to the annotations and reprints of papers in reference [2] for getting familiar with all the terminologies in this area. Having recorded diffraction patterns, the first step is to index them. For indexing, we need a set of basis vectors that are integrally independent. All three dimensionally periodic solids need three integrally independent basis vectors to index their diffraction patterns. The minimum number of integrally independent basis vectors is known as rank of any solids possessing long range translational orders [3-4]. Thus, all 3d periodic solids have rank equals to three. Remember, one uses four basis vectors for hexagonal crystals in Miller-Bravais (MB) scheme but all of them are not independent. As a consequence of this, assignment of indices to a diffracted spot in MB scheme is made unique by invoking condition that sum of indices along the three planar basis vectors is zero. We shall deliberate on this aspect further while discussing indexing of decagonal quasicrystals (DQC).

All Quasiperiodic structures possess rank greater than three. This is true even for incommensurate structures. However, quasicrystals diffraction patterns are invariant under non-crystallographic point groups that are incompatible with periodic lattice translations. We do not propose to discuss this aspect further and interested readers may refer to international tables on crystallography [5] for subtle distinction between these phases. IQC are three dimensional (3d) quasicrystals whose diffraction patterns are invariant under icosahedral point group. For the purpose of indexing, one requires six integrally independent basis vectors parallel to the vertex vectors of an icosahedron. Thus, 3d IQCs have rank equals to three. We shall demonstrate that all the diffraction spots can be indexed with the help of sextuplets of indices. For spanning the entire reciprocal space one has to follow the rules of addition and subtraction. This is true for all reciprocal lattices. In addition to this, Quasiperiodic reciprocal lattices require inflation and deflation of vectors. We shall show this as we progress. The Quasiperiodic lattice constant is determined with the help of first strongest diffraction spots along the basis vectors. Owing to quasiperiodicity, Quasiperiodic lattice constant is scale dependent. The concepts of inflation/deflation and scale dependent lattice constant are reflection of the fact that we are dealing with a new state of order. In fact both these concepts are related to golden mean $\tau = \frac{\sqrt{5}+1}{2}$ which is one of the roots of the equation $x^2 - x - 1 = 0$. Please note that the golden mean has many interesting properties like $\tau^2 = \tau + 1, \left(\frac{1}{\tau}\right) = \tau - 1, \tau^3 = 2\tau + 1$ *and so on.* IQC phases are related to such a number. This is due to the presence of 5-fold rotational symmetry. There are three 6d Bravais lattices for IQC and they are expected to give rise to three distinct icosahedral phases. These are (i) Simple IQC (SI), (ii) Body centered IQC (BCI) and (iii) Face centered IQC (FCI). Both SI and FCI phases are reported in literature but we shall restrict our discussion on SI only. The need of 6d cubic lattice in the structural description of IQC has been found to be essential and convenient in higher dimensional crystallography. It is therefore essential that we get familiarize with the concept and methodology related to higher dimensional crystallography. We shall demonstrate all the necessary aspects pertaining to this as we proceed.

Following the discovery of IQC, a new class of quasicrystals was found. This class of quasicrystals has 2d quasiperiodicity and 1d periodicity. Four popularly known DQCs have 2,4,6 and 8 layer periodicities along 10-fold axis and are designated as T_2, T_4, T_6 and T_8 phases respectively. The diffraction pattern displayed the presence of a unique 10-fold symmetry axis. Hence, this class of 2d QC was christened as decagonal quasicrystals. The discovery of decagonal phases [6-7] posed novel problems to their structural description and indexing. The rank of such a solid is five. Four basis vectors oriented with respect to each other by 72^0 and the fifth one perpendicular to this plane are sufficient to map the entire reciprocal space. However, a set of vectors related by five -fold symmetry may not possess quintuplet of indices that are permuted. These aspects are dealt extensively in references [8-9]. To attain permuted indices to refer to a set of vectors related by 5-/10-fold symmetries one uses six basis vectors [10]. The additional vector required to preserve symmetry in the indices for a family of directions/planes during indexing gives rise to the problem of redundancy. This refers to the non-unique assignment of indices to a diffraction spot. This can be surmounted by putting condition on indices akin to those of MB scheme for hexagonal crystal. We have recently discussed many novel aspects pertaining to this [9] by revisiting the MB indexing schemes of hexagonal [10-11], decagonal [8] and their related phases with the help of higher dimensional approach. For indexing diffraction patterns, one has to follow the approach of Copernican crystallography developed in Fourier space [3, 4].

In contrast, the canonical cut and project scheme [12-17] is capable of providing information about the atomic positions as well as the intensity of Quasiperiodic structures. We will not be deliberating on these in this chapter excepting succinctly dealing with the concepts of higher dimensional crystallography in the next section. Higher dimensional crystallography utilizes a mathematical construct for structural modeling of Quasiperiodic phases in a convenient way.

2. Elements of quasicrystallography

We need to remember that diffraction patterns offer the best way to define a lattice. We may understand this in the following way. If we ignore the intensity variation and replace all the diffracted spots by a point (for example, selected area electron diffraction pattern from a single grain) then we get arrangement of points in 2d section of a 3d reciprocal lattice. The prefix reciprocal before lattice appears because the distances are measured in terms of $Å^{-1}$. For describing atomic arrangements of solids one has to express distances in terms of Å and the underlying lattice is known as direct lattice. A lattice has a set of basis vectors whose integral linear combination helps us reach any diffracted spot with respect to the transmitted beam. Similar concept will hold for direct primitive lattice. The minimum number of integrally independent basis vectors required to accomplish this task is known as the rank of solids. Please note that the way we have introduced lattice does not demand anything other than the existence of diffraction patterns. Latter is a hard verifiable experimental fact. Selected area diffraction is the most convenient way to record single crystal like diffraction patterns from a polycrystalline specimen. Developing crystallography with the help of reciprocal space was central philosophy advocated earlier [18]. This concept was emphasized and extended further by N.D.Mermin (1992) after the discoveries of IQC, DQC and many other related phases [3-4]. If we observe three dimensional periodicities in diffraction patterns then we are dealing with the crystals of rank 3. The observation of incommensurate structures in 1977 [12] and their interpretation demanded four basis vectors instead of three. Incommensurate structures display crystallographic point group symmetry but their diffraction patterns possess aperiodicity in one of the three directions. In contrast, quasicrystals discovered by Shetchman et al. in 1984 displayed icosahedral point group symmetry [1] and aperiodicity of a special type. We call this as quasiperiodicity. In general, the existence of aperiodic order in a direction, demands at least two length scales that are relatively incommensurate. This is said in other words means the ratio of two fundamental lengths cannot be expressed as a rational number. We may understand this in a mathematical way as given below.

Let us consider a point on the line whose distance from an arbitrary origin is X_n. We define this by

$$X_n = a\left(n + (1/\tau)\lfloor (n/\tau) \rfloor\right) \qquad (1)$$

Where the floor function $\lfloor x \rfloor$ of any number x is defined by the number x minus the fractional part of x. This means that floor function is basically the lowest possible integer of that number. As a consequence of this we realize that the two terms in the above equation are periodic independently with a period of a and (a/τ) respectively. H.Bohr has shown that equation (1) generates points that are almost periodic or Quasiperiodic [19-20]. The floor function also ensures that the two consecutive points are not coming arbitrary close to each other. This is important for a real solid. There has to be a minimum distance of separation between two atoms.

Thus, we may write a general expression having aperiodicity as

$$X_{m,n} = a\lfloor m + n\tau \rfloor \tag{2}$$

Where, m and n are integers.

A simple calculation of X_n based on equation (1) generates points on a line that is well known Fibonacci series for τ = 1.618…….(golden mean). This is a quadratic irrational. As mentioned earlier, this is one of the two roots of equation $x^2 - x - 1 = 0$. Golden mean τ has a continued fraction representation: τ = 1+1/1+…..=1/1;2/1;3/2;5/3;8/5;13/8;21/13 and so on. These are the successive approximants of τ. One way to generate periodic structures of varying periods is by replacing τ by one of the above rational values. Such structures are known as rational approximants. We proceed to give alternative discussion of equations (1) and (2) in terms of higher dimensional crystallography.

Let us consider a set of points on one dimension of the type that we have been discussing so far. If we have to index each of the points on the lattice generated by equation (1) then we will do it conveniently by specifying two integers (m,n). Thus the rank of such a lattice is 2. Please note that the points are lying on a line. If this happens for a three dimensional solids in all the three directions then we will require 6 indices to indicate a point for such cases. Thus, the rank of such a solid will be 6. This is the case for IQC phase. Let us translate this discussion in terms of basis vectors. We shall first do it for the one dimensional Quasiperiodic case.

Let us consider two vectors V_1 and V_2 along a direction with unit vector x^{par} such that $V_1 = \cos\theta\, x^{par}$ and $V_2 = \sin\theta\, x^{par}$. Any vector on the line terminating at a lattice point is given by:

$$R = a(m_1\ V_1 + m_2\ V_2) \quad \text{and} \quad G = (1/a)(n_1\ V_1 + n_2\ V_2) \tag{3}$$

Where R and G are direct and reciprocal space lattice vectors; m_1; m_2 and n_1; n_2 are set of integers and a is spacing in Å. For direct space there has to be a minimum separation between two points as every lattice point is a probable location of atom. For reciprocal space this is not at all essential. If we choose $(\cos\theta / \sin\theta) = \tau$, then we have

$$R = a\sin\theta(m_1\tau + m_2) \quad \text{and} \quad G = (1/a)Sin\theta(n_1\tau + n_2) \tag{4}$$

The form of set of equations now resembles with those of equations (1) and (2). If we choose integers that are a combination of positive and negative both then there is an important difference between periodic solids and aperiodic ones. The set of points generated on the line is uniformly and densely filled. For R, this is obviously inconsistent. We commented earlier about the minimum distance of separation between two points. We also know from the experimental observations of IQC that in reciprocal space, we do observe sharp Bragg peaks at discrete locations just like periodic crystals. Let us attempt to understand this through the language of higher dimensional crystallography. We construct a projection matrix (P) through the dot products of the basis vectors.

$$P=\{P_{ij}\}=\{V_{ij}\}=\{V_i.V_j\} \text{ where } P_{11}=\cos^2\theta;\ P_{22}=\sin^2\theta; P_{12}=\cos\theta\sin\theta$$

$$P=\begin{bmatrix}\cos^2\theta & \cos\theta\sin\theta \\ \cos\theta\sin\theta & \sin^2\theta\end{bmatrix} \tag{5}$$

Equation (5) satisfies $P^2=P$. This is an important property of a projection matrix. Please note that the determinant of this matrix is zero and the trace is equal to 1. We can also define a matrix Q by the following relation:

$$Q=I-P=\begin{bmatrix}\sin^2\theta & -\sin\theta\cos\theta \\ -\sin\theta\cos\theta & \cos^2\theta\end{bmatrix} \tag{6}$$

Where I is an identity matrix of order 2. Please see that $Q^2=Q$ and $PQ=0$. All other properties of P are displayed by Q also. The space generated by this matrix is orthogonal to that of P.

If we choose another set of basis vectors W_1 and W_2 then we are in a position to define these on a line such that they are anti-parallel or at 180° with each other. Please note that in parallel space V_1 and V_2 are parallel. From the learning of matrix representation theory, we call the two spaces to be orthogonal. In terms of a unit vector x^{perp} we may write $W_1=\sin\theta\ x^{perp}$ and $W_2=-\cos\theta\ x^{perp}$.

We may define a orthonormal basis such that $V_i+W_i=e_i$ where i =1 to 2 and $e_ie_j=\delta_{ij}$ with $\delta_{ij}=1$ for i=j and $\delta_{ij}=0$ for $i\neq j$. The two basis vectors e_1 and e_2 are orthonormal and clearly represent the bases for a two dimensional space. With the help of two matrices given in equations (5 and 6), we can also write $V_i = P_{ij}\, e_j$ and $W_i = Q_{ij}\, e_j$ for i, j to vary from 1 to 2 only. The two dimensional direct lattice vector R^2 and reciprocal lattice vectors G^2 can now be written as:

$$R^2 = a\,(m_1e_1+m_2e_2) \text{ and } G^2=(1/a)(n_1e_1+n_2e_2) \tag{7}$$

Where a can be identified as two dimensional lattice parameter of a square lattice. Let us designate now R and G as the components of Equation (7) in "V" space or "physical space" or "par space" and S and H as the components of Equation (7) in "W" space or "pseudo space" or "complementary space" or "perp space". All these terminologies are used in Quasicrystalline literature and we need not get frightened by them. They are given by

$$R = a\,(m_1V_1+m_2V_2) \quad\text{and}\quad G=(1/a)(n_1V_1+n_2V_2) \tag{8}$$

$$S = a\,(m_1W_1+m_2W_2) \quad\text{and}\quad H=(1/a)(n_1W_1+n_2W_2) \tag{9}$$

We get after substituting for V_i and W_i the following expressions:

$$R = a\sin\theta\,(m_1\tau+m_2)\,x^{par} \tag{10a}$$

$$G = (1/a)\sin\theta\ (n_1\,\tau + n_2)\ X^{par} \tag{10b}$$

$$S = a\sin\theta\ (m_1 - m_2\tau)\ x^{perp} \tag{11a}$$

$$H = (1/a)\sin\theta\ (n_1 - n_2\tau)\ x^{perp} \tag{11b}$$

We observe that the terms within the parentheses of equations (10a) and (10b) are of the same form as we have written earlier in equations (1) and (2). Are we getting any extra information by following the path that we have adopted in the latter part of the discussion? Yes, we can ensure minimum separation between two points in the direct lattice if we put a condition on the set of indices m_1 and m_2 with the help of equation (11a). Please recall that the floor function in equation (1) was helping us accomplish this task. The matrix formulation presented here can be generalized without any difficulty for two and three dimensional aperiodic structures or Quasiperiodic structures for our discussion.

As mentioned above, we demonstrate the method of ensuring minimum distance of separation between two points in parallel or physical direct space. We have a square 2d lattice (we call them as hyperlattice) having basis vectors e_1 and e_2 with lattice parameter a. We attach a line element to each of the lattice points. The length of the line element is L^{perp}*(say)*. Please note that (L^{perp}/a^2) is the linear density of points. We choose this based on the density of the points required for the purpose. This can be arrived at by referring to the density of the points needed for the related approximant structures. As stated earlier, structures that are generated by substituting τ by its successive approximants are termed as rational approximant structures. The α-Al-Mn(Fe)-Si cubic structure with space group Pm3 having lattice parameter of ~12.68Å and 138 atoms in the unit cell is one such example. This is described as (1/1) approximant of icosahedral structures in Al-Mn system. Please see annotations in reference [2] and also consult [21] for further details.

In cut and project scheme, the length of such a line element is taken as L^{perp} = a (sinθ + cosθ). This is the external boundary of the shape in W-space obtained by projecting all the vertices of the higher dimensional unit cell on to it. We place it on the lattice point such that it is parallel to *perp.space* and is symmetric around it. This means the line element is extending both sides from +($L^{perp}/2$) *to* –($L^{perp}/2$). We may also work with the asymmetric setting by placing the line element having its extent from 0 to L^{perp}. The physical space line will cut this line element selectively and minimum separation between the two consecutive points can be ensured [14-16]. Now we would like to understand the *Fourier Transform (F.T.)* of the structure that we have generated. It is a combined effect of two distinct entities: the 2d square hyperlattice and the line element L^{perp} that is serving as motif in crystallographic parlance. Hence the *F.T.* of the 1d quasicrystals in physical space will be given by the convolution of the two functions. The *F.T.* of the 2d lattice will be a delta function and the weight of the delta function will be modulated by the *F.T.* of the line element. We therefore write:

F.T.of the 1d Quasicrystals = F.T. of the 2d lattice × F.T. of the line element

$$F.T.(G) = 1 \times (1/L^{perp}) \int_{-(L^{perp}/2)}^{(L^{perp}/2)} Exp(-2\pi i H.S)\,dS \tag{12}$$

This reduces to

$$F.T.(G) = (\sin \pi H L^{perp}) / (\pi H L^{perp}) \quad (13)$$

For a given G, there is an unique H and are given by indices of reflection

$(n_1\ n_2)$. The intensity is given by the following equation:

$$\mathrm{I(G)} = \mathrm{F.T.(G)} \times \mathrm{F.T.(G)}^* \quad (14)$$

Where $F.T.(G)^*$ is the complex conjugate of $F.T.(G)$. Please note that G and H are implicitly and uniquely related. The right hand side of the equation (13) will try to attain maximum value for those G for which H is close to zero. In experimental condition, the instrument has a cut off limit below which it will merge with the background. Hence most of the intensities may not be significant. The observed G with reasonable intensity shall correspond to only those G's for which n_1/n_2 is nearing to τ. This is the reason why we observe sharp Bragg peaks at discrete locations in physical or parallel reciprocal space under experimental condition. Having explained the philosophy for 1d case, we come back to the indexing of diffraction patterns again-the primary aim of this paper. We reproduce the two expressions again:

$$G = (1/a)\sin\theta\ (n_1 \tau + n_2)\ x^{par} \text{ and } H = (1/a)\sin\theta\ (n_1 - n_2\tau)\ x^{perp}$$

Please note that $(n_1 - n_2\tau) \rightarrow 0$ *when* $(n_1/n_2) \rightarrow \tau$. Thus all the diffracted vectors having significant observable intensities will be indexed as *(1,1);(2,1);(3,2);(5,3)* and so on. If we follow similar arguments for 3d or 2d Quasiperiodic solids, we may conclude that icosahedrally related structures will display strong intensity spots separated by τ or its power dependent on the symmetric direction(s). All those solids having τ or its successive approximants playing a role in the diffraction patterns are important for icosahedrally related structures. We advise readers to consult books and reviews available in this area through various scientific databases for further details on this aspect.

3. Icosahedral quasicrystals

As mentioned above, we have 3d quasiperiodicity for IQC and we need 6 integrally independent basis vectors to index the diffraction patterns. The rank of this class of solids is six. Recall that icosahedron is one of the five platonic solids having 20 equilateral triangles giving it a convex shape. It has 6 (or 12) five-, 10 (or 20) three- and 15 (or 30) two- fold axes. They are respectively the vertex, face and edge vectors of an icosahedron. The angular relationships amongst them can be known by referring to stereographic projection of icosahedral point group symmetry given in Figure 1. The icosahedral point group is $(m\bar{3}\ \bar{5})$ and the order of the group is 120. If we ignore inversion then it is of order 60. Please note that number of vertex vectors and rank are the same. Hence six vertex vectors of an icosahedron can serve as the six integrally independent basis vectors. All the diffracted spots are indeed seen to be expressible in terms of six integers. We shall designate such a sextuplet of indices as $(n_1n_2n_3n_4n_5n_6)$ and six basis vectors (vertex vectors) by V_1, V_2, V_3, V_4, V_5 *and* V_6. We may express these in relation to three Cartesian bases (x^{par}, y^{par} and z^{par}) in physical or parallel space by the following matrix relation.

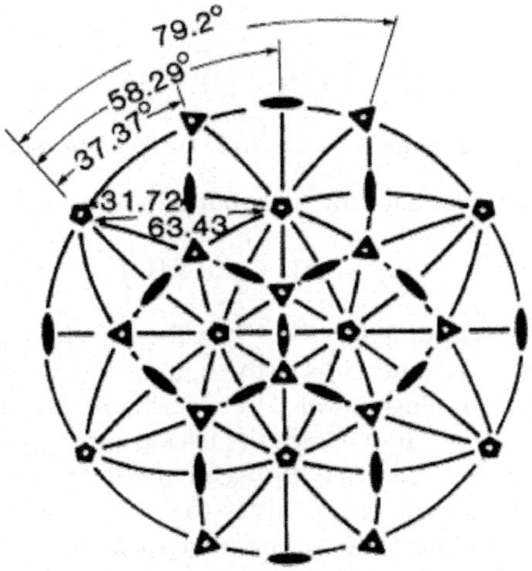

Fig. 1. Stereographic projection of the symmetry elements of the icosahedral point group

$$\begin{pmatrix} V_1 \\ V_2 \\ V_3 \\ V_4 \\ V_5 \\ V_6 \end{pmatrix} = 1\Big/\sqrt{2(1+\tau^2)} \begin{bmatrix} \tau & 1 & 0 \\ 0 & \tau & 1 \\ -1 & 0 & \tau \\ 0 & -\tau & 1 \\ \tau & -1 & 0 \\ 1 & 0 & \tau \end{bmatrix} \begin{pmatrix} x^{par} \\ y^{par} \\ z^{par} \end{pmatrix} \tag{15}$$

It is obvious that the magnitudes of all the vertex vectors are equal. $|V_1| = |V_2| = |V_3| = |V_4| = |V_5| = |V_6| = (1/\sqrt{2})$. V_6 is the polar vector in this setting and $\sum_1^5 V_i = (2\tau - 1)V_6$. Thus a vector represented by the sextuplets (111112) is a five-fold vector parallel to V_6 with a magnitude of τ^3 times that of V_6. There are five distinct set of equivalent choices for choosing a set of three Cartesian bases as there are fifteen two fold vectors in icosahedron. The orientation and magnitude of vectors are known directly by the projection matrix P^{icos}. One may write the elements of this matrix by dot or scalar product of basis vectors (Cf. section 2) and is given by

$$P^{icos} = \left(1\Big/\sqrt{20}\right) \begin{pmatrix} \sqrt{5} & 1 & -1 & -1 & 1 & 1 \\ 1 & \sqrt{5} & 1 & -1 & -1 & 1 \\ -1 & 1 & \sqrt{5} & 1 & -1 & 1 \\ -1 & -1 & 1 & \sqrt{5} & 1 & 1 \\ 1 & -1 & -1 & 1 & \sqrt{5} & 1 \\ 1 & 1 & 1 & 1 & 1 & \sqrt{5} \end{pmatrix} \tag{16}$$

Please note all the diagonal elements are equal and $V_i.V_i = (1/4)$ and the trace of the matrix is 3. The projection matrix P^{icos} has the property as mentioned in section 2. The angle between V_i (i=1 to 5) with V6 is $cos^{-1}\left(1/\sqrt{5}\right)$. The matrix elements in Equation (16) has all the information about the metrical and symmetrical properties of basis vectors. We shall extensively utilize this equation for knowing a direction of SADP of IQC. We can also write the projection matrix Q^{icos} in the complementary or pseudo or perpendicular space as:

$$Q^{icos} = I - P^{icos} = \left(1/\sqrt{20}\right)\begin{pmatrix} \sqrt{5} & -1 & 1 & 1 & -1 & -1 \\ -1 & \sqrt{5} & -1 & 1 & 1 & -1 \\ 1 & -1 & \sqrt{5} & -1 & 1 & -1 \\ 1 & 1 & -1 & \sqrt{5} & -1 & -1 \\ -1 & 1 & 1 & -1 & \sqrt{5} & -1 \\ -1 & -1 & -1 & -1 & -1 & \sqrt{5} \end{pmatrix} \quad (17)$$

Please note that trace of Q^{icos} is 3 and we can see that polar vector W_6 (complementary vector of V_6 in perp.space) is inverted with respect to other vectors in perp. space. We designate the perp.space basis vectors by W_1, W_2, W_3, W_4, W_5 *and* W_6. We may express these in relation to three Cartesian bases (x^{perp}, y^{perp} and z^{perp}) in physical space by the following matrix relation.

$$\begin{pmatrix} W_1 \\ W_2 \\ W_3 \\ W_4 \\ W_5 \\ W_6 \end{pmatrix} = 1/\sqrt{2(1+\tau^2)}\begin{bmatrix} 1 & -\tau & 0 \\ 0 & 1 & -\tau \\ \tau & 0 & 1 \\ 0 & -1 & -\tau \\ 1 & \tau & 0 \\ -\tau & 0 & 1 \end{bmatrix}\begin{pmatrix} x^{perp} \\ y^{perp} \\ z^{perp} \end{pmatrix} \quad (18)$$

The magnitude of all the vectors (W_1, W_2, W_3, W_4, W_5 *and* W_6) is same and is equal to $(1/\sqrt{2})$. This is just like that of *par* space basis vectors mentioned earlier. V- and W- spaces possess vector like and non-vector like representations respectively. Any 5- fold rotation around V_6 takes $V_1 \to V_2 \to V_3 \to V_4 \to V_5 \to V_1$ whereas that around W_6 takes $W_1 \to W_3 \to W_5 \to W_3 \to W_4 \to W_1$. Such spaces are said to be orthogonal in matrix representation [see annotations in reference 2]. Please note the way, we have defined the projection matrices here, permits us to write the decomposition of the 6d hyperspace orthonormal basis vectors e_i *(i=1 to 6)* in terms of two complementary 3d spaces basis vectors as explained in section 2. This is given by

$$e_i = Vi + Wi \quad (19)$$

$$V_i = P^{icos}_{ij} e_j \text{ and } Wi = Q^{icos}_{ij} e_j \quad (20)$$

$$P^{icos}_{ij} + Q^{icos}_{ij} = I_{ij} \quad (21)$$

Where, I_{ij} is an identity matrix of order 6.

The 6d direct hypercubic lattice vector is given by

$$R^6 = t^{icos} \sum_{1}^{6} m_i e_i \tag{22}$$

The 6d reciprocal lattice vector is written as

$$G^6 = \left(1 / t^{icos}\right) \sum_{1}^{6} n_i e_i \tag{23}$$

Where m_i and n_i are integers and t^{icos} is the 6d hypercubic lattice parameter. The physical or parallel and pseudo or perpendicular space components are written in a straight forward manner owing to the identity given by equation (19). They are

R^{icos} = parallel space component of R^6

$$= t^{icos} \sum_{1}^{6} m_i V_i \tag{24}$$

S^{icos} = perpendicular space component of R^6

$$= t^{icos} \sum_{1}^{6} m_i W_i$$

G^{icos} = parallel space component of G^6

$$= \left(1 / t^{icos}\right) \sum_{1}^{6} n_i V_i \tag{26}$$

H^{icos} = perpendicular space component of G^6

$$= \left(1 / t^{icos}\right) \sum_{1}^{6} n_i W_i \tag{27}$$

For indexing, the relevant expression is given by equation (26) and any diffracted spot or peak is represented by the sextuplet of indices ($n_1n_2n_3n_4n_5n_6$). Equation (26) can also be written with the help of matrix given in equation (15) in the following way:

G^{icos} = parallel space component of G^6

$$= (1/t^{icos})\ \left(1/\sqrt{2(1+\tau^2)}\right)\ (G_x\, x^{par} + G_y\, y^{par} + G_z\, z^{par}) \tag{28}$$

Where,

$$G_x = \tau\,(n_1+n_5) + (n_6-n_3)$$

$$G_y = \tau\,(n_2-n_4) + (n_1-n_5) \tag{29}$$

$$G_z = \tau\,(n_3+n_6) + (n_2+n_4)$$

Please note that a vector (000001) is parallel to V_6. Also, a vector (111112) is τ^3 times that of V_6. These are known as τ^3 inflation rule along the five- fold axis. The inflation matrix can be written for this purpose as

$$\begin{pmatrix} 2 & 1 & -1 & -1 & 1 & 1 \\ 1 & 2 & 1 & -1 & -1 & 1 \\ -1 & 1 & 2 & 1 & -1 & 1 \\ -1 & -1 & 1 & 2 & 1 & 1 \\ 1 & -1 & -1 & 1 & 2 & 1 \\ 1 & 1 & 1 & 1 & 1 & 2 \end{pmatrix}$$

We may write the deflation matrix by inverting it. We also note that for the two fold and three fold directions, the following inflation relations hold:

2-fold vector: (001001) = τ (010100)

3-fold vector: (011001) = τ^3 $(1001\bar{1}0)$

We may write the indices of any reflections by inflation and deflation along any direction with respect to first strongest spot that one encounters with respect to transmitted beam in SADP. The principal zone axes (5,-3- and 2-fold directions) can be easily recognized by looking at rotational invariance of diffraction pattern. Please see Cahn et al. (1986) for this purpose [22].

We enlist steps that may be followed for indexing any SADP pattern of IQC. We show three SADPs of IQC in Figure 2. Figure 2a is invariant under five- fold rotation. We call this pattern belonging to a five-fold zone. We consider a 72° sector of this SADP. We observe two spots of same intensities at same reciprocal distance from the transmitted beam along the two 2-fold directions oriented at 72°. All the vertex vectors are 5-fold vectors. We therefore may designate this zone by any one of these. Let us designate this zone as [000001]. We would like to know a set of vectors that are orthogonal to this. All the 2-fold vectors $[1\bar{1}0000], [10\bar{1}000], [100\bar{1}00], [1000\bar{1}0], [01\bar{1}000], [010\bar{1}00], [0100\bar{1}0], [001\bar{1}00], [0010\bar{1}0]$, *and* $[0001\bar{1}0]$ are orthogonal to [000001]. One may verify this by looking at the dot products from equation (16). Hence, all these 20 (10 negatives of these two fold vectors) are lying in the zone. Out of these five 2-fold vectors $[100\bar{1}00], [0100\bar{1}0], [0\bar{1}0100], [00\bar{1}010]$, *and* $[\bar{1}01000]$ are oriented with respect to each other at 72° in this zone. Owing to scaling property, assignment of indices to the first strongest spot may be done by recalling the nature of appearance of spots. Along the two fold axes, the location of spots follows τ-scaling. The inflation matrix corresponding to τ-scaling is given by

$$(1/2)\begin{pmatrix} 1 & 1 & -1 & -1 & 1 & 1 \\ 1 & 1 & 1 & -1 & -1 & 1 \\ -1 & 1 & 1 & 1 & -1 & 1 \\ -1 & -1 & 1 & 1 & 1 & 1 \\ 1 & -1 & -1 & 1 & 1 & 1 \\ 1 & 1 & 1 & 1 & 1 & 1 \end{pmatrix}$$

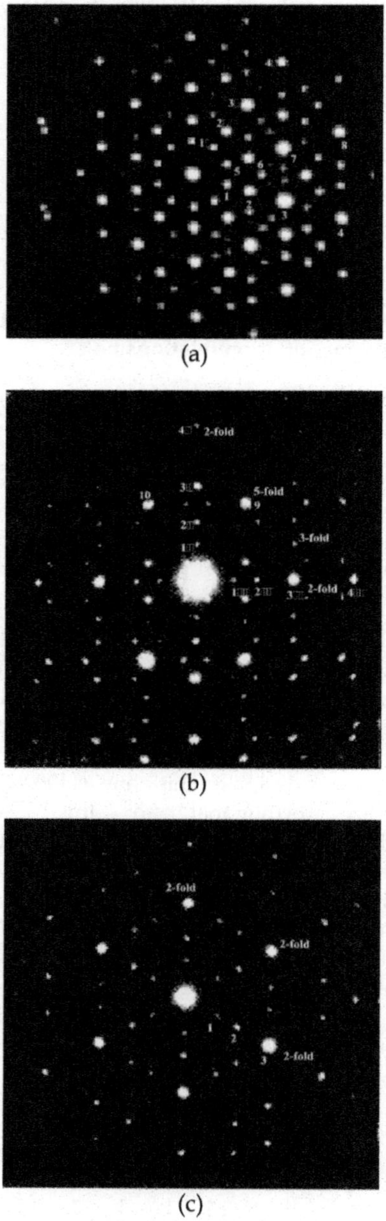

Fig. 2. Selected area diffraction pattern from simple icosahedral phase under (a) 5-fold (b) 2-fold and (c) 3-fold zones. Indices of the important spots are given in Table 1.

The first spot along one of the aforesaid five 2-fold directions can be indexed as $[001\bar{1}00]$ and if we take help of above matrix then next spot after inflation by τ will be $[0100\bar{1}0]$. Likewise we may index spots that are lying along a two-fold direction $[0\bar{1}0100]$ at 72° to $[0100\bar{1}0]$. Having done this we may utilize the rules of vector addition to index spots that are lying within 72° sector. We tabulate the indices of spots based on these steps in table 1a.

Let us try to index spots in a 2-fold zone now (see Fig. 2b). Consider a 2-fold zone [001001]. Please note that this is parallel to Z-axis and all the diffraction vectors having only X-Y components will be contained in this zone. The sextuplets of indices corresponding to 2-fold $[00\bar{1}001]$ *and* $[100010]$ are parallel to X-axis and are related by τ-scaling. Similarly, the 2-fold vectors $[1000\bar{1}0]$*and* $[010\bar{1}00]$ are parallel to Y-axis. Latter is τ time more than the former. We see strong spots at $(1/2)\cos^{-1}(1/\sqrt{5})$ ~31.7° orientation with respect to these 2-fold directions that are lying along 5-fold directions. From the geometrical relationships of 3-fold and 5-fold vectors, we identify a direction in the 2-fold zone at 37.37°. The 3-fold directions $[110\bar{1}00]$*and* $[00\bar{1}0\bar{1}1]$ are parallel to each other and are related by τ^3-scaling. The dot product of these 3-fold vectors with [001001] is zero. Hence such a 3-fold direction is contained in this zone. The indices of diffracted spots in 2-fold zone are given in table 1b. Having done this exercise, the indexing of spots in 3- fold zone is straight forward as it contains 2-fold vectors and their addition/subtraction (see Fig. 2c). Please note that spots indicated along one of the two fold directions in this figure have the similar spot designations akin to those of Figure 2a. As a consequence of this, their indexing remains the same. It is important to mention here that the 2-fold zone axis from IQC is important as it contains 2-fold, 3-fold and 5-fold directions. If one is able to get this zone at zero tilt position while recording SADP then with the help of tilt, one will be able to visit all the principal zone axes by recalling orientation of various symmetry directions with the help of Fig. 1.

(a) 5-fold zone axis [000001]					
Spot No.	Indices	Spot No.	Indices	Spot No.	Indices
1	$001\bar{1}00$	1′	$0001\bar{1}0$	5	$\bar{1}10000$
2	$0100\bar{1}0$	2′	$\bar{1}01000$	6	$0010\bar{1}0$
3	$011\bar{1}\bar{1}0$	3′	$\bar{1}011\bar{1}0$	7	$\bar{1}110\bar{1}0$
4	$021\bar{1}\bar{2}0$	4′	$\bar{2}021\bar{1}0$	8	$\bar{1}120\bar{2}0$
(b) 2-fold zone axis [001001]					
1″	100010	1‴	$010\bar{1}00$	9	$21\bar{1}\bar{1}11$
2″	$10\bar{1}011$	2‴	$110\bar{1}\bar{1}0$	10	$1\bar{1}\bar{1}121$
3″	$20\bar{1}021$	3‴	$120\bar{2}\bar{1}0$	3-fold (τ^3 inflated)	$44\bar{1}\bar{4}\bar{1}1$
4″	$30\bar{2}032$	4‴	$230\bar{3}\bar{2}0$		

Table 1. Sextuplets of indices assigned to various spots for Icosahedral phase

How to identify the first vector in five-fold direction? As explained earlier, we do not have a fixed magnitude of the vector to identify owing to the nature of expressions given in equation (29). However, this can be settled by identifying the appropriate approximant for a particular type of quasicrystals [13]. For example, the approximant corresponding to Al-Mn type IQC, is the cubic α-Al-Mn-Si structure with space group Pm3 and lattice parameter ~12.68Å. Reflections having h+k+l=odd, are weak. This is due to the large motif (having icosahedral symmetry) in the unit cell. We observe a strong spot corresponding to a plane at a distance of ~2Å having indices {530}. These are oriented at an angle of ~31° (=Cos-1(5/ $\sqrt{34}$)) with respect to cubic axes. As mentioned earlier, the angle between the 2-fold and 5-fold is = (1/2) $\cos^{-1}(1/\sqrt{5})$ ~31.7°. Thus, the first strongest reflection corresponding to inter-planar spacing of ~2Å along the 5-fold axis in IQC for this alloy system can be indexed as (000001) or its τ^3 inflated value (111112). This problem is unlike that of a crystal [cf. 22]. We have deliberated on this while introducing the subject with the help of 1d example. As mentioned earlier, the intensity of any reflection will be governed by the Fourier Transform of the motif lying in perpendicular space. The symmetry of the motif must conform to the point group of the diffraction patterns. Thus, analysis of diffraction patterns is the first step towards structural modeling.

4. Decagonal quasicrystals

It has been emphasized earlier by us [17, 23-24] that a distorted icosahedral basis vectors are the best to establish relation between the icosahedral and two dimensional Quasiperiodic structures having 10-fold and 5-fold symmetries. A distortion along one of the six five-fold axes of an icosahedron preserves a five-fold symmetry along it. For continuity, the basis vectors utilized by us earlier are reproduced below:

$$V_i = |V_1| [\sin\theta_D T^{i-1} X^{||} + \cos\theta_D Z^{||}] \tag{30}$$

$$V_6 = |V_6| Z^{||}$$

where V_i (i = 1 to 5) are parallel to the vertex vector of an icosahedron and V_6 is the sixth vertex vector along which distortion can be given by taking value of $\cos\theta_D$ different from $\frac{1}{\sqrt{5}}$. For this choice, an ideal icosahedral basis vectors are recovered in physical space [14-15]. A symmetric projection matrix can be constructed through the dot product of vectors given in equation (30). The matrix P has the following form

$$P = \begin{bmatrix} P_{11} & P_{12} & P_{13} & P_{13} & P_{12} & P_{16} \\ P_{12} & P_{11} & P_{12} & P_{13} & P_{13} & P_{16} \\ P_{13} & P_{12} & P_{11} & P_{12} & P_{13} & P_{16} \\ P_{13} & P_{13} & P_{12} & P_{11} & P_{12} & P_{16} \\ P_{12} & P_{13} & P_{13} & P_{12} & P_{11} & P_{16} \\ P_{16} & P_{16} & P_{16} & P_{16} & P_{16} & P_{66} \end{bmatrix} \tag{31}$$

where $$P_{11} = V_1 \cdot V_1 = \frac{2}{5\sin^2\theta_D};$$

$$P_{12} = P_{11}\left[\frac{1}{2\tau}\sin^2\theta_D + \cos^2\theta_D\right]; \quad P_{13} = P_{11}\left[-\frac{\tau}{2}\sin^2\theta_D + \cos^2\theta_D\right];$$

$$P_{16} = |V_1|\,|V_6| \cos\theta_D\,;$$

$$P_{44} = V_6 \cdot V_6 = \frac{5}{2}\,|V_1|^2\,(1-3\cos^2\theta_D) \text{ and } \tau = \frac{\sqrt{5}+1}{2}.$$

This matrix permits 6d orthonormal basis to define the physical or parallel (| |) space bases as given in equation (30). For details, readers are referred to our earlier work [17]. The projection matrix Q in the complementary or pseudo or perpendicular (⊥) space is given by Q = I–P where I is an identity matrix of order 6. The corresponding basis vectors in ⊥ space are written as

$$W_i = |W_1|\,[\sin\phi\, T^{\langle 2i-1\rangle}\, X^{\perp} + \cos\phi\, Z^{\perp}] \tag{32}$$

$$W_6 = |W_6|\, Z^{\perp}$$

where $|W_1|^2 = 1 - |V_1|^2$; $|W_6|^2 = 1 - |V_6|^2$; $\cos\phi = \left[\dfrac{1 - 3\cos^2\theta_D}{3 - 5\cos^2\theta_D}\right]^{1/2}$; and

$i = 1$ to 5 and $\langle 2i-2\rangle$ is modulo 5. The 6d reciprocal lattice vector G^6 in terms of orthonormal basis vectors (e_i for i = 1 to 6) for orthogonal cell [17] is written as

$$G^6 = \frac{1}{t_1}\sum_{1}^{5} N_i e_i + \frac{1}{t_6} N_6 e_6 \tag{33}$$

where t_1 and t_6 are hyperlattice parameters for the 6d orthogonal cell.

The 3d parallel (| |) and perpendicular (⊥) spaces are denoted here by $G^{||}$ and $G^{\perp}$ respectively. They are given by following equations

$$G^{||} = \frac{1}{t_1}\sum_{1}^{5} N_i V_i + \frac{1}{t_6} N_6 V_6 \tag{34}$$

$$G^{\perp} = \frac{1}{t_1}\sum_{1}^{5} N_i W_i + \frac{1}{t_6} N_6 W_6$$

where N_i's are indices of reflections.

Similarly, the 6d direct space lattice vector (R^6) is written as

$$R^6 = t_1 \sum_{1}^{5} M_i e_i + t_6 M_6 e_6 \tag{35}$$

The parallel and perpendicular space components of R^6 are designated here by $R^{||}$ and $R^{\perp}$ respectively. They are depicted in the form of equations below.

$$R^{||} = t_1 \sum_1^5 M_i V_i + t_6 M_6 V_6 \tag{36}$$

$$R^{\perp} = t_1 \sum_1^5 M_i W_i + t_6 M_6 W_6$$

where M_i (i = 1 to 6) are integers. The product

$$G^{||}.R^{||} = P_{11} \sum_1^5 M_i N_i + P_{66} M_6 N_6 + P_{12} N_{12} + P_{13} N_{13} \tag{37}$$

$$+ \frac{t_1}{t_6} P_{16} N_6 \sum_1^5 M_i + \frac{t_6}{t_1} P_{16} M_6 \sum_1^5 N_i$$

where
$$N_{12} = N_1(M_2+M_5) + N_2(M_1+M_3) + N_3(M_2+M_4) + N_4(M_3+M_5) + N_5(M_1+M_4)$$

$$N_{13} = N_1(M_3+M_4) + N_2(M_4+M_5) + N_3(M_5+M_1) + N_4(M_1+M_2) + N_5(M_2+M_3)$$

For the icosahedral phase $\cos\theta_D = \frac{1}{\sqrt{5}}$ and $t_1 = t_6$, hence $P_{11} = \frac{1}{2}$; $P_{12} = \frac{1}{2\sqrt{5}}$; $P_{13} = -\frac{1}{2\sqrt{5}}$; $P_{16} = \frac{1}{2\sqrt{5}}$ and $P_{66} = \frac{1}{2}$. This reduces equation (37) as

$$2\sqrt{5}\, G^{||}.R^{||} = \sqrt{5} \sum_1^6 M_i N_i + (N_{12} - N_{13} + N_{16}) \tag{38}$$

where,

$$N_{16} = \left[N_6 \sum_1^6 M_i + M_6 \sum_1^5 N_i \right]$$

The right hand side of equation (38) has rational and irrational parts. Hence, left hand side cannot be equated to zero to recover exact zone rule that is applicable for crystals. However, this can be made to accept values nearer to zero and for special set of $(N_1 N_2 N_3 N_4 N_5 N_6)$ corresponding to chosen symmetric direction (like 2-fold, 3-fold and 5-fold), it may display exactly zero. We have already seen example of latter in section 3. Thus, quasiperiodic phases will have the notion of exact and nearly exact zone axes. The purpose of this section is to discuss the structural characteristics of 2d-quasiperiodic structures belonging to two distinct classes(Viz .$P10_5/mcm$;$P10/mmm$).

The planar pentagonal scheme [25-28, 8], in the model of Mandal and Lele [17] corresponds to $\theta_D = 90^\circ$. This leads to $P_{11} = 2/5$; $P_{12} = \frac{1}{5\tau}$; $P_{13} = -\tau/5$; $P_{16} = 0$ and $P_{66} = 1$. Substituting these values in equation (37) gives

$$5\,G^{||}.R^{||} = \left[2\sum_{1}^{5} M_iN_i + 5\sum_{1}^{6} M_6N_6 - N_{12}\right] + \tau\,(N_{12} - N_{13}) \qquad (39)$$

The absence of N_{16} term for this case is due to the fact that $P_{16} = 0$ (cf. equation 31). Hence, for this case $\sum_{1}^{5} V_i = 0$. This condition gives rise to problem of non-uniqueness in indexing. The uniqueness can be ensured by imposing $\sum_{1}^{5} N_i = -2$ to +2. Equation (39) is satisfied for diffraction patterns corresponding to 10/m symmetry. It is clear that such bases are devoid of group-subgroup relationship with the icosahedral phase in view of severe distortion of the icosahedron. This aspect has been dealt while discussing interfaces and twinning in quasiperiodic structures [29-31].

There are experimental observations of two separate classes of decagonal phases [32-36]. They are having space groups $P10_5/mcm$ and $P10/mmm$. The presence of screw in the former case necessarily demands preservation of $\bar{5}$ symmetry in their bases. Such a choice will generate structures that are maintaining group-subgroup relationship with the icosahedral phase. This will also include $P10_5/mcm$. Please note that for this case, the bases will satisfy the following general condition

$$\sum_{1}^{5} V_i = 5\cos\theta_D\ |V_1|\ Z^{||} \qquad (40)$$

A choice of $\cos\theta_D = \frac{1}{2}$ has been shown to conform to the experimentally observed structures having $P\,10_5/mcm$ space group [23]. The structural details of this class in terms of 5d Patterson map can be found elsewhere [37-38].

Owing to equation (40), the indexing would require imposition of condition on the sextuplet rather than the quintuplet corresponding to $\theta_D = 90^\circ$. It is to be remembered here that different nature of group-subgroup relationships that the two classes of decagonal phases maintain with those of icosahedral structures are expected to reflect upon the condition of uniqueness on indices. Such a condition for $\theta_D = 60^\circ$ can be achieved by recalling that $G^{||}(22222\bar{5}) = 0$ and $R^{||}(11111\bar{4}) = 0$ [23-24]. Hence, the conditions on indices are

$$\sum_1^6 N_i = 0 \text{ modulo } 5$$
$$= -2, -1, 0, 1, 2$$

and

$$\sum_1^6 M_i = 0$$

For $\cos\theta_D = \frac{1}{2};\ P_{11} = \frac{8}{15};\ P_{12} = \frac{3\tau-1}{15};P_{12} = P_{13};P_{16} = \frac{1}{3}\sqrt{\frac{2}{5}};\ P_{66} = \frac{1}{3};$ $(t_1 / t_6) = 2\sqrt{2/5}$.

Hence, Equation (37) for this case reduces to

$$30\, G^{||}.R^{||} = \left[16\sum_1^5 M_iN_i + 10M_6N_6 - 2(N_{12}-N_{13}) + 8N_6\sum_1^5 M_i + 5M_6\sum_1^5 N_i\right] + 6\tau\,[N_{12}-N_{13}] \quad (41)$$

Equation (41) is depiction of zone rule for $P10_5/mcm$ structure corresponding to $\cos\theta_D$ = 1/2. The right hand side of equation (41) has quadratic irrational. For those zone axes for which term containing quadratic irrational is absent, one will fulfill exact zone condition. Recording diffraction patterns corresponding to such zones is essential for determination of point group symmetry of the reciprocal space. For this, the quality of thin foil has to be good so that tilt experiment can be conducted during transmission electron microscopic studies of quasicrystals both in imaging and diffraction modes. We are making these specific statements to clarify that quasicrystals are normally brittle materials and for making thin foil one should devote time. Having achieved this, we have to conduct diffraction and imaging experiments to uniquely settle its existence to rule out the possibility of irrational twins. Kindly recall while arriving at zone rule in equation (39) for $\theta_D = 90^\circ$ (P10/mmm structure) the ratio of (t_1 / t_6) was not required. This indicates (t_1 / t_6) may not be fixed for P10/mmm phase. In contrast, quantification of zone rule for $P10_5/mcm$ types of structure requires a priori prescription of (t_1 / t_6) ratio. Thus the two Quasiperiodic structures having periodicity in one dimension of similar type [32-35] demand different types of hyperlattice parameters. This distinction has to be kept in mind while discussing these structures [9].

Having discussed zone rules of various Quasiperiodic structures, we proceed to index SADPs of decagonal phase. The unique 10-fold pattern containing 2-fold axes can be indexed by noting two distinct set of 2-fold directions. These are termed as P- and D-direction respectively. They are oriented with respect to each other at 18^0 intervals. Let us try to identify them with the help of two characteristics patterns viz. P- and D- patterns. We show them in Figures 3a and 3b. Both these patterns have 10-fold direction. We see strongest

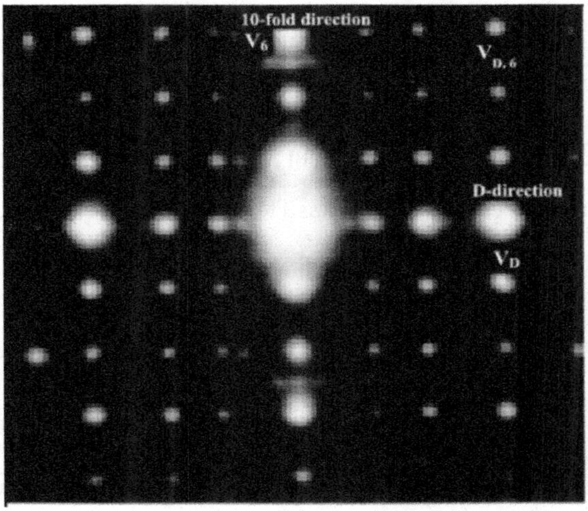

(a)

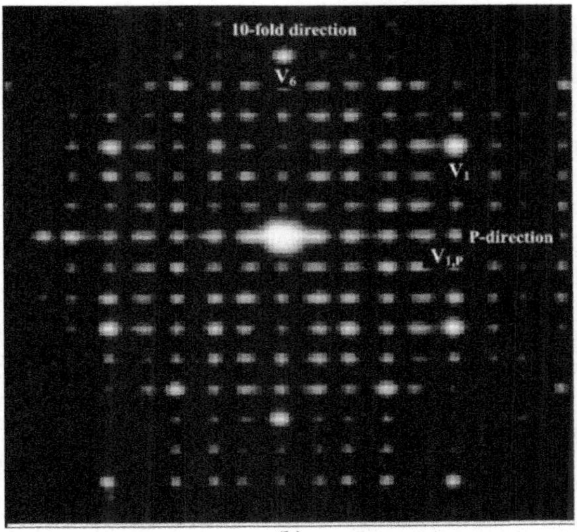

(b)

Fig. 3. Selected area diffraction patterns from decagonal phase (T_6) under two 2-fold zones (a) P-pattern and (b) D-pattern. The indices of important spots are given in Table 2.

spots (V_6) along this. There are 3 and 6 intervals between transmitted spots and V_6 respectively in P– and D– patterns. In P-pattern, odd rows have streaking while those in D-pattern are weak spots. We expect nearly absence of odd rows in P-pattern due to 10_5 screw. However, this may not be the case for D-pattern owing to presence of diffraction vectors of the type V_1. The strongest spot along 10-fold axis getting divided into six equal intervals permits us to infer that we are dealing with T_6 phase. The Z-direction of $G^{//}$ (physical reciprocal lattice vector) for $\cos\theta = ½$ is given by

$$G_z^{II} = \frac{1}{t_1}\sqrt{8/15}\left[\sum_{i=1}^{5} N_i + 2N_6\right] \quad (42)$$

Any vector having $N_1 = N_2 = N_3 = N_4 = N_5$ will have G''_{xy} equal to zero. Thus, first spot along 10-fold (periodic) direction can be indexed as [$11111\bar{2}$]. However, $\sum_1^6 N_i$ is equal to 3. This summation will be brought to -2 by subtracting zero vector [$22222\bar{5}$] or its integral multiple to above. Hence, first spot along 10-fold may be indexed as [$\bar{1}\bar{1}\bar{1}\bar{1}\bar{1}3$]. This direction is periodic and V_6 can be assigned [$\bar{6}\bar{6}\bar{6}\bar{6}\bar{6}18$]. However, $\sum_1^6 N_i = -12$ and we have proposed $\sum_1^6 N_i = -2$, -1, 0, 1, 2 for uniqueness. This can be achieved by adding zero vector [$44444\bar{1}\bar{0}$] to this. Hence, the sixth spot V_6 along 10-fold will be indexed by [$\bar{2}\bar{2}\bar{2}\bar{2}\bar{2}8$] with $\sum_1^6 N_i = -2$. We note that spot V_1 in D-pattern is at 60^0 to V_6. This is the reason why we have chosen cosθ = ½. This spot can be indexed as [300000]. To make this unique, we need to add zero vector [$\bar{2}\bar{2}\bar{2}\bar{2}\bar{2}5$]. Hence, indexing of spot V_1 is [$1\bar{2}\bar{2}\bar{2}\bar{2}5$]. The diffraction vector $V_{1,P}$, will not have z-component. We may take out z-component from V_1 and index it by [$2\bar{1}\bar{1}\bar{1}\bar{1}1$]. All other spots along P-direction can be indexed by inflation / deflation matrix given by us [23,24]. The indexing of spot V_D would again require absence of z-component. This can be achieved by assigning [$003\bar{3}00$] to this. The indexing of $V_{D,6}$ can be achieved by adding 6-layer height in [$003\bar{3}00$]. The sextuplet of indices thus becomes [$\bar{2}\bar{2}1\bar{5}\bar{2}8$]. We have listed them in Table 2. As mentioned earlier 10-fold diffraction pattern will contain $V_{1,P}$ and $V_{D,6}$ directions. All other spots can be indexed by inflation and deflation matrix [23,24] as we have indicated while indexing diffraction patterns of IQC. We have mentioned earlier that DQC have four types of phases T_2, T_4, T_6 and T_8. The indexing of spots of V_D type in P-pattern will accordingly be given by [$00n\bar{n}00$] for T_{2n} phase. The indexing of V_1 type spot in D-pattern will be given by [100000], [200000], [$1\bar{2}\bar{2}\bar{2}\bar{2}5$] and [$2\bar{2}\bar{2}\bar{2}\bar{2}5$] for T_{2n} phases respectively. For T_6 and T_8, we have differently looking indices due to uniqueness condition of $\sum_1^6 N_i = 0, \pm1, \pm2$.

P-pattern			D-pattern		
Spot designation	Indices	After [23]	Spot designation	Indices	After [23]
V_6	$\bar{2}\bar{2}\bar{2}\bar{2}\bar{2}8$	000003	V_6	$\bar{2}\bar{2}\bar{2}\bar{2}\bar{2}8$	000003
V_D	$003\bar{3}00$	$003\bar{3}00$	V_1	$1\bar{2}\bar{2}\bar{2}\bar{2}5$	300000
$V_{D,6}$	$\bar{2}\bar{2}1\bar{5}\bar{2}8$	$003\bar{3}03$	$V_{1,P}$	$2\bar{1}\bar{1}\bar{1}\bar{1}1$	$0\bar{3}\bar{3}\bar{3}\bar{3}6$

Table 2. Sextuplets of indices assigned to various spots for decagonal phase (T_6-6 layer periodicity along 10-fold axis)

5. Conclusions

We have discussed reciprocal lattices of icosahedral and decagonal quasicrystals with the help of projection formalism utilized in higher dimensional structural description of such phases. The necessary mathematical details pertaining to higher dimensional crystallography were understood with the help of one dimensional aperiodic case (cf. section 2). It was also shown that we need expression of physical reciprocal lattice vector for mapping the entire diffraction space and indexing of diffraction spots. Such a viewpoint was extended to a six dimensional cubic lattice for the icosahedral phase (cf. section 3). Having got the expression of physical reciprocal lattice vector we have spelt out necessary steps for indexing of diffraction patterns. We have shown the solution of important spots in terms of sextuplets of indices and given them in a table for the readers. We have put forward our generalized formalism based on 6d orthogonal cell for decagonal phases in section 4 of the chapter. A general expression of zone rule as a dot product of physical reciprocal and direct lattice vectors has been given. Two distinct cases corresponding to decagonal phases having 10/m and 10_5 screw axis have been given. We have adopted a six index notation for both these structures. The conditions for uniqueness for them are discussed in terms of the ratio of two lattice parameters of the orthogonal cell. The indexing of important diffracted spots corresponding to two distinct two-fold patterns (P and D patterns) has been accomplished for T_6 phase. We have indicated methods of indexing of diffraction patterns for ten-fold zone. At the end of section 4, we have also given the indices of some of the equivalent diffracted spots expected in other decagonal phases.

6. Annexure A

This note to the readers has been added after the announcement of Nobel Prize to Dan Shechtman in Chemistry (2011) for the discovery of quasicrystals in the rapidly solidified Al-Mn alloys [1]. It is important for all of us to know and understand the path of discovery of a truly seminal nature. We need to recall some of the historical facts in this regard. Steurer and Deloudi [2], while reviewing various aspects of quasicrystals in the silver jubilee year of its first report, quoted the legendry materials scientist John Cahn who later co-authored paper [1]. It is mentioned [2] that when Dan went to NIST, USA for sharing his excitement, Cahn said "Go away, Dany. These are twins and that's not terribly interesting". However, when Cahn got convinced, he led the discovery by defending its tantalizing nature from the front. One may substantiate by remembering the chronological order of the lively debate that has gone on. One may recall that when Linus Pauling put forward twinning models [3, 4] for interpreting icosahedral diffraction patterns, Cahn and Gratias supported Shechtman by taking a firm position [5] about the new state of atomic order in the solid state. At this crucial juncture, a leading role was played by a versatile genius, Mackay [6] who demonstrated that diffraction results reported in [1] cannot be understood in terms of fundamental axiom of conventional 3d-crystallography. It is against this background that Senechal [7] titled the discovery of long range aperiodic order with sharp diffraction patterns [1] as the demise of a paradigm. The demise pertains to foundation of classical 3d crystallography where we get the impression that diffraction patterns cannot display invariance under non-crystallographic point symmetry. The selected area diffraction patterns that remain invariant under icosahedral point group have given rise to a new paradigm that led to a change in the definition of crystal by International Union of Crystallography in 1992 [2].

The profound impact of the emergence of new paradigm can be understood by going through the foreword by Freeman Dyson [8]. While writing foreword for the book [8] entitled "The Mathematical Century", he remarked discovery of quasicrystals as one of the three jokes of nature of the last century. The other two jokes are (a) appearance of imaginary quantity in the solution of Schrodinger wave equation in quantum mechanics and (b) possible states in quantum mechanics forming a linear space, which have much wider ramification and deeper connection with the fundamental laws of nature governing mechanics of quantum particles. Thus, experimental observation of Quasiperiodic translational order is truly revolutionary and has led to a veritable change in our age old belief of classical crystallography. The paintings in ancient art perhaps inspired generations of mathematicians to systematically develop mathematics for tilings, coverings and packings that are not periodic. In this context, it is important to mention contributions of Penrose [9] and Mackay [10] prior to the discovery of icosahedral quasicrystals. Intellectually stimulating exercises of many such authors helped experimentalists to visualize images of possible underlying atomic arrangements and their Fourier transforms in reciprocal space.

There are many reviews and books summarizing developments in this fascinating area. Some of these are given in this chapter. For Indian contributions, we refer the readers to a report by international union of crystallography [11]. Indian experimental investigations prior to the first report of diffraction patterns with non-crystallographic point group symmetry are seen to have patterns akin to D-patterns of decagonal phases [12]. We are able to recognize this only in retrospect after we have the necessary knowledgebase on quasicrystals. In contrast, Shechtman and his colleagues' perseverance to determine the invariance of diffraction patterns under icosahedral point group symmetry and bold announcement to the scientific community are testimony to the fact that discoverers are genius and courageous men/women of all times. Hence, the discovery being a class of its own for which Shechtman as the chief architect has rightly been chosen for the Nobel Prize.

[1] D. Shechtman, I. Blech, D. Gratias, & J.W. Cahn, Phys. Rev. Lett. 53, 1951 (1984).

[2] W. Steurer and S. Deloudi, Acta Cryst. A64, 1 (2007)

[3] L. Pauling, Nature 317, 512(1985)

[4] L. Pauling, Proc. Natl Acad. Sci. 86, 8595 (1989)

[5] J.W. Cahn, D. Gratias and D. Shechtman, Nature 319, 102 (1986)

[6] A.L. Mackay, Nature 319, 104 (1986)

[7] M. Senechal, Quasicrystals and Geometry, Cambridge University Press (1996)

[8] P. Odifreddi, The Mathematical Century, University Press (India) Pvt. Ltd. (2004)

[9] R.Penrose, Bull.Inst.Math.appl., 10, 266(1974)

[10] A.L.Mackay, Physica, A114, 609(1982)

[11] Report on Crystallography in India by Int.Union of Crystall. (2007)

[12] G.V.S.Sastry, C.Suryanarayana, M.van Sande and G.van Tendeloo,Mat.Res.Bull., 13, 1065 (1978)

7. References

[1] D. Shechtman, I. Blech, D. Gratias, & J. W. Cahn, Phys. Rev. Lett. 53, 1951 (1984).

[2] P. J. Steinhardt, and S. Ostlund, annotation in the Physics of Quasicrystals, World Scientific (1987)

[3] N. D. Mermin, Phys. Rev. Lett., 68, 1172 (1992).

[4] N. D. Mermin, Rev. Mod. Phys., 64, 3 (1992).

[5] International tables for Crystallography, vol B (2006)

[6] K. Chattopadhyay, S. Lele, S. Ranganathan, G. N. Subbanna and N. Thangaraj, Curr. Sci., 54, 895(1985).

[7] L. Bendersky, Phys. Rev. Lett., 55, 1461 (1985).

[8] S. Ranganathan, E. A. Lord, N. K. Mukhopadhyay and A. Singh, Acta Cryst., A63, 1 (2007).

[9] R. K. Mandal, arXiv:1010. 2643V1 (October 2010)

[10] F. C. Frank, Acta Cryst., 18, 862 (1965).

[11] A. L. Mackay, Acta Cryst., A33, 212 (1977).

[12] P. M. de Wolff, T. Janssen, and A. Janner, Acta Cryst., A37, 625 (1981)

[13] C. L. Henley, M. de Boissieu, and W. Steurrer, Phil. Mag., 86, 1131(2006)

[14] V. Elser and C. L. Henley, Phys. Rev. Lett., 55, 2883 (1985).

[15] M. Duneau and A. Katz, Phys. Rev. Lett., 54, 2688 (1985).

[16] P. Bak, scripta Metall. 20, 1199 (1986)

[17] R. K. Mandal and S. Lele, Phys. Rev. Lett., 62, 2695 (1989).

[18] A. Bienenstock and P. P. Ewald, Acta Cryst, 15, 1253 (1962).

[19]H. Bohr, Almost-periodic functions, Chelsea (1947)

[20]N. G. de Bruijn, Proc. K. Ned. Akad. Wet. A84 38 (1981)

[21]C. Janot "Quasicrystals a Primer" Oxford Univ. Press (1998).

[22] J. W. Cahn, D. Shechtman and D. Gratias, J. Mater. Res., 1, 13 (1986).

[23] R. K. Mandal and S. Lele, Phil. Mag., B63, 513 (1991).

[24] R. K. Mandal, A. K. Pramanick, G. V. S. Sastry, and S. Lele, Mater. Res. Soc. Sym. Proc. 805, LL1. 6. 1 (2004).

[25] J. D. Fitz Gerald, R. L. Withers, A. M. Stewart and A. Calka, Phil. Mag., B58, 15 (1988).

[26] Singh, A. & Ranganathan, S., Phil. Mag., A74, 821-840 (1996).

[27] Singh, A. & Ranganathan, S., Phil. Mag., A74, 841-859 (1996).

[28] N. K. Mukhopadhyay and E. A. Lord, Acta Cryst., A58, 424 (2002).

[29] R. K. Mandal, S. Lele and S. Ranganathan, Phil. Mag. Lett., 67, 301 (1993).

[30] R. K. Mandal, Phil. Mag., B79, 157 (1999).

[31] R. K. Mandal and S. Lele, Mater. Sci. Engg., A294-296, 813 (2000).

[32] K. Edagawa, M. Ichihara, K. Suzuki and S. Takeuchi Phil. Mag. Lett., 66, 19 (1992).

[33] K. Edagawa, H. Tamura, S. Yamaguchi, K. Suzuki and S. Takeuchi, Phys. Rev., B50, 12413 (1994)

[34] Ritsch, S., Beeli,C., Nissen, H. U. & Luck, R. (1995) Phil. . Mag., A71, 671

[35] Ritsch, S., Beeli,C., Nissen, H. U., Godecke, T., Schefler, M. & Luck, R. (1998) Phil. Mag. Lett., 78, 67

[36] A. K. Pramanick, R. K., Mandal, and G. V. S. Sastry, J. Non-Cryst. Solids, 334-335, 234 (2004).

[37] Steurer, W. and Kuo, K. H. Acta Cryst., B46, 703 (1990).
[38] Steurer, W., Haibach, T. and Zhang, B. Acta Cryst., B49, 661 (1993).

2

Conventional Transmission Electron Microscope Observation of Electric and Magnetic Fields

Katsuhiro Sasaki[1], Hidekazu Murata[2], Kotaro Kuroda[1] and Hiroyasu Saka[1]
[1]Department of Quantum Engineering, School of Engineering, Nagoya University,
[2]Department of Electrical and Electronic Engineering, Meijyo University,
Japan

1. Introduction

Transmission electron microscopy (TEM) has played an increasingly important role in the successful development of the micro-electronics industry by characterizing electric and/or magnetic device structures with decreasing dimensions. In particular, characterizing the dynamic behavior of electric and magnetic fields within and surrounding a device is important to understand and control its properties (Wang et al., 2004).

Two well-known methods have been used to observe the electric and magnetic field distribution in a TEM. The first one is electron holography (Tonomura, 1987), which requires a highly coherent electron source and an extensively modified transmission electron microscope. Though electron holograph gives a quantitative measurement of electromagnetic potential, it offers only a limited view due to the limitation of the width of the interference region between the object and the reference wave.

The other one is Lorentz microscopy. Though this method can be performed in a conventional TEM, the image in Fresnel mode is obtained in defocused condition and, the image in Foucault mode is not suitable for quantitative analysis (Cohen, 1968; Fisher and Blades, 1972; Chapman, 1984). Some efforts have been made to extract quantitative results from Lorentz images, for example, the transport intensity equation (TIE) approach by Bajit et al. (2000) and, Graef and Zhu (2001). However, this approach requires post image processing to form the images representing the electric or magnetic fields.

There is the third method called the shadow method that was developed in the early days of the development of transmission electron microscopy (Marton and Lachenbruch, 1949; Blackman and Grunbaum, 1957; Jakubovics, 1964; Lazzari et al., 1971; Keyser, et al., 1975; Wade, 1976). However, the development of this method has been suspended and neglected compared to the rapid development of the other two methods.

In our previous work (Sasaki, et al, 2000, 2003) on in-situ electron holography experiments, we found a similar phenomenon, that is the distortion of the shadow image of the selected area diffraction (SAD) aperture while applying an external electric potential to the specimen. In the experiment, the observation was performed in very low magnification mode, called

"Low Mag" mode (hereinafter referred to as "LM" mode). As a practical method of rough alignment of the beam tilt in the LM mode, the shadow image of the SAD aperture was observed. The specimen is a tip of GaP Light Emitting Diode (LED), whose shape observed in a focused ion beam microscope is shown in Fig. 1a. The schematic view of the specimen and the geometrical relationship of the electron beam, when it is observed in a TEM, is shown in Fig. 1b. During the alignment procedure, as shown in the LM mode images in the TEM (Fig. 2), a local distortion of the edge shape of the SAD aperture near the reverse biased p-n junction of the specimen was observed that did not affect the image of the specimen at all. The continuous circular edge shape of the aperture was observed when there was no external voltage, and when a forward bias was applied as shown in Fig. 2a and b, respectively. However, a discontinuity in the edge shape appeared when reverse bias was applied as shown in Fig. 2c and d. This distortion increased with increasing applied voltage as shown in the superimposed schematic drawing of the images in Fig. 2e.

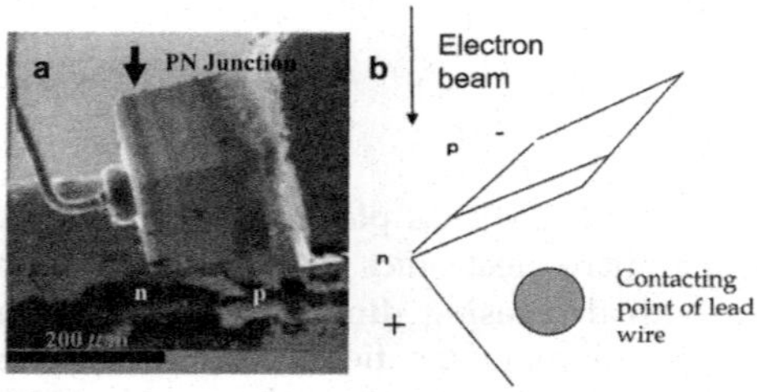

Fig. 1. A scanning ion microscope (SIM) image (a) and the schematic drawing (b) of the GaP LED chip. The arrow indicates the direction of the electron beam when the specimen is observed in a TEM.

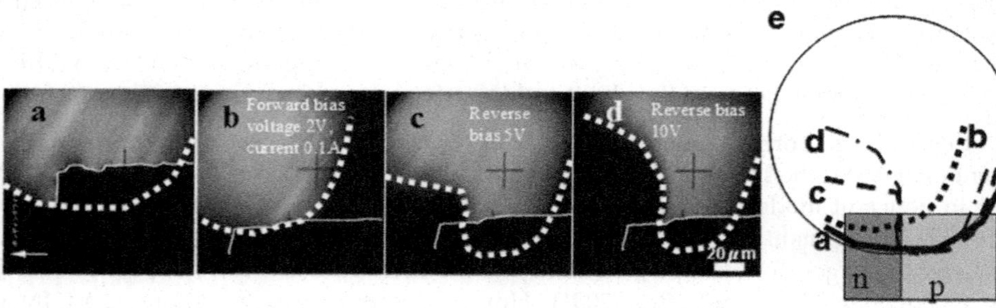

Fig. 2. Low magnification image of GaP chip and the shadow image distortion of the SAD aperture at (a) 0 V, (b) 2 V of forward, (c) 5 V and (d) 10 V of reverse bias. (e) the superimposed schematic drawing of the shadow images of the aperture.

In the above mentioned work, we discussed the geometrical optics for this phenomenon and found that the distortion of the shadow image of the SAD aperture was proportional to the deflection of the electron beam by the electric field around the position of the specimen (Sasaki and Saka, 2005; Sasaki, et. al. 2006). We also suggested that quantitative measurement not only of electric fields but also magnetic fields could be accomplished using this phenomenon. Analogous to electron optics, our method can be viewed as a shadow method (Marton and Lachenbruch, 1949; Jakubovics, 1964; Wade , 1976). However, our method, and the benefit thereof, is applicable in a modern conventional TEM without extensive modification. We call our method, temporally, the shadow image distortion (SID) method.

In this chapter, we will discuss further details of the geometrical electron optics in this phenomenon and the method for determining the optical parameters of a TEM system, which are important for observing the electric and/or magnetic fields using this phenomenon. Some examples of applications which demonstrate the quantitative measurement of the electric field using our method are introduced.

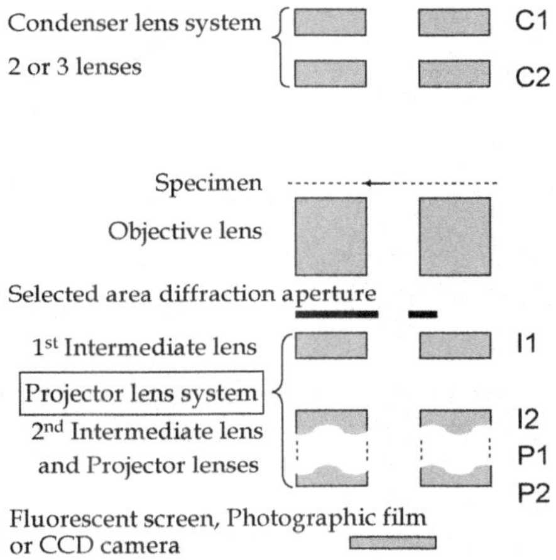

Fig. 3. Typical lens system of a conventional TEM.

2. Geometrical optics analysis of shadow image distortion

A conventional TEM consists of three lens systems as shown in Fig. 3. The first is the condenser lens system which focuses the electrons coming from the electron gun and illuminates the specimen. The second is the objective lens which, in ordinary mode, forms the image of the specimen for the following lens system. The third is the projector lens system which projects the enlarged specimen image onto the fluorescent screen, the photographic film or, contemporary, charge coupled device (CCD) camera. For example, in the H-9000NAR (Hitachi High Technologies, Tokyo, Japan) used in our work, the condenser lens system consisted of two lenses, i.e., C1 and C2, and the projector lens system consisted of four lenses, i.e., two intermediate lenses, i.e., I1 and I2, and two projector lenses, i.e., P1 and P2. We consider the behavior of the lenses, which are the condenser lens C2, the objective lens and the intermediate lens I1, to analyze the movement of the shadow image of the beam obstructer placed at the position of the SAD aperture that is between the objective lens and intermediate lens I1.

In the LM mode, the image of the specimen is formed by the intermediate lens, and the condenser lens system locates the electron beam onto the specimen for illumination. The objective lens which has a lesser effective part in this mode is usually turned off or only weakly activated. As mentioned above, we consider one condenser lens C2 as a condenser lens system, objective lens and first intermediate lens I1 only, which are shown in the ray diagram of the electron microscope in Fig. 4.

First of all, we consider the simplest condition in which an ideal parallel beam illuminates the specimen when the objective lens is turned off (Fig. 4a). Then the behavior of C2 condenser lens is taken into account. In the second condition, the objective lens is completely turned off in the divergent beam illumination (Fig. 4b). In the third condition, the objective lens is weakly activated in the parallel beam illumination (Fig. 4c). We found that the objective lens does not affect the imaging of the specimen. However, it can control the sensitivity of the field detection. In this last case, which can be performed in a real operation, the objective lens is weakly activated in the divergent beam illumination (Fig. 5).

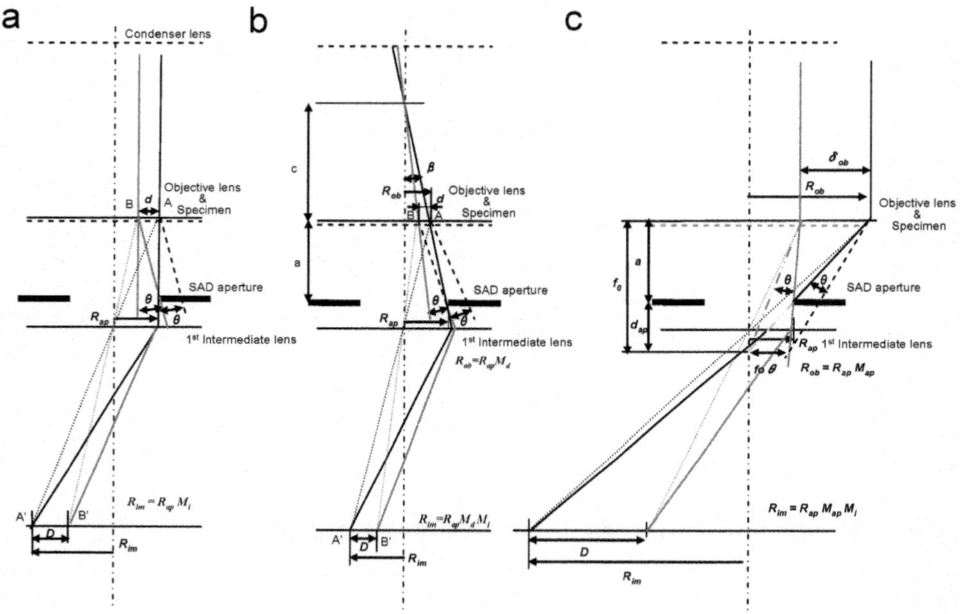

Fig. 4. Ray diagrams of (a) 1st, (b) 2nd and (c) 3rd condition.

The ray diagram in the first condition is shown in Fig. 4a. The SAD aperture is located between the specimen and the intermediate lens. Here we assume that the divergence of the electron beam is relatively small, for example on the order of 10^{-4} rad. In the parallel beam illumination, the shadow image of the SAD aperture (R_{ap} in radius) is projected onto the image plane with a radius $R_{im} = R_{ap} M_i$, where M_i is the magnification of the intermediate lens. When the electron beam is deflected by the electric field of the specimen at an angle θ, the electron beam from the specimen between A and B is obstructed by the SAD aperture. The electron beam obstructed by the SAD aperture has been imaged between A' and B' on the image plane of the intermediate lens. This means that the edge of the shadow image of the SAD aperture will move from A' to B'. This is observed as a shift of the aperture edge in the image. The distance of A-B, i.e. d, can be calculated by the deflection angle θ and the distance a between the SAD aperture and the specimen. The distance d will be enlarged by the intermediate lens to $D = a\,\theta\,M_i$ at the image plane and then enlarged or reduced to be imaged on the fluorescence screen by the following lens system with a magnification M_p (Sasaki and Saka, 2005).

In the TEM that we usually used, the condenser lens system focused the electron beam once before the specimen, then illuminated the specimen with a certain divergence angle of electron beam as shown in the ray diagram in Fig. 4b. In the second condition, we take into account the divergence of the beam due to the condenser lens (Sasaki et al., 2008a). In this condition, the beam direction has a different vector as a function of distance from the optical axis of the electron microscope. The angle β between the optical axis and electron beam direction will be $\beta=r/c$, where r is the distance from the optical axis and c is the distance from the specimen surface to the focal point of the electron beam. Though the electron beam has a divergence angle, the image is formed at the same position on the image plane of the intermediate lens in the magnification M_i as shown in the Fig. 4b. However, the area on the specimen that corresponds to the area where the electron beam can pass through the SAD aperture with radius R_{ap} is reduced due to the electron beam divergence. The radius R_{ob} of the area on the specimen plane that corresponds to the area on the SAD aperture plane with radius R_{ap} is written as $R_{ob} = R_{ap}\, c/(a+c)$. Namely, the radius of the shadow image of the SAD aperture will be reduced by magnification M_d, where $M_d = c/(a+c)$. When the electron beam is deflected by the electric field of the specimen at an angle θ, the electron beam at the edge of the SAD aperture is shifted $a\,\theta$ and obstructed by the SAD aperture. The shift $a\,\theta$ on the aperture plane corresponding to the distance A and B on the specimen plane will be reduced to $a\,\theta\, M_d$ where this corresponds to the image between A' and B'. The distance between A' and B' which corresponds to the shift of the edge of the shadow image of aperture D is calculated to be $D = a\,\theta\, M_d\, M_i$ with M_i the magnification of the intermediate lens. This result indicates that the shift of the shadow image of the SAD aperture is reduced by M_d using the electron beam with the divergence angle.

In the third condition, particularly, of the H-9000 NAR, the objective lens is weakly activated in the LM mode. The ray diagram in this condition is shown in Fig. 4c. In the LM mode with the weakly activated objective lens, the back focal plane of the objective lens is located slightly lower than the front focal plane of the intermediate lens. For simplicity, the objective lens plane is considered to be approximately in the plane of the specimen, because a very low current is applied to the objective lens as for example, in the case of H-9000 NAR, the current is one twentieth of the ordinary imaging mode, then it will work as a transfer lens and does not affect the image formation of the lens which has the object plane at the specimen position. Then we consider a parallel beam condition again to discuss the effect of the objective lens in the geometrical optics. When the objective lens is in the same plane as the specimen, the image of the specimen is formed at the same magnification as if there was no objective lens. However, the electron beam passing through the specimen will be focused to one point on the back focal plane of the objective lens and then diverge. The area on the specimen through which the electrons pass, and then pass through the SAD aperture hole with radius R_{ap} is the reverse projection of the SAD aperture on the specimen plane with radius R_{ob} as indicated by the arrows in Fig. 4c. The radius R_{ob} is calculated to be $R_{ob} = R_{ap}\, f_o\, /\, d_{ap}$, where R_{ap} is the radius of the SAD aperture, f_o is the focal length of the objective lens and d_{ap} is the distance between the back focal plane of the objective lens and the SAD aperture. The shadow image radius R_{im} of the SAD aperture in the image plane of the intermediate lens is the magnified image of this reverse projection by the intermediate lens; it is given by: $R_{im} = R_{ap}\, M_{ap}\, M_i$, where M_{ap} is $f_o\, /\, d_{ap}$.

When the electron beam is deflected by the electric field at an angle θ, the focal point of the electron beam on the back focal plane of the objective lens is displaced by $f_o\,\theta$. On the surface of the specimen, the area on the specimen through which pass the electrons passing through the SAD aperture hole will be changed. The shift of the aperture edge δ_{ob} is calculated to be $\delta_{ob} = f_o\, \theta\, a\, /\, d_{ap}$. The shift δ_{ob} is enlarged onto the image plane of the intermediate lens with magnification M_i. Thus the shift of the aperture edge D is calculated as $D = a\, \theta\, M_{ap}\, M_i$.

In the forth condition, that could be obtained in real operating conditions, the optics will be divided into two cases, i.e., whether the relations of the length between c and f_o are either $c > f_o$ or $c < f_o$.

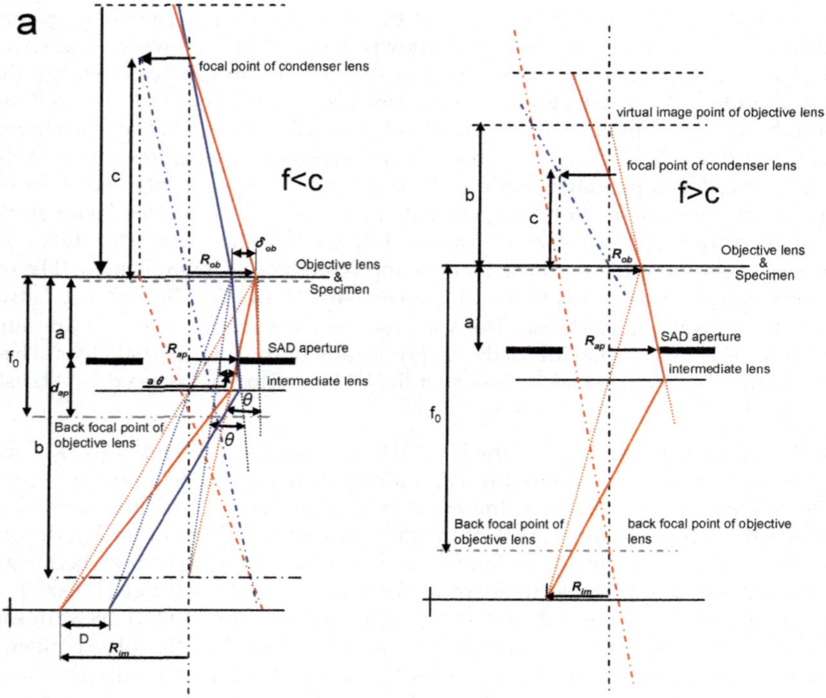

Fig. 5. Ray diagrams of 4th condition in (a) f<c and (b) f>c.

If $c > f_o$, as shown in Fig. 5a, the electron beam diverges from the focal point of the condenser lens and is focused again at the image plane of the objective lens at b. The length b can be calculated from the relation $1/b + 1/c = 1/f_o$, where c is the distance between the focal point of the electron beam after the condenser lens and the position of the objective lens, then b can be rewritten to be $f_o\, c/(c - f_o)$. The area on the specimen through which pass the electrons, passing through the SAD aperture hole is, again, the reverse projection of the SAD aperture on the specimen plane. The radius R_{ob} of the reverse projected SAD aperture is calculated as $R_{ob} = R_{ap}\, b/\,(b - a)$ in this case, where R_{ap} is the radius of the SAD aperture, a is the distance between the specimen and/or objective lens, and the SAD aperture. Replacing $f_o c/(\,c - f_o)$ into b gives

$$R_{ob} \,/\, R_{ap} = f_o c \,/\left[f_o \left(a + c \right) - ac \right]$$

M_d and M_{ap} can be written as $c/(a+c)$ and $f_o/(f_o-a)$, respectively. Rewriting R_{ob}/R_{ap} with M_d and M_{ap} gives

$$R_{ob} / R_{ap} = M_d M_{ap} / \left(M_d + M_{ap} - M_d M_{ap}\right)$$

Analogous to the above mentioned discussion in the third condition gives

$$D = a\,\theta M_d M_{ap} M_p M_i / \left(M_d + M_{ap} - M_d M_{ap}\right) \tag{2.1}$$

In the second case, $c < f_o$, the focal point of the condenser lens can be assumed to be the virtual image at the position $-b$, which is the object plane of the weakly activated objective lens. The length b can be calculated from the relation $1/c = 1/f_o + 1/b$ in this case, then b can be rewritten to be $f_o c/(f_o - c)$. The radius R_{ob} of the reverse projected SAD aperture is calculated as $R_{ob} = R_{ap}\, b/(b+a)$ in the latter case. Replacing $f_o c/(f_o - c)$ into b, gives $R_{ob}/R_{ap} = f_o c/[f_o(a+c) - ac]$. This result gives the same equation as (2.1).

If M_{ap} is nearly equal to unity, the above equation can be simplified to

$$D = a\theta M_d M_{ap} M_p M_i \tag{2.2}$$

As discussed above, the results show that the displacement D of the shadow image of the beam obstructer, which is placed at the position of the SAD aperture, will be proportional to the deflection angle θ of the electron beam at the position of the specimen. The differences in the deflection angle θ at different positions on the specimen causes different displacements of the shadow, which forms the distortion of the shadow image of the beam obstructer placed at the position of the SAD aperture (hereinafter referred to as the "shadow aperture").

3. Experimental confirmation of geometrical optics analysis

3.1 Measurement of optical parameters of the magnetic lens system in a TEM

As discussed in section 2, several magnification parameters, i.e., M_d, M_{ap}, M_p and M_i, play an important role in measuring the electric and/or magnetic field of the specimen in the optical system of the TEM. In order to confirm the above mentioned geometrical electron optics analysis and measure the optical parameters, we performed the following experiments.

The electron microscopy observation was done in a TEM H-9000NAR equipped with an ordinary LaB_6 thermal emission electron gun at an accelerating voltage of 300 kV. The observation was performed in the LM mode.

Uniformly shaped lattice grids of known size were simultaneously placed at the position of the specimen and the SAD as the shadow aperture. The magnification of the shadow image of the grid placed at the same position as the SAD aperture was compared to the magnification of the image of the specimen allowing evaluation of the optical parameters.

Three types of lattice shaped grids were employed as the specimen and the shadow aperture in this experiment. The first is the Nickel screen mesh with spacing of 16.94 μm. The second is the holey carbon films (Quantifoil R2/2: Quantifoil Micro Tools GmbH, Jena, Germany) with holes of 2 μm diameter arranged in a square lattice pattern with edge-to-edge spacing of 4 μm. The third is the square lattice on a gold thin foil with thickness 10 μm fixed to a 100 mesh Cu grid, which was formed in the shape of a square lattice with 1 μm x 1 μm square holes with lattice spacing of 2.46 μm by using Focused Ion Beam (FIB) apparatus FB-2000 (Hitachi High Technologies, Tokyo, Japan).

In order to evaluate the optical parameters M_d, M_{ap}, M_p and M_i, the Nickel screen mesh or Quantifoil R2/2 was used. Figure 6 shows an example of the image and shadow image when the Quantifoil R2/2 was used, in which the holes in the carbon film with same as size in 2 μm diameter, were observed in different diameter.

First of all, the magnification of the specimen image, the value of $M_p \times M_i$ was determined directly on the photographic film. We used the standard setting of the projector lens system of the microscope with zero objective lens current, which corresponds to the 1st and 2nd condition discussed in section 2, therefore giving the optical parameter M_{ap} as unity. The magnifications of the condition assigned to 0.2 K, 0.3 K and 0.5 K of LM mode was 230, 320 and 586 with the zero objective lens current, respectively.

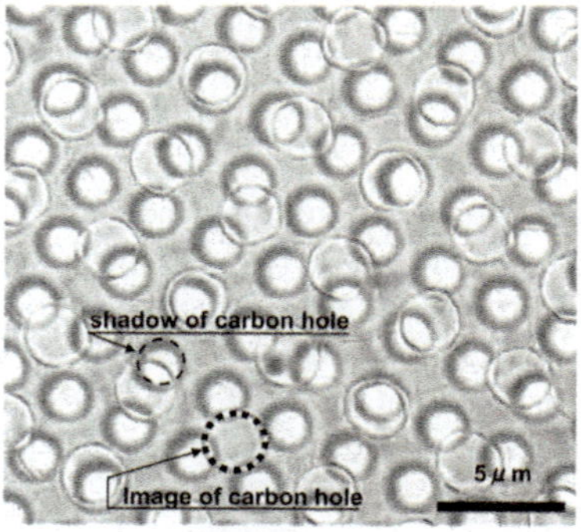

Fig. 6. Observed image and shadow image of Quantifoil 2/2.

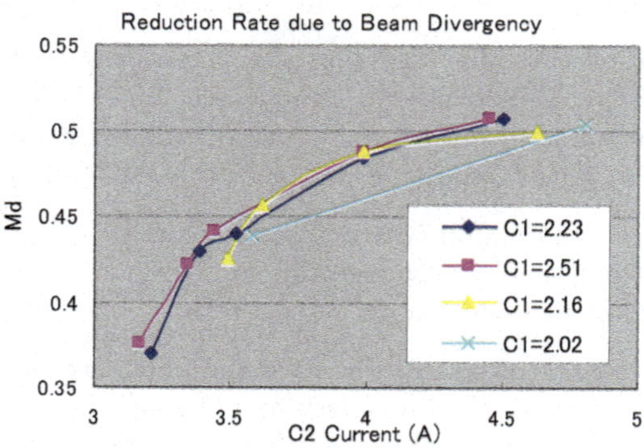

Fig. 7. Measured M_d as a function of C2 condenser lens current at different C1 condenser lens currents.

Figure 7 shows the ratio of the distance between the holes on the image and shadow image as a function of the current I_{c2} which was applied to the C2 condenser lens when the objective lens was turned off. Some data were obtained using the lattice of the Nickel screen mesh and others using the holes of Quantifoil R2/2. The results correspond to the value of M_d, which showed an initial rapid increase and gradual saturation as a function of the current applied to the C2 condenser lens. The measurements were taken at different currents applied to the first condenser lens C1. However, the results show that the current applied to C1 does not affect the value of M_d. This result supports our optical model which states that only C2 is appropriate to understanding our optical system.

As shown in Fig. 4b, M_d can be written as $c/(c+a)$. where a is the geometrically determined constant value. The value of c can be changed according to the focal length f_c of lens C2, which is determined as $c = L_{c\text{-}o} - f_c$, where $L_{c\text{-}o}$ is the distance between C2 and the objective lens. The focal length f_c of the magnetic lens is given as $f_c = k_c/I_{c2}^2$ (Liebmann 1955; Spence 1981), where I_{c2} is the current applied to lens C2 and k_c is the constant determined by the lens structure. Replacing c using $L_{c\text{-}o}$ and k_c/I_{c2}^2 gives

$$1/M_d = 1 + a/(L_{c\text{-}o} - k_c/I_{c2}^2)$$

Rewriting this equation gives

$$aM_d/(1-M_d) = -k_c/I_{c2}^2 + L_{c\text{-}o} \tag{3.1}$$

This result indicates that if the value of $a\,M_d/(1\text{-}M_d)$ is drawn as a function of $1/I_{c2}^2$, a liner relation will be given, and then k_c and $L_{c\text{-}o}$ can be given as the inverse slope and the y-intercept of the liner relation, respectively.

Figure 8 shows $a\,M_d/(1\text{-}M_d)$ as a function of $1/I_{c2}^2$ for C2 current between 3.42 and 4.45 A while keeping other lens conditions constant. The graph shows a liner relation and gives the value of k_c and $L_{c\text{-}o}$ to be 915 A^{2}mm and 218 mm, respectively, where a is given as 155 mm on the blueprint of the microscope. This result indicates that the above geometrical optics analysis can explain the optical system in our microscope very well.

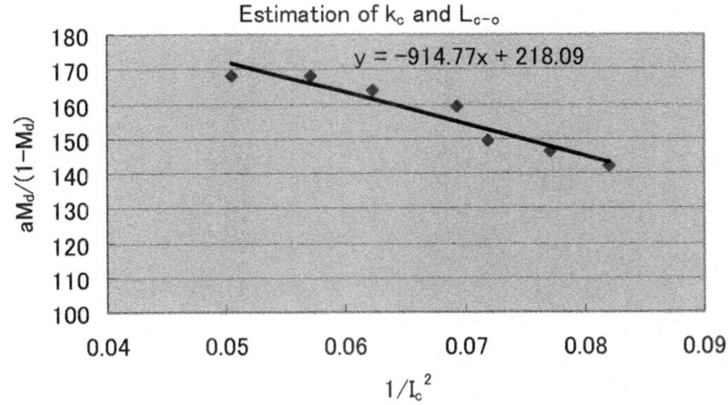

Fig. 8. Estimation of the distance between C2 and objective lens $L_{c\text{-}o}$ and the lens constant of objective lens k_c.

When current is applied to the objective lens, the parameter M_{ap} is not unity. The value M_{ap} might be estimated as a function of the objective lens current I_o. In real conditions in the electron microscope, a completely parallel electron beam can not be obtained. Under this limitation, M_d can not to be unity. Therefore, the multiplied value of M_{ap} and M_d can be obtained as the ratio M_{dap} of the distance between the lattice on the image and the shadow image, which can be directly measured on the recorded image in the same way shown above, and will be a function of both M_{ap} and M_d. Regarding equation (2.1), M_{dap} could be written as

$$M_{dap} = M_d M_{ap} / \left(M_d + M_{ap} - M_d M_{ap}\right) \tag{3.2}$$

Replacing $M_{ap} = f_o / d_{ap} = f_o / (f_o - a)$ into (3.2), which is derived from the definition, gives

$$M_d / M_{dap} = f_o - M_d a / f_o \tag{3.3}$$

The f_o can be written in the same way as for the case of f_c as $f_o = k_o/I_o^2$ where k_o is the constant of the objective lens. Placing this relation into equation (3.3) gives k_o as

$$k_o = a I_o^2 M_{dap} M_d / \left(M_{dap} - M_d\right) \tag{3.4}$$

The value of k_o plotted as a function of I_o^2 for several conditions (Fig. 9) gives a constant value of about 58 A^2mm except at low I_o. For I_o less than 0.2 A, the difference between M_{dap} and M_d is lower by three orders of magnitude than the value of M_{dap} and M_d, which is less than the error in the measurement of M_{dap} and M_d. This is the reason why k_o is not constant at I_o less than 0.2 A. This result suggests that geometrical optics can explain the behavior of the electron microscope in our condition. As a result, all the parameters M_d, M_{ap} and $M_p \times M_i$ in equation (2.1) can be determined using the currents I_{c2} and I_o, which are those applied to the 2nd condenser and the objective lens, respectively.

Similar measurements have been performed in two types of high voltage electron microscopes (HVEM), H-1250ST (Hitachi High Technologies, Tokyo, Japan) and JEM-1000k RS (JEOL Ltd., Tokyo, Japan) as shown in Figs. 10 and 11. Both types of HVEM followed equations (3.1) and (3.4) very well, proving that our analysis can be applied to any type of TEM. Higher penetration power of a higher accelerating voltage will allow application of our method to observe the inside of materials.

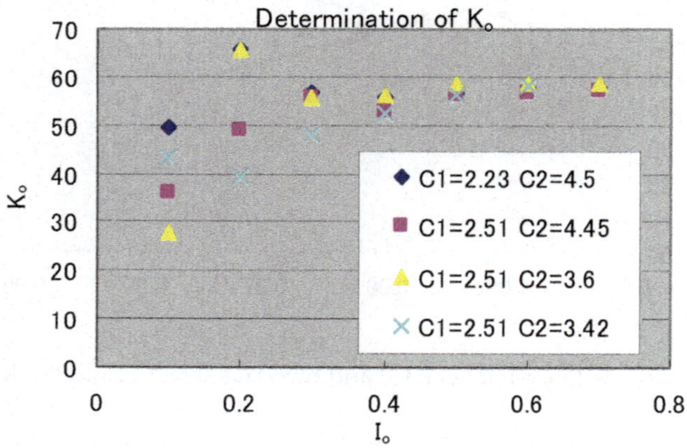

Fig. 9. Estimation of objective lens constant k_o

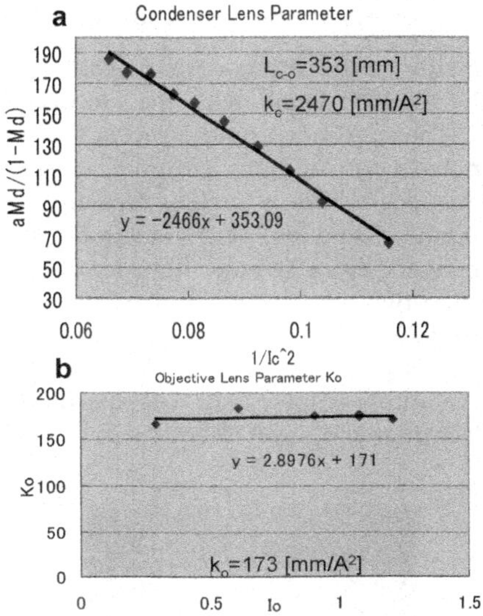

Fig. 10. The measurement of optical parameter (a) $L_{c\text{-}o}$ and k_c and (b) k_o of H-1250ST at an accelerating of 1000 kV.

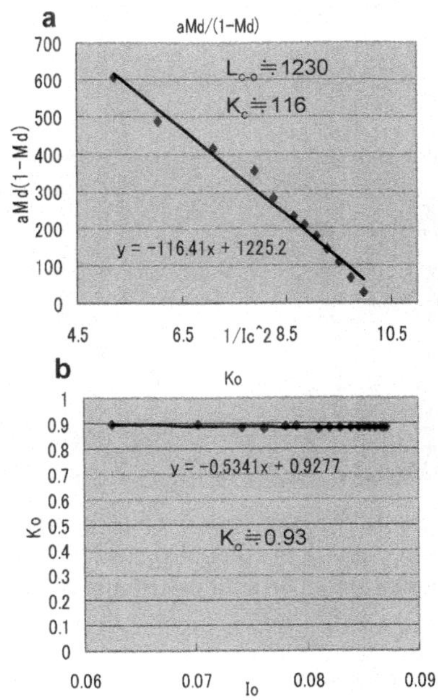

Fig. 11. The measurement of optical parameter (a) $L_{c\text{-}o}$ and k_c and (b) k_o of JEM-1000k RS at an accelerating of 1000 kV. The values are shown as the ratio of maximum current of the lenses.

3.2 Measurement of the shadow image shift in a uniform electric field

The amount and proportionality of the shadow image shift was examined by applying a uniform electric field to the electron beam with the electrodes placed at the specimen position as shown in Fig. 12. The specimen holder used in this experiment was the Kamino holder. The Kamino holder that was originally designed by Kamino and Saka (1993) as a heating holder has two electrodes where a heating element is usually fixed. One of the electrodes is connected to ground and the other electrode, through which the external voltage can be applied, is insulated from ground. In order to create a uniform electric field between the narrow gap of the two electrodes, a pair of electrodes consisting of Al plates with thickness of 13 μm were fixed to the respective electrodes of the specimen holder and carefully positioned with a spacing of 10 μm as shown in the optical microscope image and schematic drawing in Fig. 12a and b, respectively. The spacing between the electrodes was measured precisely on an electron micrograph to be 10.6 μm before the experiment.

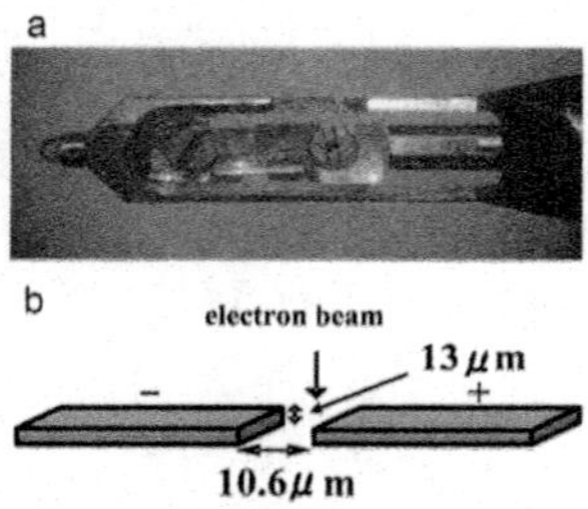

Fig. 12. The two electrodes fixed on the specimen holder (a) and the schematic drawing of them (b).

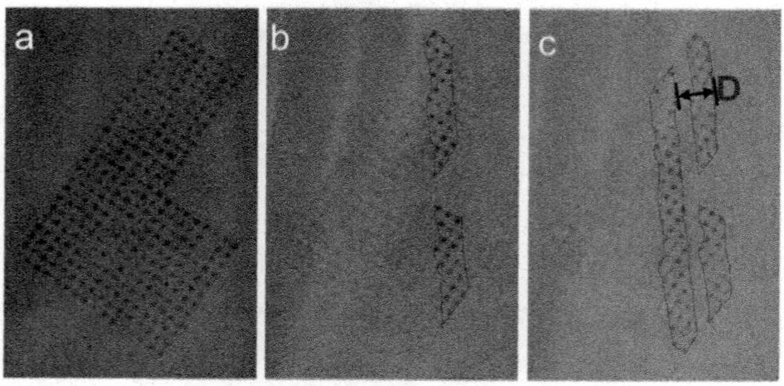

Fig. 13. The shadow image of (a) the FIB fabricated grid, (b) after the electrodes placed at the specimen position without applying voltage, i.e., 0 V, and (c) the shadow image applying 5 V with the superimposed image of 0 V. The shallow image of (a) is also superimposed on (b) and (c) as a guide of eyes.

The FIB fabricated grid as mentioned above was positioned at the SAD aperture as shown in Fig. 13a. This grid is well suited to avoiding misreading of the position in the repeated pattern because of minor irregularities in the shape caused by the unstable fabrication conditions during the home made fabrication processes. This type of shadow aperture was used to measure the shift of shadow image in a uniform electric field. The move of the shadow image of the grid observed between the image of two electrodes (Fig. 13b and c) was recorded on TEM films as a function of the applied voltage between these two electrodes.

The shift of the shadow image of the FIB fabricated grid was measured up to 2 V on the fluorescence screen. The value of deflection of the electron beam calculated from the shift of the shadow image is shown as a function of applied field strength between the electrodes as shown in Fig. 14 for the condition a = 155 mm, $M_i \times M_p$=514, M_d=0.632, M_{ap}=1.57. Because the deflection angle θ of the electron beam is proportional to the integral of the x-y component of the electric field E along the length t of the electron passage at the specimen, the shift D of the aperture image is given by

$$D = kM_d M_{ap} M_i M_p Et \tag{3.5}$$

where k is the sensitivity coefficient.

Comparison of equations (2.2) and (3.5) gives,

$$\theta = k\, Et \,/\, a \tag{3.6}$$

In the uniform electric filed, the deflection angle of the electron is given by (Merli et al., 1975)

$$\theta = eEt \,/\, mv^2 \tag{3.7}$$

and in the case of magnetic field B is

$$\theta = eBt \,/\, mv^2$$

where e is the electron charge (1.602×10^{-19}C), and m and v are the relativistic mass and velocity of the electrons. Substituting equation (3.7) into (3.6) gives

$$k = ea \,/\, mv^2 \tag{3.8}$$

At an accelerating voltage of 300 kV, $m=1.587\times m_0$, where m_0 is the rest mass of the electron, 9.109×10^{-31}kg and $v=2.330\times10^8$m/s (Williams and Carter, 1996) gives $k=3.06\times10^{-7}$m/V.

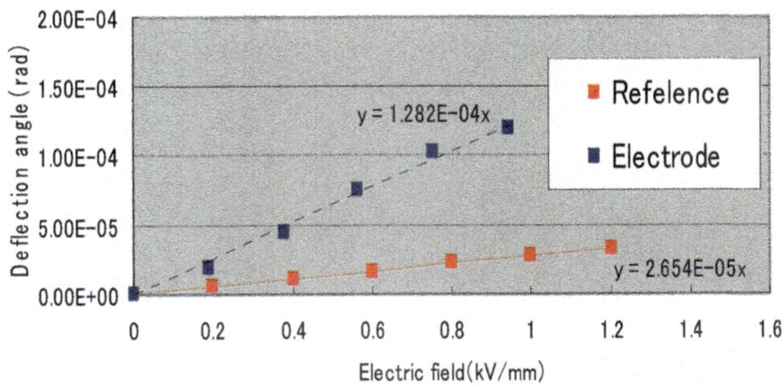

Fig. 14. The deflection angle of the electron beam as a function of applied electric field. The reference means the value when the field is restricted within the thickness of electrodes.

The experimental results showed good proportionality to the voltage applied to the electrodes. However, the discrepancy of *k* between the value derived from the equation (3.7) and the measured value was 4.8 times. In order to resolve this discrepancy, we will discuss the electric field distribution in our experiment, which in reality was not uniform and restricted to the gap between the two electrodes.

3.3 Numerical calculation of the electric field distribution along the electron trajectory

A numerical calculation has been performed to evaluate the distribution of the electric field around the two thin Al plate electrodes used in our experiments. An ordinary finite difference method was employed to calculate the 2D electric potential distribution on the cross section perpendicular to the edge of the electrodes and along the electron beam trajectory. The rectangular area of 2 mm and 1 mm was divided into 200 times 100 lattices with 10 μm mesh, and the potential $U_{x,y}$ on each lattices was calculated, where x=0 to 200 and y=0 to 100. The schematic drawing of the calculated lattice is shown in Fig. 15

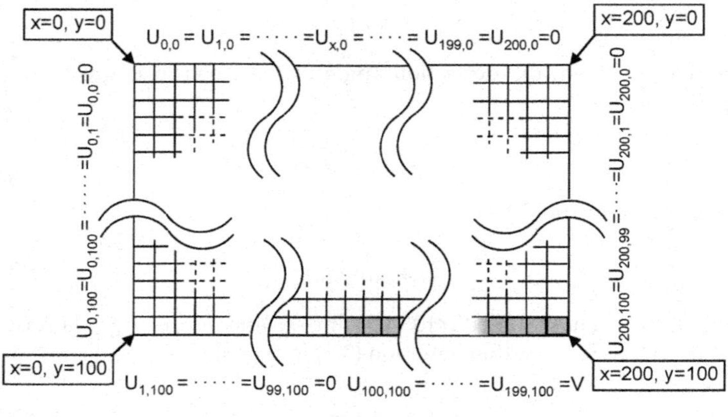

Fig. 15. The lattice model used for the estimation of the distribution of the field

The potentials on the boundaries were set on the top, i.e., $U_{x,0}$, and both side, i.e., $U_{0,y}$ and $U_{200,y}$, of the lattice at 0 Volt. The boundary values of the left bottom side in Fig. 15, which corresponds to the electrode connected to the ground earth, i.e., $U_{1,100}$ to $U_{99,100}$, were also set to 0 Volt. The boundary values of the right bottom side, i.e., between $U_{100,100}$ and $U_{199,100}$, were set to the applied voltages to the electrode. The difference equation on each lattice point was

$$U_{x,y} = \left(U_{x-1,y} + U_{x+1,y} + U_{x,y-1} + U_{x,y+1}\right) / 4$$

The matrix of $U_{x,y}$ for x=1 to 199 and y=1 to 99 were calculated using Gauss-Zeidelle method until the condition $\Sigma\Delta U_{xy}/\Sigma U_{xy}<10^{-3}$ is satisfied.

Electric field was calculated taking the difference of adjacent potentials. The electron deflection was calculated from the integration of the field in the x-y plane assuming the trajectory of the electrons was along the Z axis during their passage through the field.

Figure 16a shows a schematic drawing of the calculated model around the electrodes using the finite differential method. The brighter contrast corresponds to higher potential in the calculation result in Fig. 16b. As shown in Fig. 16b, the potential of the electrodes diverged until about 0.5 mm away from the electrodes. Figure 16c shows the horizontal part of the electric field derived from the differentiation of the potential along the line running on the vertical centre of Fig. 16b, where the electron beam is running. The integral of the horizontal part of the electric field (Fig. 16d) giving a value 4.8 times of the value if the electric field is restricted to the thickness of the electrodes along the electron passage. This value corresponds to the difference of the sensitivity factor k between the experimental value and the calculated value using equation (3.8). This means that our observation is really measuring the deflection of the electron by this applied electric field.

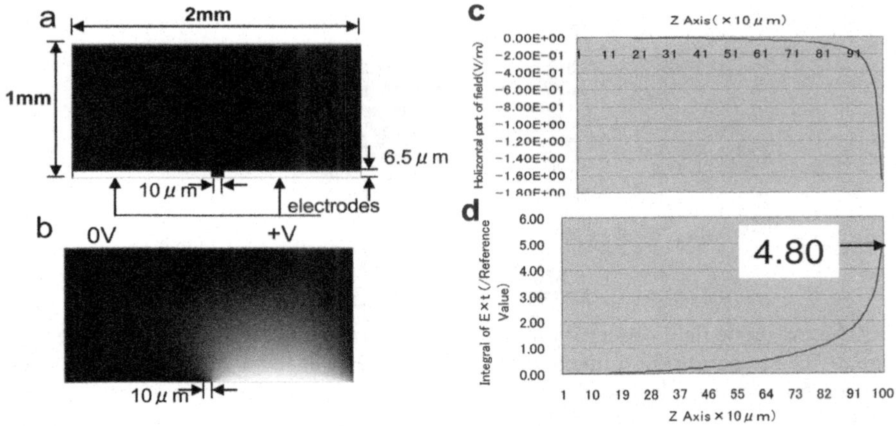

Fig. 16. The results of numerical calculation. (a) The schematic drawing of the calculated model and (b) gray scale map of the potential. (c) The horizontal part of electric field and (d) the integral along the line running vertical centre of (b).

We need the correction of the field distributed in the upper and lower part in which we are interested. However, this kind of problem could happen in the other method. For example, in the case of electron holography, the electromagnetic field is usually distributed over a wide area in and/or around the specimen whereas the interference region is limited to a bundle like shape. The reference electron wave will be affected by the electromagnetic field in the region that it is traversing, which is undetectable. The electric field observation in the previous work is just showing the difference between the region of interest and the region where no significant distribution of the field is assumed. However, our calculation revealed that the distribution of the field is hard to limit near the specimen and is usually distribute in the order of mm. The reliability of quantitative measurement will be discussed again in the next section.

4. Applications

The geometrical optics analysis and the measurement of a simple electric field showed us that the method we developed allow not only qualitative but also quantitative measurement of electric and/or magnetic fields. In this section, we describe three examples of applications. Two of them are examples of qualitative measurement showing the facility of our method, and the other is an example of quantitative measurement.

As a preliminary experiment before the detailed experimental studies which use the electric field, we will show the magnetic field observation around a particle of commercially available ferrite magnet. Figure 17a shows the superimposed image of the specimen onto the shadow image of the Ni screen mesh when the specimen has not been inserted, i.e., without the distortion. An apparent distribution of the shadow image distortion was observed after the insertion of the specimen. In Fig. 17b, the magnetic line of force running perpendicular to the arrows indicates the direction and amount of shadow image shift. The figure demonstrates that the method can observe a magnetic field. In our method, the specimen is free from the magnetic field of the lens, because the objective lens is basically turned off or only weakly activated, which could be a benefit for the application of magnetic field observation.

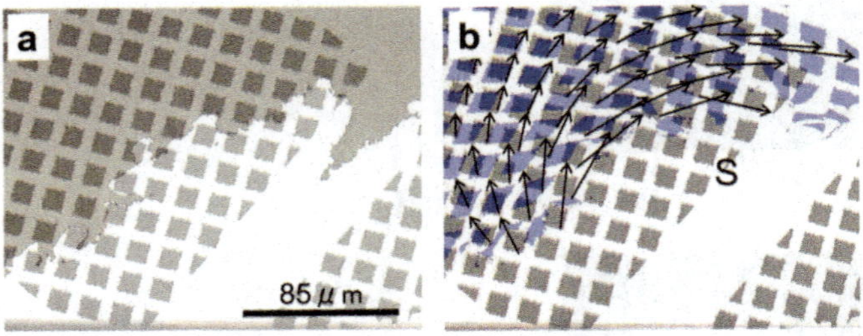

Fig. 17. The magnetic field observation of around a particle of commercially used ferrite magnet. (a) The superimposed image of the specimen and shadow image of the Ni screen mesh when the specimen has not been inserted. (b) The superimposition of the shadow images with the field (blue) onto without the field (gray). The arrows indicate the deflection of the electron beam. The magnetic line of force runs perpendicular to the arrows. The magnetic south pole of the particle marked by S.

In the following detailed experimental studies, we used the electric field only, because it was easy to create and control.

4.1 Qualitative measurement of a non-uniform electric field embedded in a uniform field

This example is the measurement of a non-uniform electric field embedded in a uniform field. The electric field far away from the electrodes will show a moderately changing feature with an average value compared to the field near the electrodes, which will show a rapid change. In order to observe the contribution of these two fields onto the shadow image, we prepared a specimen as shown in Fig. 18. Figure 18a is a schematic drawing of the specimen configuration. The half cut Cu grid with a carbon film is fixed to the electrode and a voltage applied to the counter electrode. As shown in Fig. 18b, the electric lines of force (the solid line arrows) and the electron beam trajectories (the red line arrows) are drawn, schematically, in the cross sectional view of the specimen. In the crescent of the metal, there will be a non-uniform rapidly changing field distribution. The electrodes supporting the Cu grid with the large gap will make a moderately changing field distribution in a large area.

Figure 19 shows the superimposed shadow images of FIB shaped grid with and without external potential, which is the same as the one used in section 3.2. The applied external voltages to the counter electrode in Fig. 19a and b were 20 V and 10 V, respectively. The averaged shifts of the grid correspond to the external potential as shown by the dotted line arrows in the figures. The narrow and thick line arrows in each figure indicate the unequal shift of the lattice holes from the position of without and with external potential to the electrodes, respectively. In Fig. 19a, there is systematic distortion in the position of the holes, and almost no distortion was observed in Fig. 19b. The carbon film in the metal crescent has been removed during specimen preparation for Fig. 19a. The local non-uniform distribution of electric field is formed in the metal crescent. However, the carbon film in the metal crescent will have the same potential as the metal electrode, which means that there is uniform potential distribution in the area shown in Fig. 19b; therefore, there is no distortion of the shadow of the lattice. This result shows that the rapid change of electric field near the specimen can be detected separately from the moderately changing electric field in the wide area, if the local distortion is taken into account in the hole shift of the shadow image of the lattice. Of cause, the complete separation and quantitative evaluation of the local field requires comparison between numerical calculation and measured value.

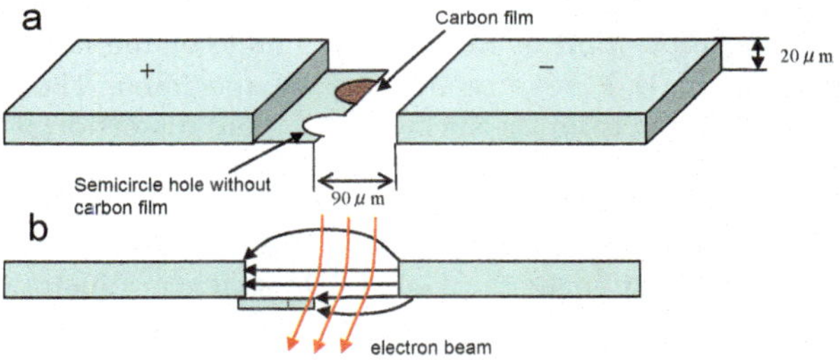

Fig. 18. Schematic drawing of (a) the specimen configuration and (b) the cross sectional view of the specimen. The electric lines of force (arrows of sold line) and the electron beam trajectories (arrows of red line) are drawn, schematically.

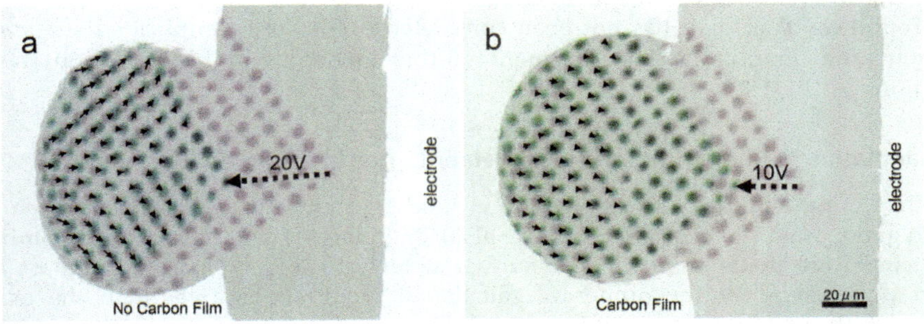

Fig. 19. The 2D-maps showing the difference of the local field distribution in the semicircle holes (a) without and (b) with carbon film indicated by the solid line arrows. The shallow pink and green grids correspond to the shadow image before and after applying potential. The conductive carbon film normalized the field in the hole (b). The whole shifts of the shadow image of the grid (the dotted line arrows) correspond to the field strength in the upper part of the specimen.

4.2 Two-dimensional field distribution observation of the reverse biased p-n junction

The second example is the two-dimensional field distribution around a reverse biased p-n semiconductor junction, which we mentioned in section 1.

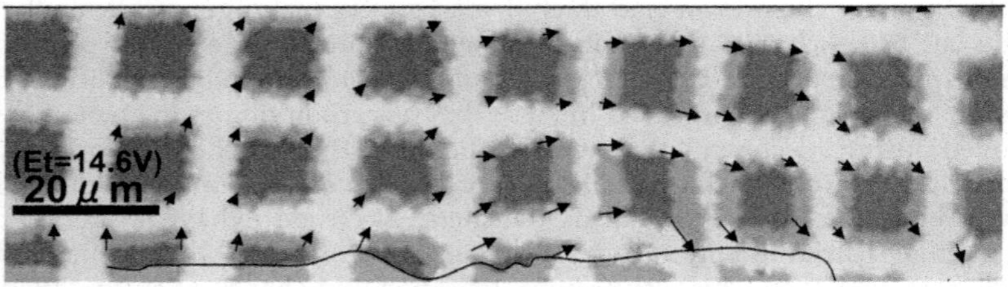

Fig. 20. The electric field observation using the square lattice shaped aperture. The direction and the length of arrows indicate the direction and the strength of the integrated electric field along the electron passage (*Et*). The arrow length of 20 μm in the image corresponds to 14.6 V of *Et*.

Figure 20 shows the shadow image distortion of the square lattice shaped Nickel sheet mesh, which is the superimposition of the shadow image of the lattice without reverse bias onto the image with 5.01 V reverse bias to the specimen. The two images were aligned at a position far enough from the specimen to avoid distortion. The corresponding positions in the two images are connected by arrows. The direction and length of the arrows indicates the direction and strength of the integrated electric field along the electron path (*Et*). The arrows in the figure show the two-dimensional distribution of electric field near the p-n junction in the GaP LED, which is illustrated in Fig. 1. The direction of the field is opposite in the p- and n-regions along the edge of the specimen. In vacuum, a few microns away from the specimen edge, a continuous potential gradient is observed from the p- to the n-region. The result is, qualitatively, in good agreement with the electron holography observation (Sasaki, et. al., 2000; Wang, et al., 2003). At present, the separation of the contribution of the field which is near the specimen from the field away from the specimen is not available, because the detailed numerical calculation of the field around the specimen has not been completed due to the complicated shape of the specimen. The quantitative measurement of this method will be shown in the next subsection.

4.3 Quantitative field distribution measurement around a field emitter needle

The third example is the field distribution around an emitter needle of the field emission electron gun (Sasaki, et al., 2008b; Sasaki, et al 2010). In this system, the materials comprising the specimen are made of metal. The surface potential could be specified almost as the applied potential. Then the numerical calculation would be simpler than the example discussed in the section 4.2.

The two dimensional distribution of the measured value was compared with the calculated value using the surface charge method (Murata, et al., 2004). The numerical calculation model of the specimen was constructed based on the optical image shown in Fig. 21 and TEM observation, though the asymmetry of the needle was not taken into account in the first approximation.

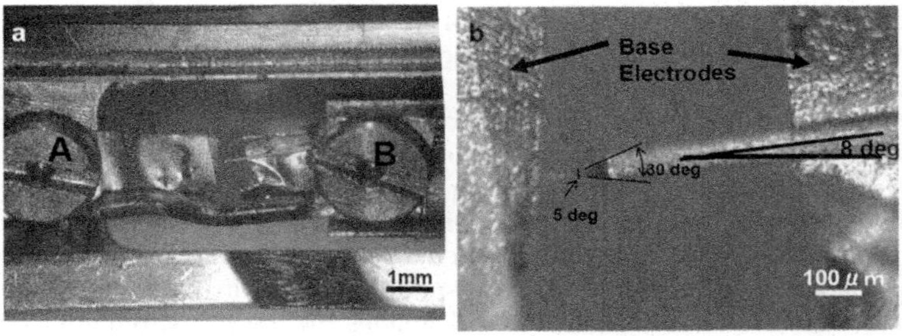

Fig. 21. An optical micrograph of the specimen (a). The enlarged image between the needle tip and the grounded electrode is shown in (b).

In this experiment, two different types of shadow aperture were used. The first one is the holey carbon film Quantifoil R2/2 as shown in Fig. 22a, which was placed at the position of the SAD aperture. This aperture allows us easy identification of the shift of the shadow image of holes by eye. Another type of shadow aperture has been utilized in order to modify the spatial resolution and to allow an automated analysis. The home made carbon film with the rough surface was fixed onto the 200 mesh Cu grid. The surface of the film was decorated by Au deposited to a thickness of about 10 nm. The carbon film showed randomly distributed granular contrast as shown in Fig. 22b. It was hard to identify the distortion of the shadow image of this aperture by eye. However, the dedicated software using the pattern matching method was employed to detect the distortion in the shadow image.

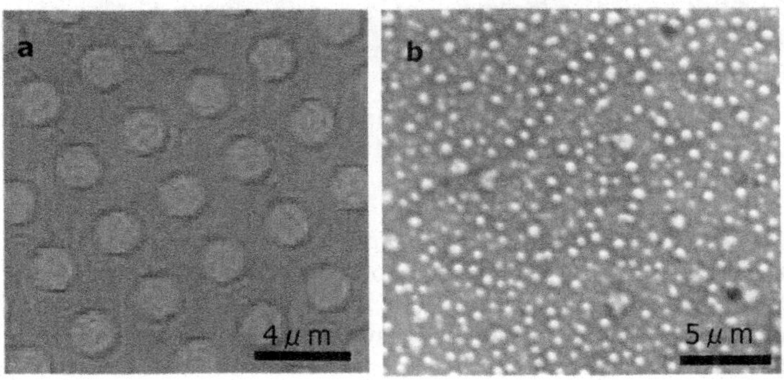

Fig. 22. Two different types of shadow image aperture. (a) Square lattice patterned holy carbon film. (b) Au decorated carbon film with randomly distributed granular contrast.

TEM images of the needle and the shadow images of the Quantifoil R2/2 before applying external voltage and after applying +30 V are shown in Fig. 23a and b, respectively. The shadow image of the aperture is fixed clearly in an area larger than 100 μm×100 μm. In the center of Fig. 23a and b, the image of the needle was observed as a dark shadow because the needle was too thick to be transparent to the electron beam. The square lattice arrangement of the circles corresponds to the shadow image of Quatifoil 2/2. The image of the edge of the ground electrode is seen as the horizontal line in the dark area in the upper part of Fig. 23a and b. The tilted square of dark area surrounding the figures corresponds to the shadow image of the grid mesh which is supporting the holy carbon film of Quatifoil 2/2. There is a discontinuity in the shadow image of the grid mesh at the bottom of the needle in Fig. 23b, which could be interpreted as the difference of the field direction on the left and right hand side of the needle. The vector of the field on the left and right hand side of the needle has the X part toward left and right hand side, which will shift the shadow image of the grid mesh to left and right hand side, respectively. This counter move of the shadow image causes the discontinuity at the edge of the grid mesh. The images after applying external voltage were superimposed onto the image before applying external voltage carefully aligning the needle images at the same position. The displacement and distortion of the holes of the shadow image of the aperture were observed, particularly near the needle tip in Fig. 23b. The displacement of the shadow image of corresponding holes in the shadow aperture was measured for each hole in the superimposed images. In the measurement, X and Y axes which are parallel and perpendicular to the edge of the ground electrode, respectively, with the origin at the needle tip were determined as shown in Fig. 23a. The displacement *D* was converted to the field strength *Et* using equation (3.5).

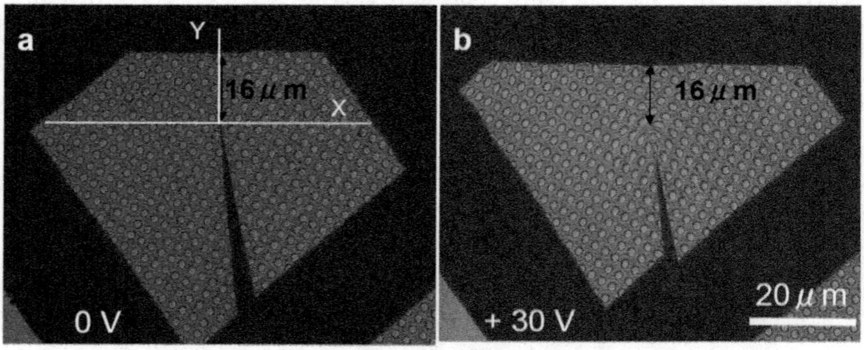

Fig. 23. Shadow images of the Quantifoil R2/2 aperture (a) before and (b) after applying external voltage at +30 V. The displacement and the distortion of the aperture holes are shown in (b).

Figure 24 shows measured values of the electric field at an applied voltage of 30 V. The closed circles in Fig. 24 show the Y component of the electric field, which is perpendicular to the edge of the ground electrode, along the line parallel to the ground electrode as a function of the distance X from the needle. The solid lines in the figure show calculated values using the surface charge method. The calculated values were the integration of the horizontal part of the electric field along the trajectory of the electron above and below the specimen until the value became saturated. The thickness taken into account in the calculation was 2 mm above and below the specimen. Figures 24a, b, c, d, e and f show the Y component of measured and calculated values of the electric field along the line parallel to the ground electrode at Y distance from the tip of the needle for 3 μm, 5 μm, 7 μm, 9 μm, 11 μm and 13 μm, respectively. Due to not only the limitation of the hole position in the shadow aperture but also the tilt of the row of holes to the X axis as shown in Fig. 23, the Y distance measured position could have a repeated systematic increase and sudden decrease of errors in the range of ± 1 μm from the given positions. Such systematic error in the measured position causes a systematic error in field strength of about ± 1 V, which could account for the shallow pseudo shoulder in the field distribution at the discontinuity of error. Taking into account the estimated error, it was found that the measured values showed good agreement with the calculated values with an error of less than 10%. In the range of this error, the measured value of the field decreases monotonically as a function of the X distance from the needle. However, in Fig. 24f, the dispersion of the value became wider than the estimated error. This is due to the field disturbance caused by the irregularity of the shape of the ground electrode and the charging up of dust at the edge of the electrode, which could be observed by the irregular distortion of the shadow image of holes along the ground electrode in Fig. 23b. The agreement was better on the left hand side of the figures. Particularly in Fig. 24a, a rapid increase of the value near the needle tip showed good agreement with the calculated value. In contrast to the good agreement between the measured and the calculated values on the left hand side, the results on the right hand side in the figures showed comparably poor agreement, perhaps, because of the asymmetry of the field due to the tilt of the needle at an angle of 8 degrees from the line perpendicular to the edge of the ground electrode. Further detailed calculation taking into account the asymmetry of the needle will be required. However, the asymmetry could be understandable qualitatively as follows. When the needle is tilted counter-clockwise as in Fig. 23, the vector of the field will follow the tilt. In the left hand side of the needle, the vector of the field will turn away from the perpendicular line to the ground electrode and then the Y component of the vector will decrease. However, the field attracted to the ground electrode could have the tendency to tilt back to the perpendicular line to the ground electrode, which will reduce the decrease of Y component of the vector. In the right hand side of the needle, the vector of the field will be tilted and positioned along the perpendicular line to the ground electrode, which will increase the Y component of the vector. However, the compensation of the field rotation could not work compared to the left hand side of the needle, and then the increase of the Y component of the vector in right hand side of the needle will remain.

The spatial resolution using the shadow aperture which has a square lattice pattern should be limited by the spacing of the pattern. In order to improve the spatial resolution of the method, a modified shadow aperture which gives the shadow images of randomly dispersed dots and dedicated software which allows automatic detection of the shadow image distortion have been applied to measure the field distribution at the tip of the needle. Figure 25 shows the main window of the developed software. The shadow images before and after applying an external field by applying 4 V to the electrode are shown in Fig. 25a and b. It was hard to identify the difference between the two images by eye. However, the pattern matching method (Szeliski, 2006) revealed the distortion in the shadow image. The image was divided into a series of cell-like images with size SX and SY. The template images were selected with size CX and CY in each cell of the image before applying external voltage. The sum of the squares differences (SSD) was calculated between the template and the corresponding cell-like image after applying external voltage shifting the position up and down to AX and AY. At the minimum of the SSD found in the area of AX and AY, the shift parameters in X and Y direction were determined to be the displacement vector of the template image in the image after applying the external voltage. The displacement vectors were displayed as the solid line which originates at the center of the circle indicating the left upper corner of the template as shown in Fig. 25c. The processed image using this software showed the two-dimensional field distribution with spatial resolution less than 1 μm. In the area where there was no rapid change of the field, the spatial resolution of the software was modified up to 100 nm. However, in the area near the needle tip, the simple SSD function gave the wrong result due to the large distortion in the shadow image. Further modification in the pattern matching method will be required to obtain nanometer spatial resolution without exception.

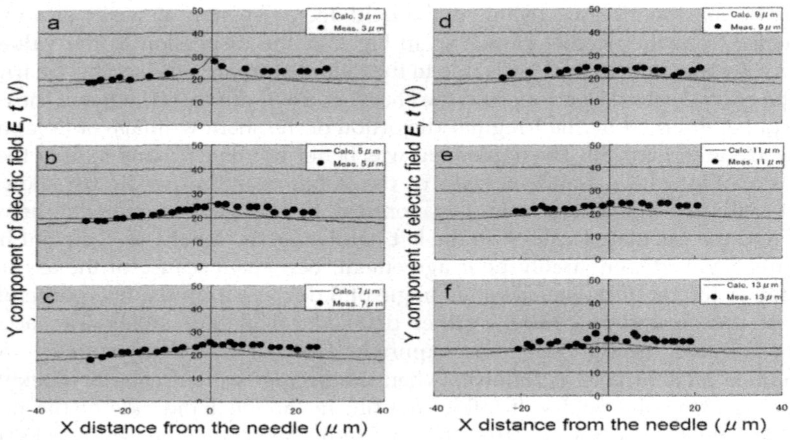

Fig. 24. Y component of the electric field as a function of the distance X from the needle along the line parallel to the ground electrode with distance from the tip of the needle at (a) 3 μm, (b) 5 μm, (c) 7 μm, (d) 9 μm, (e) 11 μm and (f) 13 μm, respectively.

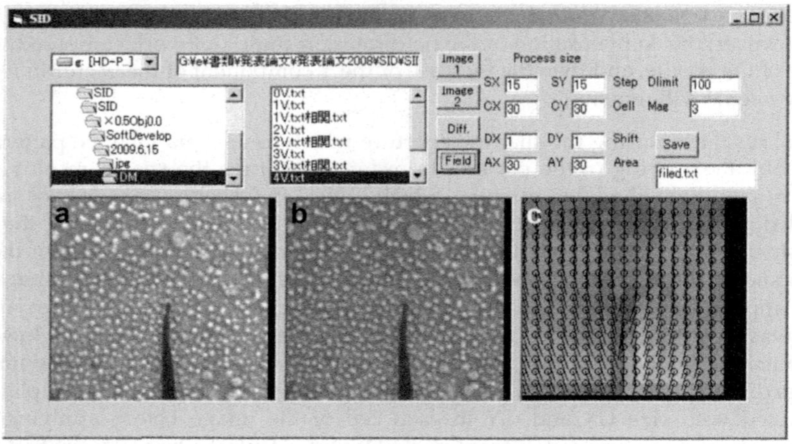

Fig. 25. The main window of the developed software. The shadow image (a) before and (b) after applying external voltage at 4 V, and (c) the detected image distortion corresponding to the electric field distribution.

5. Conclusion

We have developed a simple and quick method to measure the electric and magnetic field around a specimen using a conventional TEM. In this chapter, we have discussed the geometrical optics analysis of our method in a modern conventional transmission electron microscope. The results suggested that the shift of the shadow image of the beam obstructer placed at the position of the selected area diffraction aperture is proportional to the deflection angle around the specimen. It was found that the magnifications of the specimen image $M_i \times M_p$ and the shadow image of the beam obstructer M_d and M_{ap} on the fluorescent screen have an important role in measuring the electric and magnetic fields at the specimen. The magnification $M_i \times M_p$ can be measured obtaining the lattice image of the specimen with known spacing. The method to evaluate the optical parameters M_d and M_{ap} using the current applied to the second condenser lens and objective lens has been discussed. The quantitative measurement of the electric field has been tested in a uniform field. The comparison of the experimental result with the numerical calculation of a long range field distribution showed the possibility of quantitative measurement of the electric field using this method.

Some preliminary applications have been introduced. Two examples of qualitative observation of the two dimensional projection of the electric field along the electron beam direction were investigated involving a qualitative separation of the field near the specimen from the moderately distributing field away from the specimen.

An example of the quantitative measurement of the electric field has been shown using the electric field distribution at the tip of a narrow needle. The measured value was confirmed as the integrated value of the horizontal part of the electric field along the electron trajectory. The result showed not only good agreement with the calculated value with error of less than 10% but also the possibility to reveal further detail of the electric field distribution. The dedicated shadow aperture and software allow us to map the electric field distribution with spatial resolution on the order of 100 nm, particularly in the area where the field is changing moderately. We believe that the method can measure the two-dimensional distribution of the field over an area more than 100 μm with submicron spatial resolution.

Our method is limited to the LM mode. However, it is possible to detect the electromagnetic field distribution in a large area at the specimen level, since the method simply detects the deflection of the electron beam at that level. As shown in section 3, the electromagnetic field is usually distributed over a wide area in and/or around the specimen. In electron holography, the reference electron wave will be affected by the electromagnetic field in the region that it is traversing, whereas the interference region in the image of an electron hologram is limited to a bundle like shape. Our method can also be performed in a field emission gun transmission electron microscope (FEG-TEM) equipped with an electron bi-prism because the electron bi-prism is placed at the same position as where the SAD aperture was placed. The distortion of the shadow image of the bi-prism had been observed in an electron holography observation (Sasaki, et al., 2000). The narrow and sharp image of the bi-prism is more suitable than the irregular edge shape of the SAD aperture for quantitative measurements. Similar phenomenon has been studied by Pozzi (1975, 1980 and 1982) and Frzzini et al., (2006). In order to estimate and/or correct the phase disturbance in the reference wave of electron holography, using both methods simultaneously would be advantageous.

The Lorentz microscopy uses contrast modification of the specimen image due to the field. Therefore, it cannot obtain any information in the area of vacuum. Our method is suitable to observe the field distribution in vacuum rather than in the material which reduces the intensity of the electron beam. Simultaneous use of both methods could make them complementary to each other.

Some other shadow methods (Marton and Lachenbruch, 1949; Jakubovics, 1964; Ishba et al., 1974) which have optics similar of the Foucault mode of Lorentz microscopy, i.e., the beam obstructer is placed after the imaging lens, have already been developed, however, the methods have never been used in a wide variety of applications. Presumably, the optics has a problem, because the imaging lens which has the strongest magnetic field will be changed and adjusted during imaging in these methods. This causes a change of major optics parameter, which gives instability to the optics and a loss in quantitative ability of the measurement. Our method avoids this problem by using the 1st intermediate lens as an imaging lens, which is placed after the beam obstructer.

In future development, observation of the field distribution inside materials will be required, and employment of the HVEM looks promising. Improvement of the spatial resolution in a medium voltage electron microscope to better than 10 nm has been obtained through higher magnification (Mori, 2009). Though the complete quantitative comparison with other methods, for example, electron holography (Oikawa et al. 2007), will be required to generalize this method, we can foresee that our method will give more fluent results, not only if it is used with other methods, i.e., electron hologram or the Lorentz microscopy but also as a standalone method .

6. Acknowledgement

This work was supported by Grants-in-Aid for Scientific Research B18360303 and B21360307.

7. References

Bajit, S., Barty, A., Nugent, K. A., McCartney, M., Wall, M. and Paganin, D., (2000). Quantitative phase-sensitive imaging in a transmission electron microscope. *Ultramicroscopy*, 83, pp. 67-73.

Blackman, M. and Grunbaum, E (1957). An investigation into the effect of magnetic domains in cobalt on an electron beam. *Proc. Roy. Soc.* A241, pp. 508-521.

Cohen, M., (1968) Recent developments in Lorentz electron microscopy. *IEEE Trans. Mag.*, MAG4, 1, pp. 48-50.

Chapman, J. N., (1984). The investigation of magnetic domain structures in thin foils by electron microscopy. *J. Phys.*, D17, pp. 623-647.

Fisher, R. D., and Blades, J. D., (1972). Recording gap fields by Lorentz shadowgraphs and characteristics of single crystal MnZn ferrites. *IEEE Trans. Mag.*, MAG8, 2, pp. 232-238.

Frzzini, P. F., Ortolani, L., Pozzi, G. And Ubaldi, F., (2006). Interference electron micrography of one-dimensional electron-optical phase objects. *Ultramicrocsopy*, 106, pp. 620-629.

Graef, M. De. and Zhu, Y., (2001). Quantitative noninterferometric Lorentz microscopy. *J Appl. Phys.*, 89, pp. 7177-7179.

Ishba, T., and Suzuki, H., (1974). Measurements of magnetic field of magnetic recording head by a scanning electron microscope. Jpn. J. Appl. Phys., 13, 3, pp. 457-462.

Jakubovics, J. P., (1964). The effect of flux closure at the edges of thin cobalt films on electron microscope images. *Philos. Mag.* 10, pp. 675-694.

Kamino, T. and Saka, H., (1993). Newly developed high resolution hot stage and its application to material science. *Microsc. Microanal. Microstruct.*, 4, pp. 127-135.

Keyser, U., Schaerpf, O and Schwink, Ch., (1975). Electron optic investigation of magnetic stray field on (110) planes of Ni crystals oriented with high accuracy. *Physica* 80B, pp. 389-396.

Lazzari, J. P. and Wade, R. H., (1971). Electron probe measurements of field distributions near magnetic recording heads. *IEEE Trans. Mag.*, MAG7, pp. 709-704.

Marton, L. and Lachenbruch, H., (1949). Electron optical mapping of electromagnetic fields. *J. Appl. Phys.* 20, pp. 1171-1182.

Merli, P. G., Missiroli, G. F. and Pozzi, G.. (1975). *Phys. Stat. Sol.*, A30, pp. 699-711.

Mori, H., (2009). *Master thesis*, Nagoya University, Nagoya, Japan (in Japanese).

Murata, H., Ohye, T., and Shimoyama, H., (2004). High accuracy calculation of electric field in composite dielectric system by improved 3-D boundary charge method. *Nucl. Inst. Meth. Phys. Res.*, A519 pp. 184–195.

Oikawa, T., Kim, J. J., Tomita, T., Park, H. S., and Shindo, D., (2007). Measurement of electric potential distributions around FEG-emitters by electron holography. J. Electr. Microsc. , 56, 5, pp. 171-175.

Pozzi, G., and Vanzi, M., (1982). Interpretation of electron interference images of reverse-biased p-n junctions. *Optic*, 60, 2 pp. 175-180.

Pozzi, G., (1980). Asymptotic approximation of the image wavefunction in interference electron microscopy. II. Extension to the biprism edges. Optic, 56, 3, pp. 243-250.

Pozzi, G., (1975). Asymptotic approximation of the image wavefunction in interference electron microscopy. Optic, 42, 1, pp. 97-102.

Sasaki K, Kato N, Miyashita K, Wang Z, Hirayama T, Saka H. (2000). Development of FIB Techniques for the *In-Situ* Observation of Living Semiconductor Devices. *Electron Microscopy*, 35 Supplment 1, p. 190.

Sasaki, K., Wang, Z., Hirayama, T., Yaguuchi, Y., Saka H. (2003). Electron holography observation of p-n junction in a semiconductor. *Electron Microscopy*, 38, pp. 216-218 (in Japanese).

Sasaki, K., and Saka, H., (2005). A Simple method of the electric/magnetic field observation by a conventional transmission electron microscope. In: *Materials Science Forum 475-479*, Zhong, Z. Y., Saka, H., Kim, T. H., Holm, E. A., Han, Y. F. and Xie, X. S., Trans Tec Pub., Zurich, pp. 4029-4034.

Sasaki, K., Wang, Z., Fukunaga, K., Hirayama, T., Kuroda, K. and Saka, H., (2006). In-Situ TEM Observation of Electromagnetic Field in Some Real Materials. Materials Research Society Symposium Proceedings 907E, 0907-MM04-02, pp. 1-6, Boston, USA, Dec., 2006.

Sasaki, K., Kubo, Y. and Kuroda, K., (2008a). Shadow Image Distortion in a Conventional Transmission Electron Microscope. *Korean Journal of Microscopy* 38, Supplement, pp. 33-34.

Sasaki, K., Tanaka, N., Murata, H., Morita, C., Shimoyama, H. and Kuroda, K., (2008b). Electric Field Observation around a FEG-Emitter Tip Using a Conventional TEM. *Int. J. Adv. Microsc. Theor. Calc. Lett* 1 pp. 98-99.

Sasaki, K., Mori, H., Tanaka, N., Murata, H., Morita, C., Shimoyama, H. and Kuroda, K., (2010). Measurement of electric field distribution using a conventional transmission electron microscope. *J. Electr. Microsc.*, 59, pp. S89-S94.

Szeliski, R., (2006). Image Alignment and Stitching. In: *Handbook of Mathematical Models in Computer Vision*, Paragios N, Chen Y and Faugeras O, pp. (273-292), Springer, New York.

Tonomura, A., (1987). Applications of electron holography. *Rev. Mod. Phys.*, 59, pp. 639-669.

Wade, R. H., (1976). The measurement of magnetic microfields. *IEEE Trans. Mag.*, MAG12, pp. 34-38.

Wang, Z., Kato, N., Sasaki, K., Hirayama, T. and Saka, H., (2004). Electron holographic mapping of two-dimensional doping areas in cross-sectional device specimens prepared by the lift-out technique based on a focused ion beam. *J. Electr. Mmicrosc.* 53, pp. 115-119.

Wang, Z., Sasaki, K., Hirayama, T., Yabuuchi, Y. and Saka, H., (2003). Observing a p-n junction in a reverse-biased GaP light-emitting diode by combining electron holography and focused-ion-beam milling. *Microsc. Microanal.*, 9, supplement, pp. 772-773 .

Williams, D. B. and Carter, C. B., (1996). *Transmission Electron Microscopy, Basics I.* Plenum Press, New York, p. 1356

3

Advanced Techniques in TEM Specimen Preparation

Jian Li
CANMET-Materials Technology Laboratory, Hamilton, Ontario, Canada

1. Introduction

Since it was invented about 30 years ago, focused ion beam microscopes (FIB) has mostly been used in semiconductor industry for mask repair, lithography, doping and circuit edit (Melngailis, 1987). The number of FIB systems increased significantly in recent years, and their applications have extended into materials and biological sciences (Phaneuf et al.,, 1998, 1999; Li et al., 2001, 2006a).

The recent development of Dual-Beam^{TM} or Cross-Beam^{TM} FIB systems has gradually taken over the traditional single beam FIB systems. A typical FIB column contains a liquid metal ion source that produces a finely focused Ga ion beam. The primary Ga ion beam is accelerated by 30-50 kV, and directed towards the features of interest on the specimen. The incident ion beam will sputter atoms from the specimen surface, producing both secondary ions and secondary electrons. Depending on the application, the beam current can be set to high (e.g. over 60 nA) for rapid ion beam milling. Aside from electron beam imaging, operator can perform high-resolution ion beam imaging (up to 2.5 nm imaging resolution in some FIB systems). Site-specific micro-depositions (e.g., Pt, W, C, and SiO2) and micro-etching can also be achieved by the interaction of the primary ion beam (or electron beam) with the deposition (or etching) gas introduced into the system. Figure 1 shows a schematic diagram of ion-specimen interaction in FIB microscopes.

Similar to a typical SEM, FIB microscopes can be used to produce high-resolution images directly from almost any kind of samples including fine powder particles. The primary Ga ion beam can produce images with enhanced crystallographic orientation contrast using secondary electrons (SE) generated from the specimen surface. The ion beam induced secondary electron yield is strongly dependant on the crystallographic orientation of the specimen. This is attributed to the very small ion beam-specimen interaction volume. The secondary ion (SI) particles can also be collected by the detector to produce secondary ion images that can sometimes provide valuable information related to the local chemistry. In addition to high-resolution imaging on sample surfaces, ion beam cross sectioning is the most important feature of FIB microscopes. The stress-free FIB cross-sections produce valuable subsurface information. In dual-beam FIB systems equipped with EDS and EBSD, subsurface chemistry and crystallographic orientation of specific subsurface features can be obtained.

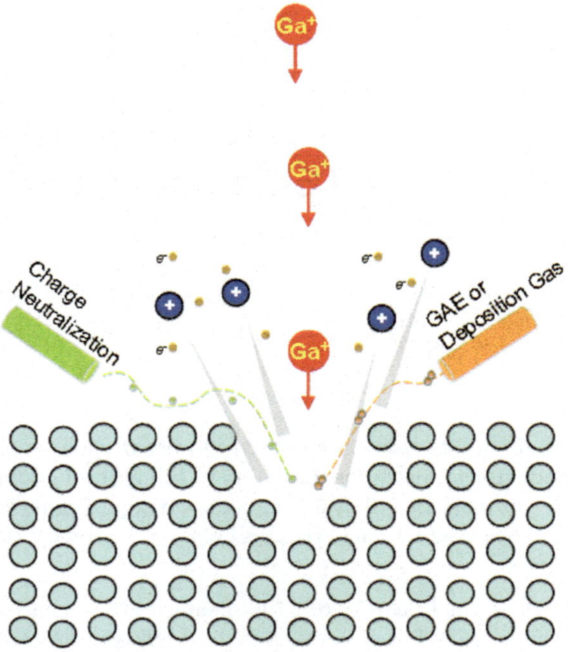

Fig. 1. Schematic diagram of a typical FIB system

Another major application of modern FIB microscopes is to prepare TEM specimens. FIB-TEM specimen preparation techniques have been reported (Giannuzzi & Steve, 1999; Anderson, 2002; Shankar et al., 2003). In this chapter, we will briefly review and compare the currently available techniques, and provide examples using our practical experiences.

2. Focused-ion beam techniques and associated TEM specimen preparation routines

Modern FIB systems are used to perform the following functions: high-resolution imaging using electron- and ion-beam, TEM specimen preparation (where STEM imaging can be performed in-situ), micro-machining, micro-deposition, and microanalysis using EDS and EBSD. Most of FIB microscopes are mainly used to prepare TEM specimens using semi-automatic routines provided by microscope vendors, while the unique imaging capability using the primary Ga ion beam has frequently been overlooked. In addition, the high-resolution SEM column tends to take over the "imaging job" while the Ga ion beam in the FIB columns are often regarded as the dedicated "milling machine". Although the modern FEG SEM columns could, no doubt, achieve higher ultimate imaging resolution, the unique FIB images are still beneficial in many aspects (Li et al., 2004, 2006a). The heavy Ga ions, although accelerated by 30 kV, can only penetrate a few nanometers into the specimen (depending on the material's properties). This makes FIB imaging extremely surface sensitive. Strong crystallographic contrast can be obtainable directly from the metallographic polished surface.

2.1 FIB imaging and cross sectioning

Similar to conventional SEMs, FIB microscopes can produce high-resolution secondary electron and secondary ion images. Chemical etching, although necessary to reveal metallurgical features in many microscopy routines, are not preferred since it can introduce many forms of artifacts. A typical SEM study would require metallographic etching that result in a modified surface topography. The FIB secondary ion imaging could provide both chemical and crystallographic details from the as-polished surface. Careful metallographic polishing is required to obtain high-quality FIB images (Figure 2). Apart from potential surface/subsurface mechanical damage introduced during sample polishing, any surface oxidation or contamination will also have a significant effect on FIB imaging. Some FIB systems are equipped with plasma cleaners to remove surface contaminants and thin oxide films in-situ. In addition, the gallium ions can also be used to sputter off the surface oxide film prior to imaging. Figure 3 shows a stress corrosion crack (SCC) tip imaged using an optical microscope, a SEM and a FIB (ion beam induced secondary electron image) microscope. The optical and SEM images were taken from a chemically etched surface, while the FIB image was taken from an as-polished surface. The significance of these images was published elsewhere (Li et al., 2008a, 2008b).

Fig. 2. FIB images of an IF steel (a) annealed and (b) deformed. Images taken from a metallographic polished surface

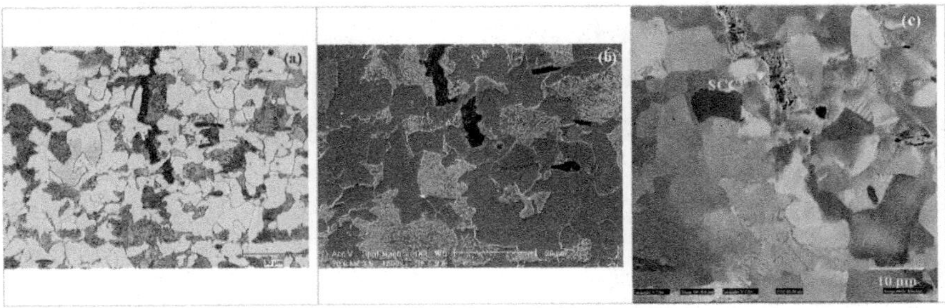

Fig. 3. Images taken from the same crack tip (a) optical, (b) SEM secondary electron image and (c) FIB secondary ion image

Focused-ion beam microscopes are also used to produce high quality cross sections. Microscopic features appearing on the sample surfaces can be cross-sectioned using the primary Ga ion beam. High-resolution images of the cross sections can be obtained by tilting the sample in single beam FIB systems, or more conveniently using the electron beam in modern dual-beam systems. The ion beam milling process can be monitored in real-time using images generated by the electron beam. Prior to FIB sectioning, a strip of coating is deposited in the FIB to prevent the surface features from ion beam erosion during milling. Figure 4 shows the corrosion of a 316L stainless steel test coupons under supercritical water condition (Li et al., 2011a, 2011b). The typical sizes of FIB cross-sections are usually less than 100 μm. Very small cross-sections usually suffer from re-deposition, while large FIB sections not only need extensive FIB milling time but are also generally difficult to produce high-quality cross-sections. The primary reason for this difficulty is the relatively coarse ion beams that are frequently used to minimize milling time usually result in cross-sections with a "curtain" effect. Smaller beam current can result in good cross-sections, but when milling in large scale which needs the extensive milling time, the system instability (beam- and stage-drift) could diminish the advantage. The largest cross-section accomplished by the author was 500 μm in width on a galvanealed steel sheet to assess the quality of the Zn coating. In this case, the major milling job was performed using high current (40 nA) and consecutive small sections of ~100 μm in width were polished using a lower beam current (1.5 nA) in a relatively old Micrion-2500 FIB system owned by FIBICS Inc. If larger cross-sections are needed, a combination of mechanical polishing and FIB imaging could be used. Topics related to FIB sectioning and imaging have been published extensively in the past years (Su et al., 2002; Schmidt et al., 2011; Li et al., 2006c).

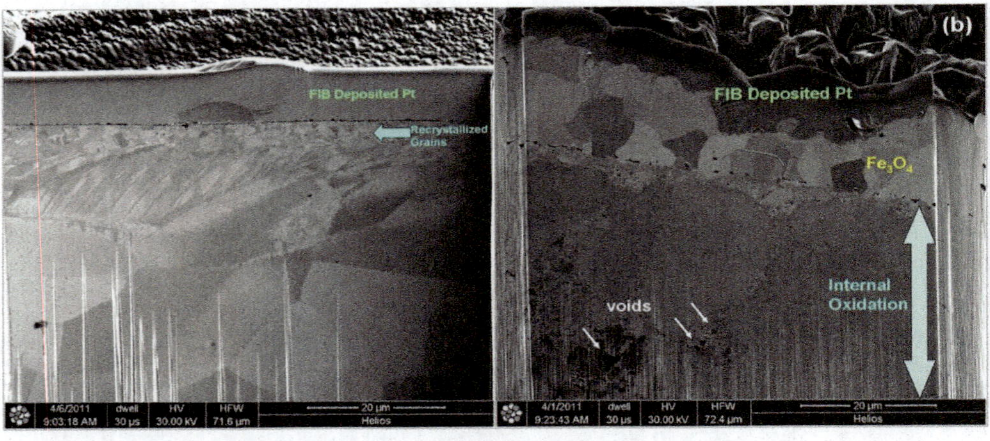

Fig. 4. SCW corrosion of stainless 316L. (a) layer of fine recrystallized grains prevented extensive corrosion, (b) severe corrosion due to lack of Cr_2O_3 passive film formation

2.2 TEM sample preparation

FIBs are only used to produce TEM specimens in many laboratories. FIB microscopes have become powerful tools in TEM specimen preparation (Phaneuf, 1999; Giannuzzi & Steve, 1999; Anderson, 2002; Shankar et al., 2003), and the techniques have evolved rapidly. The "lift-out" technique has shown significant advantages over the "original" H-bar technique that TEM specimens can be made directly from a bulk specimen (Giannuzzi and Steve, 1999). The original lift-out technique does not allow foil re-thinning once prepared. The lift-out technique was later improved that much thicker specimens (typically 5 µm in thickness) are lifted out from the bulk and then transferred to TEM grids using a micromanipulator (in-situ or ex-situ) followed by final FIB thinning. This technique can be used to produce almost all types of TEM specimens. However they are ineffective in some special cases. In this section, FIB TEM specimen preparation techniques are reviewed and compared using practical examples in our recent research.

2.2.1 Conventional H-bar technique

The conventional H-bar technique was the only available FIB TEM preparation method in the mid-1990s (Longo et al., 1999). Figure 5 shows a schematic diagram of the H-bar technique, which includes the following steps:

1. Small samples, containing the feature of interest, (about 2.5x1.0x0.5 mm in size) are cut out from bulk specimens using a precision diamond cut-off wheel.
2. Both sides of the small specimen are then carefully polished using a "tripod polisher" (Anderson et al., 1997) in order to produce parallel surfaces and minimize the mechanical damage introduced during diamond saw cutting. The polished sample should have a thickness of less than 100 µm.
3. The specimens are then mounted onto a half TEM grid using epoxy adhesive and allowed sufficient time to cure.
4. The mounted specimens are loaded into a FIB microscope for precise ion-beam thinning.

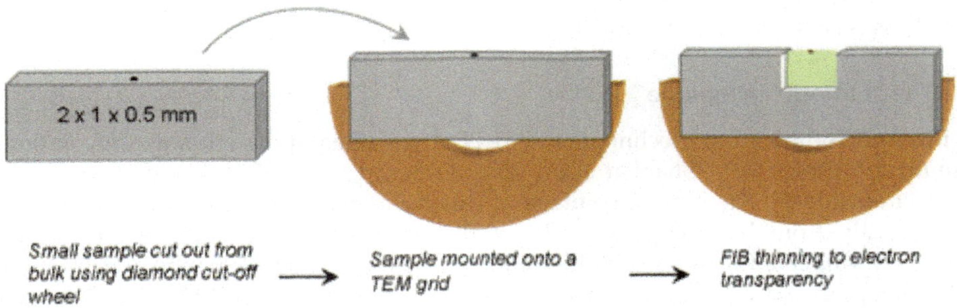

Fig. 5. Schematic diagram of conventional FIB-TEM specimen preparation technique

Some thought should be given as to the sample's thickness before FIB thinning. The thinner the sample, the less the FIB milling time will be needed. However, if the sample is too thin, it may have insufficient mechanical strength, and residual mechanical damage from the mechanical polishing could lead to problems and artifacts. This technique has found many applications, especially for non-site-specific TEM specimens. Once the sample is mounted onto a TEM grid, FIB sectioning and imaging can be used to identify the area of interest, as shown in Figure 6. An electron-transparent TEM sample can subsequently be made by thinning the backside of the specimen. Care must be taken during diamond saw cutting and various stages of mechanical polishing to ensure the integrity of the area of interest. With the invention of lift-out technique, especially the perfection of in-situ lift-out instruments, the H-bar FIB-TEM specimen preparation technique has become obsolete. However, it remains as a powerful method that can still be used in some special cases.

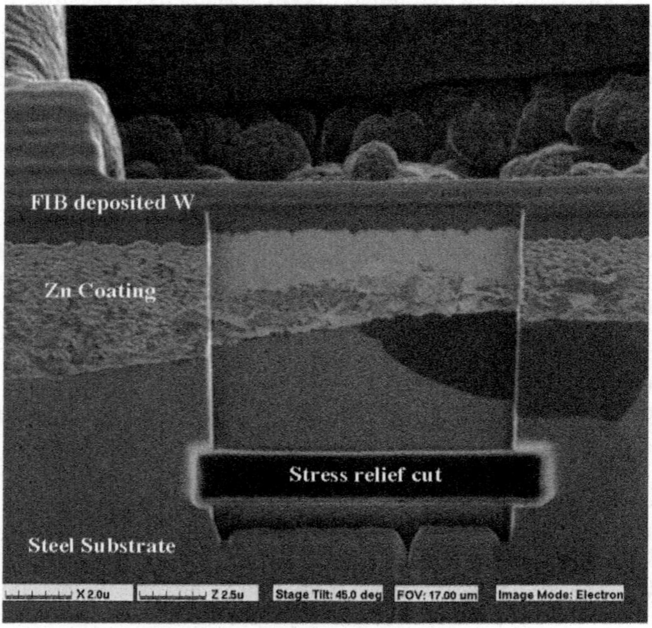

Fig. 6. An example of the H-bar technique to make a TEM specimen from galvannealed steel. Stress relief cut could be made for specimens with high residual stress.

2.2.2 The "Lift-out" technique

The invention of the lift-out technique in FIB-TEM specimen preparation was revolutionary (Giannuzzi & Steve, 1999; Shankar et al., 2003; Prenitzer et al., 1998; Langford et al., 2002; Gasser et al., 2004). The main advantage of this technique is that TEM specimens can be made directly from bulk samples, without the need of cutting and polishing. This is particularly valuable for those samples that are sensitive to contamination and fragile that cannot be prepared by any mechanical means. The traditional lift-out processes are illustrated in Figure 7, and described as follows:

1. Once the feature of interest is identified on the surface, a protective metal (e.g. tungsten or platinum) is used to cover that local area (using FIB deposition).
2. FIB milling around the targeted area to create an electron transparent membrane with a typical thickness of less than 100 nm depending on the material, type of analysis and the TEM.
3. This thin-membrane (TEM foil) containing the feature of interest is then cut free from the bulk using the ion beam.
4. The thin membrane is transferred to carbon-coated TEM grid using an in-situ or ex-situ micromanipulator for TEM observation.

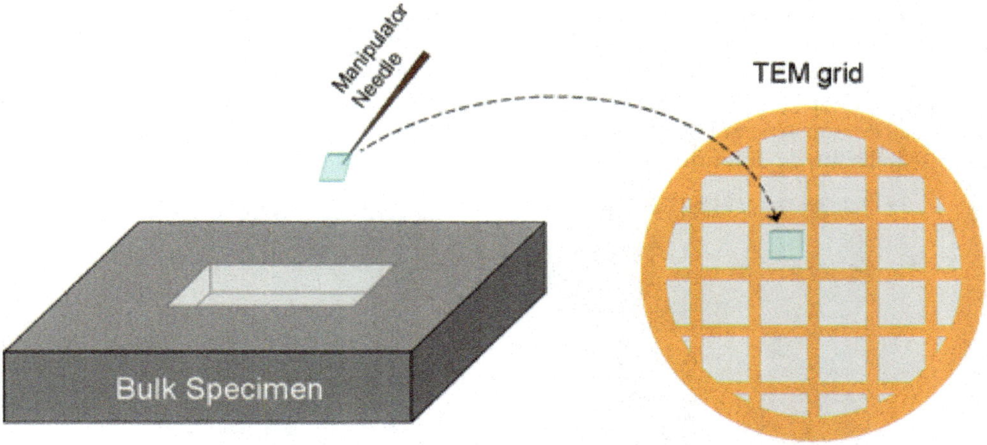

Fig. 7. Schematic diagram of traditional lift-out TEM specimen preparation technique

The preparation of a lift-out TEM specimens affect only a relatively small local region typically on the order of 50 x 50 x 50 μm in size, where other part of the surface is virtually unaffected. The lift-out technique enables rapid TEM specimen preparation with minimal mechanical damage. The TEM specimens produced by this technique have reduced spurious bulk X-ray signal due to significantly reduced bulk material around the electron-transparent membrane. This is particularly beneficial for TEM investigations of magnetic materials as there is less mass of the specimen to be attracted to the TEM pole pieces. In addition, samples that are fragile or sensitive to contamination can be prepared using this technique. In the early stage of the development of lift-out technique, we undertook a challenging case of precision specimen preparation - preparing TEM sections through controversial features in ALH84001, the NASA Mars meteorite that reportedly contains evidence of former Martian life. On that small and extremely fragile surface of the meteorite, small features of 20-30 nm (Figure 8) in the smallest dimension was lifted out (using an ex-situ lift-out instrument), and studied in TEM (Li et al., 2003). The lift-out process was challenging and rewarding.

The conventional lift-out technique proved to be an efficient and versatile TEM specimen preparation method where the conventional H-bar technique can not be used. However, TEM foil thickness cannot be precisely measured in FIB systems, and the degree of final FIB thinning is very dependant on the operator's experience. Once the TEM specimen is made and transferred onto the TEM grid, it is not possible to perform any further FIB re-thinning.

2.2.3 The "H-bar lift-out" and "Plan-view lift-out" techniques

The lift-out process described in the previous section can be very risky. The potential risk of losing a TEM specimen is quite high. The electrostatic charge at the tip of the glass needle (used in ex-situ lift-out process) could repel the tiny TEM foil and cause it to "fly-off" the needle tip. Also, too much charge at the needle tip could make it very difficult to "unload" the specimen onto the TEM grid.

Patterson combined the "H-bar" and "lift-out" techniques (Patterson et.al., 2002). Using this combined method, a much thicker "slab" on the order of ~5 μm in thickness is transferred to a TEM grid (similar to that shown in Figure 7) before final FIB thinning to electron transparency. This is by far the most versatile and widely used lift-out technique under almost all circumstances. Figure 9 shows a relatively thick specimen cut out from a Mg alloy particle to be lifted out for final FIB thinning.

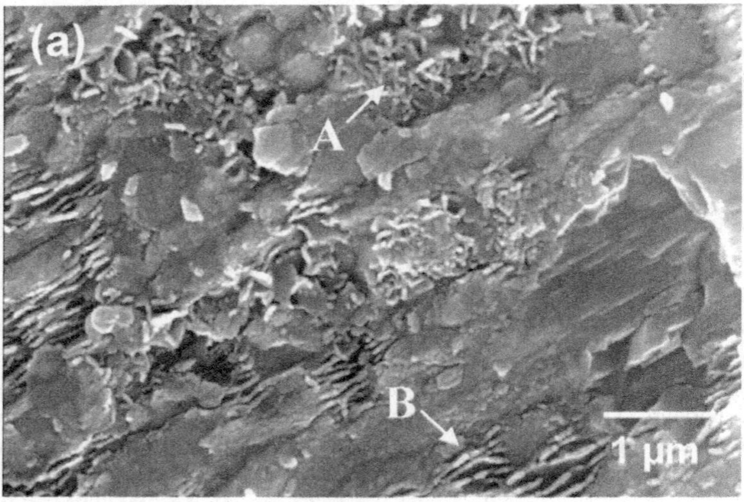

Fig. 8. High-resolution SEM image showing the small feature of interest found on a piece of Martians meteorite, from which a TEM specimen was made.

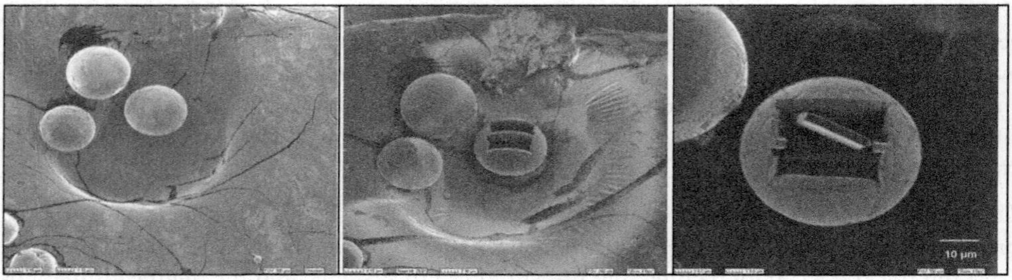

Fig. 9. H-bar lift out TEM sample preparation from a fine Mg alloy particle.

The H-bar lift-out technique is particularly powerful when used to prepare plan-view TEM specimens. In some TEM investigations, plan-view samples are desired to characterize areas of interest. For example, a TEM foil containing the crack tip would be extremely valuable to understand the mechanism of SCC (Li et al. (2004, 2005, 2006d; Elboujdaini et al. 2004). Preparing site-specific plan-view TEM samples are tedious and time consuming. Figure 10 shows a baddeleyite (ZrO_2) grain identified on a small piece of Martian meteorite, which is to be used to date the mineral. The largest challenge posed by baddeleyite in Martian meteorites is their concentration and grain size, with grains typically < 1 to 30 microns in longest dimension. The identified baddeleyite island was lifted out (planview), and fully characterized in TEM (Li et al., 2010; Herd et al. 2010).

2.2.4 The "Direct Lift-out" technique for ultra-fine specimens

The commonly used FIB-TEM specimen preparation techniques present difficulties when dealing with small and delicate samples, such as fine powders and fine and/or fragile fibers. Once the sample dimensions approach the size of the usual lift-out specimen, the application of conventional lift-out techniques becomes difficult. Early work by Cairney and Munroe (Cairney & Munroe, 2001) prepared TEM specimens from fine FeAl and WC powders by embedding the powder into Epoxy resin. A TEM specimen was prepared using the conventional H-bar technique by treating the hardened resin (embedded with the powder particles) as a bulk specimen. TEM specimens prepared using this method contains significant amount of epoxy, and can be problematic in many ways during TEM examination. In addition, residual stress introduced to the particles during the resin-curing process could be unfavorable.

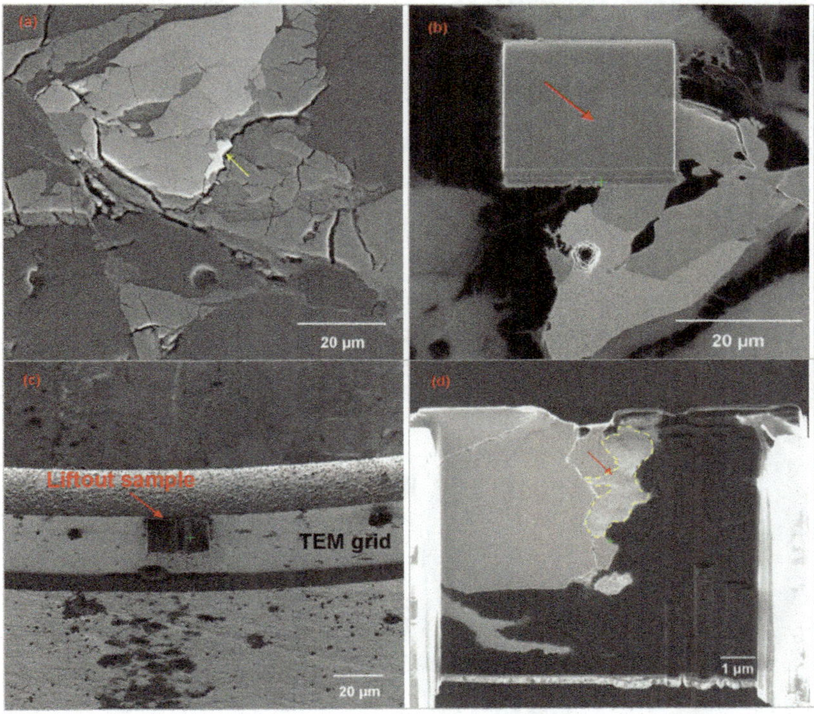

Fig. 10. Planview lift out of a baddeleyite phase from a polished section of Martian meteorite. (a). baddeleyite phase identified on the polished thin section, (b) FIB deposited W to protect the baddeleyite phase from ion beam damaging, (c) The protected zone containing the baddeleyite phase cut free, lifted out and mounted on TEM grid, (d) FIB thinning specimen showing the retained baddeleyite phase.

In a recent study (Li et al. 2003, 2006b), the author needs to evaluate the coating integrity of nickel-coated carbon fibers (about 10 μm in diameter). The entire cross-section of the fiber was to be made electron transparent. The project was to investigate the coating integrity produced using various coating processes. Some of the coatings were extremely fragile or even flaky. The "resin embedding" technique described in the previous section could have caused unacceptable mechanical damage in resin solidification process. None of the lift-out techniques described in the previous section was deemed likely to provide artifact-free TEM samples from these coated fibers. Even the very versatile technique of diamond-knife ultramicrotomy, normally excellent for cross-sectional TEM specimens, would have produced mechanically damaged or "shattered" cross-sections, and coating delamination would have been highly likely. The authored discovered the fine glass needle used for ex-situ lift-out process can pick up much larger pieces than previously reported (Li et al., 2006b). The fibers were first carefully cut to 4-5 mm in length and transferred onto the inner edge of a TEM copper grid using an ex-situ micromanipulator. A minimal amount of low-shrinkage epoxy was used to stabilize the fiber to the grid. The fiber that is fixed onto the grid was then directly cut and thinned in a FIB microscope. Figure 11 shows the FIB and

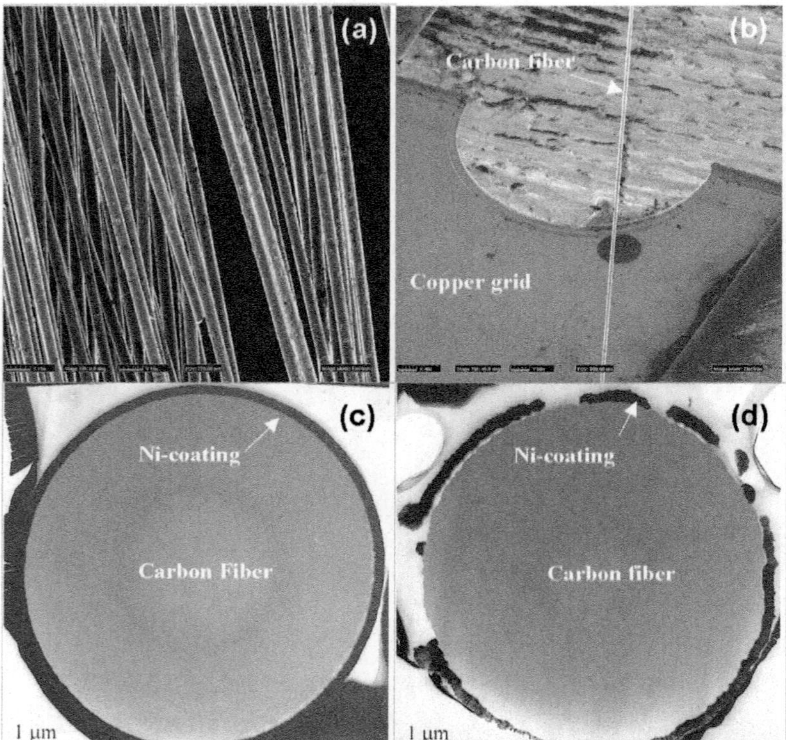

Fig. 11. FIB and TEM images of successfully prepared TEM specimens using the direct lift-out technique. (a) FIB low magnification image of a bundle of fine fibers. (b) FIB image of a mounted fiber on a copper grid. (c) A low-magnification TEM image of a fiber with good coating quality. (d) Low-magnification TEM image showing a fiber with poor coating quality.

TEM images of the fiber made using the direct lift-out technique. There are several advantages of this new technique:

1. The entire cross-section of the fiber can be made electron transparent for TEM observation.
2. Minimum mechanical damage is introduced during the TEM specimen preparation process.
3. Lower risk of mechanical damage by the micromanipulator compared to other lift-out processes.

This lift-out technique has also been used to make TEM samples from fine powders. The powder particles are spread onto a clean surface, and small particles are lifted-out and mounted directly onto the edge of a copper grid (as that shown in Figure 12). These particles can be made electron transparent in the FIB with minimal milling effort.

The conventional H-bar FIB-TEM specimen preparation technique is a simple and straightforward technique. With the invention of lift-out technique, this relatively "old" technique becomes unpopular. The lift-out technique not only provides the capability to prepare TEM samples with minimum mechanical damage and minimal contamination, but is also capable of producing site-specific TEM specimens. TEM specimens can be prepared either perpendicular or parallel to the sample surface (plan-view). The "direct lift-out" technique further facilitates TEM sample preparation of small and/or fragile specimens such as fine fibers and powders.

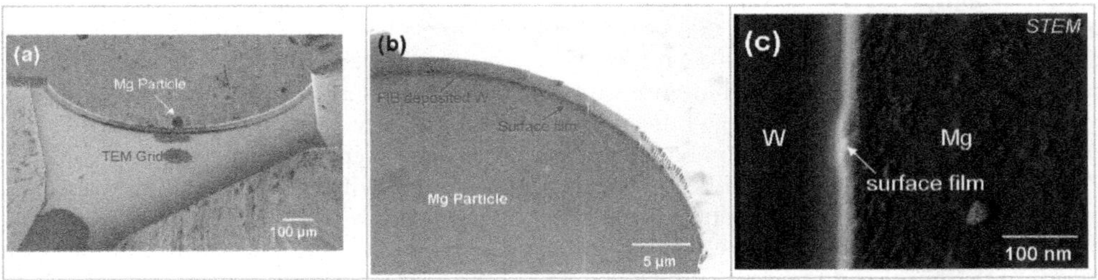

Fig. 12. Surface oxidation of pure Mg particles, (a) A Mg particle lifted out, (b) FIB image of showing thin oxide film on the surface, and (c) STEM work confirming thin MgO layer of about 20 nm in thickness.

2.3 Potential artifacts

In the past years, almost every major micro-beam vendors started to offer FIB systems. The number of FIB owners and users increased exponentially, and applications of FIB techniques extended into various field of materials engineering. FIB users ranging from university students to reputable research institutes start to produce TEM specimens like in massive production mode without thorough understanding of potential artifact. Throughout the years in the author's materials research, in both private sectors and the Canadian national materials research lab, the author encountered a wide range of engineering materials and processes. His FIB experience started about 15 years ago in the early days when FIB was just beginning to be noticed by materials researchers. The author has gradually experienced, learned, and start to understand and tried hard to cope with most of the commonly occurring FIB-induced artifacts, and in many cases, after hard lessons learnt.

Common FIB induced artifacts include:

1. "curtain effect": due to insufficient FIB metal deposition and improper milling parameter.
2. low melting point gallium phase formation on sample surface,
3. beam damage on FIB prepared TEM specimens,
4. beam induced grain growth in nano-crystalline materials,
5. beam damage to most of the HCP materials,
6. materials redeposition during milling,
7. surface amphorization.

The root causes of FIB-generated artifacts are often complex. Aside from the nature of the materials and the type of FIB work to be performed, the proper control of the Ga ion beam is of paramount importance. However, the control of the ion beam has frequently been overlooked by many FIB users. Different FIB systems are designed with different milling routines. It is nearly impossible to create a universal recipe that is optimal for both efficient milling and minimize artifacts. In general, controlling the ion beam includes: acceleration voltage, beam current, beam dwell time, pixel spacing (beam overlap), re-trace, refresh time etc. In addition, one of the most of important factors is the actual beam shape. When performing fine milling especially during the final stage of the TEM specimen preparation, the beam shape should be checked carefully (with each aperture change). One should definitely eliminate "beam tail". A tight, stable and well-aligned beam is essential to reduce most of the FIB-induced artifacts on the TEM foil. Figure 13 shows examples of FIB hole-drills on silicon wafer using (a) a nearly perfectly aligned, and (b) a poorly aligned ion beam. A well-aligned beam with minimum beam spread is essential to produce high-quality ion beam milling and polishing, while a beam tail of the poorly aligned beam will result in a noticeable "curtain effect". In some materials with hexagonal crystal symmetry (e.g. Zn and Zr), artifacts resulting from ion-beam milling are almost inevitable. Figure 14 shows a FIB cross-section of a galvanealed Zn coating on an interstitial-free (IF) steel. Ion-beam damage appears as dark speckles in the coating microstructure. It seems that fine grains at the Fe-Zn interface (Γ phase) and the Zn-rich phase (ξ phase) near the coating surface are less prone to ion-beam damage and can be resolved as shown in Figure 14(a). The majority of the δ phase in the coating suffers noticeable damage. A few more imaging passes using a very small ion beam current (only 32 pA) worsens the damage, as shown in Figure 14(b).

FIB IC chip circuit modification of Cu-based interconnect is generally problematic (Malis et al., 2002; Herschbein et al., 1998; Casey et al., 2000) due to the significantly different sputter rate of Cu metal interconnects. The sputter rate variation is related to Cu grain crystallographic orientations. Measurements of FIB sputter rates on single crystal Cu specimens (Phaneuf et al., 2003) show a sputter rate variation of about four times between fast milling orientations, such as (111), and slow milling orientations, such as (110). This difference in sputter rate is not limited only to Cu, but has also been observed in FIB milling of Au and other alloys with cubic structures. The slow sputtering of the Cu (110) orientation is attributed to the of Ga ion beam channelling, and the formation of an anomalous metal-gallium (M_xGa_y) phase during FIB milling under conditions in which the incident FIB beam hits the specimen at angles far from glancing and closer to normal incidence. During FIB milling of Cu, some grains become dark in color; these dark islands grow and spread as the ion dose increases (Shuman et al., 2001; Phaneuf et al., 2002). The appearance of such Ga-rich phase is also found to appear on low-carbon steel as shown in Figure 15. There have been little reports on how to reduce Ga phase formation during FIB milling. It is not known if changing milling parameters (reduce beam energy etc.) will have any positive effect. However, one should always be aware of these gallium phases in any FIB milling.

Fig. 13. FIB drills on silicon using a 670 pA beam. (a) a well-aligned beam, (b) a poorly aligned beam.

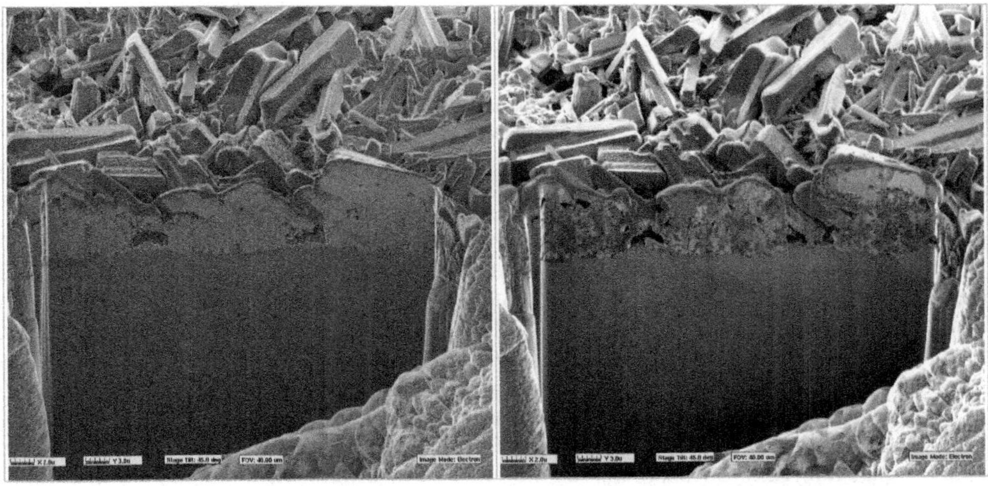

Fig. 14. FIB cross-section of a galvanealed Zn coating on IF steel. (a) polished with a well-aligned beam of 210 pA and imaged with a 32 pA beam, (b) a few more imaging passes using 32 pA causes more damage to the coating.

Fig. 15. Fe_xGa_y phase formation on low carbon steel during FIB imaging. (a) first image pass, (b) multiple imaging passes under 1.5 nA beam current, (c) multiple passes under 6 nA beam.

3. Conclusion

Advances in materials engineering depend heavily on high-level microstructure characterization using a combination of state-of-the-art instruments. New characterization techniques and methodologies are critical in understanding the fundamentals of materials properties. High-quality TEM work is heavily dependent on the preparation of TEM foil with minimum artifacts. FIB lift-out techniques are versatile, and can produce almost any kind of TEM specimen (provided they are both vacuum and beam stable). The operator should always try to optimize milling parameters in order to reduce potential artifacts.

4. Acknowledgement

The author gratefully acknowledges the efforts of the following individuals for their contributions of research and consultations leading to the completion of this book chapter: Tom malis, Pei Liu, Catherine Bibby, Wenyue Zheng and Darren Bibby (Materials Technology Laboratory, Natural Resources Canada), Michael Phaneuf (Fibics Incorporated), Chris Herd (University of Alberta) and George Kiporous (Dalhousie University).

5. References

Anderson, R. (2002). Comparison of FIB TEM Specimen Preparation Methods, *Proceeding of Microscopy and Microanalysis*, 2002, vol.8, pp. 44-45.

Anderson, R. & Klepeis, S.J. (1997). Combined Tripod Polishing and FIB Method for Preparing Semiconductor Plan View Specimens, *Materials Research Society Symposium Proceedings*, Pittsburgh, 1997, vol. 480, p. 187.

Cairney, J.M. & Munroe, P.R. (2001). Preparation of transmission electron microscope specimens from FeAl and WC powders using focused-ion beam milling, *Materials Characterization*, 2001, 46, pp. 297-304.

Casey, J.D., Noll, K.E., Shuman, R., Chandler, C., Megordan, M., Phaneuf, M. and Li, J (2000). Copper Device Editing Strategy for FIB Milling, *LSI Testing Symposium*, 2000, Osaka, Japan.

Gasser, P., Klotz, U.E., Khalid, F.A. and Beffort, O. (2004). Site-specific specimen preparation by focused ion beam milling for transmission electron microscopy of metal-matrix composites. *Microscopy and Microanalysis*, 2004, vol.10, pp. 311-316.

Giannuzzi, L.A. and Stevie, F.A. (1999). A review of focused ion beam milling techniques for TEM specimen preparation, *Micron*, 1999, vol. 30, pp. 197-204.

Elboujdaini, M., Li, J., Gertsman, V., Gu, G., Revie, W., Gao, M. and Katz, D. (2004). *NACE International Conference, CORROSION*, 2004, March, Paper#04554; New Orleans.

C.D.K.Herd, R.A.Stern, E.L.Walton, Li,J. and Bibby, C. (2010). TEM and SEM-CL Analysis of the Baddeleyite in NWA 3171: Geochronological Implications for Martian Meteorites, *Lunar and Planetary Science Conference XLI*, Paper#2280, March 1-5, 2010, Houston, Texes.

Herschbein, S., Fisher, L.S., Kane, T.L., Tenney, M.P. and Shore, A.D. (1998). Focus Ion Beam Process for Removal of Copper, *Proc. ISTFA*, 1998, p. 127.

Langford, R.M., Petford-Long, A.K. and Gnauck, P. (2002). Preparation of Transmission Electron Microscopy Cross-section Specimens Using Focused ion Beam Milling, *Journal of Cacuum Science and Technology A*, 2001, vol. 9, pp. 2186-2193.

Li, J., McMahon, G.S. and Phaneuf, M.W. (2001). *Proceedings of the Microscopically Society of Canada, 2001, vol. XXVIII*, pp. 26-27.

Li, J., Gertsman, V.Y. and Lo, J. (2003). Preparation of Transmission Electron Microscope Specimens from Ultra-fine Fibers by a FIB Technique, *Microscopy and Microanalysis*, 2003, vol.9 (2), pp. 888-889.

Li, J., Elboujdaini, M. and Gertsman, V.Y., Gao, M. and Katz, D.C. (2004). Advanced characterization of a stress corrosion crack from an in-field serviced X-52 steel line pipe, *In: Aerospace Materials and Manufacturing: Development, Testing and Life Cycle Issues*, ed. P.C. patnaik, 2004, pp. 515-523.

Li, Jian, Elboujdaini, M., Gertsman, V. Y., Gao, M. and Katz, D. C. (2005), Characterization of a Stress Corrosion Crack with FIB and TEM, *Canadian Metallurgical Quarterly*, 2005, vol.44 (3), pp. 331-338.

Li, J. (2006a). Focused Ion Beam Microscope – More than a Precision Ion Milling Machine, *Journal of Metal*, 2006, vol.58 no.3, pp. 27-31.

Li, J., Dionne, S. and Malis T. (2006b). Recent Advances in FIB-TEM Specimen Preparation Techniques,*Materials Characterization*, 2006, vol. 57, pp. 64-70.

Li, J., Elboujdaini, M. and Liu, P. (2006c). Crack Tip Plastic Zone Formation during Pipeline Hydrostatic Testing,*Proceeding of Microscopy and Microanalysis 2006*, Chicago, August, 2006, vol.12, pp. 1306-1307.

Li, J., Elboujdaini, M., Fang, B., Revie, R.W. and Phaneuf, M.W. (2006d). Microscopy Study of Intergranular Stress Corrosion Cracking of X-52 Line Pipe Steel, *Corrosion*, 2006, vol. 62 (4), pp. 316-322.

Li, J., Elboujdaini, M., Gao, M. and Revie, R.W. (2008a). Formation of Plastic Zone at the SCC Tip in a Pipeline Experienced Hydrostatic Testing, *Materials Science and Engineering A*, 2008, vol. 486, pp. 496–502.

Li, J. (2008b). The Detection of Local Plastic Strain in Microscopic Scale, *Materials Letters*, 2008, vol. 62, issu.6-7, pp. 804-807.

Li, J., Herd, C. & Malis, T. (2010). Investigation of Baddeleyite Phase Found in Martian Meteorites, *Microsc. Microanal.* 2010, vol. 16 (Suppl 2), pp. 218-219.

Li, J., Zheng, W., Cook, W., Toivonen, A., Penttilä, S., Guzonas, D., Woo, O.T., Liu, P., and Bibby, D. (2011a). Effect of Coating and Surface Modification on the Corrosion Resistance of Selected Alloys in Supercritical Water, *32nd Annual CNS Conference*, Niagara Falls, ON, June 5-8, 2011.

Li, J., Zheng, W., Kuyucak, S., Miles, J., Cook, W., Woo, O.T., Zhou, Z. & Ge, C. (2011b). Corrosion Resistance of Experimental Alloys and Coatings Under SCW Conditions, *5th International Conference on SCWR (ISSCWR-5)*, Vancouver, BC, March 13-16, 2011.

Longo, D.M., Howe and Johnson, W. C., Experimental method for determining Cliff-Lorimer factors in transmission electron microscopy (TEM) utilizing stepped wedge-shaped specimens prepared by focused ion beam (FIB) thinning, *Ultramicroscopy*, 1999, vol. 80, (2), pp. 69-84.

Malis,T., Carpenter, G.J.C., Dionne, S., Botton, G.A. and Phaneuf, M.W. (2002). Industrial Applications of Electron Microscopy, ed. Li, Zhigang R., Marcel Dekker Inc. New York, 2002.

Melngailis, J. J. (1987). Focused ion beam technology and applications, *Vac. Sci. Technol.* B, 1987, vol. 5(2), pp. 469-495.

Patterson, R.J., Mayer, D., Weaver, L. and Phaneuf, M.W. (2002). FIB-TEM Preparation for Ex- Situ Cross- Sectional and Plan-View FIB Specimen Preparation, *Proceeding of Microscopy and Microanalysis*, 2002, vol.8, pp. 566-567.

Phaneuf, M. W.; Li, Jian and Malis, T. (1998). FIB techniques for analysis of metallurgical specimens, *Microscopy and Microanalysis*, 1998, vol. 4, pp. 492-493.

Phaneuf, M. W. (1999). Imaging, spectroscopy and spectroscopic imaging with an energy-filtered field emission TEM, *Micron*, 1999, vol. 30, pp. 277-288.

Phaneuf, M.W., Li, J. & Casey, J.D. Jr. (2002). Gallium Phase Formation in Cu and Other FCC Metals During Near-Normal Incidence Ga-FIB Milling and Techniques to Avoid this Phenomenon, *Microscopy and Microanalysis*. 2002, vol. 8 (2), pp. 52-53.

Phaneuf, M.W., Li, J., Shuman, R.F. Noll, K. and Casey, J.D. Jr. (2003). Apparatus and method for reducing differential sputter rates, US patent #6,641,705, issued November 04, 2003.

Prenitzer, B.I., Giannuzzi, L.A., Newman, K., Brown, S.R., Irwin, R.B., Shofner, T.L. and Stevie, F.A. (1998). Transmission Electron Microscope Specimen Preparation of Zn Powders Using the Focused Ion Beam Lift-out Technique, *Metallurgical and Materials Transactions A*, 1998, vol. 29A, pp. 2399-2406.

Schmidt, F., Uhbacher, M., Kyriakopoulos, A. , Schubert, H. & Zehbe, R. (2011). From 2D slices to 3D volumes: Image based reconstruction and morphological characterization of hippocampal cells on charged and uncharged surfaces using FIB/SEM serial sectioning, *Ultramicroscopy*, (111), 2011, pp. 259-266.

Shankar, S. Riddle, Y.W. & Makhlouf, M.M. (2003). Focused ion beam milling: A practical method for preparing cast Al-Si alloy samples for transmission electron microscopy, *Metallurgical and Materials Transactions A*, 2003, vol. 34A, pp. 705-707.

Shuman, R.F., Noll, K., Casey, J.D. Jr. (2001). Copper Device Editing: Strategy for Focused Ion Beam Milling of Copper, U.S. Patent 6,322,672 B1, issued November 2001.

Su, X., Bouchard, R., Li, J. & Tyson, W.R. (2002). Identification of Cleavage Origins Using Focused Ion Beam (FIB) Sectioning, *Microsco. Microanal.* 2002, 8 (Suppl. 2) , p548.

4

Orientation Microscopy in the Transmission Electron Microscope - Investigations of Small Orientations Changes by Means of Orientation Mapping in TEM

M. Bieda, K. Sztwiertnia, A. Korneva and J. Kawalko
Institute of Metallurgy and Materials Science PAS, Krakow, Poland

1. Introduction

Orientation Microscopy (OM) is a technique for determining the crystallites orientations in the automatic way using systems for acquisition and indexing diffraction patterns in grid of points refers to sample coordinate system. In this way, plenty of additional information about examined material can be provided. As a conventional method of orientation visualization, color code map is used; however, data behind each pixel is applicable for further computing. Quantifying parameters and characteristics based on the sets of measured orientations provide essential information about grains, grain boundaries and about local crystallographic lattice deformation.

OM was introduced in the Scanning Electron Microscope (SEM) in the 1990s [Adams, 1993]. Nowadays OM or Orientation Mapping technique is a very well known and common technique in the SEM. There exist several commercial systems for creating and processing data obtained in the SEM. In order to improve spatial and angular resolution introducing the OM in the Transmission Electron Microscope (TEM) was considered.

2. Crystallographic orientation and measurements methods

The study of the microstructure can be carried out using many techniques eg. light and electron microscopy. Most of the classical methods allow only the qualitative analysis. The use of crystallographic information enables more advanced qualitative and quantitative description of microstructure. Of particular importance are information obtained from systematic measurements of crystallographic orientation of the sample in defined areas of the investigated material. Knowledge of the orientation distribution can be linked with the properties characterizing the polycrystalline material.

Crystallographic orientation can be defined as function of coordinates (x,y,z) correlated with points in the sample of investigated material. Since measurements are usually carried out in the plane of the sample, the function depends on coordinates x, y. Then crystallographic orientation at each point with coordinates x, y can be defined as the rotation transforming the coordinate system centered at the point (x,y) associated with the sample of the material in a coordinate system associated with a single crystal [Bunge, 1982].

To describe the orientation can be used a number of parameterization. The basic form of presentation of crystallographic orientation is a matrix of rotation. In addition, the orientations can be expressed through the axis and angle of rotation, Euler angles, Rodrigues parameters, quaternions et al. [Bunge, 1982; Morawiec, 2004, Frank, 1988]. The most commonly used are the Euler angles parameterization, where the orientation is expressed by three angles $\varphi1$, Φ, $\varphi2$, respectively: $0<\varphi1<2\pi$, $0<\Phi<\pi$, $0<\varphi2<2\pi$ (the representation introduced by Euler in 1775) [Wenk, 2004; Morawiec, 2004]. The term crystallographic orientation is also related to the concept of crystallographic texture. Texture is called crystallographic orientation distribution in polycrystalline material [Kocks, 1998]. As defined by the texture of the material can be determined unambiguously from the measured single crystallographic orientation. Collection of individual crystallographic orientation can be used to determine the microstructural characteristics such as orientation distribution functions (ODF), misorientation distribution functions (MDF) of grain boundaries having the same or different phases. Based on the data obtained it is possible to analyze the stereological parameters of microstructures and phase distribution in the material. Proper selection of the data obtained using different measurement methods can provide the full characterization of the microstructure and determine its impact on the important, from a technological point of view, material properties.

To measure the crystallographic orientation can be used various techniques, from optical methods to diffraction methods [Wenk, 2004]. The most commonly used methods are diffraction: X-ray diffraction, neutrons and electrons.

The phenomenon of diffraction on the crystal can be described by the Braggs' equation.

X-ray diffraction methods are widely used to global texture measurements. To advantages of this method belong:

- Simple methods of sample preparation,
- Large area of investigation, global texture can be obtained,
- Nondestructive method.

Spatial resolution of diffraction methods by means of the diffractometer depends on the used method and ranging from 50 μm to 0.1 mm [Schwarzer, 1998].

In order to increase spatial resolution instead of the traditional lamp as a source of the x ray radiation can be applied synchrotron. In that way can be obtained spatial resolution about 5 μm. Using the synchrotron radiation it is possible to obtain 3D microstructural characteristic 3DXRD (*Three-dimensional X-ray Diffraction*). This method was developed in Risø National Laboratory (Denmark) [Poulsen, 2004].

For texture characterization of the large volume of bulk materials the most proper is using neutron diffraction. But the access to the neutron diffractometer is limited.

New intensively developed methods are methods of local oreintation measurements by means of electron microscopes.

Due to the progress in equipment and fast data collecting computers methods of orientation measurements in scanning and transmission electron microscopy have been developed rapidly in the recent years. Especially in the case of SEM orientation measurements become almost as widely used as X-ray methods. The basic idea of the local orientation measurements is of recording diffraction images, and then finding a crystallographic orientation corresponding to the image.

2.1 Orientation microscopy in SEM

Scanning electron microscope has become in the recent years versatile tool to observe microstructure. It can be applied for visualization of the topography of the surface of the sample for analyzing chemical composition or for crystallites orientation determination in specific area of the sample.

Exist several types of difraction patterns in SEM eg.:

- MKXD (*Micro Kossel X-ray Diffraction*),
- ECP (*Electron Channeling Pattern*), SAECD (*Selected Area Channeling Diffraction*),
- EBSD (*Electron Back-Scatter Diffraction*).

Currently, most developed and most commonly used diffraction patterns for the measurement of local crystallographic orientation in the SEM is EBSD technique. Diffraction pattern is formed under Braggs' law. Kikuchi lines or pairs of lines originating from different crystallographic planes on the pattern.

Automating the processes of acquisition and indexing diffraction images has allowed the creation maps by systems to measure the orientation in the SEM. The development of such systems was associated with the development of modern research equipment including cameras CCD (Charge Coupled Device) for recording microscopic images, and numerical methods allow for automatic indexing diffraction images based on supplied data. It initiated the development of a new field of electron microscopy, called the orientation microscopy [Adams, 1993]. OM or OIM™ acronym was introduced in the 90s of last century and concerned the distribution of crystallographic orientation in the material in terms of the coordinates of the sample by means of SEM. The best known commercially available systems for measuring the crystallographic orientation is OIM™ TSL - EDAX using the method of indexing diffraction images created by Wright and Adams [Wright, 1992] and HKL Channel 5 - Oxford Instruments based on indexing algorithm developed by Schmidt et al. [Schmidt, 1989]. Own system, named ACOM (Automated Crystal Orientation Mapping) was developed by Schwarzer [Schwarzer, 1997]. The principle of operation of all systems is based on the automatic recording of diffraction images obtained from the particular sample area, indexing and mapping of orientation. The orientation map is a graphical representation of the distribution of orientation on the sample surface. Mapping the orientation is the acquisition of diffraction images in a specific grid of points with a given step [Randle, 2000]. Received orientations are represented on the map by colors matched according to previously defined key colors. In the EBSD system grid step is done by controlling the electron beam position or the stage position. The introduction of a new FIB technique (Focused Ion Beam) in scanning electron microscopes equipped with gallium gun for cutting thin film of a desired size and thickness allowed to extend to three dimensions of orientation maps. With combining FIB and EBSD measurements, it is possible to cut successive samples (with an area of approximately 50x50 microns and a thickness of about 150 nm) of bulk material, and then make orientation measurements [Zeafferer, 2005;

Konrad, 2006]. In this way, based on measured sequence of two-dimensional maps one can obtain three-dimensional image of the microstructure. Modern commercial systems used in SEM, are also equipped with the possibility of a combined analysis of EDS (Energy Dispersive X-ray Spectrometry) and EBSD allowing obtaining information about chemical composition, crystallographic phase and orientation in a certain area of the sample. An important parameter for mapping is the spatial resolution of the method. The spatial resolution can be defined as possibility of distinguishing between two points in which orientation is different. In scanning electron microscopy measurement spatial resolution depends on investigated material and the type of electron source and is limited by the volume at which the beam interacts with the sample. In case of field emission gun (FEG) spatial resolution is the best and is about 30 nm. However, in the case of standard measurements to obtain the resolution of less than 100 nm is very difficult. While the angular resolution of EBSD images is above 0.5 ° [Schwarzer, 1998]. The advantages of EBSD systems are speed, full automation possible and very good quality maps for single-phase materials. Difficulties occur in the analysis of multiphase materials. Due to the slope of sample (>60 °) during the EBSD measurements, the important role plays the good preparation of the sample surface [Wenk, 2004]. Multiphase material, with different susceptibility to polishing and chemical reagents for the various phases may cause difficulties in obtaining equal sample surface, which affects the quality of the diffraction images, and thus the ability to automatically detection lines in the image. In case of materials after severe plastic deformation with nano and submicrocrystalline grains obtain the good quality of orientation map in SEM is challenging. These methods are also not effective for the analysis of local, small changes in orientation (<1 °) on the border of the angular-resolution diffraction EBSD.

2.2 Orientation microscopy in TEM

There is a strong need for development the method of orientation investigation of nanomaterials with angular and spatial resolution better that can be offered by EBSD/SEM.

This can be attained by means of transmission electron microscopy (TEM) Where the spatial resolution of the measurement is more limited by thickness of thin foils then the parameters of beam (diameter of electron beam can be lower then 2 nm) [Wu ,2009 Zaefferer, 2011]. This allows for analysis materials with ultra crystalline grains.

Automation of the systems in transmission electron microscope (TEM) is difficult. There exist several systems but most of them are not commercial available.

In TEM there are several different types of diffraction patterns [Carter & Williams, 1996]. The most common and wide used for phase discrimination are spot patterns (fig 1.).

Taking into account only geometry of spot patterns, orientation can be determined with accuracy ~5°. Higher accuracy (up to ~0.5°) can be assured by taking into consideration intensity on the diffraction patterns.

Investigation used local orientation measurements in TEM were performed in 1983 by Sztwiertnia and Haessner [Haessner, 1983]. Based on manual collected spot patterns were calculated local partial orientation and misorientation distribution function in cold rolled copper.

Fig. 1. Spot pattern in TEM.

Exemplary applications of spot patterns for investigation of the materials after severe plastic deformation are presented in [Zeafferrer, 2000]. In this work both geometry and intensity of spot diffraction patterns are used for determination crystallographic orientation. Acquisition of images is performed manually by microscope operator.

Another system based on spot pattern was presented by Rauch [Rauch, 2006]. Further development of this system resulted in commercially available system called Astar offered by NanoMEGAS company [Rauch, 2008].

In this case, the main goal of indexing is to match acquired spot pattern to the previously calculated database of patterns.

(a) (b)

Fig. 2. Kikuchi type diffraction patterns: CBED a) and TKP b).

Better angular resolution has Kikuchi types diffraction patterns (Fig. 2). They allow to specify orientation with accuracy 0.1°.

To creating orientation map can be applied Kikuchi types patterns obtained in transmission beam (TKP - Transmission Kikuchi Diffraction Patterns) and convergent beam (CBED - Convergent Beam Electron Diffraction Patterns).

Method for automatic measurement of local crystallographic orientations in the TEM in defined areas of thin film using diffraction CBED images developed as one of the first Haessner and Sztwiertnia [Sztwiertnia, 1991; Pospiech, 1991; Haessner, 1992]. This method has been used to study recrystallization textures in the alloys with cubic symmetry. The first semi-automatic system for mapping the orientation of the TEM was created by Schwarzer and Zeafferer [Zeafferer, 1994] and then developed by Zeafferer [Zeafferer, 2000]. System operation is based on direct analysis of the collected diffraction images: spot patterns and Kikuchi type patterns. Besides the automatic indexing of images of diffraction system also contains procedures for computer processing of images and to determine the crystallographic dependencies such as: disorientation between grains, Burgers vector, slip systems and twinning systems.

Since systems for measuring the orientation maps in transmission microscopy are much harder to automate, most of these systems are created without the automation of image acquisition diffraction [Zeafferer, 1994; Schwarzer, 1997a]. System relies on recording diffraction in automatic way was built at the University of Metz [Funderberger, 2003; Fundenberger, 2003a]. These systems are not commercially available and must be individually adapted to a specific laboratory.

In addition to the different diffraction image analysis technique in TEM orientation map are created using Dark Field Image technique [Dingley, 2006]. This technique was introduced as a commercial system in 2000 by TSL-EDAX [Dingley, 2000]. For the measurements are used large series of dark field images recorded at different diffraction conditions. Intensity analysis is made on each dark field image. This allows the reconstruction of diffraction images and calculation the orientation at that point. The resolution of such measurements is dependent on the magnification used and the resolution of the microscope in dark field images [Zeafferer, 2002]. The advantage of this system is a fast measurement time of about 15 - 20 minutes. However, there are problems associated with drift of the sample and the need to use a large range of intensity (required for high-quality CCD camera).

All these methods and the system built in the IMIM PAS, whose description will be presented in the following subchapter, are used in various aspects of research in materials science. Together with X-ray difractometry and EBSD measurements techniques are complementary methods that allow for a detailed description of the microstructure of materials.

3. Elements of the system

In IMIM PAS the system for orientation mapping on TEM was built [Bieda, 2008]. The main idea of this system is automatic recording of diffraction images and save them on computer hard disk in the form of graphic files using a digital CCD camera from a few to tens thousand of diffraction images and their subsequent indexing and analysis of the crystallographic orientation set.

The system (Fig. 3 and 4) includes the following elements:

- Philips CM20 microscope - transmission electron microscope with a maximum accelerating voltage 200 kV and spatial resolution (the possibility of distinguishing between two points on the microscopic image) value of ~ 0.2 nm. The microscope is connected to a PC via the serial port (RS232-C).
- CCD camera Gatan Dual Vision Company - 12 bit (4095 gray scale) digital CCD camera with a resolution of 1300 x 1030 pixels and exposure time from 0.01 to 10 s, mounted on the side of the microscope column. The camera is connected to the computer using a cable card and DMA enabling fast data transfer.
- Software Gatan Digital Micrograph - the software used for communication between a computer and a camera and a microscope. Allows to save images on a computer, has a built-in tools for image analysis, and allows to create own programs in developing capabilities of Digital Micrograph.
- Software EP for indexing Kikuchi diffraction patterns developed by A. Morawiec [Morawiec, 1999], as the result of cooperation between the IMIM PAS and the University of Metz (France).
- Software for analysis of data containing information about the distribution of orientation in a sample of material such as Tango (part of Channel 5 system for the analysis data from EBSD in SEM), programs to analyze the components of the texture, to calculate the ODF, etc.

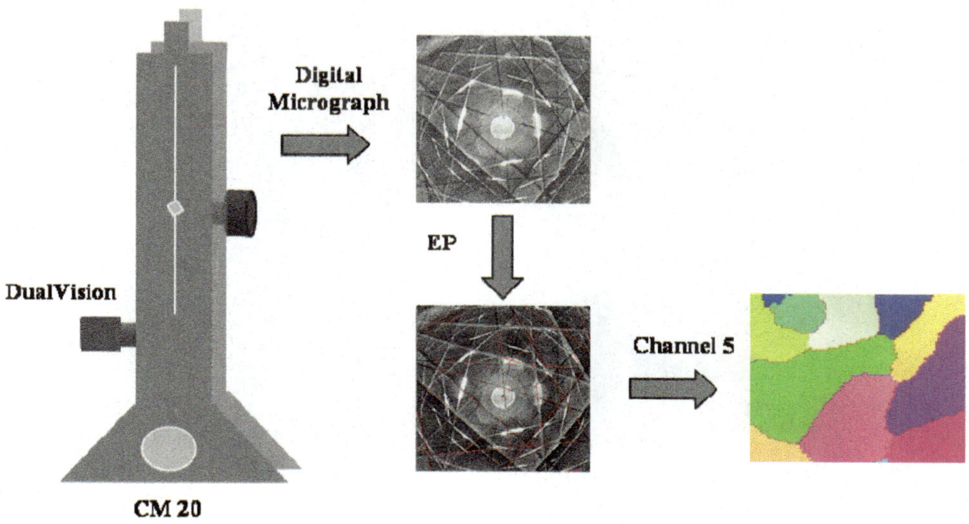

Fig. 3. Elements of orientation mapping system.

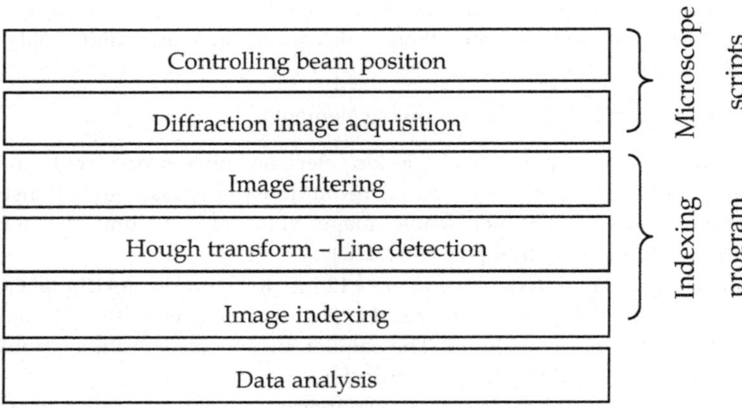

Fig. 4. Schema of orientation mapping system.

4. Application of the system

One of the possible applications of the OM measurements in the TEM are presented. Deformed pure (4N) aluminum was chosen as an exemplary material. The new quality and quantity of supplied data made possible better microstructural characterization and observation of very small orientation changes produced characteristic subgrains structure (fig. 5.).

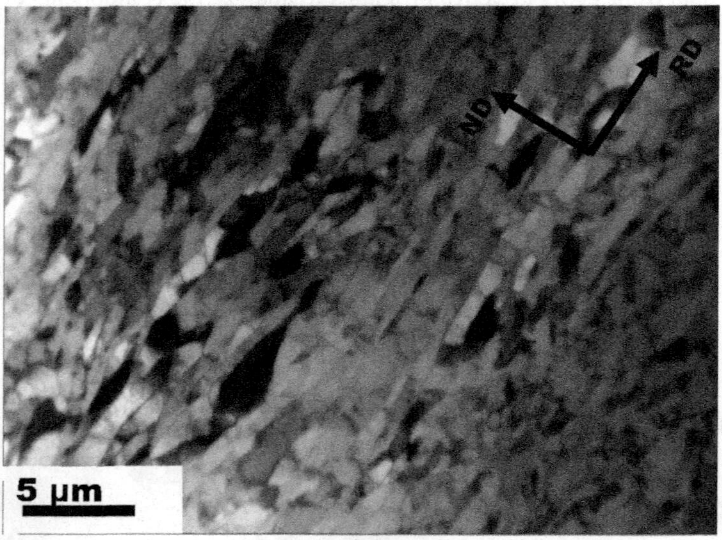

Fig. 5. TEM bright field image of cold-rollling up to 90% pure alluminium

Measurements of the local crystallographic orientation were carried out in aluminum with a purity of 99.99%. Thin foils were prepared from cross-rolled up to 90% sheet of metal. That represented characteristic layered microstructure for the deformed material consisting of grains and subgrain bands lying almost parallel to the rolling plane, Figure 5. In previously published works such as [Pospiech, 1998; Sztwiertnia, 2006a] was presented the study of the microstructure of aluminum rolled on the basis of that single crystallographic orientation manually recorded along the direction normal to the plane of rolling (ND). The results obtained (about 1000 crystallographic orientation) were analyzed due to the contribution of individual components of texture in the plane perpendicular to the direction of the transverse plate (TD). An example of the orientation distribution function calculated from the measured orientation is presented in Figure 6. Pure aluminum after cold rolling to 90% deformation texture is characterized by copper with fiber type α and β. In the main β fiber are S and Copper components, and in the fiber α can be distinguished Brass and Goss components.

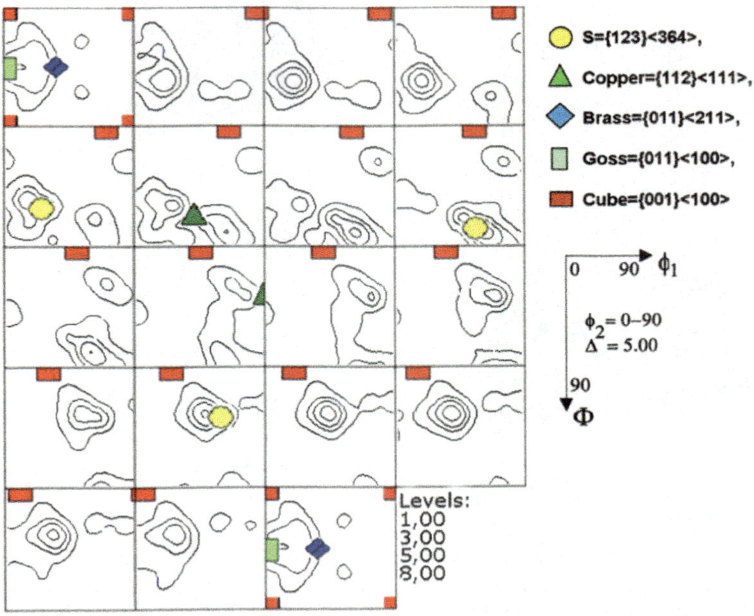

Fig. 6. Exemplary local orientation distribution function (ODF) for cold rolled up to 90% pure aluminium counted on the base of 1106 of orientation with marked texture components.

5. Results and discussion

Local crystal orientation measurements using the system to create orientation maps, allowed the registration of several thousand orientation at one measurement with a fixed step controlling electron beam displacement. This allowed obtaining the orientation maps in areas with a width from a few to several microns. With the ability to measure diffraction at the measuring points distant from each other by several nanometers, could be observed subtle changes in the crystallographic orientation in very small distances.

Orientation maps were measured for the longitudinal section (perpendicular to transverse direction TD) in such a way as to include as much information about the boundaries lying parallel to the plane of the aluminum sheet, Figure 7, 8 and 9. The analyzed areas of the samples were above 10 microns of thickness (along normal direction ND). In such areas piles of elongated subgrains in rolling direction RD were observed, which boundaries in the ND were characterized by disorientation angles generally less than 5 °.

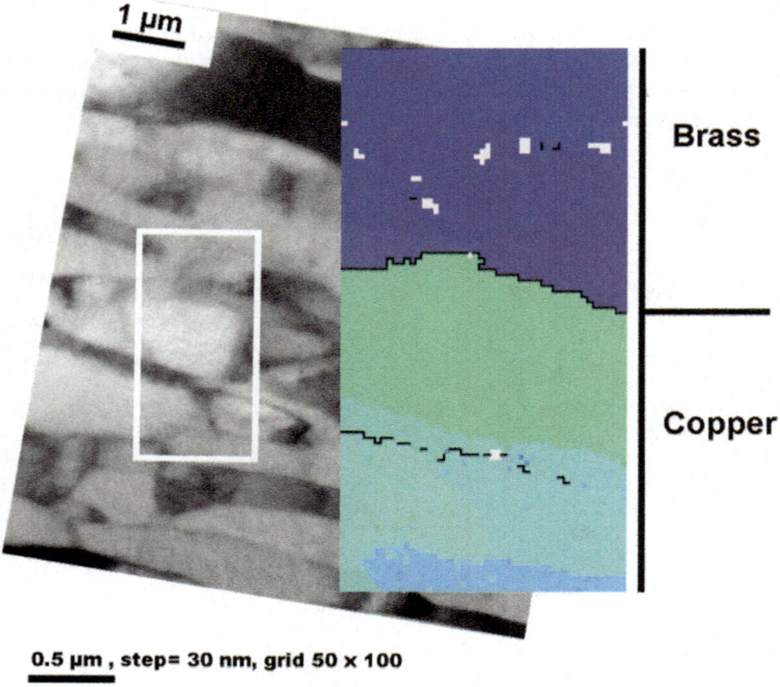

Fig. 7. Bright field image (TEM) of aluminum cold rolled up to 90% and orientation map with colors from inverse pole figure, the black lines are misorientations angles > 5° and corresponding texture components. Unresolved diffraction patterns are presented as white points.

Cumulative disorientation angles, less than 5 °, occurred between the orientations of subgrains, corresponding to one component of the texture. In Fig.7 on orientation map is marked boundary misorientation angle of > 12 ° occurs between the areas representing the type of texture components, respectively Copper and Brass. In the area of Copper component, we can distinguish the boundary disorientation angle of approximately 5 °. Since the border between the two components of the texture observed a gradual shift in orientation towards the ideal position of component-type Copper. Larger changes in the angle of disorientation identify the extent of occurrence of individual components of the texture (Fig. 7 and 10), while only minor changes were observed within the same component.

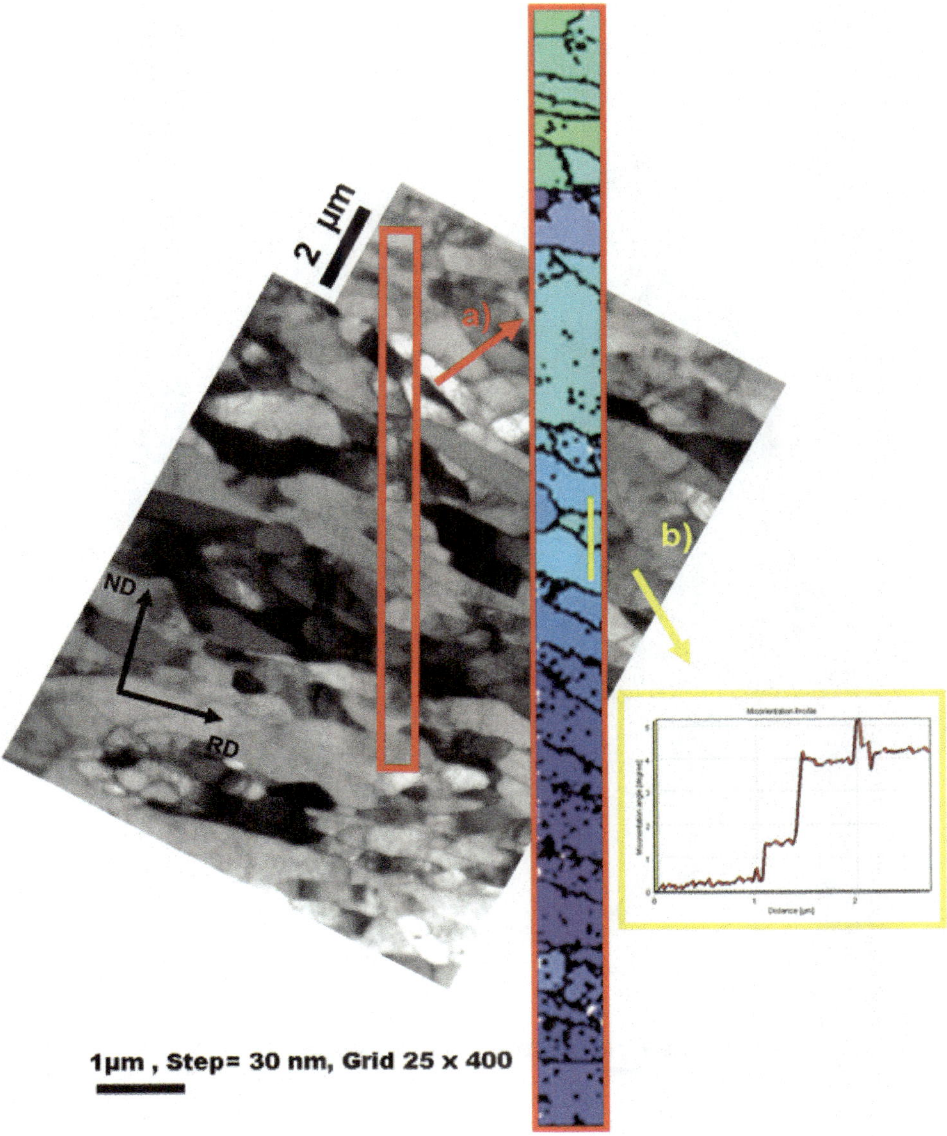

Fig. 8. Bright field image (TEM) of aluminum cold rolled up to 90% and a) orientation map with colors from inverse pole figure, the black lines are misorientations angles > 1°. Unresolved diffraction patterns are presented as white points and b) misorientation profile across three subgrains.

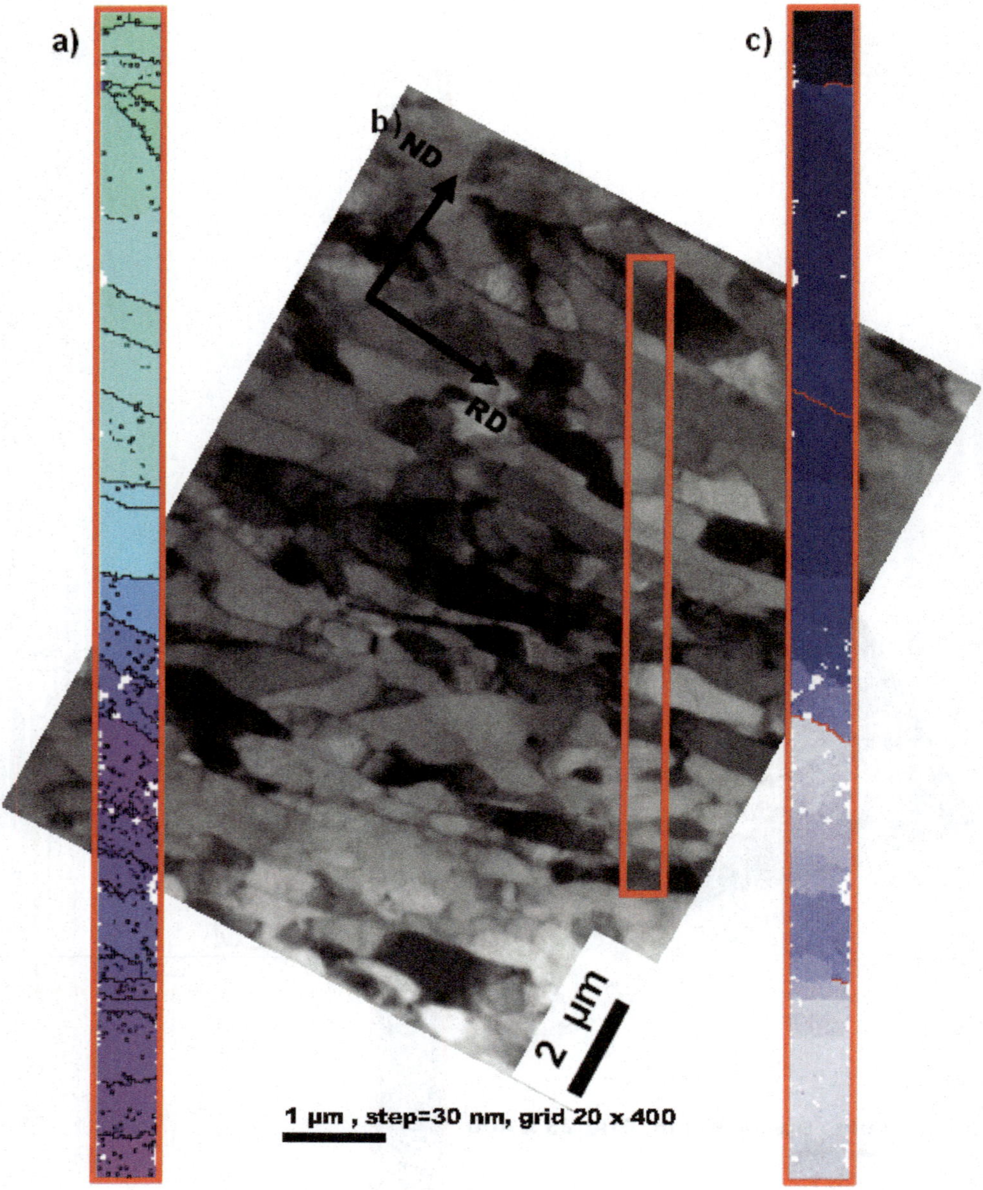

Fig. 9. Bright field image (TEM) of aluminum cold rolled up to 90% b) and orientation map with colors from inverse pole figure a), the black lines are misorientations angles > 1°. Unresolved diffraction patterns are presented as white points and c) color map where gradient of color means deviation from the first orientation up to 20°, the red lines are misorientations angles > 5°.

Bands shown in the orientation maps (Fig. 8 and 9) are composed of bundles of parallel, highly elongated in the RD subgrains, between which there are relatively small (between 1 ° and 10 °), cumulative disorientation angles. In addition, in Figure 9c, is showed that the color gradation is related to a gradual change from the initial orientation. On this map red are marked boundaries, where the orientation changes were> 5 °. Typical course of changes in the angle of disorientation between adjacent subgrains along a line drawn parallel to the length of the orientation map is shown in Fig. 10a. Again, it can be seen that the differences in orientation are often very small on the boundaries of two subgrains, and may be cumulative, with the angle of disorientation between the first and last subgrains of the band may exceed 20 °. For example, total disorientation angle across the transition band width of about 12 microns, Fig. 10c, is ~ 20 °. Identifying the orientation of the corresponding components of the texture, it appears a transition from a stable orientation of S in the direction of another stable orientation Brass (Fig. 10).

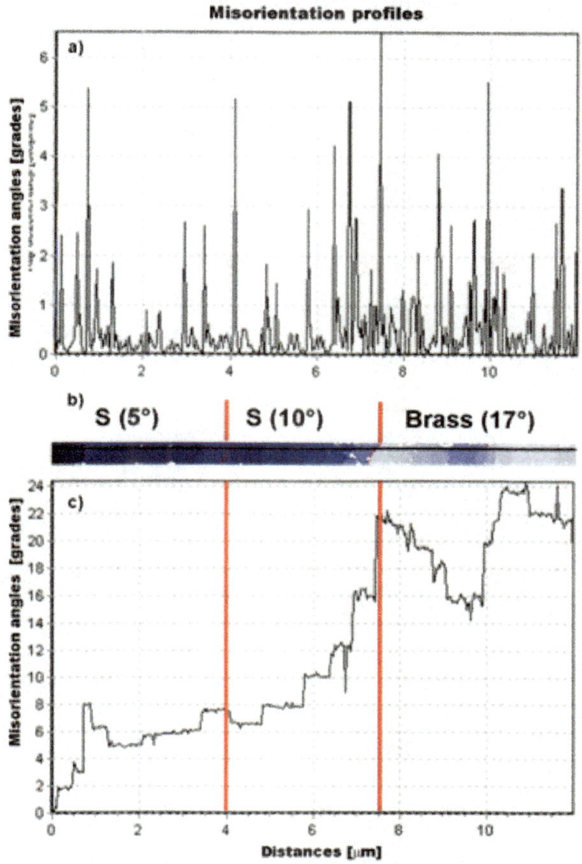

Fig. 10. Misorientation profiles across the line marked on the orientation map b) between particular grains a) and relative to the first orientation c) with marked texture components, the numbers in bracket means deviation from ideal orientation.

Accumulation of disorientation angles between subgrains across the spectrum is characteristic for the transition band [Hjelen, 1991], which distinguishes it from other elements of the microstructure of similar morphology. Transition band separates the fragments of the original grains, in which, during deformation, act various combinations of slip systems as a result of what they turned toward different, stable end orientations [Hu, 1963a]. With increasing strain - as the fragments of deformed grains are close to the stable end orientation - total disorientation angle increases, e.g. [Hjelen, 1991]. It should be emphasized that the transition band is often observed in deformed single crystals. In polycrystalline materials, their identification has been difficult because there were no methods to study small changes in orientation at short distances. In earlier work, in which orientation measurement was performed [Pospiech, 1998; Sztwiertnia, 2006a], the transition bands in deformed polycrystalline microstructure could not be noticed, because the diffraction at certain points of measurement were recorded only when followed their apparent change in that way it was easy to "lose" their subtle changes within the band.

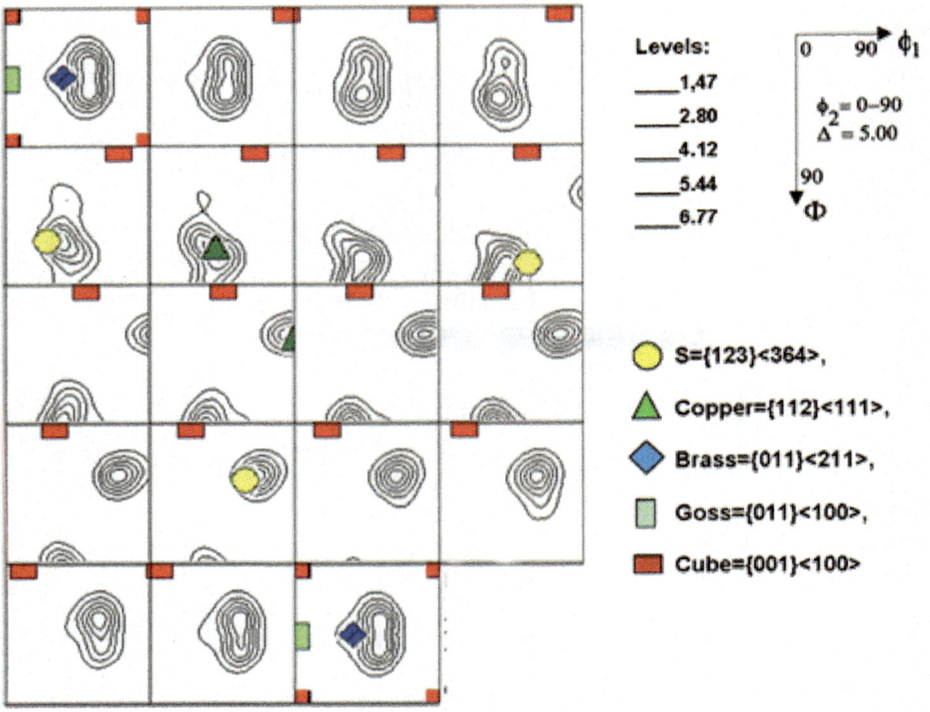

Fig. 11. Local orientation distribution function (ODF) for cold rolled up to 90% pure aluminium counted on the base of 56 000 of orientation measured in TEM with marked texture components.

On the basis of 56 000 measured orientation orientation distribution function (ODF) was calculated (Fig.11) and texture components as S, Brass and Copper were identified. Their respective percentage share was: 35.3%, 29.7% and 24%. Transitional areas between the bands in which there are no large misorientation are called deformation bands [Dillamore, 1972]. An example of such spectrum is shown in Fig. 12. Here we do not observe a smooth change of orientation angle combined with the accumulation of misorientaton angles, but only the fluctuations around a certain position. Thickness of the deformation bands having the same orientation was a few micrometers, and the angles of misorientation between adjacent bands were not large. Distribution of orientation along the ND was not accidental. Cells or subgrains were often concentrated in clusters containing up to 15 crystallites having the same type of orientation.

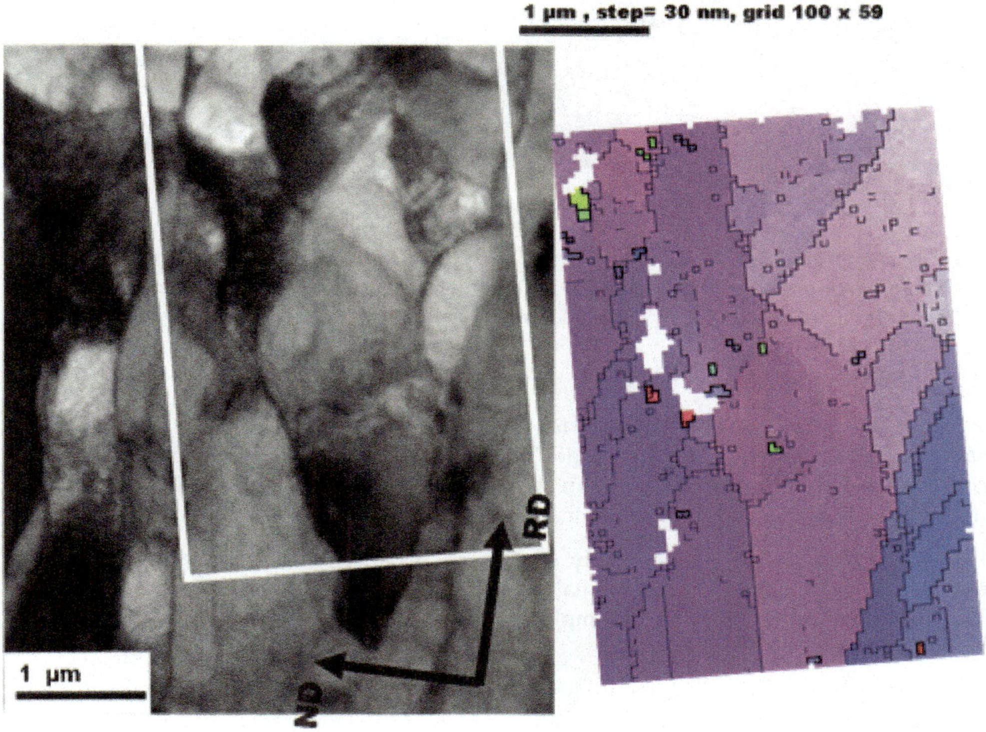

Fig. 12. Bright field image (TEM) of aluminum cold rolled up to 90% and orientation map with colors from inverse pole figure, the black lines are misorientations angles > 1°. Unresolved diffraction patterns are presented as white points.

Detailed maps obtained at higher magnification allowed to show the microstructure of subgrains in bands around 100 nm in widith (Fig. 13) with an angle of disorientation ~ 1 °.

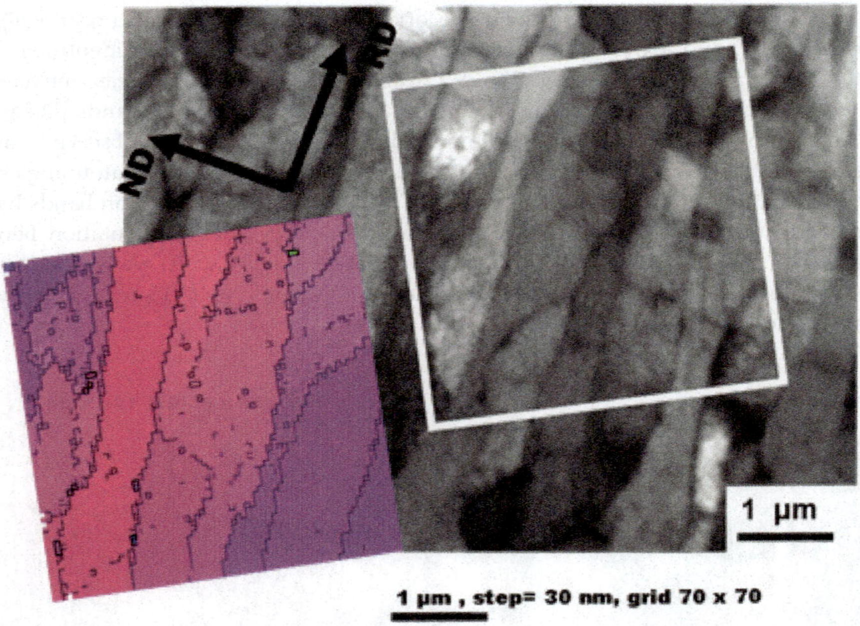

Fig. 13. Bright field image (TEM) of aluminum cold rolled up to 90% and orientation map with colors from inverse pole figure, the black lines are misorientations angles > 1°. Unresolved diffraction patterns are presented as white points.

6. Conclusions

Principles of diffraction patterns used for orientation measurements in the TEM and short survey of existed systems for orientation mapping in the TEM was introduced. Pros and cons of TEM systems and comparison with SEM systems was presented in this chapter including spatial and angular resolution.

Difficulties in automation and variety of diffraction patterns used in the TEM made the existing systems, with some exceptional, hard to commercialize. Generally, methods in the TEM and the SEM can be used as complementary methods for investigating microstructure in nano and micro scale.

In spite of better spatial resolution in the TEM than in the SEM, the more important is better angular resolution in the TEM (about 0.1°), which can be assured of using Kikuchi type diffraction patterns and make possible investigation of small orientation changes in material in micro and nano-scale.

7. References

B. L. Adams, D. J. Dingley Materials Science Forum 157-62 p. 31 (1994)
M. Bieda PhD. thesis IMIM PAS (2008)

M. Bieda, K. Sztwiertnia, A. Korneva, T. Czeppe, R. Orlicki Solid State Phenomena Vol. 163 p.13 (2010)
H.J. Bunge Buterworths Londyn (1982)
D.J. Dingley, S.I. Wright. 7th Asia-Pacific Electron Microscopy Physical Sciences, Singapore, Times Printers Pte. Ltd. (2000)
D. Dingley Proceedings of the Electron Crystallography School (2005)
I.L. Dillamore, P.L. Morris, C.J.F. Smith, W.B. Proc. Roy. Soc., 329A, s.405 (1972)
F.C. Frank Metall. Trans. A 19 s.403 (1988)
J.J. Fundenberger, A. Morawiec, E. Bouzy, J.S. Lecomte Mater. Chem. Phys. 81, p.535 (2003)
J.J. Fundenberger, A.Morawiec, E. Bouzy, J.S. Lecomte Ultramicroscopy 96 p. 127 (2003)
J. Hansen, J. Pospiech, K. Lücke Springer-Verlag (1978)
F. Haessner, J. Pospiech, K. Sztwiertnia Materials Science and Engineering 57 s.1 (1983)
F. Haessner, K. Sztwiertnia, Scripta metall. mater.27 s. 1545 (1992)
J. Hjelen, R. Ørsund, E. Nes Acta metall. Mater. 39 s. 1337 (1991)
H. Hu J. Wiley and Sons New York s. 311 (1963)
U.F. Kocks, C.N. Tome, H.R. Wenk, Cambridge University Press (1998)
J. Konrad, S. Zaefferer, D. RaabeActa Materialia 54 s. 1369 (2006)
N. C. Krieger Lassen Proc. of 16th Riso Int. Symp. on „Microstructural and Crystallographic Aspects of Recrystallization" Riso National Laboratory, Roskilde, Denmark s. 405 (1995)
N.C. Krieger Lassen Journal of Microscopy 190 s.375 (1998)
A. Morawiec J. Appl. Cryst. 32 p. 788 (1999)
A. Morawiec Springer-Verlag (2004)
A. Morawiec, J.J. Fundenberger, E. Bouzy, J.S. Lecomte J. Appl. Cryst. 35 s. 287 (2002)
A. Morawiec J. Appl. Cryst. 32 p. 788 (1999)
A. Morawiec Springer-Verlag (2004)
A. Morawiec, M. Bieda Arch. Metall. Mater. 50 s. 47 (2005)
J. Pospiech, K. Lücke, K. SztwiertniaTextures and Microstructures 14-18 s. 103 (1991)
J. Pospiech, K. Sztwiertnia, J. Jura Advanced Light Alloys and Composites Springer s. 361 (1998)
H. Poulsen Ph D thesis Riso Natinal Laboratory Dania (2004)
V. Randle, O. Engler CRC Press (2000)
E.F. Rauch, L. Dupuy Arch. Metall. Mater. 50 p.87 (2005)
E. F. Rauch, M. Véron, J. Portillo, D. Bultreys, Y. Maniette and S. Nicolopoulos Microscopy and Analysis 22(6):S5-S8 (EU) (2008)
N-H. Schmidt, N.Ø. Olenson Canadian Mineralogist 28 s. 15 (1989)
R.A. Schwarzer Ultramicroscopy 67 p.19 (1997)
R.A. Schwarzer Micron 28 s.249 (1997)
R.A. Shwarzer Materials Science Forum 287-288 s.23 (1998)
K. Sztwiertnia, F. Haessner Textures and Microstructures, 14-18 s.641 (1991)
K. Sztwiertnia, F. Haessner Mater. Sci. Forum, 157-162, s.1069 (1994)
K. Sztwiertnia, M. Bieda, G. Sawina Arch. Metall. Mater. 51 s.55 (2006)
H.R. Wenk, P. Van Houtte „Texture and anisotropy" Rep. Prog. Phys. 67 s.1367 (2004)
D.B. Williams, C.B. Carter Plenum Press Nowy Jork (1996)
G. Wu, S. Zaefferer Ultramicroscopy 109 p. 1317 (2009)
S.I. Wright, B.L. Adams Met. Trans. A 23 s. 759 (1992)

S. Zaefferer, R. Schwarzer Z. Metallkd. 85 (1994)
S. Zaefferer J. Appl. Cryst. 33 s.10 (2000)
S. Zaefferer Advanced in imaging and electron physics 125 s.355 (2002)
S. Zaefferer Advanced engineering materials 5(10) s.745 (2003)
S. Zaefferer Cryst. Res. Technol., 1-22 (2011)

5

Low-Dose Imaging Techniques for Transmission Electron Microscopy

David B. Carlson and James E. Evans
Department of Molecular and Cellular Biology, University of California at Davis, USA

1. Introduction

Transmission electron microscopes have the ability to visualize almost any specimen at the nanoscale. Unfortunately, modified imaging protocols are needed when samples are composed predominantly of low atomic number elements. In this chapter we review several methods that are currently employed to mitigate the adverse effects caused by electron irradiation to materials at the interface of biology and nanotechnology. We also highlight future technological advancements to the microscope platform that may further enhance the quality of imaging for beam sensitive samples.

2. Need for low-dose imaging with transmission electron microscopy

Although material scientists are able to routinely image inorganic specimens at atomic resolution, the same is not true for organic and biological specimens. This divergence in attainable resolution is largely a result of varying tolerance to electron beam exposure. Organic and biological specimens exhibit increased sensitivity to high-energy electron irradiation as compared to inorganic materials. For example, metals and monocrystals of silicon can handle very large doses of radiation without showing significant signs of damage. However, radiation damage, in the form of bond breakage and mass loss, from the electron beam is both deleterious and unavoidable for organic samples made of low atomic number elements such as carbon, nitrogen, and oxygen (Thach & Thach, 1971). The damage response of biological specimens is dependent on the cumulative exposed dose and was estimated using spot fading diffraction experiments on two-dimensional (2D) crystals of the amino acid L-valine. In those experiments, a dose of 6 $e^-/Å^2$ at 80 keV resulted in the complete disappearance of diffraction spots and this value is typical, within an order of magnitude, of most biological specimens (Glaeser, 1971). Aliphatic polymers such as polyethylene are equally sensitive to electron irradiation and display similar critical doses to that of proteins (Revol & Manley, 1986). However, more aromatic polymers such as polystyrene, anthracence and poly-xylene can withstand higher irradiation doses due to the stability of the phenyl group (36, 42 and 120 $e^-/Å^2$ respectively) (Kumar & Adams, 1990; Williams & Carter, 2004). Importantly, it should be noted that although zeolites and ceramics are technically inorganic specimens consisting of high atomic number elements, they exhibit intermediate sensitivity to electron irradiation with critical doses of approximately 100 and 600 $e^-/Å^2$, respectively (Pan & Crozier, 1993).

2.1 Primary damage mechanisms

There are two main types of primary damage (radiolysis and knock-on damage) that occur in samples following exposure to high-energy electrons (Ugurlu et al., 2011). Both knock-on damage and radiolysis are examples of inelastic scattering in which a portion of the incident energy of an imaging electron is lost due to a collision event. Knock-on damage occurs when an incident electron directly interacts with the nucleus and transfers enough energy to eject the atom from the specimen. Thus, knock-on damage is manifested as the displacement of atoms within a sample and results in the formation of point defects in a periodic lattice. Radiolysis damage arises due to sample ionization by incident electrons interacting with valence electrons and breaking chemical bonds (Williams & Carter, 2004).

Knock-on damage can be quite substantial for inorganic specimens but is generally less significant for biological samples. Although the use of higher incident electron energies (≥300 keV) increases the likelihood of knock-on damage it also decreases the frequency of radiolysis (Csencsits & Gronsky, 1987). At lower incident electron energies, knock-on damage can be completely mitigated if the energy transferred during inelastic scattering events is below the threshold energy for displacement of a particular atom. When imaging graphene sheets, an accelerating voltage less than 80 keV has been shown to prevent knock-on damage to individual carbon atoms during continuous imaging (Girit et al., 2009).

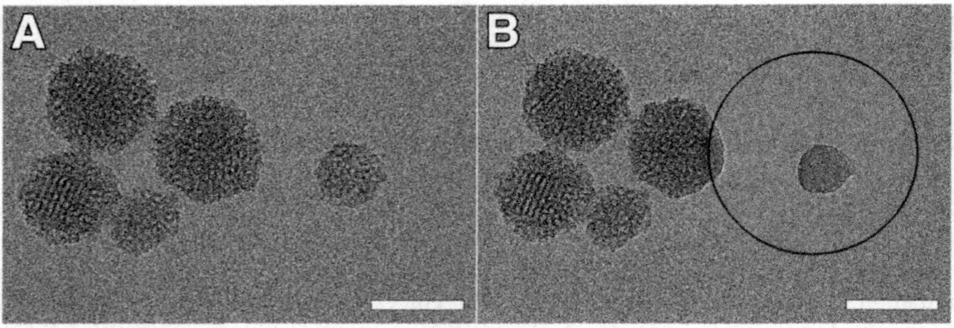

Fig. 1. Amorphization of a mesoporous silica nanoparticle following intense electron irradiation. A) Low-dose image of 5 mesoporous silica nanoparticles clearly showing lattice fringes and porosity. B) A second low-dose image of the same area as (A) following selective irradiation of a single nanoparticle with a focused electron beam. The outline of the 100 nm diameter electron beam is seen as a black circle. Scale bars represent 50 nm.

However, radiolytic damage is still detected at low and intermediate incident electron energies (80-300 keV) and can affect both organic and inorganic specimens (Csencsits & Gronsky, 1987; Frank, 2006). For example, previous work has shown that ceramic materials can be amorphized due to radiolytic cleavage of chemical bonding and subsequent rearrangement of the atoms (Blanford & Carter, 2003). This process is demonstrated in Figure 1, which shows a mesoporous silica nanoparticle losing internal structure due to radiolysis. Five mesoporous silica nanoparticles are clearly evident using standard low-dose and high-resolution imaging conditions (Fig. 1A). Following selective irradiation of the right-most nanoparticle with a total dose of 600 $e^-/Å^2$, a second low-dose image was acquired that clearly shows the same nanoparticle underwent amorphization and no longer displays any lattice fringes or porous structure.

2.2 Secondary and tertiary damage mechanisms

In addition to the primary damage mechanisms of radiolysis and knock-on damage, one must also consider secondary and tertiary damage. Secondary damage results from inelastic collision events which form phonons, free radical species and secondary electrons that can each propagate through the sample and heat the sample or cause further damage via chemical reaction or secondary scattering (Williams & Carter, 2004). Tertiary damage effects arise from the evolution and buildup of gas (generally hydrogen and oxygen) following the decomposition of organic specimens via radiolysis (Meents et al., 2010).

For specimens that are good thermal conductors, phonon induced heating is usually negligible (Williams & Carter, 2004). However, the majority of biological and organic specimens (as well as ceramics) are thermal insulators and therefore such heating can be considerable, sometimes to the point of melting the sample. Simulations of proteins frozen and imaged at liquid nitrogen temperature (77 K) suggest that this heating may be insignificant if the dose rate is less than 50 $e^-/Å^2$ per second (Karuppasamy et al., 2011).

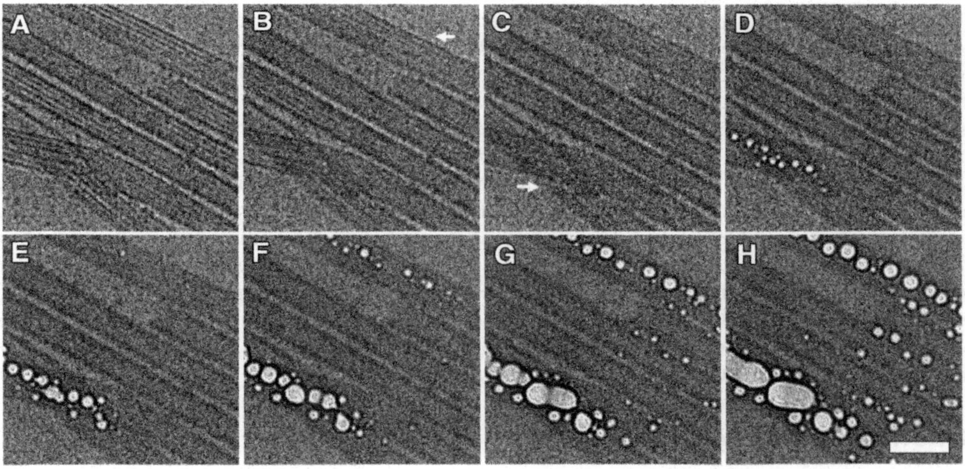

Fig. 2. Time course of radiolysis damage and hydrogen gas evolution imaged as a function of cumulative dose. A) Initial low-dose cryo-EM image of microtubules at a dose of 4 $e^-/Å^2$. The protein content of the microtubules appears as black lines while the background represents the vitrified ice substrate. B-H) Serial low-dose images with increasing cumulative doses of an additional 4 $e^-/Å^2$ per frame. The arrow in (B) indicates the initial blurring of the microtubule structure due to the onset of radiolysis damage that becomes more severe with continued imaging. The arrow in (C) indicates the first detection of gas bubble formation corresponding to the area of highest protein density as two microtubule filaments can be seen overlapping. With continued irradiation, the gas bubbles form in other locations along the microtubules and eventually merge together. Scale bars are equivalent for all images and represent 50 nm.

In addition to radiolysis damage causing primary and secondary damage it also leads to tertiary damage to organic specimens. Beyond the typical critical electron dose for proteins (10 $e^-/Å^2$), cumulative radiolysis damage causes high-resolution spots to fade as bonds stabilizing the native structure are broken (Chiu & Jeng, 1982). Since hydrogen bonding is highly prevalent in biological specimens, radiolysis damage results in the release of hydrogen atoms from the protein backbone (Dubochet et al., 1982; Leapman & Sun, 1995). Due to the sample being frozen, released hydrogen and other trapped atoms cannot easily diffuse and therefore gas bubbles begin to form. Figure 2, shows the preferential evolution of gas bubbles in areas of high protein density. An initial blurring of the protein density (black lines) can be seen due to radiolysis damage to high-resolution components. After a cumulative dose of 12 $e^-/Å^2$ small gas bubbles can be seen forming in areas occupied by protein filaments. Finally, with continued irradiation the gas bubbles continue to expand.

2.3 Benefits of low temperature imaging

There are two major benefits to TEM imaging at low temperatures. First, vitrified samples can be imaged in a frozen hydrated state allowing optimal sample preservation of the native structure (Adrian et al., 1984; Taylor & Glaeser, 1976). Second, lower temperatures enable an increased tolerance to ionizing radiation damage (Chiu et al., 1987).

While the process of plunge freezing dynamically fixes the sample to maintain atomic resolution for organic specimens, the ultimate benefit of low temperature imaging is the reduction in radiation damage at temperatures below ~170 K (Frank, 2006). This radiation protection is a direct result of dynamically immobilizing the sample through embedment into vitrified ice (Knapek & Dubochet, 1980). By caging the sample in a frozen environment, free radicals generated from inelastic scattering events are unable to diffuse through the sample and cause secondary damage (Knapek & Dubochet, 1980). In addition, the freezing also constrains the movement and degrees of freedom for the atoms of a molecule after a bond is broken thereby limiting the structural rearrangement produced during irradiation (Frank, 2006). As a result, embedment in vitrified ice and imaging at liquid nitrogen temperature improves radiation tolerance 2 to 6 fold over room temperature imaging (Chiu et al., 1987; Frank, 2006). Further cooling to liquid helium temperature (4 K) has shown an additional two-fold increase in radiation tolerance of vitrified two-dimensional (2D) membrane protein crystals to 20 $e^-/Å^2$ (Fujiyoshi, 1998).

3. Low-dose imaging techniques

3.1 Conventional low-dose imaging

Low-dose imaging is a standard technique and is ubiquitously used in nearly all cryogenic-Electron Microscopy (cryo-EM) applications. The main goal of low-dose imaging is to locate an area of interest on the grid surface and accurately determine the appropriate focus without significantly pre-exposing the location to be imaged at high-resolution. If done correctly, the effective pre-exposure dose to the sample is less than 0.2 $e^-/Å^2$ (Sun & Li, 2010).

Most modern microscopes come with pre-installed low-dose software. JEOL microscopes have a Minimal Dose System (MDS) interface that stores three different sets of lens values for efficient swapping between imaging modes. SEARCH mode is generally a low-magnification overview image used to identify areas of interest while PHOTO mode is used for actual data collection at high-magnification. FOCUS mode is set at the same or higher magnification as PHOTO mode, but the beam and image is shifted to an adjacent area to avoid pre-exposing the area of interest identified for high-resolution imaging in PHOTO mode. Reliable swapping between the three modes is possible since the last lens settings of each mode are automatically saved and recalled whenever transferring modes.

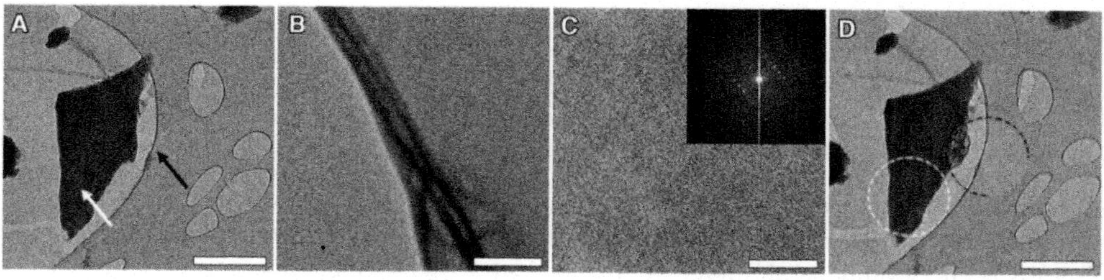

Fig. 3. Demonstration of the areas used for low-dose cryo-EM imaging. A) SEARCH mode: Low-magnification overview of the sample grid used to identify areas of interest. Black arrow points to the area that will be used for FOCUS mode while the white arrow indicates the sample area of interest that will be used to record the PHOTO mode data. B) FOCUS mode: High-magnification image of an adjacent area used to manually find eucentric height and determine optimal defocus parameters. C) PHOTO mode: High-magnification image of the sample area of interest (catalase crystal). The inset shows the corresponding FFT with lattice spots verifying the presence of the protein crystal. D) Same area as (A) but after acquiring (B) and (C). The white dotted circle represents the electron beam diameter used for imaging in PHOTO mode. The black dotted ¾ circle indicates the irradiated area while working in FOCUS mode. Note that the location imaged for FOCUS mode shows damage to the edge of the catalase crystal as well as a thinned circular region of vitreous ice. However, the adjacent area used for PHOTO mode remained unaltered and pristine for data collection and the single low-dose image did not cause any detectable damage. Scale bars represent 2 micrometers in (A&D) and 100 nanometers in (B&C).

The MDS interface is accessed by selecting "MDS" under the TEMCON "Alignment" tab and is activated upon clicking on the PHOTO mode button. To ensure the best high-resolution imaging capabilities, the microscope should be aligned in PHOTO mode. This can be done at the beginning of any microscope session. Once the microscope is optimally aligned in PHOTO mode, the user should select the appropriate magnification and brightness settings for the desired experiment. Next, all lens values should be recorded in a notebook for later reference. At this point the remaining modes can be setup. Ideally, the only differences between the three modes should be magnification, brightness, image shift and beam shift. Therefore, the first step in setting up the SEARCH and FOCUS modes is to guarantee that all three modes are initially perfectly aligned. To do this, the PHOTO mode lens values recorded in the laboratory notebook can be duplicated for both SEARCH and FOCUS modes. Now that all three modes are equivalent, SEARCH mode can be set to a low-magnification overview and perfectly centered with PHOTO mode by using a combination of beam shift and image shift. It should be noted that some researchers use a defocused diffraction pattern rather than low-magnification TEM for SEARCH mode. In this case, a focused diffraction pattern is defocused and projects an image of the sample onto the viewing screen that can be perfectly centered with PHOTO mode using the projector lens shift. In both cases, the beam diameter used for SEARCH mode should be large enough to minimize the electron dose during screening but bright enough to detect small or low contrast features. After successfully aligning SEARCH mode, the last step is to align FOCUS mode. Here, the magnification is usually the same or higher than that used for PHOTO mode to ensure the accuracy of defocus measurements. Using beam shift and image shift,

FOCUS mode should be moved to an area adjacent to that illuminated by PHOTO mode. Additionally, the beam diameter should be condensed to prevent any unwanted pre-exposure of the area of interest for PHOTO mode. If using a holey carbon cryo-EM grid, FOCUS mode is usually positioned directly over the carbon support nearby a hole to be imaged in PHOTO mode.

Now that MDS is properly setup, SEARCH mode can be used to find the area of interest while imparting less than 0.2 $e^-/Å^2$ (Fig. 3A). Once an interesting sample is located, the electron beam should be shuttered or blanked and MDS should be switched to FOCUS mode. While imaging in FOCUS mode (Fig. 3B) the actual focus should be determined and any astigmatism corrected using a Fourier Transform (FT) of the CCD image. The electron beam should once again be blanked or shuttered to prevent unwanted exposure and MDS switched to PHOTO mode. At this point the viewing screen should be raised and a single image can be collected on the nearly unperturbed sample at a dose below the damage threshold (Fig. 3C). Successful integration of this approach when imaging radiation sensitive samples can reliably provide high-resolution images prior to the onset of significant damage (Fig. 3D).

3.2 Spot scan imaging

In addition to the damage processes described earlier, exposure to a high-energy electron beam can cause additional imaging artifacts such as beam-induced movement and charging. Both artifacts produce image blur that inherently limits the attainable resolution during data collection (Downing & Glaeser, 1986). Electron beam induced charging can be a significant problem for flood beam illumination of organic and biological specimens since these tend to be insulating materials. The interaction of high-energy electrons with the insulating sample can cause a charge separation between sample and substrate that is further exacerbated by tilting the sample (Gyobu et al., 2004). This is of particular importance for low temperature imaging where carbon tends to become an insulator as temperatures approach that of liquid helium (Frank, 2006).

To mitigate such movement, spot scan imaging was developed using a condensed electron beam to image the area of interest as a composite of individual sub-areas (~ 100 nm diameter) in a grid like pattern (Downing & Glaeser, 1986; Zemlin, 1989). This technique reduces beam-induced movement and charging effects since each sub-area is spatially distinct and surrounded by an unexposed sample region. Generally, CCD or film acquisition times are set to several seconds allowing multiple exposures of adjacent sub-areas to be collected in a single image (Downing & Glaeser, 1986). Although this procedure may not be beneficial to improving resolution for single particle analysis, where each complex adopts a random orientation, spot scanning has been used in many studies to improve the resolution of radiation sensitive 2D crystalline arrays (Frank, 2006).

3.3 Electron crystallography

One method to attain high-resolution data under low-dose conditions is the use of electron diffraction. Analogous to X-ray diffraction of three-dimensional (3D) crystals, electron diffraction can be used to image 2D crystals and uses an incredibly small dose of electrons,

0.5 e-/Å2, to generate enough signal to accurately describe high-resolution data (Schenk et al., 2010; Unwin & Henderson, 1975). This is possible because electron crystallography amplifies the redundant unit cell information within the crystal (Unwin & Henderson, 1975). The more unit cells imaged, the greater the signal for a given dose, as seen by bright spots corresponding to repetitive crystal spacings within the Fourier transform or diffraction image (Downing & Li, 2001). Currently, the highest resolution structure determined by electron microscopy was solved using electron diffraction (Gonen et al., 2005).

While electron diffraction suffers from the same lack of phase information as X-ray diffraction and requires some method of phase retrieval, there are several advantages over real space imaging. First, diffraction patterns are not effected by sample movement resulting from beam-induced charging or drift. Second, and arguably more important to resolution gain, much larger crystals can be imaged and used to generate more accurate diffraction spots through improved statistics (Downing & Li, 2001). Alternatively, gathering low-dose real-space images of 2D crystals provides both the phases and amplitudes of the diffraction spots in the FT. However, the Fourier transforms are subjected to information loss when the contrast transfer function equals zero, therefore, at least two images are needed to recover any missing information (Unwin & Henderson, 1975). Additonally, diffraction patterns do not exhibit the resolution-dependent falloff of spot amplitude generated during real-space imaging due to the contrast transfer function.

3.4 Single shot dynamic TEM

Single shot dynamic TEM (DTEM) has opened up an exciting new avenue for electron microscopy. Since high-energy electrons accelerated to 200 keV travel at approximately 2/3 the speed of light, only a single electron is within the microscope column at any given time for conventional TEM. However, for DTEM an ultrafast laser pulse illuminates a photocathode source causing upwards of 1 billion electrons to be photoemitted as a single packet on the nanosecond timescale (LaGrange et al., 2006). Since the number of electrons in a single pulse are fairly constant for a given laser power, the dose is determined by the beam diameter used during imaging. For example, 1 billion electrons spread to a beam diameter of 5.0 micrometers results in a dose of 0.5 e-/Å2 per pulse while a beam diameter of 1.2 micrometers would yield a dose of 8.8 e-/Å2. In addition, DTEM is compatible with MDS, and allows screening and focusing the sample in SEARCH and FOCUS modes using continuous wave laser pulses and then acquiring the final image with single shot pulsed mode.

Currently the major advance with this technology is the ability to study the transient behavior of materials such as reaction boundaries looking at deformation and phase transformations in inorganic materials (LaGrange et al., 2006). However, with brighter and faster pulses, DTEM may allow for "diffract then destroy" imaging of beam sensitive samples, in which the packet of electrons interacts with and passes through the sample before damage propagates and affects imaging (Reed et al., 2009). If this type of ultrafast imaging does "outrun" the mechanisms of damage, drift and movement, then doses far greater than 10 e-/Å2 could be used yielding greater contrast and higher resolution for cryo-EM structure determination of proteins. Second generation DTEMs are currently being installed to evaluate such ultrafast imaging capabilities for organic material research (Evans et al., 2011).

4. Solving the attainable contrast versus resolution tradeoff

Contrast, by definition, is the difference in intensity of two adjacent pixels. Our eyes have difficulty detecting contrast at levels less than 10% and are unable to detect differences at 5% (Williams & Carter, 2004). The Rose Criterion was originally developed to estimate the visibility or detectability of an object within an X-Ray image based upon irradiation dose. However the Rose Criterion is also applicable to TEM imaging and suggests that to truly resolve a feature at any resolution, the signal needs to be 3-5 times the value of the average background or noise. Since the amount of signal is dependent on the sampling dose, there is a direct relationship between irradiation dose and attainable resolution (Yakovlev & Libera, 2008).

4.1 Identifying location of specimen within image due to low contrast

Low signal-to-noise ratios are the major limiting factor for high-resolution structure determination of beam sensitive samples (Evans et al., 2008; Okamoto, 2008). For organic specimens, the limited contrast is caused by a combination of the maximum dose allowed to maintain high-resolution structures (roughly 10 $e^-/Å^2$), and the low scattering cross-section of low atomic number elements that constitute the sample. In certain cases, the low signal-to noise may be so low that single proteins and 2D crystals may be indiscernible from the surrounding vitreous ice layer in cryo-EM at perfect focus. Therefore, to locate a given protein, it has been necessary to acquire significantly under-focused images at values ranging from 10-30 times Scherzer defocus (Evans et al., 2008) to enhance contrast of low-resolution components. Unfortunately, with the typical defocus used for cryo-EM, the contrast transfer function (CTF) exhibits many contrast reversals at high-resolution producing information gaps where the CTF equals zero and limiting the directly interpretable image information to the point resolution (Fig. 4A).

4.2 Phase shifting technology

Without advances in imaging technology, it is necessary to compromise resolution for contrast. One solution to this problem is the introduction of a phase plate in the back focal plane of the objective lens (Nagayama, 2011). With a phase device installed in the microscope, imaging can take place at or near Gaussian defocus (Fig. 4D) as compared to the 1-4 microns underfocus used for cryo-EM 3D reconstructions of single protein molecules (Fig. 4A). The resulting in-focus image has increased contrast at low-resolution while greatly flattening CTF reversals, yielding less data loss at higher frequencies (Figure 4). Ultimately, such increased contrast will permit more accurate angle assignments and alignment of single particles and 2D crystals for averaging and 3D reconstructions.

Unfortunately, the direct interaction of the electron with the phase plate is a major hurdle associated with current phase plate designs. In addition to changing the phase of the electron, the interaction between the phase plate material and the electron can result in secondary scattering that results in the loss of information and effectively reduces the signal to noise ratio. This is because the electron's interaction is not a zero energy-loss event due to the physical passing of electron through a solid medium such as thin amorphous carbon (Nagayama 2011). Thus, zero-loss electrostatic phase plates are needed to truly allow atomic resolution imaging for biological samples.

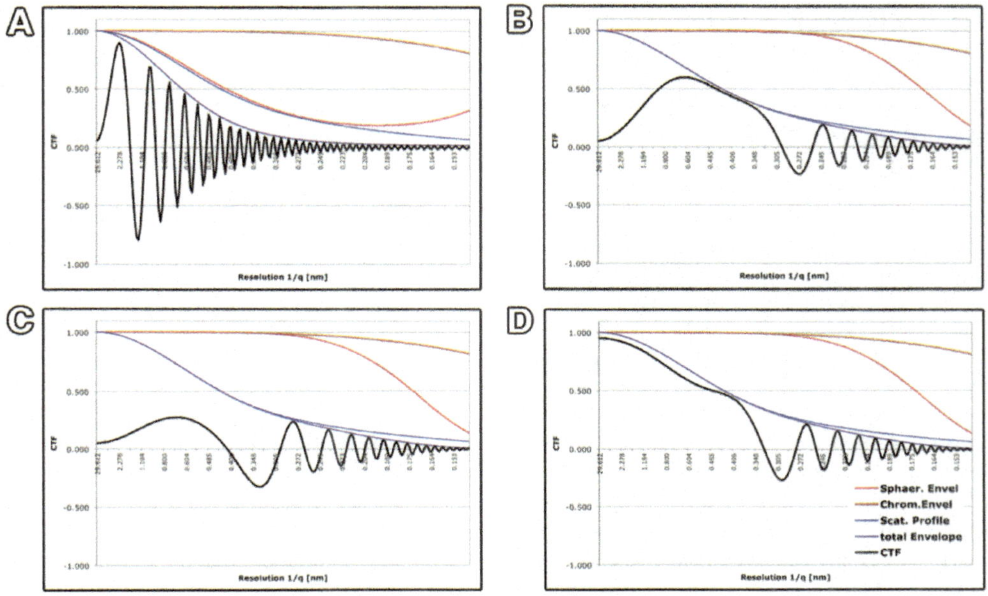

Fig. 4. Calculated CTF profiles for conventional cryo-EM with and without a zero-loss phase plate. (A) At 1000 nm underfocus, a typical defocus for biological samples, low-frequency amplitudes are greatly increased improving low-resolution contrast but causing amplitudes to decay quickly at higher frequencies while introducing a large number of contrast reversals. (B) 70 nm underfocus (Scherzer defocus) typically used for material science TEM imaging. Low-frequency contrast greatly decreases; however, higher frequency contrast decays more slowly and yields fewer contrast reversals than 1000 nm defocus. 35 nm underfocus without (C) and with (D) a zero-loss 90-degree phase plate. Note the improvement of the contrast for the low-resolution frequencies. For all CTF profiles, the microscope parameters are from the specifications of the JEM-2100F used for the other experiments in this paper: Accelerating voltage of 200 keV, energy spread of 0.7 eV, Cs of 2.0 mm and Cc of 2.0 mm.

4.3 Low-dose aberration corrected electron microscopy

A major advance in TEM optics is the development of spherical aberration (Cs) correction. Using an electromagnetic hexapole element to correct for spherical aberration, the first studies significantly improved point resolution from 0.24 nm to 0.13 nm with a 200 keV microscope (Haider et al., 1998; Urban et al., 1999). Additionally, new advancements to the aberration corrected microscope platform have given rise to imaging of inorganic specimens with sub-angstrom spatial resolution.

In order to study samples at atomic resolution, it was necessary to image at relatively high accelerating voltages, generally 200 keV or higher (Haider et al., 1998). These electrons have far more energy than needed to produce knock-on damage to graphitic carbon with a critical threshold at 60 keV (Egerton et al., 2004). Therefore, carbon-based samples will succumb to ionizing damage rather quickly. Using a low voltage (80 keV) Cs corrected microscope is one method to reduce or potentially eliminate knock-on damage. In addition, lower accelerating voltages are beneficial to sample preservation and also improve phase contrast. While low voltage Cs corrected microscopes may not have the point resolution of higher voltage microscopes, the added sample stability has the potential to allow the dose necessary to study single atom dynamics and EELS (Chuvilin et al., 2008). It should be noted that while low voltage electrons produce less knock-on damage, the slower electrons interact with the sample more frequently causing increased ionization and localized heating, both of which are potentially detrimental to image quality (Blanford & Carter, 2003).

4.3.1 Low-dose aberration corrected TEM

Cs corrected TEM has been used for the past decade to resolve inorganic specimens with unprecedented detail. However, it has only been within the last four years that spherical aberration correction has been applied towards improving the imaging of organic specimens (Evans et al., 2008). Part of the delay in applying Cs correction to beam sensitive materials was due to a limitation in the design of the microscope platform. On JEOL microscopes, the image shift coils typically utilized during setup of MDS are removed to accommodate the Cs corrector. While the Cs corrector itself has image shift coils, the lens values are not stored by the MDS software and thus automated switching between modes is impossible. Instead, a partially manual approach is necessary in which MDS is setup normally except no image shift is applied through the microscope controls. Then, whenever switching to or from FOCUS mode, the IShift lenses of the Cs corrector are adjusted by a known amount through the CEOS interface (Evans et al., 2008). This modified low-dose methodology permitted the imaging of Silicon (Fig. 5A) and a two-dimensional crystal of the organic molecule paraffin (Fig. 5B) to 0.11 and 0.16 nm respectively (Fig. 5).

Although Cs correction by itself does not enhance the phase contrast of low-resolution spatial frequencies, it does permit improved contrast within the 0.7 - 0.1 nm range. Thus, in addition to most inorganic specimens, organic samples with primary lattice spacings smaller than 7 angstroms (most self-assembled monolayers of polymers) would immediately benefit from low-dose Cs corrected imaging. This is because the extension of the point resolution and single continuous phase across this resolution range (Fig. 5C) makes the information in such images directly interpretable for thin samples. Currently, the rapid oscillations of the CTF for a non-corrected microscope require CTF correction and full image simulation to understand the structural details. For biological imaging of protein molecules, the future coupling of aberration correction and zero-loss phase plate technologies promise improved point resolution AND contrast for all spatial frequencies (Fig. 5D) thereby allowing more reliable particle identification and structure determination.

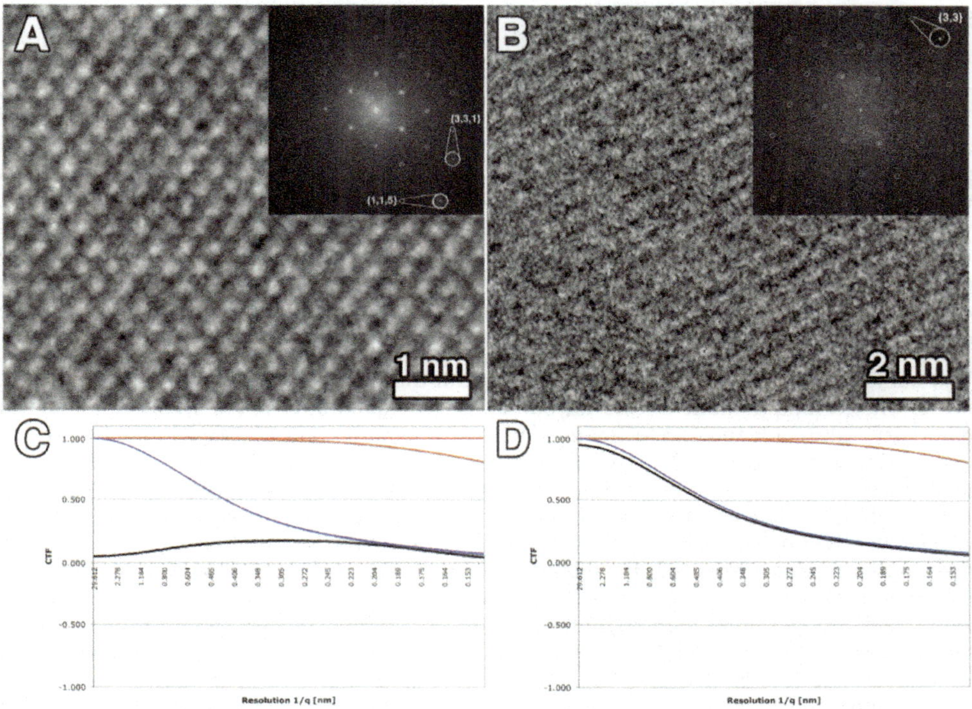

Fig. 5. Analysis of low-dose aberration corrected TEM performance on inorganic Si (110) and organic paraffin crystal specimens recorded under low-dose cryo-conditions at 10 e-/Å2. (A) Under low-dose conditions, atomic fringes within the crystal lattice of Si (110) are directly visible. The Fourier transform in the upper right corner shows lattice spacings to 0.11 nm resolution. (B) Paraffin crystals, consisting of low atomic number elements carbon and hydrogen, directly show lattice fringes with Cs corrected optics to 0.16 nm resolution. (C&D) CTF simulations of a 200 keV Cs corrected microscope without (C) and with (D) a 90-degree phase shift from a zero-loss phase plate.

4.3.2 Low-dose aberration corrected STEM

Currently, Cs corrected scanning TEM (STEM) can achieve higher spatial resolution and better contrast for robust inorganic samples than Cs corrected TEM. The first description of low-dose Cs corrected STEM used a dedicated 200 keV Cs corrected JEOL 2100F/Cs retro-fitted with a CEOS Cs corrector. By reducing the dwell time for each pixel to 1.0 microseconds per pixel and reducing the beam current to about 1 pA with a pixel size of 0.4 Å2, atomic resolution could be seen in the power spectrum of strontium titanate crystal lattices. Although real space images could not discern any structures, Fourier filter algorithms detected reflections to 0.28 nm for Cs corrected STEM images at a dose of 15 e-/Å2 (Buban et al., 2010). Those results suggested that Cs corrected STEM can be used under low-dose conditions and achieve atomic resolutions that make it amenable to studying beam sensitive materials, however, further research is needed to prove the capabilities of the technique with organic specimens.

5. Conclusion

Proteins, polymers, zeolites, and ceramics are all examples of specimens that are sensitive to exposure by high-energy electrons (Csencsits & Gronsky, 1987; Glaeser, 1971; Revol & Manley, 1986). Cumulative exposure of such samples to high-energy electrons causes structural changes that limit interpretation of the native structure via bond breakage, displacement, and gas evolution (Frank, 2006; Williams & Carter, 2004). In particular, radiolysis damage destroys high-resolution information rather quickly once exposed to even a small number of electrons, typically 10 $e^-/Å^2$ for protein in cryo-EM (Frank, 2002). Therefore, in order to generate a high-resolution image at the appropriate focus, a method for limiting pre-exposure is needed to localize the sample and allow for proper defocusing prior to image acquisition. This is done using low-dose procedures that are easily adaptable and can be used in synergy with all of the other advanced imaging techniques mentioned above.

As it stands, this paper outlines methods to retain the maximum resolution of the sample, but no one technique alone allows for sub-angstrom imaging of beam sensitive biological materials. Future advances in technique and instrumentation must be developed in order to push through to the next level of resolution while maintaining the dose requirements needed to preserve native structures and dynamics. While structural biologists are limited to looking at static images of single proteins and protein crystals, new advances in in-situ stages have the potential of opening the door into protein dynamics that may require even lower radiation doses to ensure the integrity of enzyme functionality. In addition, new ways of dealing with even lower signal-to-noise, inherent with in-situ fluid stages, will most likely need development before high-resolution dynamics can be explored. As a result, this may only be possible by implementing DTEM with spherical and chromatic aberration correction as well as zero-loss obstruction-less phase optics and single-electron CCD detectors. If realized, rapid acquisition at the microsecond timescale could allow direct viewing of intermolecular conformational changes (McPherson and Eisenberg, 2011). Ultimately, moving to the nanosecond timescale could be possible with ultrafast imaging (Reed et al., 2009) and allow exploration of protein domain movements responsible for converting between an active and inactive functional state (McPherson & Eisenberg, 2011).

6. Acknowledgements

This work was supported by NIH grant number 5RC1GM091755.

7. References

Adrian, M.; Dubochet, J.; Lepault, J. & McDowall, A. W. (1984). Cryo-electron microscopy of viruses. *Nature,* Vol. 308, No. 5954: pp. 32-36, ISBN

Blanford, C. F. & Carter, C. B. (2003). Electron Radiation Damage of MCM-41 and Related Materials. *Microscopy and Microanalysis,* Vol. 9, No. 03: pp. 245-263, 1431-9276 ISBN

Buban, J. P.; Ramasse, Q.; Gipson, B.; Browning, N. D. & Stahlberg, H. (2010). High-resolution low-dose scanning transmission electron microscopy. *Journal of Electron Microscopy,* Vol. 59, No. 2: pp. 103-112, ISBN

Chiu, W.; Downing, K. H.; Dubochet, J.; Glaeser, R. M.; Heide, H. G.; Knapek, E.; Kopf, D. A.; Lamvik, M. K.; Lepault, J.; Robertson, J. D.; Zeitler, J. D. & Zemlin, F. (1987). Cryoprotection in electron microscopy. *J. Microsc.,* Vol. 141, No.: pp. 385-391, ISBN

Chiu, W. & Jeng, T. W. (1982). Electron radiation sensitivity of protein crystals. *Ultramicroscopy,* Vol. 10, No. 1-2: pp. 63-69, 0304-3991 ISBN

Chuvilin, A.; Kkaiser, U.; Obergfell, D.; Kkhlobystov, A. & Roth, S. (2008). Application of 80kV Cs-corrected TEM for nanocarbon materials. *Proceedings of European Microscopy Conference*, Aschen, Germany, month year

Csencsits, R. & Gronsky, R. (1987). Damage of zeolite Y in the TEM and its effects on TEM images. *Ultramicroscopy*, Vol. 23, No. 3-4: pp. 421-431, 0304-3991 ISBN

Downing, K. H. & Glaeser, R. M. (1986). Improvement in high resolution image quality of radiation-sensitive specimens achieved with reduced spot size of the electron beam. *Ultramicroscopy*, Vol. 20, No. 3: pp. 269-278, 0304-3991 ISBN

Downing, K. H. & Li, H. (2001). Accurate Recording and Measurement of Electron Diffraction Data in Structural and Difference Fourier Studies of Proteins. *Microscopy and Microanalysis*, Vol. 7, No. 05: pp. 407-417, ISBN

Dubochet, J.; Lepault, J.; Freeman, R.; Berriman, J. A. & Homo, J. C. (1982). Electron microscopy of frozen water and aqueous solutions. *Journal of Microscopy*, Vol. 128, No. 3: pp. 219-237, 1365-2818 ISBN

Egerton, R. F.; Li, P. & Malac, M. (2004). Radiation damage in the TEM and SEM. *Micron*, Vol. 35, No. 6: pp. 399-409, 0968-4328 ISBN

Evans, J. E.; Hetherington, C.; Kirkland, A.; Chang, L.-Y.; Stahlberg, H. & Browning, N. (2008). Low-dose aberration corrected cryo-electron microscopy of organic specimens. *Ultramicroscopy*, Vol. 108, No. 12: pp. 1636-1644, 0304-3991 ISBN

Frank, J. (2002). SINGLE-PARTICLE IMAGING OF MACROMOLECULES BY CRYO-ELECTRON MICROSCOPY. *Annual Review of Biophysics and Biomolecular Structure*, Vol. 31, No. 1: pp. 303-319, 1056-8700 ISBN

Frank, J. (2006). *Three dimensional electron microscopy of macromolecular assemblies: Visualization of biological molecules in their native state.* (2nd), Oxford University Press, USA,

Fujiyoshi, Y. (1998). The structural study of membrane proteins by electron crystallography. *Advances in Biophysics*, Vol. 35, No. 0: pp. 25-80, 0065-227X ISBN

Girit, Ç. Ö.; Meyer, J. C.; Erni, R.; Rossell, M. D.; Kisielowski, C.; Yang, L.; Park, C.-H.; Crommie, M. F.; Cohen, M. L.; Louie, S. G. & Zettl, A. (2009). Graphene at the Edge: Stability and Dynamics. *Science*, Vol. 323, No. 5922: pp. 1705-1708, ISBN

Glaeser, R. M. (1971). Limitations to significant information in biological electron microscopy as a result of radiation damage. *Journal of Ultrastructure Research*, Vol. 36, No. 3-4: pp. 466-482, 0022-5320 ISBN

Gonen, T.; Cheng, Y.; Sliz, P.; Hiroaki, Y.; Fujiyoshi, Y.; Harrison, S. C. & Walz, T. (2005). Lipid-protein interactions in double-layered two-dimensional AQP0 crystals. *Nature*, Vol. 438, No. 7068: pp. 633-638, 0028-0836 ISBN

Gyobu, N.; Tani, K.; Hiroaki, Y.; Kamegawa, A.; Mitsuoka, K. & Fujiyoshi, Y. (2004). Improved specimen preparation for cryo-electron microscopy using a symmetric carbon sandwich technique. *Journal of Structural Biology*, Vol. 146, No. 3: pp. 325-333, 1047-8477 ISBN

Haider, M.; Rose, H.; Uhlemann, S.; Schwan, E.; Kabius, B. & Urban, K. (1998). A spherical-aberration-corrected 200 kV transmission electron microscope. *Ultramicroscopy*, Vol. 75, No. 1: pp. 53-60, 0304-3991 ISBN

Karuppasamy, M.; Karimi Nejadasl, F.; Vulovic, M.; Koster, A. J. & Ravelli, R. B. G. (2011). Radiation damage in single-particle cryo-electron microscopy: effects of dose and dose rate. *Journal of Synchrotron Radiation*, Vol. 18, No. 3: pp. 398-412, 0909-0495 ISBN

Knapek, E. & Dubochet, J. (1980). Beam damage to organic material is considerably reduced in cryo-electron microscopy. *Journal of Molecular Biology*, Vol. 141, No. 2: pp. 147-161, 0022-2836 ISBN

Kumar, S. & Adams, W. W. (1990). Electron beam damage in high temperature polymers. *Polymer,* Vol. 31, No. 1: pp. 15-19, 0032-3861 ISBN

Leapman, R. D. & Sun, S. (1995). Cryo-electron energy loss spectroscopy: observations on vitrified hydrated specimens and radiation damage. *Ultramicroscopy,* Vol. 59, No. 1-4: pp. 71-79, 0304-3991 ISBN

McPherson, A. & Eisenberg, D. (2011). Advances in Protein Chemistry & Structural Biology: Protein Structure and Diseases. In, R. Donev, pp. 189-190. Academic Press,

Meents, A.; Gutmann, S.; Wagner, A. & Schulze-Briese, C. (2010). Origin and temperature dependence of radiation damage in biological samples at cryogenic temperatures. *Proceedings of the National Academy of Sciences,* Vol. 107, No. 3: pp. 1094-1099, ISBN

Nagayama, K. (2011). Phase Plate Electron Microscopy Supramolecular Structure and Function 10. In, J. Brnjas-Kraljević and G. Pifat-Mrzljak, pp. 101-113. Springer Netherlands, 978-94-007-0893-8,

Okamoto, H. (2008). Noise suppression by active optics in low-dose electron microscopy. *Appl. Phys. Lett.,* Vol. 92, No. 6: pp. 063901, ISBN

Pan, M. & Crozier, P. A. (1993). Quantitative imaging and diffraction of zeolites using a slow-scan CCD camera. *Ultramicroscopy,* Vol. 52, No. 3-4: pp. 487-498, 0304-3991 ISBN

Reed, B. W.; Armstrong, M. R.; Browning, N. D.; Campbell, G. H.; Evans, J. E.; LaGrange, T. & Masiel, D. J. (2009). The evolution of ultrafast electron microscope instrumentation. *Microsc Microanal,* Vol. 15, No. 4: pp. 272-281, 1435-8115 (Electronic) 1431-9276 (Linking) ISBN

Revol, J. F. & Manley, R. S. J. (1986). Lattice imaging in polyethylene single crystals. *Journal of Materials Science Letters,* Vol. 5, No. 3: pp. 249-251, 0261-8028 ISBN

Schenk, A. D.; Castaño-Díez, D.; Gipson, B.; Arheit, M.; Zeng, X. & Stahlberg, H. (2010). Chapter Four - 3D Reconstruction from 2D Crystal Image and Diffraction Data. In: *Methods in Enzymology,* J. J. Grant, pp. 101-129. Academic Press, 0076-6879,

Sun, J. & Li, H. (2010). Chapter Ten - How to Operate a Cryo-Electron Microscope. In: *Methods in Enzymology,* J. J. Grant, pp. 231-249. Academic Press, 0076-6879,

Taylor, K. A. & Glaeser, R. M. (1976). Electron microscopy of frozen hydrated biological specimens. *Journal of Ultrastructure Research,* Vol. 55, No. 3: pp. 448-456, 0022-5320 ISBN

Thach, R. E. & Thach, S. S. (1971). Damage to biological samples caused by the electron beam during electron microscopy. *Biophysical journal,* Vol. 11, No. 2: pp. 204-210, ISBN

Ugurlu, O.; Haus, J.; Gunawan, A. A.; Thomas, M. G.; Maheshwari, S.; Tsapatsis, M. & Mkhoyan, K. A. (2011). Radiolysis to knock-on damage transition in zeolites under electron beam irradiation. *Physical Review B,* Vol. 83, No. 11: pp. 113408, ISBN

Unwin, P. N. & Henderson, R. (1975). Molecular structure determination by electron microscopy of unstained crystalline specimens. *Journal of Molecular Biology,* Vol. 94, No. 3: pp. 425-440, 0022-2836 ISBN

Urban, K.; Kabius, B.; Haider, M. & Rose, H. (1999). A way to higher resolution: spherical-aberration correction in a 200 kV transmission electron microscope. *Journal of Electron Microscopy,* Vol. 48, No. 6: pp. 821-826, ISBN

Williams, D. B. & Carter, C. B. (2004). *Transmission Electron Microscopy: A textbook for materials science.* (1st), Springer,

Yakovlev, S. & Libera, M. (2008). Dose-limited spectroscopic imaging of soft materials by low-loss EELS in the scanning transmission electron microscope. *Micron,* Vol. 39, No. 6: pp. 734-740, 0968-4328 ISBN

Zemlin, F. (1989). Dynamic focussing for recording images from tilted samples in small-spot scanning with a transmission electron microscope. *Journal of Electron Microscopy Technique,* Vol. 11, No. 4: pp. 251-257, 1553-0817 ISBN

6

Transmission Electron Microscopy for the Quantitative Analysis of Testis Ultra Structure

Saeed Shokri[1*], Masoud Hemadi[2] and Robert John Aitken[3]
[1]*Anatomy Department, School of Medicine, Zanjan University of Medical Sciences, Zanjan,*
[2]*Fertility and Infertility Research Center, Imam Khomeini Hospital, Faculty of Medicine, Ahvaz Jundishapur University of Medical Sciences, Ahvaz,*
[3]*ARC Centre of Excellence in Biotechnology and Development, Discipline of Biological Sciences, University of Newcastle, NSW,*
[1,2]*Iran*
[3]*Australia*

1. Introduction

1.1 Testes structure

The testes have two functions: they produce the male gametes, spermatozoa, and also produce the male sexual hormone, testosterone, which stimulates the accessory male sexual organs and causes the development of the masculine extragenital sexual characteristics. The testis is surrounded by a thick capsule, the tunica albuginea, from which a conical mass of connective tissue, the mediastinum testis, projects into the testis [1]. The tunica albuginea is covered externally by a serosa. From the mediastinum, delicate fibrous septa radiate towards the tunica albuginea and divide the parenchyma of the testis into about 300 lobuli testis, which communicate peripherally. Each lobule contains 1-4 convoluted seminiferous tubules (about 150-300 μm in diameter, 30-80 cm long). Interstitial tissue between the convoluted tubules is continuous with a layer of loose vascular connective tissue, the tunica vasculosa testis, which is found beneath the tunica albuginea. Each seminiferous tubule continues near the mediastinum into a straight tubule, a tubules rectus. The straight tubules continue into the rete testis, a labyrinthine system of cavities in the mediastinum [1, 2].

1.2 The convoluted seminiferous tubules

These tubules are enclosed by a thick basal lamina and surrounded by 3-4 layers of smooth muscle cells (or myoid cells). The insides of the tubules are lined with seminiferous epithelium, which consists of two general types of cells: spermatogenic cells and Sertoli cells (fig 3, 5).

---

* Corresponding Author

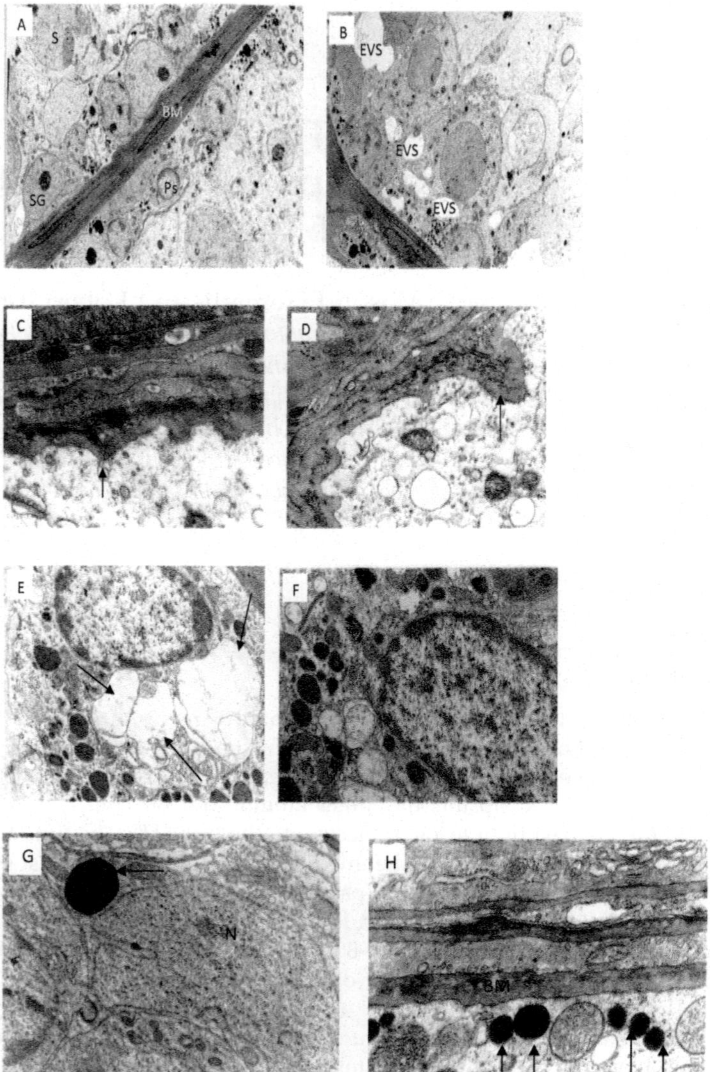

Fig. 1. Electron micrograph (TEM) of rat testes from: A, the control group showing normal basement membrane (BA), primary spermatocytes (PS), spermatogonia (SG) and Sertoli cell (S) (×4000). B, the animals in experimental group showing empty vacuolar spaces (EVS) in the Sertoli and spermatocyte cells (×5000). signs of inflammatory damage of testicular tissue in the form of irregular (C) and thickened (D) basal lamina(→) in (×11000 and ×13000), cytoplasmic vacuolation (→) in the Laydig (E) cells in experimental group (×12000) in comparision with normal Leydig cell (F) in control group (×15000), Lipid droplets (→) in Sertoli cells (N: Sertoli nucleus) in experimental group [G (×22000)] (BM: Basement Membrane) [H (×23000)].

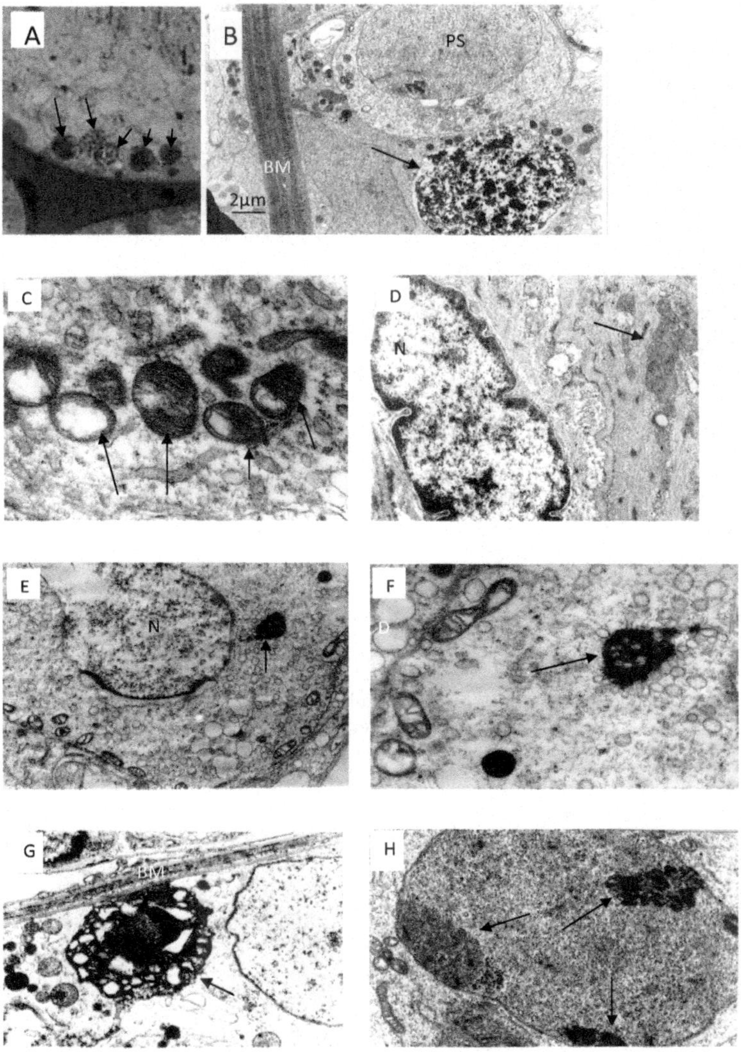

Fig. 2. Analyses of testis cross-sections from experimental group shows spermatogenic cell apoptosis (×400): The representatives TEM semithin (A) and TEM ultrathin (B) sections demonstrate an increase in germ cell apoptosis (arrows) in sections from experimental rats. BM Basement Membrane, PS Primary Spermatocyte. Vacuoles in the mitochondria (C) of Sertoli cells (→) in experimental group (×30000) in comparison with normal mitochondria (D) of Sertoli cells (N: Sertoli nucleus) in control group (×17000), dense body (E,F) in the cytoplasm of spermatid cells (N: spermatid nucleolus) in experimental group (×25000), (×45000), degenerated (G) cell (→) near the basement membrane (BM) in ND-EX group (×30000), dense clumped (H) marginal chromatin (→) in primary spermatocytes in experimental group (×35000).

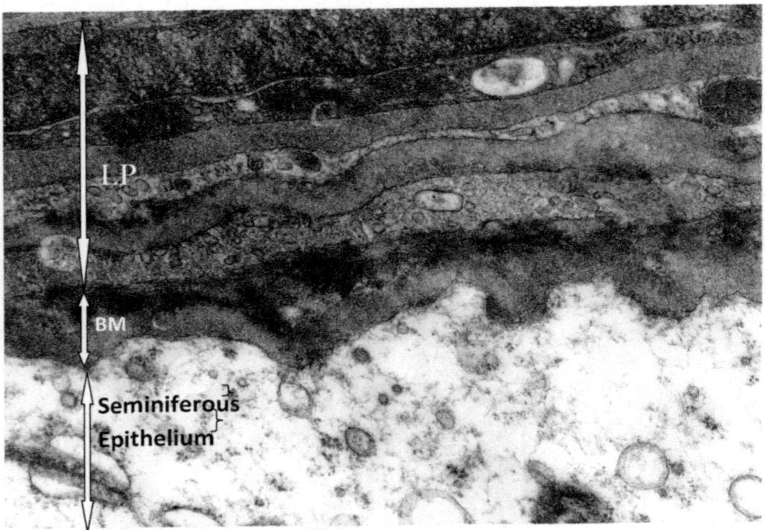

Fig3. LP:Lamina Properia, BM:Basement Membrane

Fig. 3. LP:Lamina Properia, BM:Basement Membrane

Spermatogenic cells:

Spermatogonia

Spermatogonia are the first cells of spermatogenesis. They originate from the primordial germ cells, PGC, of epiblast layer of embryo in the 8th days of feotal development and then appear in the 4th week of development in the endodermal walls of the yolk sac and migrate to the primordium of the testis, where they finally in the puberty differentiate into spermatogonia. They are always in contact with the basal lamina of the tubule. Two types of spermatogonia can be distinguished in the human seminiferous epithelium: Type A spermatogonia have a rounded nucleus with very fine chromatin grains and one or two nucleoli. They are stem cells which divide to form new generations of both type A and type B spermatogonia (fig 4).Type B spermatogonia have rounded nuclei with chromatin granules of variable size, which often attach to the nuclear membrane, and one nucleolus. Although type B spermatogonia may divide repeatedly, they do not function as stem cells and their final mitosis always results in the formation of primary spermatocytes which lie in the cell layer luminal to the spermatogonia. They appear larger than spermatogonia. They immediately enter the prophase of the first meiotic division, which is extremely prolonged (about 22 days!). A large number of primary spermatocytes are always visible in cross-sections through seminiferous tubules. Cell divisions, from the formation of primary spermatocytes and onwards, to the production of the spermatocytes, are incomplete [2]. The cells remain connected by bridges of cytoplasm (fig 4). The completion of the first meiotic division results in the formation of

Secondary spermatocytes,

which are smaller than primary spermatocytes. They rapidly enter and complete the second meiotic division and are therefore seldom seen in histological preparations. Their division results in the formation of

Spermatids,

Which lie in the luminal part of the seminiferous epithelium, they are small (about 10 µm in diameter) with an initially very light (often eccentric) nucleus (fig 4). The chromatin condenses during the maturation of the spermatids into spermatozoa, and the nucleus becomes smaller and stains darker [2].

The terminal phase of spermatogenesis is called spermiogenesis and consists of the differentiation of the newly formed spermatids into

Spermatozoa

The mature human spermatozoon is about 60 µm long and actively motile only following epididymal maturation. It is divided into head, neck and tail.

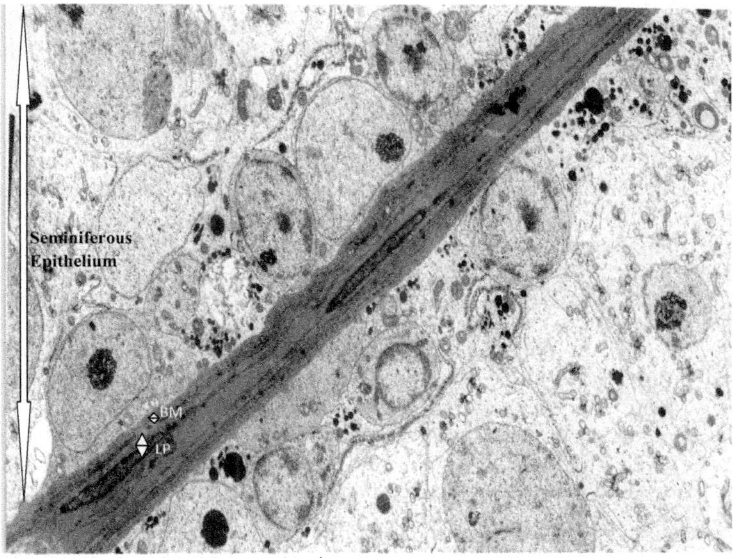

Fig4. LP:Lamina Properia, BM:Basement Membrane

Fig. 4. LP:Lamina Properia, BM:Basement Membrane

The head (flattened, about 5 µm long and 3 µm wide) chiefly consists of the nucleus (greatly condensed chromatin!). The anterior 2/3 of the nucleus is covered by the acrosome, which contains enzymes important in the process of fertilization. The posterior part of the nuclear membrane forms the so-called basal plate.

The neck is short (about 1 µm) and attached to the basal plate. A transversely oriented centriole is located immediately behind the basal plate. The neck also contains nine segmented columns of fibrous material, which continue as the outer dense fibers into the tail.

The tail is further divided into a middle piece, a principal piece and an end piece. The axoneme (the generic name for the arrangement of microtubules in all cilia) begins in the middle piece. It is surrounded by nine outer dense fibers, which are not found in other cilia. In the middle piece (about 5 µm long), the axonema and dense fibers are surrounded by a sheath of mitochondria. The middle piece is terminated by a dense ring, the annulus. The principal piece is about 45 μm long. It contains a fibrous sheath, which consists of dorsal and ventral longitudinal columns interconnected by regularly spaced circumferential hoops. The fibrous sheath and the dense fibers do not extend to the tip of the tail. Along the last part (5 μm) of the tail, called the end piece, the axoneme is only surrounded by a small amount of cytoplasm and the plasma membrane.

It takes about 48 days from the time cells enter meiosis until morphologically mature spermatozoa are formed. Depending on the length of reproduction of spermatogonia (which is not precisely determined) it takes approximately 64 days to complete spermatogenesis.

Spermatogenesis is regulated by follicle stimulating hormone (FSH), which in males stimulates the spermatogenic epithelium, and luteinizing-hormone (LH), which in males stimulates testosterone production by Leydig cells in the interstitial tissue.

Sertoli cells

are far less numerous than the spermatogenic cells and are evenly distributed between them. Their shape is highly irregular - columnar is the best approximation. Sertoli cells extend from the basement membrane to the luminal surface of the seminiferous epithelium. Processes of the Sertoli cells extend in between the spermatogenic cells (cell limits are therefore not clearly visible in the light microscope [LM]). The nucleus of Sertoli cells is ovoid or angular, large and lightly stained and often contains a large nucleolus. The long axis of the nucleus is oriented perpendicular to wall of the tubule [2]. A fold in the nuclear membrane is characteristic for Sertoli cells but not always visible in the LM (well ... actually ... it's not that difficult to find, but not that easy either ....). Lateral processes of Sertoli cells are interconnected by tight junctions, which are likely to be the structural basis for the blood-testis barrier. Spermatogonia and primary spermatocytes are located in the basal compartment; other cellular stages of spermatogenesis are located in the adluminal compartment. Tight junctions may temporarily open to permit the passage of spermatogenic cells from the basal into the adluminal compartment. Sertoli cells provide mechanical and nutritive support for the spermatogenic cells. Sertoli cells also secrete two hormones - inhibin and activin - which provide negative and positive feedback on FSH secretion from the pituitary, respectively.

Interstitial tissue

Leydig cells (15-20 μm), located in the interstitial tissue between the convoluted seminiferous tubules, constitute the endocrine component of the testis. They synthesis and secrete testosterone. Leydig cells occur in clusters, which are variable in size and richly supplied by capillaries. The cytoplasm is strongly acidophilic and finely granular. The nucleus is large, round and often located eccentric in the cell [2].

2. Fine structure of seminiferous tubule (ST) wall

The wall of seminiferous tubule (ST) has formed of two distinct concentrical layers: basal membrane (BM) and Lamina Propria (LP) [3]. However, some authors consider that BM is a component of the LP, being its most inner layer, in apposition with the seminal epithelium [4, 5]. The BM is in contact with the seminiferous epithelium and contains laminin and collagen type I and IV, synthesized by Sertoli cells [5]. The LP consists of three concentric zones [6]. The middle and external ones have a total of three to seven layers of oblonged flattened secreting peritubular cells [7]. The inner zone (lamina interna) is adjacent to the BM, containing large amounts of collagen with lamellar disposition [6]. The middle zone (myoid layer) consists of the first three to five incomplete layers of modified smooth muscle cells called myoid cells. These cells reside in the extracellular matrix among the fibrillary lamellar structures which, in the normal testis, are made of both collagen and laminin [3, 4, 5, 8, 9]. Myoid cells contain numerous actin, myosin and desmin filaments [10, 11]. In contrast to tubules, which underwent sclerosis, they show an intense positive reaction for actin and desmin but only focal positive reaction for vimentin [12,13]. The outer zone consists of the last two cell layers made of fibroblasts and fibrocytes containing characteristic collagen, elastin and vimentin cytoskeletal filaments [6, 11, 14]. Landon and Pryor (1981) found values over 5–7 µm within a narrow variation interval. On the other hand, Trainer (1997), found that the thickness of the two layers varied between 0.3 and 6 µm. Moreover, the thickening of the LP with age is considered to be rather relative as it depends on the ST decrease in length and diameter and represents the result of the reduction in volume of the seminiferous epithelium (fig3, 5) [15–19].

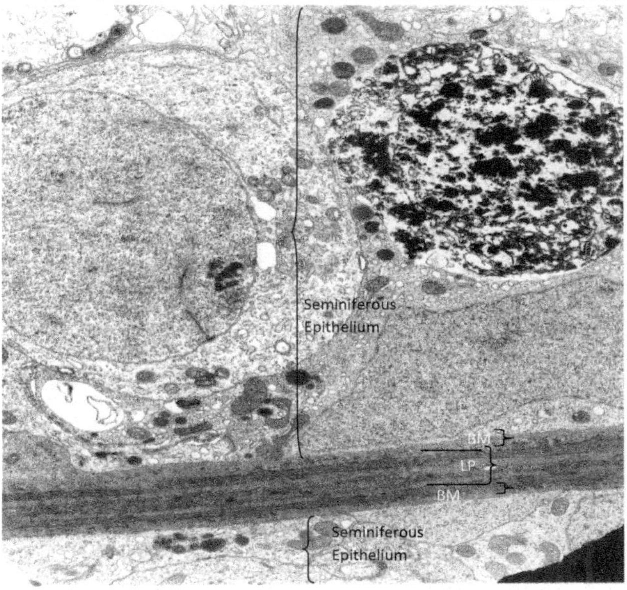

Fig. 5. LP:Lamina Propria, BM:Basement Membrane

2.1 Evaluation of the tubular basement membrane

For evaluation of the tubular basement membrane, we used OT POP et al pattern [49]. The dimensions of the basement membrane were evaluated on TEM samples. For each case, 10 tubules were randomly selected and photographed using the same magnification [49].

The algorithm for evaluating the basement membrane thickness was as follows:

For each tubule, five random measurements were performed and were assigned *g1 CIV, g2 CIV, …, g5 CIV.*

The mean thickness/tubule was calculated using the formula:

$$\frac{(g1\,CIV + g2\,CIV + g3\,CIV + g4\,CIV + g5\,CIV)}{5}$$

and the values were assigned the following symbols: *G1 CIV, ..., G10 CIV.*

The mean thickness per case was calculated using the formula:

$$\frac{(G1\,CIV + G2\,CIV + ... + G30\,CIV)}{10}$$

and the values were assigned the following symbols for each case: *GM1 CIV, ..., GMn CIV,* "*n*" being the number of the last case from each age group.

The mean thickness for each age group was calculated using the formula:

$$\frac{\left(GM1\,CIV + GM2\,CIV + ... + GMn\,CIV\right)}{n}$$

And the values were assigned the following symbols:

GM 01 CIV, ..., GM07 CIV.

2.2 Evaluation of the lamina propria thickness

The evaluation of the lamina propria thickness was performed on the sets of serial sections:

for each case, 10 tubes were randomly selected and photographed using the same magnification of the microscope [49].

The algorithm for evaluating the lamina propria thickness was as follows: For each tubule, five random measurements were performed and were assigned *g1, g2, ..., g5*. The mean thickness/tubule was calculated using the same formula:

$$\frac{(g1 + g2 + g3 + g4 + g5)}{5}$$

And the values were assigned the following symbols: G1, G2, G3, ..., G10.

The mean thickness per case was calculated using the formula:

$$\frac{(G1 + G2 + G3 + ... + G10)}{10}$$

And the values were assigned the following symbols for each case: *GM1, GM2, ..., GMn,* "*n*" being the number of the last case from each age group. The mean thickness for each age group was calculated using the formula:

$$\frac{(GM1 + GM2 + ... + GMn)}{n}$$

And the values were assigned the following symbols: *GM 01, GM 02, ..., GM07.*

3. Evaluation of ultrastructure of seminiferous tubule epithelium

To distinguish different testicular cell types, the following ultrastructural morphological characteristics were used. Spermatogonia are located on the basal lamina of the seminiferous tubules [2]. Type A spermatogonia were identified as cells with an ovoid nucleus with the nucleoli close to the nuclear membrane. The electron-dense cytoplasm contained a small Golgi apparatus, few mitochondria and many free ribosomes. Type B spermatogonia were identified by having a more rounded nucleus and heavily stained chromatin masses attached to the nuclear membrane or to the nucleoli, located at the centre of the nucleus. Sertoli cells were recognized by their location on the basal laminae of the tubuli, by their extension to the lumen of the tubule, and by their large deeply indented nucleus with a homogeneous nucleoplasm and a prominent nucleolus [2]. The cytoplasm contained oval mitochondria, a small Golgi apparatus, an agranular endoplasmic reticulum, lipid droplets and primary and secondary lysosomes. The EM sections were studied in the following categories: (1) Seminiferous tubules: Basement membrane (BM); Spermatogenic series: various stages of maturation, sperm count, and sperm morphology; and Sertoli cells: number, size, and organelles; (2) Extracellular matrix (ECM) and Leydig cells. Features of degeneration were fragmentation of the cell membrane, disorganization of the organelles, intracellular of edema, disruption of the endoplasmic reticulum (ER), and fractionation of mitochondria (fig1, 2) [2].

Groups	*Control rats*	*Sham rats*	*EX1 rats*	*EX2 rats*	*EX3 rats*
Seminiferous tubules					
1. Basement membrane thickening	0	0	2 (Gr. 2)	1 (Gr. 2)	7(Gr. 2) [a]
2. Spermatogonia					
Decreased numbers	0	0	1 (Gr. 1)	1 (Gr. 1)	5(Gr. 2) [a]
Degeneration	0	0	2 (Gr. 1)	1	3(Gr. 2) [a]
Vacuolation	0	0	1 (Gr. 1)	0	3(Gr. 2) [a]
Apoptosis	4(occasional)	3(occasional)	6(single) [a]	3(single)	7(single) [a]
3. Spermatocytes					
Decreased numbers	0	0	1 (Gr. 1)	1 (Gr. 1)	3(Gr. 1) [a]
Degeneration	0	0	2 (Gr. 1) [a]	1	5(Gr. 3) [a]
Vacuolation	0	0	1 (Gr. 1)	0	1(Gr. 2)
Apoptosis	4 (occasional)	3(occasional)	6(clumps) [a]	5(single)	7(clumps) [a]
4. Spermatids					
Decreased number	2(mild)	2(mild)	4(moderate)	2(mild),	5(moderate) [a]
Abnormal morphology	0	0	2(tail/head) [a]	0	4(head) [a]
5. Sertoli cells					
Increased cell number	2	1	1	2	2
Increased cell size	0	0	5[a]	0	7[a]
Increased lipid droplets	2	2	5[a]	1	7[a]

[a]Wilkinson rank-sum test; P value <0.05 statistically significant. EX: experimental group. EX1rats: injecting nandrolone decanoate 10mg/kg/weekly for 8 weeks, EX2 rats: swimming exercise 1hour/day for 8 weeks and EX3 rats: ND + Exercise [2]

Table 1. Ultrstructural changes in testes

The vacuolar and degenerative changes were graded according to the percentage of cells involved as follows: grade 0: < 1% of cells; grade 1: 1-25%; grade 2: 25%-50%; grade 3: > 50% of cells affected. Apoptosis was similary graded as follows: grade 0: < 1%; grade 1: 1-5%; grade 2: 6%-10%; grade 3: > 10% of cells. Cell numbers were graded as follows: grade 0:

normal; grade ±1: 25-50% increase or decrease; grade ±2: 50-75% increase or decrease; grade ±3: > 75% increase or decrease. Thickening of the BM was graded as 0: BM thickness in control rats (taken as normal range); 1: 25%-50%; 2: 50%-75%; 3: >75% increase [2].

Groups	*CO*	*EX1*	*EX2*	*EX3*
Seminiferous tubules				
Basement membrane thickening	normal	Diffuse, 2+fractionation	Mild, focal, 1+Fractionation	Diffuse, 2+ to 3+
Spermatogenesis	normal	Focal Degeneration, singl cell apoptosis, sperm count mildly decreased	Focal Degeneration, singl cell apoptosis, *sperm* count mildly decreased	Marked degeneration, Clumped apoptosis Cells, All counts reduced
Sertoli cells	normal	Increased size Increased lipid droplets (LD)	Mild increase in size, Mild increased LD	Incresed size Increased LD

EX1: injecting nandrolone decanoate 10mg/kg/weekly for 8 weeks, EX2: swimming exercise 1hour/day for 8 weeks and EX3: ND + Exercise [2]

Table 2. Changes observed in rat testes by transmission electron microscopy

4. The ultrastrucure of Leydig cells

In 2002 Redins CA et al measured diameters D1 and D2 from 20 nuclei with prominent nucleoli per animal. They used millimeter-ruled transparent plastic scale, which was placed on the photographs, twenty photographs per group with a same magnification (×2500) [48]. Therefore, they used the formula $R = \sqrt{D1 \times D2}$. Researchers can use the geometric formula of the sphere's volume (Nvol =16 4/3πR3) for determining the nuclear volume (Nvol). Incident point counting methods is useful for counting the fractional volumes of nuclei and cytoplasm of Leydig cells [20]. In this method 414 computerized points placed on photographs, twenty photographs per animal with a same magnification (×2500). These points randomly distributed on a net-like transparent test overlay. According to Bassi *et al.* (1992) and Ferreira *et al.* (1994) calculated the ratio between cytoplasm and nucleus (FCvol/FNvol) and the cytoplasmic volume (Cvol) of the cells (Cvol = Nvol × FCvol/FNvol) [21, 22]. Incident point counting methods was used for counting the cytoplasmic organells of Leydig cell as described for counting Leydig cells fractional nuclear volumes and cytoplasm. Ten 18 × 24 cm photographs at a same magnification (×12500) were examined per animal.

5. Pathological changes of testes ultrastructure

Some studies show a remarkable decrease in the number and size of the Leydig cells and depletion of intact cells in the experimental animals. Feinberg et al. (1997) found pyknotic and severe depletion of Leydig cells following treatment by anabolic androgenic steroids [23]. Close relationship between Leydig cells and blood vessels suggests that these cells are at high risk of exogenous toxicants and multivacuolated Leydig cells are probably a form of cell involution. Leydig cells are known to have receptors for LH that stimulates these cells to produce testosterone. Both LH and testosterone are responsible for normal spermatogenesis in male rats [24]. Therefore, depletion of LH receptors and decreases in peripheral LH by exogenous testosterone administration result in the reduction of testosterone secretion [25].

Some results suggest a reduction of synthetic activity in Leydig cells in some animals. The fractional volumes of 0.6% and 0.3% for the rough endoplasmic reticulum (RER) and Golgi complex, respectively, observed in mouse Leydig cells and reported by Mori *et al.*(1982), were somewhat lower than those in the experimental groups [26]. According to Palade (1975) the RER and the Golgi complex are played a role in synthesis of secretory proteins for export from the cell [27]. The Golgi complex of Leydig cells participates in the secretory pathway of glycoproteins [28]. Based on some reports, that protein synthesis may be diminished in Leydig cells of animal treated with some drugs or toxicants (fig1, 2).

An important role of basement membrane is maintaining the integrity of tissues [29]. It can stable the structure of tissue and send signals to cell [30]. Therefore, alteration of basement membrane structure can impair the severe function of testis [29]. Several proteins including laminin, type IV collagen, various heparin sulfate proteoglycans and ectatin/nidogen is collected in the basement membrane [31]. Type IV collagen is a major constituent of basement membrane which is secreted by Sertoli cells and myofibroblasts [32] and it has been localized in the inner and the outer extracellular matrix (ECM) layers of the basement membrane of seminiferous tubules [30].

Some reports have showed that some exogenous stimulants may induce myoid cells to produce more collagen and ECM that are responsible for basal lamina thickness [33]. Therefore, it may increase thickness of basement membrane and change its appearance to irregular wavy multilaminar shape (Fig 3, 5).

The essential interactions for spermatogenesis are thought to be between Sertoli cells, myoid cells, Leydig cells, and germ cells. These cells must interact together by ECM of the basement membrane.

Many reports have showed that overexpression of type IV collagen correlates with thickened basement membrane and it is related to spermatogenic dysfunction in mammals [31,33, 34]. Some studies demonstrate an increase in size of the basement membrane, others do not indicate significant changes in senescence [36–40].

Different studies that regarding testicular parenchyma, rarely mentioned changes of LP and include only general qualitative assessments. Thus, the gradual thickening of the lamina propria is parallel to tubular involution [16, 17]. Moreover, it is considered that LP thickening depending on the length and diameter decrease of the seminiferous tubules, and being the result of the disproportionate reduction in the volume of the seminiferous epithelium [16–18, 40].

Whitin the testicular epithelium observed two major changes in the Sertoli cells and in their vicinity were noticed. One was the presence of cytoplasm vacuolization, vesicular-like crista of the mitochondria, numerous lipid droplets and lysosome and phagolysosome in Sertoli cells of experimental rats. These are in agreement with the reports showing that exogenous stimulants may cause progressive apoptosis of the Sertoli cells, which affect spermatogenesis and sperm parameters [41]. The second major change was the empty vacuolar spaces between Sertoli cells that are regarded to be the place where spermatogonia and spermatocytes should be located.

In addition, the results of some studies showed that apoptosis occurred in all germ line cells especially in spermatogonia and spermatocyte. The presence of apoptotic germ cells is supported by Blanco et al. (2002) finding that has described apoptosis in hamster testis following treatment with anabolic androgenic steroids [42]. Spermatogenesis is a complex and dynamic process that results in the continual production of spermatozoa in mammals. The Sertoli cells are largely responsible for orchestrating the germ cells through sequential phases of mitosis, meiosis, and differentiation. The Sertoli cells accomplish this task by providing hormonal, nutritional, and physical support. Apoptosis of germ cells that occurs in the testicular epithelium serves as a mechanism to reduce the germ cell population to the level that the Sertoli cells can support. Some drugs and toxicants injure or disrupt the function of Sertoli cells and can effectively reduce their supportive role, resulting in an increase in the elimination of the germ cell numbers via apoptosis [43]. Also it has been described that apoptosis in the germ cells is related to the Fas signaling system that is activated by exogenous toxicants [43]. Observation of detached germ cells, amorphous head sperm, and mislocation of spermatid and spermatozoa to positions that are closely related to the basement membrane may be due to the rapid disruption of Sertoli-germ cell interaction. This physical interaction ultimately leads to the sloughing of the germ cells from the seminiferous epithelium [43]. It seems that the spermatogenesis cycle is reduced by the presence of high levels of androgens. On the other side, some researchers believed that some toxicants can result in an enhanced production of reactive oxygen species (ROS) in cells/tissues and exert oxidative stress (OS), which, in turn, increases the rates of cellular damage [44]. Pey et al. (2003) showed that prolonged stanozolol treatment as an AAS can cause an oxidative stress situation in rat liver [45]. Therefore, it seems logical that the physical/chemical-induced oxidative stress may affect the testicular antioxidant system and lipid peroxidation [46]. OS has been shown to be a major cause of male infertility; a large proportion of infertile men have elevated levels of seminal ROS. Several forms of sperm DNA damage are caused by ROS, e.g. chromatin cross-linking, chromosome deletion and DNA strand breaks (Apoptosis). Under physiological conditions, apoptosis maintains the number of germ cells within the supportive capacity of Sertoli cells. However, disturbances in this pathway can interrupt the spermatogenic cascade. High levels of apoptosis were detected at spermatogenic stages where major developmental blocks occur, and frequencies of DNA damage were higher in less mature germ cells [47] (Figs 2 and table 1,2).

6. References

[1] Shokri S, Aitken RJ, Abdolvahhabi M, Abolhasani F, Mohammad Ghasemi F *et al. Exercise and supraphysiological dose of nandrolone deconoate increase apoptosis in spermatogenic cells.* Basic Clinic. Pharmacol. Toxicol., 2010, 106:324-330.

[2] Naraghi MA, Abolhasani F, Kashani I, Anarkooli IJ, Hemadi M, Azami A *et al. The effects of swimming exercise and supraphysiological dose of nandrolone decanoate on the testis in adult male rats: A transmission electron microscope study.* Folia Morphol., 2010, 69(3):138-146.

[3] Trainer TD. *Testis and excretory duct system. In: Sternerg SS (ed), Histology for pathologists,* 2nd edition, Lippincott–Raven Publishers, Philadelphia, 1997:1019–1035.

[4] Gulkesen KH, Erdogru T, Sargin CF, Karpuzoglu G. *Expression of extracellular matrix proteins and vimentin in testes of azoospermic man: an immunohistochemical and morphometric study.* Asian J Androl., 2002, 4(1):55–60.

[5] Nistal M, Paniagua R. *Non-neoplastic diseases of the testis. In: Bostwick DG, Eble JN (eds),* Urologic surgical pathology, Mosby, St. Louis, 1997:458–544.

[6] Hermo L, Lalli M, Clermont Y. *Arrangement of connective tissue components in the walls of seminiferous tubules of man and monkey.* Am. J. Anat., 1977, 148(4):433–445.

[7] Christl HW. *The lamina propria of vertebrate seminiferous tubules: a comparative light and electron microscopic investigation.* Andrologia, 1990, 22(1):85-94.

[8] Bustos-Obregon E. *Ultrastructure and function of the lamina propria of mammalian seminiferous tubules.* Andrologia, 1976, 8(3):179–185.
[9] Bustos-Obregon E, Holstein AF. *On ultrastructural patterns of the lamina propria of human seminiferous tubules.* Z. Zellforsch Mikrosk Anat., 1973, 141(3):413–425.
[10] Ross MH, Long IR. *Contractile cells in human seminiferous tubules,* Science, 1966, 153(741):1271-1273.
[11] Virtanen I, Kallajoki M, Narvanen O, Paranko J, Thornell LE, Mietinen M, Lehto VP. *Peritubular myoid cells of human and rat testis are smooth muscle cells that contain desmin-type intermediate filaments.* Anat. Rec.,1986, 215(1):10–20.
[12] Martin R, Santamaria L, Nistal M, Fraile B, Paniagua R. *The peritubular myofibroblasts in the testes from normal men and men with Klinefelter's syndrome. A quantitative,ultrastructural, and immunohistochemical study.* J. Pathol., 1992, 168(1):59–66.
[13] Santamaria L, Martin R, Nistal M, Paniagua R. *The peritubular myoid cells in the testes from men with varicocele: an ultrastructural, immunohistochemical and quantitative study.* Histopathology, 1992, 21(5):423–433.
[14] Dekrester DM, Kerr JB, Paulsen CA. *The peritubular tissue in the normal and pathological human testis. An ultrastructural study.* Biol Reprod, 1975, 12(3):317–324.
[15] Landon GV, Pryor JP. *The blood-testis barrier in men of diverse fertility status: an ultrastructural study.* Virchows Arch. A Pathol. Anat. Histol., 1981, 392(3):355–364.
[16] Johnson L. *Spermatogenesis and aging in the human.* J. Androl., 1986, 7(6):331–354.
[17] Paniagua R, Nistal M, Amat P, Rodriguez MC, Martin A. *Seminiferous tubule involution in elderly men.* Biol. Reprod., 1987, 36(4):939–947.
[18] Johnson L, Petty CS, Neaves WB. *Age-related variations in seminiferous tubules in men. A stereologic evaluation,* J Androl, 1986, 7(5):316-22.
[19] Johnson L, Abdo JG, Petty CS, Neaves WB. *Effect of age on the composition of the seminiferous tubular boundary tissue and on the volume of each component in humans,* Fertil Steril, 1988, 49(6):1045–1050.
[20] Bozzola JJ. & Russel LD. *Quantitative Electron Microscopy. In: Principles and techniques for biologists.* Jones and Bartlett Publishers, Boston, 1992, 278- 304.
[21] Bassi WE, Hernandes R, Stipp ACM, Taga R. *Avaliação morfométrica das dimensões do sistema de ductos estriados de glândulas parótidas do rato.* Rev. Brasil. Ciên. Morfol., 1992, *9*(1): 15-19.
[22] Ferreira RR, de assis GF, Taga R. *Estudo morfométrico da evolução do volume nuclear e citoplasmático e da relação citoplasma/núcleo de células acinosas de glândulas parótidas do rato durante a vida pós-natal.* Rev. Brasil. Ciên. Morfol., 1994, *11*(2): 1142-1148.
[23] Feinberg MJ, Lumia AR, McGinnis MY. *The effect of anabolic-androgenic steroids on sexual behavior and reproductive tissues in male rats.* Physiol. Behav. 1997, 62 (1), 23-30.
[24] Zirkin BR. *Spermatogenesis: its regulation by testosterone and FSH, Semin.* Cell Dev. Biol., 1998, 9(4), 417-21.
[25] Ichihara I, Kawamura H, Nakano T, Pelliniemi LJ. *Ultrastructural, morphometric, and hormonal analysis of the effects of testosterone treatment on Leydig cells and other interstitial cells in young adult rats.* Ann. Anat., 2001, 183(5), 413-26.
[26] Mori H, Shimuzu D., Fukunishi Y, Christensen AK. *Morphometric analysis of testicular Leydig cells in normal adult mice.* Anat. Rec., 1982, *204*: 333-339.
[27] Palade G. *Intracellular aspects of the process of protein synthesis.* Science, 1975, 189: 347-358.
[28] Lalli MF, Clermont Y. *Leydig cells and their role in the synthesis and secretion of glycoproteins.* Anat. Rec., 1975, 181: 403-404.
[29] Richardson LL, Kleinman HK, Dym M. *Altered basement membrane synthesis in the testis after tissue injury.* J. Androl., 1998, 19(2):145-55.

[30] Martin GR, Timpl R. *Laminin and other basement membrane components.* Annu. Rev. Cell Biol., 1987, 3: 57-85.
[31] Dobashi M, Fujisawa M, Naito I, Yamazaki T, Okada H, Kamidono S. *Distribution of type IV collagen subtypes in human testes and their association with spermatogenesis.* Fertil. Steril. , 2003, 80(2):755-60.
[32] Skinner MK,Tung PS, Fritz IB. *Cooperativity between sertoli cells and testicular peritubular cells in the production and deposition of extracellular matrix components.* J. Cell Biol., 1985, 100 (6): 1941-47.
[33] Santoro G, Romeo C, Impellizzeri P, Arco A, Rizzo G, Gentile C. *A morphometric and ultrastructural study of the changes in the lamina propria in adolescents with varicocele.* BJU. Int., 1999, 83 (7):828-32.
[34] Häger M, Gawlik K, Nyström A, Sasaki T, Durbeej M. *Laminin {alpha}1 chain corrects male infertility caused by absence of laminin {alpha}2 chain.* Am. J. Pathol., 2005, 167(3):823-33.
[35] Bergman RA, Afifi AK, Heidger PM JR. *Atlas of microscopic anatomy. Section 14: Male reproductive system.* The Virtual Hospital University of Iowa College of Medicine, 1999.
[36] Sasano N, Ichijo S. *Vascular patterns of the human testis with special reference to its senile changes.* Tohoku J. Exp. Med., 1969, 99(3):269–280.
[37] Johnson L, Petty CS, Neaves WB. *Influence of age on sperm production and testicular weights in men.* J Reprod Fertil, 1984, 70(1):211–218.
[38] Neaves WB, Johnson L, Porter JC, Papker CR JR, Petty CS. *Leydig cell numbers, daily sperm production,and serum gonadotropin levels in aging men.* J. Clin. Endocrinol. Metab., 1984, 59(4):756–763.
[39] Meacham RB, Murray MJ. *Reproductive function in the aging male.* Urol. Clin. North. Am., 1994, 21(3):549–556.
[40] Plas E, Berger P, Hermann M, Pfluger H. *Effects of aging on male fertility?* Exp. Gerontol., 2000, 35(5):543–551.
[41] Gopalkrishnan K, Gill-Sharma MK, Balasinor N, Padwal V, D'Souza S, Parte P et al. *Tamoxifen-induced light and electron microscopic changes in the rat testicular morphology and serum hormonal profile of reproductive hormones.* Contraception, 1998, 57(4):261-69.
[42] Blanco A, Flores-Acuna F, Roldan-villalobos R, monterde JG. *Testicular damage from anabolic treatments with the beta(2)-adrenergic agonist clenbuterol in pigs: a light and electron microscope study.* Vet. J., 2002, 163(3):292-98.
[43] Richburg JH. *The relevance of spontaneous- and chemically-induced alterations in testicular germ cell apoptosis to toxicology.* Toxicol. Lett., 2000, 15(112-113):9-86.
[44] Davies KJ, Quintanilha AT, Brooks GA, Packer L. *Free radicals and tissue damage produced by exercise.* Biochem. Biophys. Res. Commun., 1982, 107(4):1198-205.
[45] Pey A, Saborido A, Blázquez I, Delgado J, Megías A. *Effects of prolonged stanozolol treatment on antioxidant enzyme activities, oxidative stress markers, and heat shock protein HSP72 levels in rat liver.* J. Steroid Biochem. Mol. Biol., 2003, 87(4-5):269-77.
[46] Husain K, Somani SM. *Interaction of exercise training and chronic ethanol ingestion on testicular antioxidant system in rat.* J. Appl. Toxicol., 1998, 18(6):421-29.
[47] Tesarik J, Greco E, Cohen-Bacrie P, Mendoza C. *Germ cell apoptosis in men with complete and incomplete spermiogenesis failure.* Mol. Hum. Reprod. 1998, 4(8):757-62.
[48] Redins CA, Redins GM and Novaes JC. *The effects of treatment with melatonin on the ultrastructure of mouce leydig cells: a quantitative study.* Braz. J. Biol. 2002, 62(3):517-523.
[49] Pop OT, Cotoi CG, Pleşea IE, Gherghiceanu M, Enache SD, Mandache E, Hortopan G, Pleşea RM. *Histological and ultrastructural analysis of the seminiferous tubule wall in ageing testis.* Rom J Morphol Embryol. 2011, 52(1 Suppl):241-8.

7

Transmission Electron Microscopy to Study Gallium Nitride Transistors Grown on Sapphire and Silicon Substrates

S. Lawrence Selvaraj and Takashi Egawa
Research Center for Nano-Device and System,
Nagoya Institute of Technology,
Japan

1. Introduction

Ever since Gallium Nitride based high electron mobility transistor (HEMT) operation was demonstrated (Khan, 1993), there is a tremendous interest in the design and growth of GaN based transistors for high power device applications. The nitride semiconductors have wide application in the fields of high electron mobility transistors (HEMTs), light emitting diodes, and various high power electronic devices. The peak electron velocity, electron mobility in GaN surpasses the best performance reported from Si based devices. GaN and related materials are highly attractive for high power and high temperature electronic devices owing to their large bandgap energy (3.4 eV), high breakdown field (3×10^6 V/cm) (Pearton, 2006) and excellent chemical stability. The growth of these nitrides layers require a substrate which should be cost effective and lattice matching to GaN. For over the past few decades, SiC, sapphire and silicon are the substrates commonly used for the growth of GaN for application as HEMTs, LEDs and other electronic devices. The choice of a suitable substrate for the growth of III-nitride semiconductor devices is very important as it influences the lattice mismatch and quality of GaN. Today most of the GaN based heterostructures (HSs) were grown either on *c*-plane (0001) sapphire or silicon. The GaN growths on silicon develop cracks due to high tensile stress and ends up in large number of threading dislocations. Poor matching between GaN and sapphire both in terms of lattice parameter and thermal expansion coefficient results in a high dislocation density as high as 10^9 - 10^{10} cm^{-2}.

The growth of low defect-density GaN is of great technological importance owing to enhanced device applications by overcoming the existing drawbacks such as current collapse and excess leakage through buffer and substrate. And for this transmission electron microscope (TEM) serves as an important tool to investigate the quality of GaN grown on various substrates. It sheds light on how the growth quality is improved by various growth advancing technologies. In this chapter, we will review how TEM continuous to play an important role in the evolution and commercialization of high quality GaN growth suitable for power device requirements.

2. Substrates for GaN

The substrates widely used for AlGaN/GaN HEMTs are SiC, sapphire and silicon. The GaN growth on SiC offers excellent quality as it has very low lattice mismatch of 3.5%. Good quality of GaN is also grown on sapphire substrate, but the thermal conductivity is very low which hinders during device operation. Therefore, Si is undoubtedly the most suitable choice for commercialization due its low cost and large size availability. Though higher lattice and thermal mismatches of Si generates a large number of dislocation and cracks, AlGaN/GaN HEMT devices on Si have shown attractive device performance for high-power applications. Until now, only Si offers a large size (8-inch diameter) wafer for growth. Table 1 compares the material properties of various substrates used for GaN growth. Therefore it is necessary to study the quality of GaN grown on Si substrate by TEM images to assess the growth quality.

Substrate material	SiC	*c*-sapphire	Si (111)
Lattice mismatch to GaN [%]	3.5	14	17.0
Thermal expansion [10^{-6} K^{-1}]	5.0	a: 7.5	2.6
(GaN : a ~ 5.5)		c: 8.5	
Thermal conductivity [W/cm • K]	4.9	0.3 ~ 0.5	1.5
Cost [per 3 inch]	US$ 2000-3000	US$ 500	US$ 100
Size available [inch]	2 to 3	2 to 4	3 to 6

Table 1. Substrates commonly used for the growth of GaN and the comparison of their properties.

Here we will be presenting how the quality of the GaN grown on sapphire and Si makes a difference in the growth quality as viewed by TEM. All the hetero-structures presented here were grown by Taiyo-Nippon Sanso, SR 4000 horizontal MOCVD set up on sapphire and Si substrates. These transistor structures grown on various substrates were reported to have excellent transistors device characteristics (Selvaraj & Egawa, 2006, 2007). In this chapter, AlGaN/GaN HEMT heterostructures grown on sapphire and Si substrates as shown in Figure. 1 will be used for TEM analysis. The MOCVD growth of GaN starts directly on sapphire substrate without any intermediate buffer layers due to reduced lattice mismatch of 14% between GaN and sapphire. On the contrary for HEMT on Si, in view of the large lattice mismatch for GaN on Si, it is necessary to reduce the dislocations by using intermediate buffer layers. The buffer layer starts with nucleation layers of 100 nm AlN followed by 40 nm AlGaN. Then, superlattice structures (multi pairs of GaN/AlN : 20/5 nm) were used as buffer of various thicknesses to reduce the dislocation density (Selvaraj et. al., 2009). The MOCVD growth completed with i-GaN channel and 25 nm i-AlGaN barrier layers as shown in Fig.1. The TEM images in the following sections will reveal how the dislocation density is lowered and the quality of the GaN was improved by increasing the thickness of the buffer.

For an excellent transistor operation and high power device application, all the layers grown and their growth quality plays a distinct role. Therefore, the growth quality of these layers will be analyzed using TEM images.

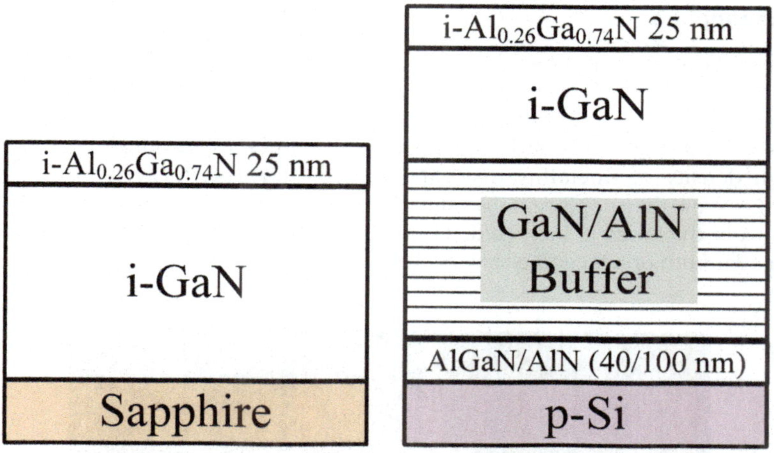

Fig. 1. The schematic description of the AlGaN/GaN heterostructures grown on Sapphire and Silicon substrates by MOCVD.

3. GaN on sapphire

Sapphire is the most extensively used substrate for growth of III-nitrides. Large area good quality crystals of sapphire are easily available at a moderate cost. They are transparent, stable at high temperature and the technology of growth of the nitrides on sapphire is quite mature. Sapphire (α-Al_2O_3) exists in four orientations namely (10-10), (0001), (2-1-10) and (11-20). The *c*-plane (0001) of sapphire has a lattice mismatch of 14% whereas the *a*-plane (11-20) has a very low lattice mismatch of 2.0% (Ambacher, 1998). The *c*-plane sapphire is widely used for GaN growth than *a*-plane sapphire. The *a*- and *c*-planes of sapphire are illustrated in Fig. 2. The growth of GaN was carried out on these two planes of sapphire namely *c*-plane and *a*-plane. With the help of TEM study, we will understand how significantly the quality of GaN changes between the these two planes of sapphire. Also the lattice mismatch of 14% and 2% respectively in the case of GaN grown on *c*-plane and *a*-plane sapphire is evident with TEM images.

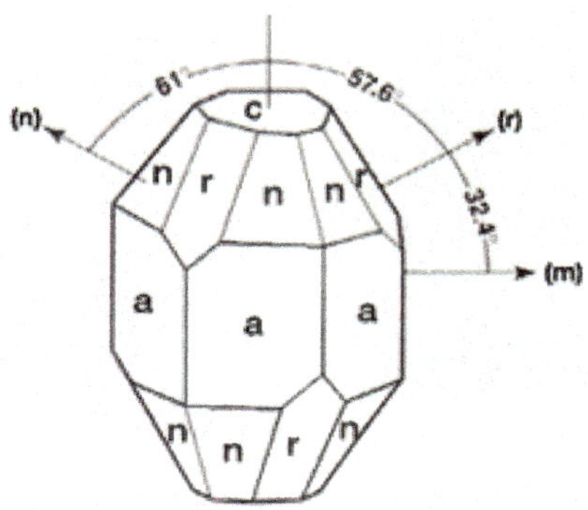

Fig. 2. Rhombohedral structure and surface planes of sapphire (Ambacher, 1998).

3.1 *c*-plane sapphire

Crystal orientations of sapphire and GaN grown on *c*-plane sapphire parallel. The *c*-plane is the commonly used orientation for the growth of GaN on sapphire. Because of the 14% lattice mismatch, a large number of dislocations are observed in the cross sectional TEM image as shown in Fig. 3. Careful observation at the interface of substrate and GaN layer shows high density of threading dislocations at the nucleation layers which begins the growth of GaN on the sapphire substrate. However, there is a decrease in the dislocation density as GaN thickness is increased. These dislocations should be minimized in view of applications for both optical and electronic devices.

Fig. 3. MOCVD grown GaN on *c*-plane sapphire showing high density of threading dislocations (Selvaraj & Egawa, 2008).

3.2 *a*-plane sapphire

The edge and screw dislocation measured from the single crystal XRD data emphasize that dislocations are less for heterostructures grown on *a*-plane sapphire (Selvaraj & Egawa, 2008). The transmission electron microscopy (TEM) images as in Fig. 4 shows that GaN

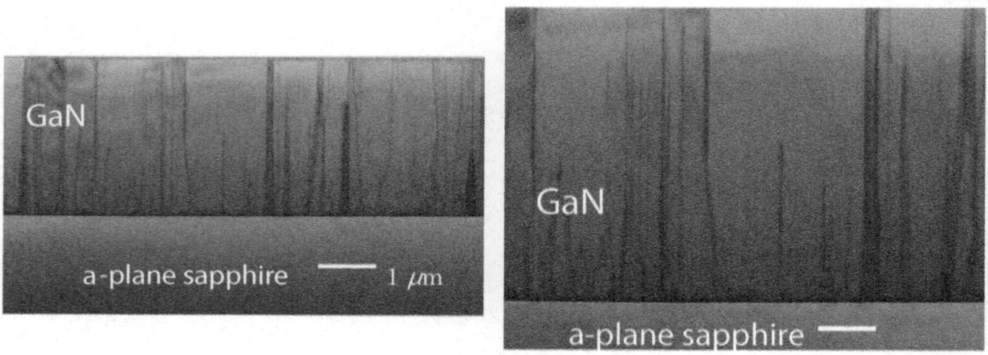

Fig. 4. MOCVD grown GaN on *a*-plane sapphire showing high quality of GaN with very low density of threading dislocations as shown in this TEM image (Selvaraj & Egawa, 2008).

grown on *a*-plane has fewer dislocations than GaN on *c*-plane as seen from Fig. 3. From the TEM images, the threading dislocation density (D_d) for GaN on *a*-plane was calculated to be lower ($1.8x10^9$ cm^{-2}) than for GaN on *c*-plane sapphire ($3.6x10^9$ cm^{-2}). These results illustrate that growth of AlGaN/GaN heterostructures on *a*-plane sapphire has good interface lattice alignment suitable for device applications.

Today *a*-plane sapphire substrate are available commercially and suitable for low cost and large scale production. The GaN grown on *a*-plane sapphire were found to have improved surface morphology with low threading dislocation density. Excellent HEMT device properties were observed with an enhanced Schottky barrier height resulting two orders of magnitude less gate leakage current and a reduced drain current collapse.

4. GaN on Si

The use of Si substrate is very promising due to low cost and large size availability. As mentioned earlier, the large lattice mismatch causes the large dislocations which affects the insulating nature of i-GaN in the heterostructure. For high power device applications, it is necessary to grow GaN on Si with low dislocations.

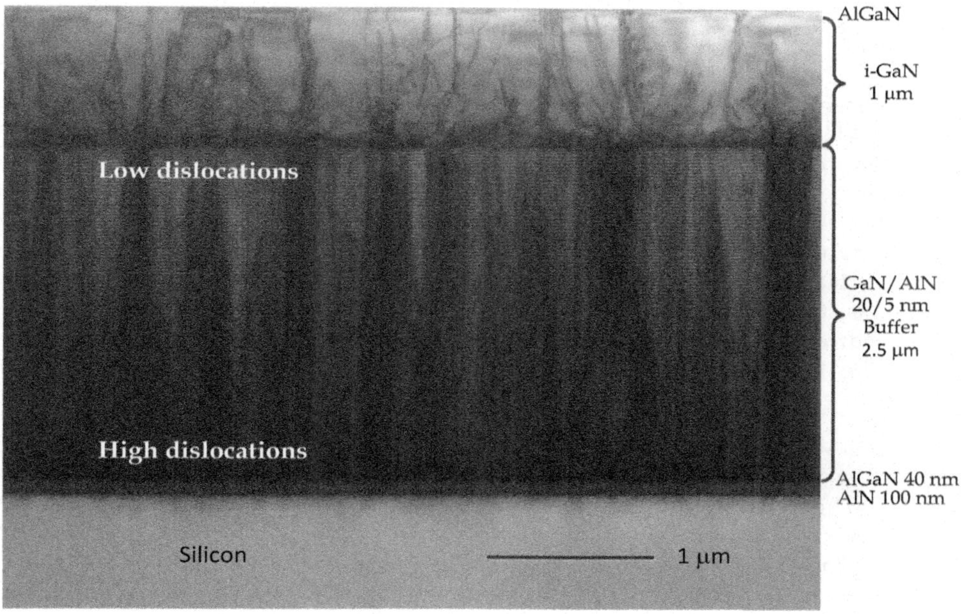

Fig. 5. TEM view of MOCVD grown GaN on silicon substrate showing improvement of GaN quality with increase in the thickness of buffer.

The GaN grown on Si with thin buffers (0.5 μm) were found to demonstrate a poor breakdown which challenges the ultimate importance of GaN for high power devices. In order to overcome the problem due to high dislocations, we have grown GaN on Si using thick buffers which improved the breakdown (Selvaraj, 2009a, 2009b). The cross sectional TEM image for GaN grown on Si using thick buffer is shown in Fig.5 which has large threading dislocations due to 17% lattice mismatch. However the dislocations shown here are lower compared to GaN grown on Si using thin buffers. In Fig. 5, it is understood that the density of threading dislocations lowered as the thickness of epi-layers is increased. The dislocations are highest at the GaN/Si interface and gradually decreases at the top layers of the wafer. It is this top surface layer that plays a major role in the horizontal conduction of electrons through the 2DEG channel forming a high electron mobility transistor. Thus the optimization of GaN growth on Si is technologically important to achieve a commercially low cost and large area fabrication and TEM plays a major role to study the quality of epi-layers grown.

The TEM figure 5 shows the various epi-layers involved in the MOCVD growth of AlGaN/GaN HEMTs on Si. The MOCVD growth on Si starts with a 100 nm AlN layer to prevent 'Ga' melt-back etching of Si substrate at high growth temperature. The 100 nm AlN was followed by 40 nm AlGaN and both these starting layers are called nucleation layers. Then follows the buffer layer which consists of multiple pairs of GaN/AlN which is also called as super-lattice structure (SLS). The thickness of the buffer is significant to compensate the strain caused by the GaN channel layer and further lowers the dislocation density of epi-layers. In Fig. 5, the buffer consists of 100 pairs of GaN/AlN giving a buffer thickness of 2.5 μm. The multiple pairs are visible here like black and white stripe lines.

A close observation of Fig.5 shows that near the Si substrate, the density of dislocations are very high and this high dislocation density is reduced gradually as the thickness of the buffer is increased. Finally in the GaN layer which is otherwise called as channel, the dislocation density is observed to be minimum. The dislocations present in the GaN have very little effect for normal biasing conditions of the transistor. However, these dislocations are very critical for reverse blocking characteristics of the transistor. When the gate is biased below the threshold voltage during OFF-state blocking measurement, the drain is strongly biased and imperfections in the device give rise to leakage currents such as source-drain leakage, substrate leakage and gate leakage. Therefore it is necessary to minimize the dislocations for achieving a transistor capable of operating at high power conditions.

The SLS multi-pairs of GaN/AlN which forms the buffer comprised major portion among the growth completed epi-layer. The TEM image in Fig. 6 shows more enlarged view of the MOCVD grown 20 nm GaN/5 nm AlN regular periods. As the thickness of the GaN layer increase, the tensile strain in the *a*-axis of the AlN template increases (Murakawa et al., 2007). The compressive stress of the 1.0 μm thick GaN channel layer is counter balanced by the stress relief through this multiple pairs of SLS buffer. The thickness of an SLS pair namely 20 nm GaN/5 nm AlN could be confirmed through the TEM image shown here.

The use of multi-pairs of GaN/AlN facilitates lowering the dislocation density. From the TEM image shown in Fig.7 we could find that a threading dislocation running through the multi-pairs of GaN/AlN is terminated towards the top-edge of the SLS buffer. In general, the advantage of using the SLS buffer reduces the dislocation density and also provides a balance for the compressive stress produced by thick GaN channel layer.

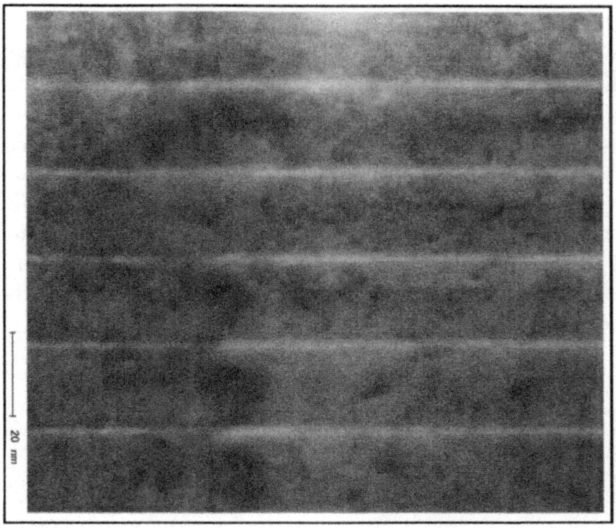

Fig. 6. The SLS buffer consisting 20 nm GaN/5 nm AlN as seen by TEM image.

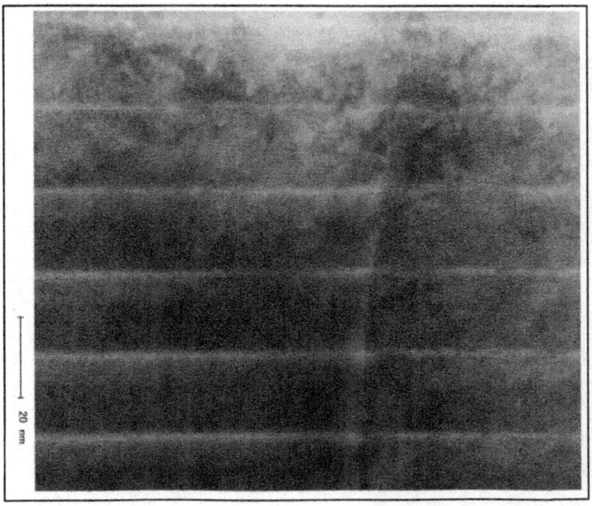

Fig. 7. Termination of a threading dislocation at the SLS buffer for AlGaN/GaN HEMTs grown on Si.

In order to find its application in the field of high power devices, we have attempted in the past to grow thick AlGaN/GaN HEMTs on Si substrate using thick GaN/AlN buffers. We found that increasing the thickness of the buffer reduces the dislocation density gradually. This was confirmed by measuring the full-width at half maximum (FWHM) from the X-ray diffraction measurement on these wafers. The screw and edge dislocation densities were calculated from the FWHM values (Metzger et al., 1998). By growing thick buffers, we achieved lowest dislocation densities of 5.8×10^{8} cm^{-2} and 2.6×10^{10} cm^{-2} respectively for screw and edge dislocation density (Rowena et al., 2011). The TEM image shown below in Fig.8 has a 2 µm i-GaN grown on 3 µm buffer. Increasing the thickness of i-GaN facilitates the growth of thick i-GaN beyond 1 µm. In comparison with Fig.5, we could observe that threading dislocations are largely reduced by increasing the thickness as shown by Fig.8. These wafers having low dislocation density showed high breakdown voltages due to low leakage currents through buffer and substrate. The buffer and substrate leakage currents which are critical for achieving a high breakdown are influenced by the dislocations and hence it is necessary to reduce the dislocations using thick buffers which compensates the lattice mismatch between Si and GaN.

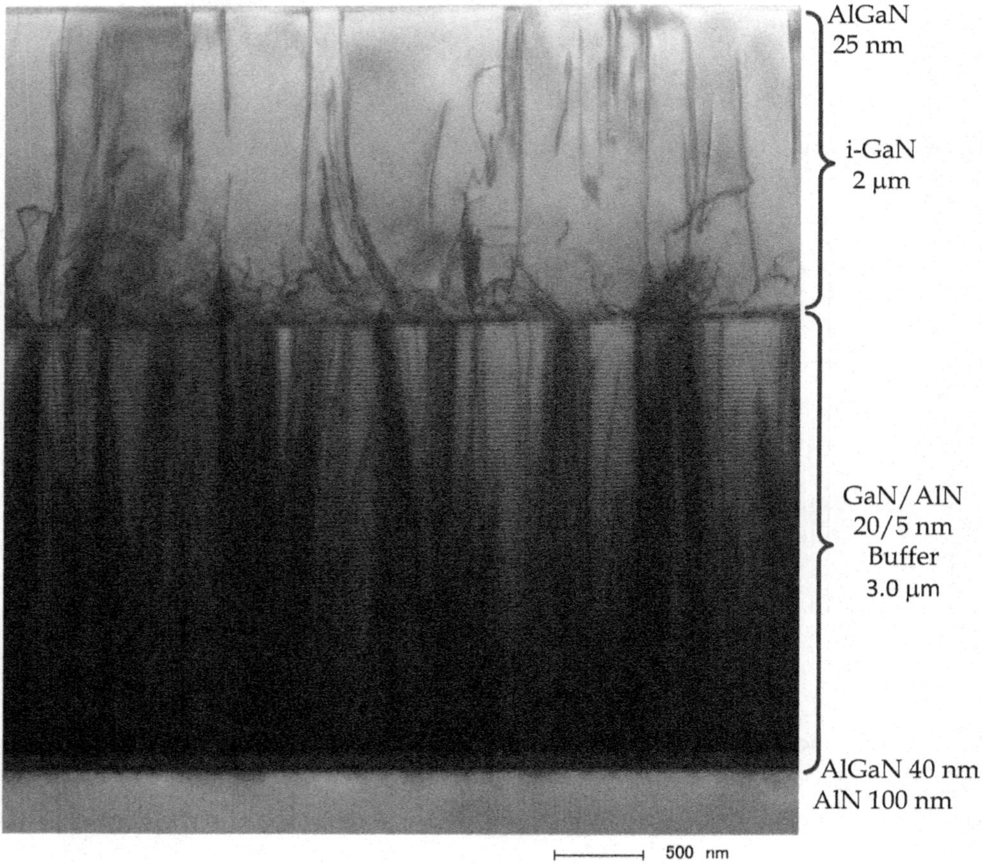

Fig. 8. A low dislocation density observed for 2 µm thick i-GaN channel layer grown on 3.0 µm thick buffer on Silicon substrate.

The 2DEG mobility of an AlGaN/GaN HEMT is around 1000 cm^2/Vs and the mobility is increased gradually all these years by improving the growth quality and other methods. One such advancement has been done by inserting a thin 1 nm AlN between the AlGaN top layer and GaN channel. This inserting 1 nm AlN layer which is called a "spacer layer", decreases the alloy disorder scattering leading to an increase in the mobility well above 1500 cm^2/Vs (Shen, 2001). A latest report demonstrate a very high mobility of 3215 cm^2/Vs which was achieved by improving the growth quality using thick buffer and using a thin AlN spacer layer at the interface of AlGaN/GaN HEMT (Selvaraj et al., 2011). The insertion of thin AlN layer produces a large discontinuity in the conduction band offset which shields the 2DEG channel from the disorder scattering. The heterostructure with 1 nm AlN layer is schematically represented in Fig.9(a). The same 1 nm AlN spacer layer is visible at the high resolution TEM image shown in Fig.9(b). Thus the highly magnified view made available using TEM helps to analyse the thin epi-layers which plays an important role in the device characteristics.

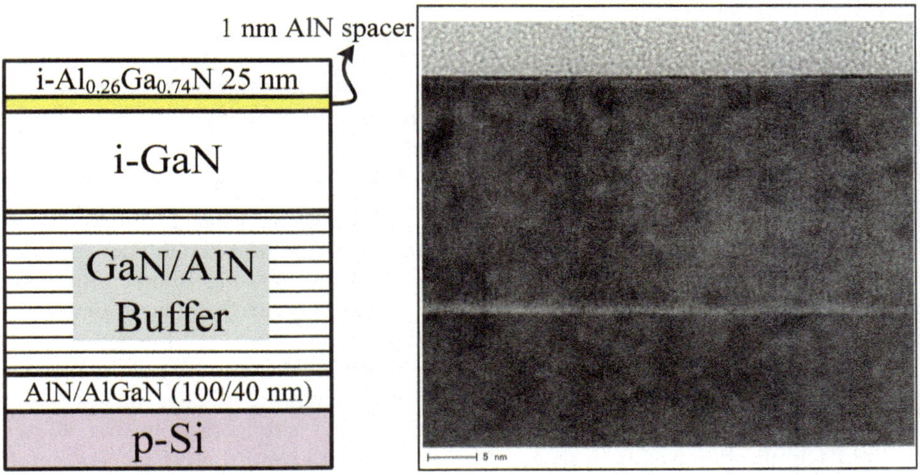

Fig. 9(a). The schematic heterostructure of an AlGaN/GaN HEMT with 1 nm AlN spacer layer; 9(b). The cross-sectional view of the 1 nm AlN spacer layer by TEM.

5. Study of defects for MOCVD grown GaN on Si

An important issue before the growth of AlGaN/GaN transistors on Si is appearance of deep-pits on surface of the growth completed wafer. These pits pose a major problem for the stability of the devices. Devices fabricated from these wafers show a high buffer and substrate leakage causing a low breakdown (Selvaraj et al., 2009). To enable the required high breakdown voltages, material quality free from bulk and surface defects is a vital concern. Deep-pits of hexagonal/polygonal shaped pyramids were observed on the on the surface of growth completed AlGaN/GaN transistors. These hexagonal pyramid shaped pits as shown by Scanning Electron Microscope (SEM) image in Fig.10 which reflects the crystal symmetry off GaN were found after growth without any intentional etching treatment. The size of these pits were around 1 to 1.5 μm in diameter.

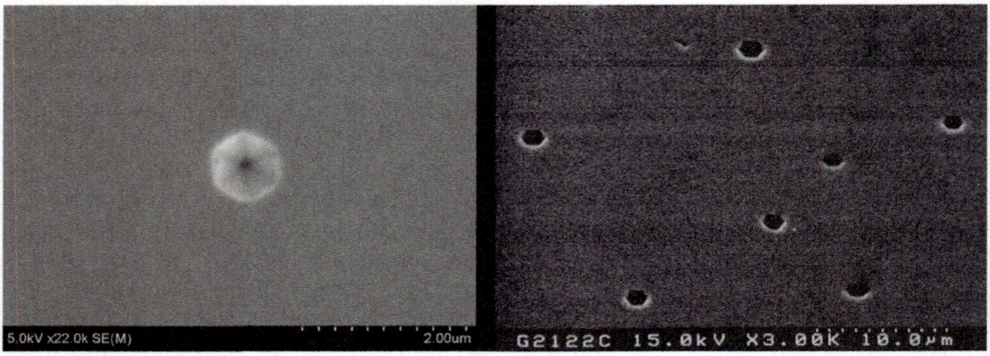

Fig. 10(a). The SEM view of the hexagonal shaped deep-pit on the MOCVD grown AlGaN/GaN layers on Si; 10(b). Another view through SEM of the deep-pits on the surface.

For AlGaN/GaN HEMTs grown on Si by MOCVD at our center, we observed density of deep-pits ranging from 900 cm^{-2} to $3x10^4$ cm^{-2}. The Fig.11 below shows the MOCVD grown AlGaN/GaN HEMTs with various density of deep-pits. It is challenging to grow a pit-free samples on Si, as the best grown wafers still have a few deep-pits. The prevention and control of deep-pits largely depend on the condition of the MOCVD growth chamber.

The HEMT devices fabricated on these wafers show an additional leakage observed through buffer and substrate finally triggering a pre-matured breakdown. When the density of deep pits are very high, there is an enormous increase in substrate leakage thereby limiting the breakdown. The three-terminal OFF breakdown voltage measured on these devices with various pit density revealed that as the pit density increased, there is a drastic reduction in the breakdown voltage. All the devices present in a single wafer with deep pits showed very poor low breakdown characteristics irrespective of the presence of deep pits at the gate-drain area. The intention of growing a thick epi-layer HEMT is to increase the breakdown so as to find its application as a high power device. But on the contrary, the presence of deep pits very much affect the device breakdown primarily due to large leakage through the buffer and substrate.

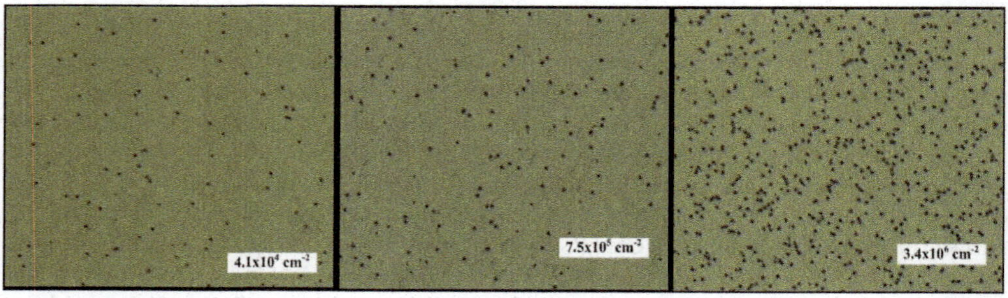

Fig. 11. Normal microscope view of the deep-pits with varying density.

The nature of these deep pits and their origin were unknown until the TEM measurement was performed on the wafers. The cross sectional images of TEM as shown in Fig. 12 revealed a hexagonal V-shaped pit originating down from the substrate and running throughout the epi-layer (Selvaraj et al., 2009). These etch pits originating at the silicon substrate run throughout the 5 μm thick epi-layer up to the AlGaN surface causing deep surface pits. And even using a thick buffer layer could not stop the deep pits reaching the AlGaN surface layer. These deep pits found on surface originate from the substrate and behave like micro-pipes for high substrate leakage flow which in turn affects the breakdown of the devices.

Fig. 12. MOCVD grown GaN on Si having a deep-pit on the surface. The TEM image reveals the origin of pit from the Si substrate causing high substrate leakage current.

Recently we also made an in-depth study on the regions around deep-pits and their limitations on the performance of MOCVD grown AlGaN/GaN HEMTs on Si substrate (Selvaraj et al., 2011). We found that for regions within 50 μm distance of deep-pits, there is a degradation observed in device performance. If the active region of the HEMTs are within the 50 μm distance from a deep-pit, then it leads to 19% decrease in the drain current and positive shift in the device threshold voltage signifying an early pinch-off. The early pinch-off for devices within the 50 μm distance of deep-pit confirms the presence of high dislocations causing dislocation scattering. These high dislocations cause 50% decrease in 3-terminal OFF breakdown voltage due to large buffer and substrate leakage current. Further, the 50 μm radius defective region around the pits were confirmed by cathodoluminescence and Raman shift measurements. Therefore, for AlGaN/GaN epi-layers grown on Si having a low density of deep-pits on the surface, the pits does not impede the normal performance of the HEMTs if the source/drain active region is located far away from the deep-pits. As the appearance deep-pits were due to melt-back etching of Si substrate by 'Ga' at high growth temperature, it is imperative that GaN growth on Si substrate is free from deep-pits for cost-cutting measures.

6. Conclusions

The TEM plays an important role in identifying the quality of GaN grown on various substrates. Further it helps in understanding the cause for failure modes arising in the devices. TEM assists in designing GaN based devices for high power device applications. In this chapter, it was discussed regarding the application of TEM for investigating the growth of Gallium Nitride on various substrates. Sapphire substrates in different planes (*c*-plane & *a*-plane) and Silicon were used for the growth of AlGaN/GaN high electron mobility transistors. The GaN grown on *a*-plane sapphire offers the best quality. Because of the use of thick buffer layers, good quality GaN were observed on Si substrate inspite of large lattice mismatch. The MOCVD grown AlGaN/GaN layers on Si had varying density of deep-pits and subsequently the devices fabricated from those wafers showed a large leakage through buffer and substrate. The reason for the leakage was understood by taking TEM images across a deep-pit. These deep-pits had an origin from the Si substrate, probably due to melt-back etching of Silicon by 'Ga'. It was with the help of TEM we found that these deep-pits are the defects behind the high leakage currents observed in the wafers. Thus, the TEM measurement provides a very important technological support for analyzing the growth of device quality epi-layers.

7. References

Khan, M.A., Bhattarai, A., Kuznia, J. N., & Olson, D. T., (1993) High electron mobility transistor based on GaN-$Al_xGa_{1-x}N$ based heterojunction, *Applied. Physics Letters*, Vol. 63, No. 9, (June 1993) 1214-1216.

Ambacher, O. (1998). Growth and applications of group III-nitrides. *Journal of Physics D: Applied Physics* Vol. 31, (June 1998), pp. (2653-2710)

Metzger, T., Hopler, R., Born, E., Ambacher, O., Stutzmann, M., Stommer, R., Schuster, M., Gobel, H., Christiansen, S., Albrecht, M., & Strunk, H.P., (1998) Defect structure of epitaxial GaN films determined by transmission electron microscopy and triple-axis X-ray diffractometry, *Philosophical Magazine* A, Vol. 77, No. 4, (1998) 1013-1025.

Shen, L., Heikman, S., Moran, B., Coffie, R., Zhang, N.-Q., Buttari, D., Smorchkova, I. P., Keller, S., DenBaars, S.P., & Mishra, U.K., (2001) AlGaN/AlN/GaN high-power microwave HEMT, *IEEE Electron Device Letters*, Vol. 22, No. 10, (October 2001), pp. (457-459).

Selvaraj, S.L., & Egawa, T., (2006), Enhancement of drain current density by inserting 3 nm Al layer in the gate of AlGaN/GaN high-electron-mobility transistors on 4 in. silicon, *Applied. Physics Letters*, Vol. 89, No. 19, (Nov 2006) pp. (193508-1-193508-3).

Pearton, S. J., Abernathy, C.R., & Ren, F., (2006), *Gallium nitride processing for electronics, sensors and spintronics,* (edition), Springer-Verlag, ISBN-10: 1852339357, London.

Murakawa, K., Niikura, E., Hasegawa, F., & Kawanishi, H., (2007), Reduction of threading dislocations in AlGaN/AlN/SiC epitaxial layers by controlled strain with (AlN/GaN) multibuffer-layer structure, *Japanese Journal of Applied Physics,* Vol. 46, No. 6A, (June 2007) pp. (3301-3304).

Selvaraj, S.L., Ito, T., Terada, Y., & Egawa, T., (2007), AlN/AlGaN/GaN metal-insulator-semiconductor high-electron-mobility transistor on 4 in. silicon substrate for high breakdown characteristics, *Applied. Physics Letters,* Vol. 90, No. 17, pp. (173506-1-173506-3).

Selvaraj, S.L., & Egawa, T., (2007) Reduced gate leakage for AlGaN/GaN HEMTs grown on a-plane (11-20) sapphire, *Extended Abstracts of Solid State Devices Materials,* (October 2007), pp. (162-163).

Selvaraj, S.L., & Egawa, T., (2007), Demonstration of AlGaN/GaN High electron mobility transistors on a-plane (11-20) sapphire, *Japanese Journal of Applied Physics,* Vol. 47, No. 5, (May 2008) pp. (3332-3335).

Selvaraj, S.L., Suzue, T., & Egawa, T., (2009) Breakdown Enhancement of AlGaN/GaN HEMTs on 4-in Silicon by Improving the GaN Quality on Thick Buffer Layers, *IEEE Electron Device Letters,* Vol. 30, No. 6, (June 2009), pp. (587-589).

Selvaraj, S.L., Suzue, T., & Egawa, T., (2009) "Enhancing the breakdown voltage by growing 9 μm thick AlGaN/GaN HEMTs on 4-in Silicon," 67th *IEEE Device Research Conference Digest*, (June 2009), pp. (283-284).

Selvaraj, S.L., Suzue, T., & Egawa, T., (2009) Influence of deep pits on the breakdown of metalorganic chemical vapor deposition grown AlGaN/GaN high electron mobility transistors on silicon, *Applied Physics Express,* Vol. 2, No. 11, (November 2009), pp. (111005-1-111005-3).

Selvaraj, Watanabe, A., & Egawa, T., (2011) Influence of deep-pits on the device characteristics of metal-organic chemical vapor deposition grown AlGaN/GaN high-electron mobility transistors on silicon substrate, *Applied. Physics Letters,* Vol. 98, No. 25, (June 2011) pp. (252105-1-252105-3).

Selvaraj, Watanabe, A., & Egawa, T., (2011) "Enhanced mobility for MOCVD grown AlGaN/GaN HEMTs on Si substrate," 69th *IEEE Device Research Conference Digest*, (June 2011) pp. (221-222).

Rowena, I. B., Selvaraj, S. L., & Egawa, T., (2011) Buffer thickness contribution to suppress vertical leakage current with high breakdown field (2.3 MV/cm) for GaN on Si, *IEEE Electron Device Letters*, Vol. 32, No. 11, (November 2011) pp. (1534-1536)

8

Determination of Aspect-Ratio Distribution in Gold Nanowires Using Absorption Spectra and Transmission Electron Microscopy Techniques

Hiroo Omi
NTT Basic Research Laboratories, NTT Corporation, Japan

1. Introduction

The surface plasmon excitation energies of noble metal nanowires are strongly dependent on the dimensions of the nanowire and the local environment, which allows for tunable optical characteristics of nanowire samples (Noguez, 2007; Nehl, C. L. & Hafner, J. H. 2008). This has great advantages for the use of nanowires in surface-enhanced Raman spectroscopy (Orendorff et al. 2006; Sztainbuch, 2006; Suzuki et al. 2004, 2005), medical imaging, disease treatment, and other applications such as highly sensitive sensors (Agarwal et al. 2007; Gibbons et al. 2006; Yu & Irudayaraj 2007; Chen et al. 2007). For this reason, the fabrication processes for nanowires and the characterization of their optical properties are of great interest in current research (Murphy et al. 2006; Gulati et al. 2006; Miranda & Ahmadi 2005; Zweifel & Wei 2005; Hu et al. 2003; Sönnichsen et al. 2002). It would be useful to have a quick and easy way to accurately determine the size distribution of nanoparticles in solution in order to analyze nanowire synthesis processes or characterize samples for applications.

In 1912, Gans extended Mie's theory of the scattering of spherical particles (Mie 1908) for the case of oblate and prolate spheroids (Gans 1912). This simple theory describes the absorption characteristics of spheroidal particles smaller than the excitation wavelength and could provide a reasonable approximation for the optical characteristics of nanowires with different geometries. Because absorption measurements are simple, fast, and can be done in situ during the growth process, they would offer an excellent basis for particle characterization.

Nanowire synthesis always leads to dispersion in the nanowire aspect ratios, commonly defined as the length of the particle divided by its width. It has been well established that the absorption spectrum of a particle is strongly dependent on this aspect ratio (Gulati et al. 2006; Link et al. 2005; Yan et al. 2003; Link & El-Sayed 2005; Prescott & Mulvaney 2006; Brioude et al. 2005; Murphy et al. 2005; Pérez-Juste et al. 2005). The absorption spectrum of a disperse nanowire solution will therefore be broadened, and information on the aspect-ratio distribution is thus contained in a measured absorption spectrum (Eustis et al. 2006).

Comparisons of experimentally determined absorption spectra and numerical approximations with Gans theory (Prescott & Mulvaney 2006; Brioude et al. 2005) suggest that there is a non-negligible dependence of the absorption spectrum on the end-cap geometry and particle size. Prescott and Mulvaney (Prescott & Mulvaney 2006) used discrete dipole approximation (DDA) calculations to determine the importance of the exact gold nanowire shape. They noticed an improvement over Gans theory by modeling the rods as cylinders. Gans theory uses perfect spheroids instead. However, they observed the presence of spherical end caps and admitted that the actual particle geometry is not described correctly by a cylinder. On the basis of their DDA model, they suggested shape factors that take into account the particle width, end-cap geometry, and aspect ratio. These shape factors are empirically derived, and there is no explanation for why the proper particle geometry (a cylinder with oblate spheroids as end caps) does not give an accurate absorption spectrum. We are therefore led to the hypothesis that there must be an additional physical quantity that is variable.

Previous papers (Yan et al. 2003; Link & El-Sayed 2005; Prescott & Mulvaney 2006; Brioude et al. 2005) have systematically reported calculated longitudinal plasmon excitation peak positions that are too low compared to various experimental results. They assumed that the dielectric constant of the medium around the particles is equal to that of the solvent ($\varepsilon_{m,water} = 1.77$). In this paper, we will use the original shape factors of Gans theory, but with the dielectric constant of the medium, ε_m, as a variable. Studies by Miranda and Ahmadi (Miranda & Ahmadi 2005) and Gulati et al. (Gulati et al. 2006) indicate that the local dielectric properties have a strong influence on the optical absorption characteristics. The presence of a surfactant around the nanowires is one of the reasons why the dielectric constant can locally be different.

We believe that the chief advantage of Gans theory is the speediness of the calculations afforded by the simplicity of its formulation. The main alternative to the Gans approximation of other particle geometries makes use of the DDA method (Prescott et al. 2006; Brioude et al. 2005; Yin et al. 2006), which allows for different particle geometries but requires a significantly more difficult implementation with long computation times. Adapting Gans theory to real nanowires and nanoparticles samples would therefore have benefits in their characterization (Amendola V. & Meneghetti M. 2009).

In 2006, Eustis and El-Sayed already fitted calculated absorption curves to measurements in order to determine the aspect-ratio distributions for polydispersed samples (Eustis et al. 2006). By analyzing high-aspect-ratio nanowires, they could limit their analysis to the longitudinal plasmon resonance absorption. In this Chapter, we extend their work to determine the aspect-ratio distributions of gold nanowire solutions from measured absorption spectra for much-lower-aspect-ratio nanowires by taking the whole UV-vis absorption spectrum into account. Our results are compared with transmission electron microscope (TEM) images to confirm their validity, and we discuss the influence of the local environment of the nanowires. We will present suitable values for the effective dielectric constant of the medium based on our calculations using Gans theory. Finally, we will show that our method is also applicable to higher-aspect-ratio nanowires through a comparison with findings in the literature.

2. Calculation methods

The absorption coefficient, κ, for elongated particles was derived by Gans (Gans 1912) and implemented by Link et al. (Link et al. 1999) as in eq 1. The results in the initial publications of Link et al. contained a calculation error that was first noticed by Yan (Yan et al. 2003) and later corrected by Link and El-Sayed (Link & El-Sayed 2005).

$$k = \frac{2\pi NV\varepsilon_m}{3\lambda} \sum_{j=A}^{C} \frac{\left(1/P_j^2\right)\varepsilon_2}{\left(\varepsilon_1 + \frac{1-P_j}{P_j}\varepsilon_m\right)^2 + \varepsilon_2^2} \tag{1}$$

where N is the number of particles per unit volume, V is the average volume per particle, ε_m is the dielectric constant of the medium, λ is the absorbed wavelength, and $\varepsilon = \varepsilon_1 + i\varepsilon_2$ is the complex dielectric function of the prolate spheroid. The dielectric constant of the medium is related to the refractive index n by $\varepsilon_m \approx n^2$. For water ($n$ = 1.33), this gives ε_m = 1.77. The product NV equals the total volume of all particles. In eq 1, the depolarization or shape factors P_j relate the plasmon excitation to the aspect ratio of the particle for the three axes j = A, B, C with lengths L_A, L_B, and L_C. For a prolate spheroid, this gives $L_A > L_B = L_C$. The usual notation for these shape factors (Gans 1912) as a function of aspect ratio $R = L_A/L_B$ is given by

$$P_A = \frac{1-e^2}{e^2}\left(\frac{1}{2e}\ln\left(\frac{1+e}{1-e}\right)-1\right)$$

$$P_B = P_C = \frac{1-P_A}{2} \tag{2}$$

$$e = \sqrt{1-\frac{1}{R^2}}$$

We use this formulation of the absorption of a prolate spheroid to approximate the absorption characteristics of gold nanowires.

Because of anisotropy, the nanowire absorption curves calculated with eqs 1 and 2 show two peaks associated with the longitudinal and transversal plasmon excitations. The position of the longitudinal absorption peak is strongly dependent on the aspect ratio of the particles as well as on the dielectric constant of the medium, ε_m.

Due to anisotropy, the absorption curves of nanorods show two peaks associated with the longitudinal (SP_{long}) and transversal (SP_{trans}) plasmon excitations. The position of the longitudinal absorption peak is strongly dependent on the aspect ratio of the particles as well as on the dielectric constant of the medium ε_m. Figures 1 and 2 show a redshift of the longitudinal absorption peak as the aspect ratio and ε_m increase. The transversal plasmon mode is also dependent on R and ε_m but the effect is much less pronounced. There is a slight blueshift when the aspect ratio increases and a redshift for higher values of the dielectric constant.

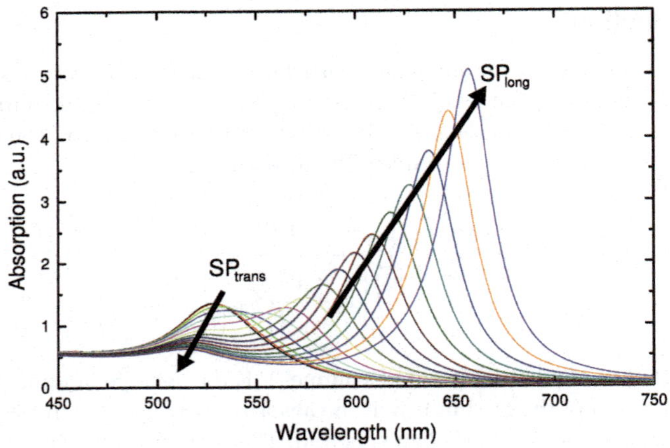

Fig. 1. Calculated absorption spectra for different aspect ratios $1 \leq R \leq 2.5$ with $\varepsilon_m = 2.1$. As the aspect ratio R increases, the transversal plasmon excitation SP_{trans} shows a slight blueshift and the longitudinal plasmon excitation SP_{long} shows a much stronger redshift.

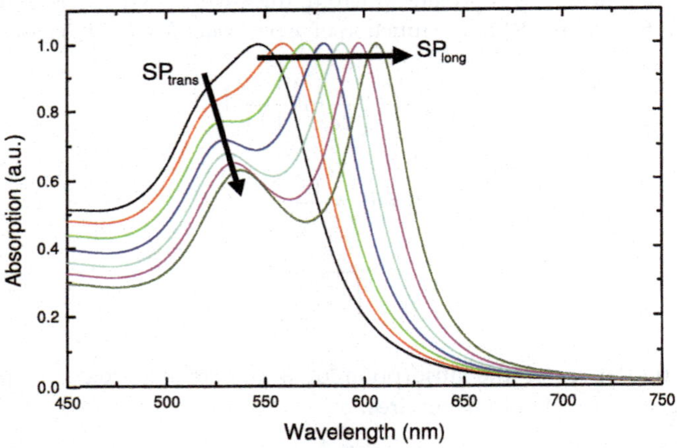

Fig. 2. Calculated absorption spectra for different dielectric constants 1.6 ≤ _m ≤ 3 with R = 1.5. There is a redshift for both SP_{trans} and SP_{long} as ε_m increases.

Because of this dependence of the absorption peak positions on the aspect ratio of the particles, we will observe a broadening of the absorption peaks when we analyze a solution containing nanowires of unequal size. By fitting the measured absorption curve with the calculated absorption curves of rods with different aspect ratios, we should be able to derive the original aspect-ratio distribution using methods similar to those employed by Eustis and El-Sayed (Eustis & El-Sayed 2006).

For a nanowire with aspect ratio R, the absorption as a function of the wavelength can be calculated using the formulation of Gans theory in eqs 1 and 2. Through the shape factors P_j, the absorption becomes dependent on the aspect ratio. Furthermore, we use the dielectric constant of the medium, ε_m, as a fitting parameter; although the factor before the summation in eq 1 is merely a constant scaling factor, the value of ε_m within the summation has a profound effect on the shape of the absorption spectrum.

We used the complex dielectric constants $\varepsilon = \varepsilon_1 + i\varepsilon_2$ for bulk gold as obtained by Johnson and Christy (Johnson & Christy 1972). Recent studies (Stoller et al. 2006) have shown that there is a discrepancy between the dielectric constants of bulk gold and that of small gold particles. However, because the difference is small and the accuracy of the new results is limited, it is deemed sufficiently accurate to use the optical constants for bulk gold. The absorption curves are calculated for a series of aspect ratios and stored in a matrix M. This reduces our problem to solving $s = Mx$ for the aspect-ratio frequencies x, with vector s being our measured absorption spectrum. For real problems, there will not be an exact solution and we have to optimize the problem for a certain norm. Given the physical constraints of positive frequencies and smoothness of the solution, finding the solution becomes nontrivial. To optimize for a least-square error, we made use of a custom optimization algorithm implemented in the SCILAB 4.1 (INRIA & ENPC. Scilab 4.1 2006) software package. The calculated aspect ratio distributions are compared with size measurements from TEM images in order to confirm the validity of our calculation method.

Because we fit both the longitudinal and transversal plasmon excitation absorption peaks, we can analyze the size distributions of much-lower-aspect-ratio nanowires than previously reported. In the paper by Eustis and El-Sayed (Eustis & El-Sayed 2006), the fitting was limited to the longitudinal plasmon mode. The authors argued that the transversal absorption peak position is only weakly dependent on the aspect ratio of the particles and could therefore be ignored. In our studies, we used particles with aspect ratios $R \leq 2.5$, whereas Eustis and El-Sayed observed nanowires with $1.5 \leq R \leq 9$. This has two important consequences. When using much-lower aspect ratios, the importance of the blueshift of the transversal absorption peak increases as the redshift of the longitudinal absorption peaks diminishes. More importantly, there will be a larger overlap between the longitudinal and transversal absorption peaks. The intensity and position of the transversal absorption peak will then have an effect on the longitudinal absorption peak, and we cannot limit our analysis to only the longitudinal plasmon excitations. Instead, we must fit both peaks in the absorption curve at once. This requirement puts additional demands on the accuracy of the theory.

Fitting both absorption peaks also allows us to determine the abundance of spherical particles. However, a smaller range of aspect ratios will require a higher resolution and greater accuracy of the fitting algorithm in order to obtain a good fit to the experimental results. As such, by fitting both the longitudinal and transversal plasmon excitation absorption peaks, we should be able to accurately determine the full aspect-ratio distribution in our samples as well as the dielectric constant of the medium.

3. Experimental methods

3.1 Materials

Hydrogen tetrachloroaurate ($HAuCl_4 \cdot 3H_2O$, 99.5%) was purchased from Merck Co. Cetyltrimethylammonium bromide (CTAB, 99+%) was obtained from Acros Organics. Silver nitrate and ascorbic acid were from Wako Chemical Co. and Tokyo Kasei Co., respectively, and were used as received. Deionized water was used throughout the experiments.

3.2 Synthesis of gold nanowires

Gold nanowires were prepared by a seed-mediated growth method (Gou & Murphy 2005). More precisely, 9.5 mL of 0.1 M CTAB was combined with 0.5 mL of 0.01 M $HAuCl_4 \cdot 3H_2O$ and 55 μL of 0.1 M ascorbic acid as well as varying amounts (20, 60, and 100 μL) of 0.01 M silver nitrate aqueous solution, under continuous stirring. A seed solution was prepared according to Nikoobakht & El-Sayed 2003, and 12 μL of this solution was injected into the mixture to initiate the growth of gold nanowires. We obtained three different samples containing gold nanowires with different aspect-ratio distributions for the silver nitrate amounts of 20, 60, and 100 μL (samples 1-3). These samples were aged for 24 h in air to ensure full formation of gold nanowires.

3.3 Instruments

The absorption spectra of the nanowire solutions were measured with a Hitachi U-3500 spectrophotometer in the 300 to 900 nm range. TEM images were made with a JEOL JEM-2100F at 200 keV after a drop of the solution had been dried on a microarray.

4. Results

4.1 Experiments

The absorption spectra, measured for three different samples with different aspect-ratio distributions, are shown in figure 3. For sample 1, the longitudinal and transversal absorption peaks overlap to form one broad absorption peak. Two distinct peaks can be seen for samples 2 and 3, corresponding to the transversal plasmon mode at 520 nm and the longitudinal plasmon excitation around 600 nm.

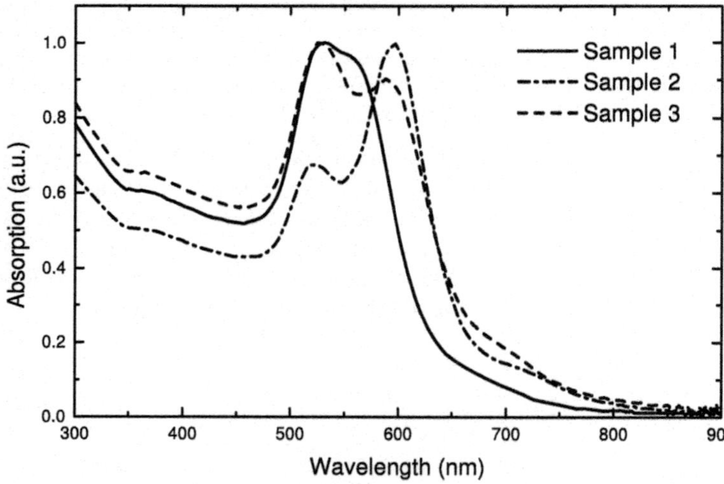

Fig. 3. Measured absorption spectra for three different aspect-ratio distributions. There is significant overlap between the longitudinal and transversal absorption peaks because of the low aspect ratios of the nanowires.

To validate our calculation methods, we have measured the size of the gold nanowires from TEM images. This enables us to later compare the calculated aspect-ratio distributions with the actual aspect ratios. Figure 4 shows the aspect ratios as determined from TEM images for more than 250 particles per sample. The lengths, L_A, and widths, L_B, were measured as the maximal length and width of the particles. For the three samples, the average particle lengths were L_A = 42, 46, and 40 nm and the average widths were L_B = 27, 25, and 22 nm. The standard deviation per sample was 15% for the widths and 12% for the lengths. The aspect ratio, R, is defined as $R = L_A/L_B$ as in the inset of figure 4. The peak aspect ratios for the rod-shaped particles were R = 1.5, 1.8, and 1.9 for samples 1, 2, and 3, respectively. The second peak in the distributions in figure 4 at $R \approx 1$ indicates the presence of spherical particles.

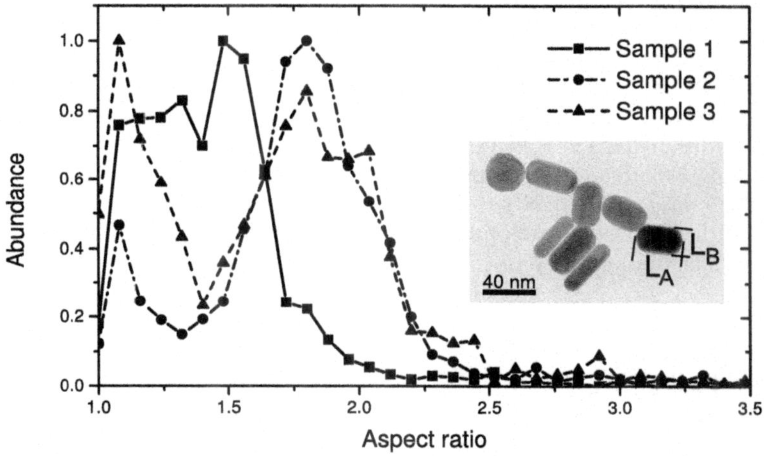

Fig. 4. Size distributions from TEM images. Curves are histograms of 253, 271, and 269 particles for samples 1, 2, and 3, respectively. The inset shows a characteristic TEM image (sample 3). The aspect ratio is given by $R = L_A/L_B$. High-aspect-ratio particles have a much smaller volume than low-aspect-ratio ones. Although the length, L_A, is constant, the width, L_B, is smaller for particles with a higher aspect ratio. An $R \approx 1$ particle can be seen on the top left.

It should be noted that the abundance in figure 4 was measured by particle volume because it is the volume of particles and not their count that defines the amount of absorption. We approximated the particle shape by using a spheroidally capped cylinder with an end-cap length of half the radius. However, the size distributions are such that the precise shape of the end caps has little influence on the total particle volume per aspect ratio. What is important is that the high-aspect-ratio particles ($R \geq 3$) are much smaller than particles with lower aspect ratios. As we can see in the inset of figure 4, particles with higher aspect ratios have lengths similar to the lower-aspect-ratio particles, but they are not as wide. The result is that the high aspect- ratio rods have a significantly smaller volume than the low-aspect-ratio ones and thus a smaller influence on the absorption spectrum. By taking the volume of the particles into account, we avoid overestimating the presence of the high aspect- ratio particles. The small shoulders at 700 nm in figure 3 should be attributed to these high-aspect-ratio particles ($R \geq$ 3). Because eq 1 is not scaled by the number of particles, N, but rather by the total volume, NV, of the gold particles of a specific aspect ratio, R, we can transform our problem from determining the count of particles with a certain aspect ratio to determining the total volume of all particles with a specific aspect ratio. Whenever we speak of abundance in this paper, we refer to the total volume of gold for all of the particles with a certain aspect ratio.

4.2 Calculations

For 100 aspect ratios between 1 and 4, we calculated the associated absorption curves using eqs 1 and 2. The measured absorption spectra of figure 3 were then fitted by a linear combination of these 100 different absorption spectra as can be seen in figure 5. For a series of ε_m, ranging from 1.7 to 2.4, the optimal fit to the absorption spectrum was determined and the fitting errors were plotted as a function of ε_m as in figure 6. The lowest fitting error then indicates the optimal dielectric constant. For samples 1-3, this gives us $\varepsilon_m = 2.09$, $\varepsilon_m = 2.18$, and $\varepsilon_m = 2.12$, respectively, which is significantly higher than the dielectric constant of the solvent $\varepsilon_{m,water} = 1.77$. This dielectric constant corresponds to a refractive index of $n = 1.46 \pm 0.02$. The quality of the fits is shown in figure 5, and it can be seen that the overall curve and the peak positions are fitted well, but there is a consistent deviation in the UV absorption. Additionally, the transversal absorption peaks are broader than Gans theory predicts. This appears to be a limitation of Gans theory. Our TEM images show the presence of several much larger, spherical particles. Because Gans theory does not include any size dependence of the absorption peak position and shape, the contribution of these particles to the absorption spectrum is ignored.

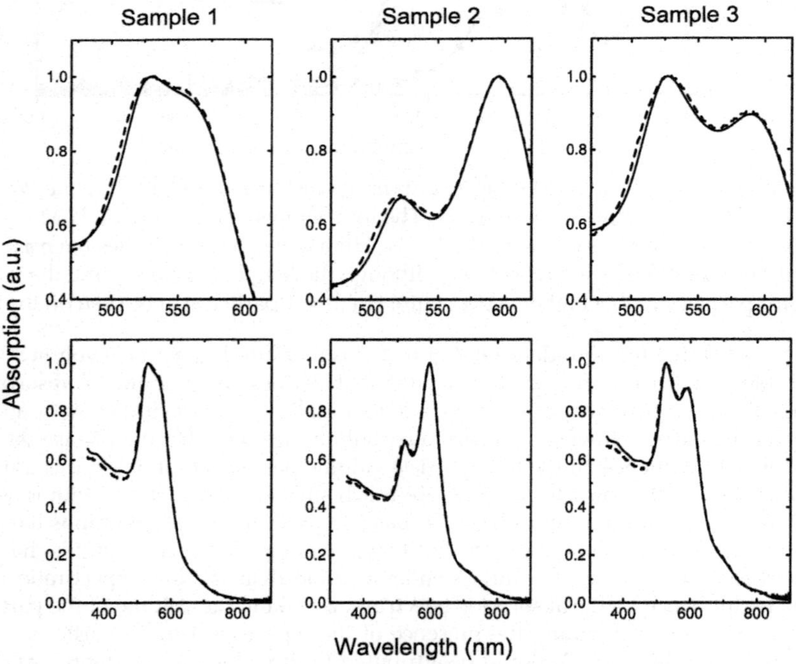

Fig. 5. Quality of Gans fitting with $\varepsilon_m = 2.14 \pm 0.05$. Measured absorption spectra (dashed lines) and simulations (solid lines). The measured absorption spectra are a linear combination of the spectra of 100 particles with $1 \leq R \leq 4$.

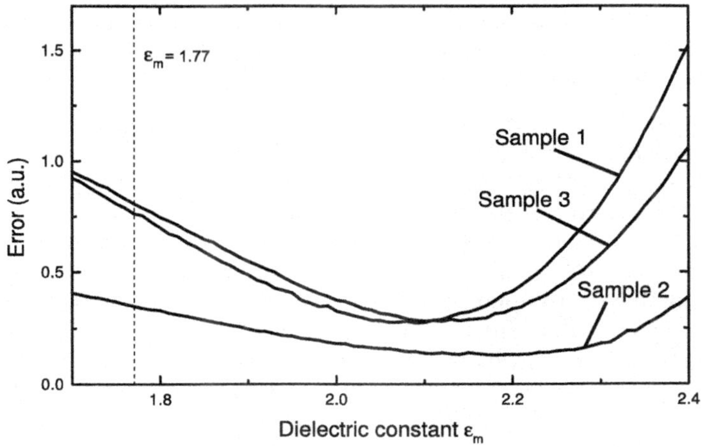

Fig. 6. Fitting errors per sample as a function of ε_m. The minimal fitting error indicates an optimal $\varepsilon_m \neq 1.77$. For the three samples, we find a minimal fitting error for $\varepsilon_m = 2.14 \pm 0.05$.

Figure 7 shows the results of the fitting using Gans theory, together with the aspect-ratio distributions as determined from TEM images. We can see that the peak aspect ratio for rodshaped particles ($R > 1.5$) matches our simulations for all three samples. Our fitting method thus gives a good approximation of the actual aspect-ratio distributions. It is clear that when we optimize the fitting of the absorption spectrum by adjusting ε_m we simultaneously optimize the fit to the measured aspect-ratio distributions. Apparently, our optimization of the medium dielectric constant in Gans theory is a necessary step in obtaining the correct aspect-ratio distributions.

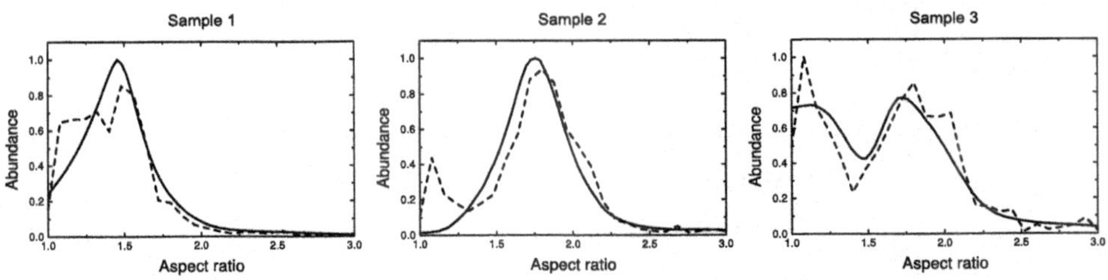

Fig. 7. Aspect-ratio distributions according to TEM images (dashed lines) and Gans theory (solid lines) for the optimized values of ε_m: $\varepsilon_m = 2.09$, $\varepsilon_m = 2.18$, and $\varepsilon_m = 2.12$ for sample 1, 2, and 3, respectively.

For samples 1 and 2, we see a deviation of the calculations from the TEM measurements for spherical particles ($R \approx 1$). The abundance of spherical particles in TEM images is larger for these samples than that suggested by our calculations. We can offer two possible explanations: either the applicability of Gans theory for spherical particles is limited or we overestimated the abundance of spherical particles in our TEM images. Nanowires that are standing perpendicular to the observation plane can be mistaken for spherical particles, possibly resulting in the disagreement for low aspect ratios for samples 1 and 2 in figure 7.

For sample 3, we do not observe such an effect. Here, the small disagreement between the TEM measurements and our fitting can be attributed to the limited number of particles measured. The simulation fits both aspect-ratio peaks at $R \approx 1$ and $R = 1.9$ to a good extent.

In the previous section, we did not need or use the TEM measurements to determine the aspect-ratio distribution or the dielectric constant of the medium. Nevertheless, to check the accuracy of the aforementioned value of ε_m and confirm the validity of our methods, we also calculated the absorption spectra with Gans theory given the particle size distributions as determined via TEM measurements. For every particle of which the aspect ratio and volume were determined from the TEM images, we calculate the absorption with eqs 1 and 2 and then sum these absorption spectra for all particles. This calculated absorption spectrum is compared with the measured absorption spectra in figure 3, while varying ε_m. Figure 8 shows the fitting results as well as the fitting errors as a function of ε_m. The optimal value of ε_m can again be determined by minimizing the fitting error. We find a good agreement between the calculated absorption spectra and the measured spectra from figure 3 for $\varepsilon_m = 2.1 \pm 0.1$. This matches the value of $\varepsilon_m = 2.14 \pm 0.05$ determined above by the inverse process.

It is possible to determine the aspect-ratio distribution of gold nanowires from an absorption spectrum using Gans theory if we optimize the dielectric constant of the medium. We can derive the optimal value of this dielectric constant from the absorption spectrum as well. Size measurements from TEM images confirm the accuracy of our method and support our hypothesis that the local environment of the particles has a significant influence on their optical characteristics.

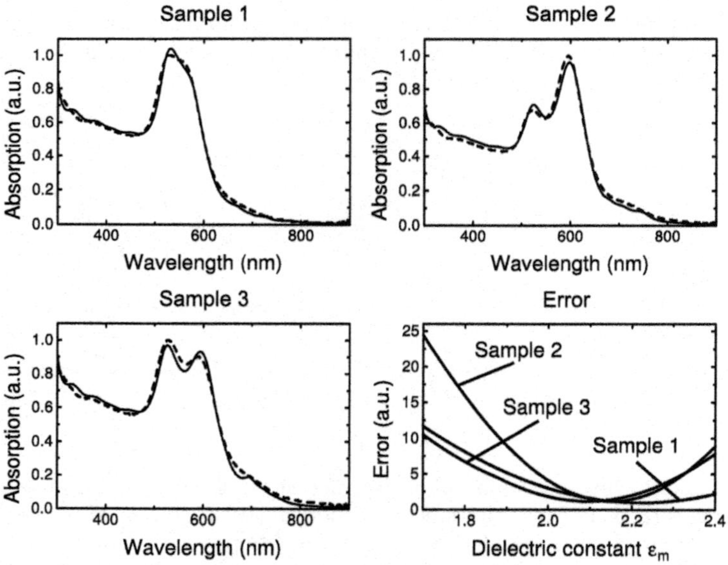

Fig. 8. Measured (dashed lines) and calculated (solid lines) absorption curves using the aspect ratios and volumes obtained from TEM images for the parameters R and NV in eqs 1 and 2. The fitting errors for the three samples are also given as a function of ε_m.

5. Comparison with other experimental results

5.1 Longitudinal peak position

Several studies use the longitudinal plasmon excitation absorption peak position to roughly determine the peak aspect ratio of nanowires in solution (Gulati et al. 2006; Zweifel & Wei 2005; Link et al. 1999; Yan et al. 2003; Link & El-Sayed 2005; Prescott & Mulvaney 2006; Brioude et al. 2005; Pérez-Juste et al. 2005). The wavelength of this absorption peak can be obtained with Gans theory by minimizing the denominator in eq 1 for the longitudinal shape factor P_A. Given that ε_2 is fairly constant in the UV-vis domain, the longitudinal absorption peak position is given by that wavelength, $\lambda_{\max}$, for which the following relation holds:

$$\varepsilon_1 = -\frac{1 - P_A(R)}{P_A(R)}\varepsilon_m \tag{3}$$

Through linearization of eq 3, one can express the wavelength of the absorption maximum, $\lambda_{\max}$, of the longitudinal plasmon resonance as a function of the aspect ratio R and the dielectric constant ε_m. Using the method by Link et al. (Link et al. 1999), we found the following relationship for $\lambda_{\max}$:

$$\lambda_{\max} = (52.58R - 41.24)\varepsilon_m + 467.31 \text{ (nm)} \tag{4}$$

This is in good agreement with the results of Yan and co-workers (Yan et al. 2003).

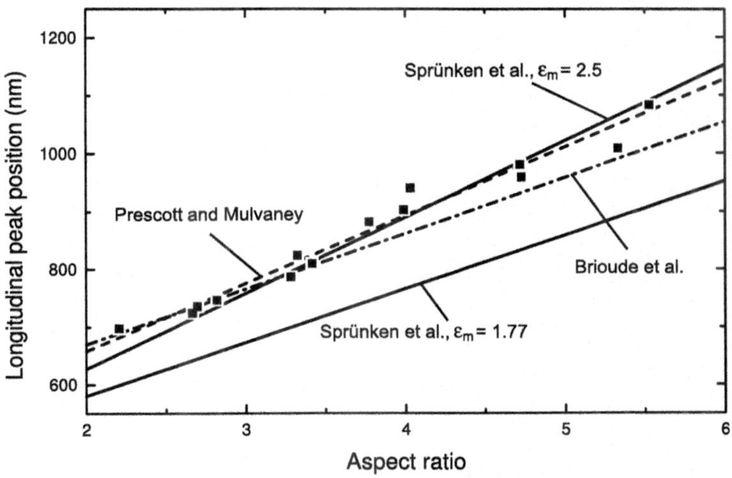

Fig. 9. Longitudinal peak positions as a function of nanowire aspect ratio. Experimental data from Pérez-Juste et al. (Pérez-Juste et al. 2005) (squares) compared with DDA calculations by Prescott and Mulvaney for a nanowire modeled as a cylinder (Figure 5 in Prescott & Mulvaney 2006) (dashed line), Brioude et al. (Figure 4 in Brioude et al. 2005) (dash-dotted line), and our simulations using Gans theory (solid line) with $\varepsilon_m = 2.5$ in eq 4. The solution of eq 4 for $\varepsilon_m = 1.77$ is given as a reference (bottom).

Figure 9 compares the experimental results of Pérez-Juste et al. (Pérez-Juste et al. 2005) with DDA calculations of Brioude et al. (Brioude et al. 2005) and Prescott and Mulvaney (Prescott & Mulvaney 2005), together with the predictions of Gans theory according to eq 4 for ε_m = 2.5. Equation 4 for ε_m = 1.77 is given as a reference. The DDA method has been used extensively in recent years to simulate gold nanowires. These studies used ε_m = 1.77 to calculate the longitudinal plasmon excitation peak positions. Figure 9 shows that, although there are some dissimilarities between the simulations of Brioude et al. and Prescott and Mulvaney, both calculations match the experimental results well. To obtain this good fit, Prescott and Mulvaney modeled the nanowires as cylinders, observing a significant dependence of the absorption characteristics on the particle geometry.

Using ε_m = 1.77, Brioude and co-workers reported (Figure 6 in Brioude et al. 2005)

$$\lambda_{\max} = 96R + 478 \text{ (nm)} \tag{5}$$

as a linear approximation of their simulation results. Equation 4 gives in this case $\lambda_{\max}$ = 93.07R + 394.3 (nm). The slope does not differ greatly, but there is a large offset of 80 nm. However, if we use ε_m = 2.5 instead of 1.77 in eq 4, we do find a good agreement between Gans theory and the experimental results of Pérez-Juste et al., as shown in figure 9. This requires a change of ε_m, but does not require changing the particle geometries. For the fittings of Brioude et al., Prescott and Mulvaney, and our simulation using Gans theory, we find sum square errors of 19×10^3, 5.7×10^3, and 9.9×10^3 (nm), respectively. Considering the experimental and calculation errors, all three fittings follow the experimental data well.

According to Prescott and Mulvaney (Prescott & Mulvaney 2006), the longitudinal surface plasmon mode is very sensitive to end-cap geometry. Their results, however, suggest that, even though the TEM images show the presence of spherical end caps (similar to our nanowires as in the inset of figure 4), one should model the nanowires as pure cylinders in order to obtain the best fit. Brioude et al. did not discuss the particle geometries used. We find that, by adjusting only the dielectric constant of the medium, we can also make Gans theory fit the experimental results of Pérez- Juste and co-workers without changing the nanowire geometries. We believe that this indicates that the dielectric constant of the medium plays as important a role as the particle geometry in the absorption characteristics of gold nanowires.

5.2 High-aspect-ratio nanowires

Eustis and El-Sayed (Eustis & El-Sayed 2006) also compared their TEM aspect-ratio measurements with Gans calculations, using ε_m = 1.77. They analyzed five samples with 1.5 ≤ R ≤ 9 and longitudinal absorption peaks at 630, 700, 850, 900, and 1000 nm (see figure 10). They only found a good match for two out of five samples, namely, for the samples with absorption peak positions of 630 and 900 nm (figure 5 in Eustis & El-Sayed 2006 and reproduced in figure 10). When one compares their measured absorption spectra with their measured aspect-ratio distributions, this is not surprising. Their two samples with longitudinal absorption peaks at 630 and 700 nm have very similar aspect-ratio distributions. The case for the samples with absorption peaks at 900 and 1000 nm is the same. When the measured absorption spectra are dissimilar but the measured aspect-ratio distributions are similar, then the aspect-ratio distribution alone does not define the absorption spectra. This means that there must be another variable involved: either the local effective dielectric constant of the medium is different or the particle geometries are dissimilar.

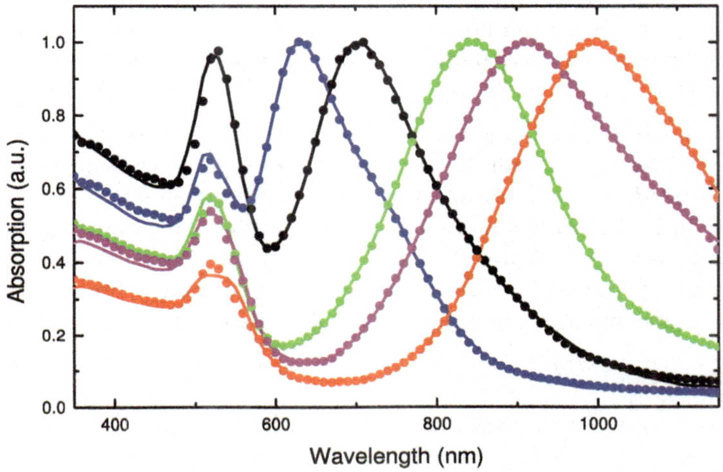

Fig. 10. Gans fitting (circles) using our method to absorption spectra (lines) by Eustis and El-Sayed from Fig. 5(c) in (Eustis & El-Sayed 2006).

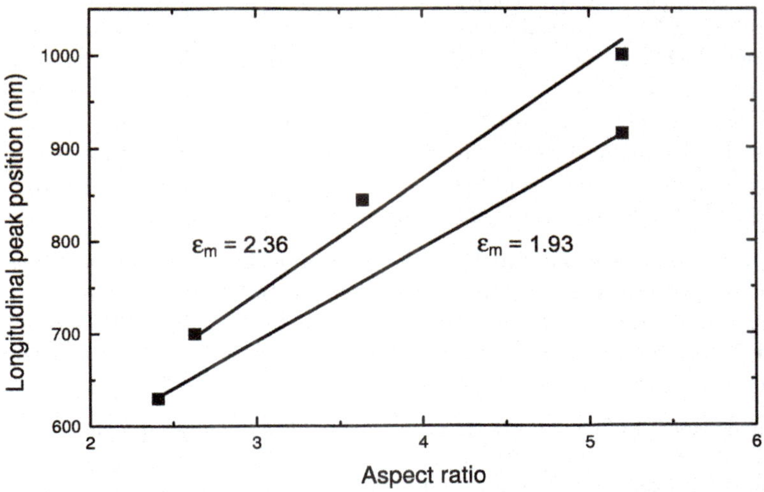

Fig. 11. Experimental data from Eustis and El-Sayed (Figure 5b in Eustis & El-Sayed 2006) (squares) fitted to eq 4 with $\varepsilon_m = 1.93$ (bottom) and $\varepsilon_m = 2.36$ (top).

Our results show that for $R \leq 2.5$ Gans theory allows for the determination of the aspect-ratio distribution of nanowires in solution, if we take the local dielectric environment of the particles into account. Eustis and El-Sayed (Eustis & El-Sayed 2006) analyzed the size distributions of samples for a much-larger aspect-ratio range: $1.5 \leq R \leq 9$. To check the validity of Gans theory for higher aspect ratios, we also applied our fitting method with variable ε_m to their results. Using the dielectric constant of the medium as a fitting parameter, we found that we can fit all five absorption spectra if we define two values for ε_m. By fitting eq 4 to the longitudinal absorption peak positions of Eustis and El-Sayed as in figure 11, we find two sets of samples that are best modeled with two distinct values of ε_m.

For the samples with absorption peaks at 630 and 900 nm, we find $\varepsilon_m = 1.93$, whereas we find an optimal fit at $\varepsilon_m = 2.36$ for the samples with absorption peaks at 700, 850, and 1000 nm. This corresponds to the results of Eustis and El-Sayed, who found a reasonable fit for the first two samples, but did not find a good fit for the other three using $\varepsilon_m = 1.77$. In fitting eq 4 to the peak positions in figure 11, only the slope of the line can be varied by changing ε_m. between the experimental results and theoretical calculations. Through linearization of the longitudinal absorption peak position in Gans theory we obtain

$$\lambda_{max} = (52.58R - 41.24)\varepsilon_m + 467.31 \text{ (nm)}. \quad (6)$$

All curves share a common point at $R = 0.78$ for which eq 4 reduces to $\lambda_{\max} = 467.31$ (nm). This means that, even for this limited dataset, we still have more data points than parameters in the curve fitting.

Figure 12 shows the aspect ratios as measured by Eustis and El-Sayed (Figure 5b in Eustis & El-Sayed 2006) compared to the fitting results for our method. Figure 12a shows the samples with absorption peaks at 630 and 900 nm, modeled with $\varepsilon_m = 1.93$. The samples with peaks at 700, 850, and 1000 nm, modeled with $\varepsilon_m = 2.36$, are shown in figure 12b. Note that these curves do not indicate the volume of particles but rather the count of particles. Eustis and El-Sayed (Eustis & El-Sayed 2006) reported that the volume of their nanowires was independent of the aspect ratio. From the good correspondence between the measured aspect ratios and our calculations in figure 12, it is reasonable to assume that we are dealing with two sets of samples that require different fitting parameters.

In other words, we can conclude that our fitting method is also applicable for high-aspect-ratio particles. Gans theory accurately describes the relation between the aspect-ratio distributions (Figure 12) and absorption spectra of Eustis and El-Sayed (Figure 5c in Eustis & El-Sayed 2006) without considering the particle geometries, when we regard ε_m as a fitting parameter.

The calculated aspect-ratio peak positions in Figure 12, as well as the shape of the curves, match the TEM measurements of Eustis and El-Sayed (Figure 5b in Eustis & El-Sayed 2006) well. It should be noted that for the sample with an absorption peak at 1000 nm (rightmost curve in figure 12b) there is a relatively large discrepancy between the experimental and theoretical results for $R > 5$. This is probably caused by the fact that for high aspect ratios and these values of ε_m the longitudinal absorption peaks fall outside of the spectral domain. The absorption spectra of Eustis and El-Sayed (Figure 5c in Eustis & El-Sayed 2006) are truncated at 1150 nm, which is lower than the longitudinal peak positions of the high-aspect-ratio nanowires in our simulations. Another possible reason is that, as in our experiments, the nanowire volume might not be independent of aspect ratio, but smaller for high-aspect-ratio rods. As such, we may underestimate the count of high-aspect-ratio particles.

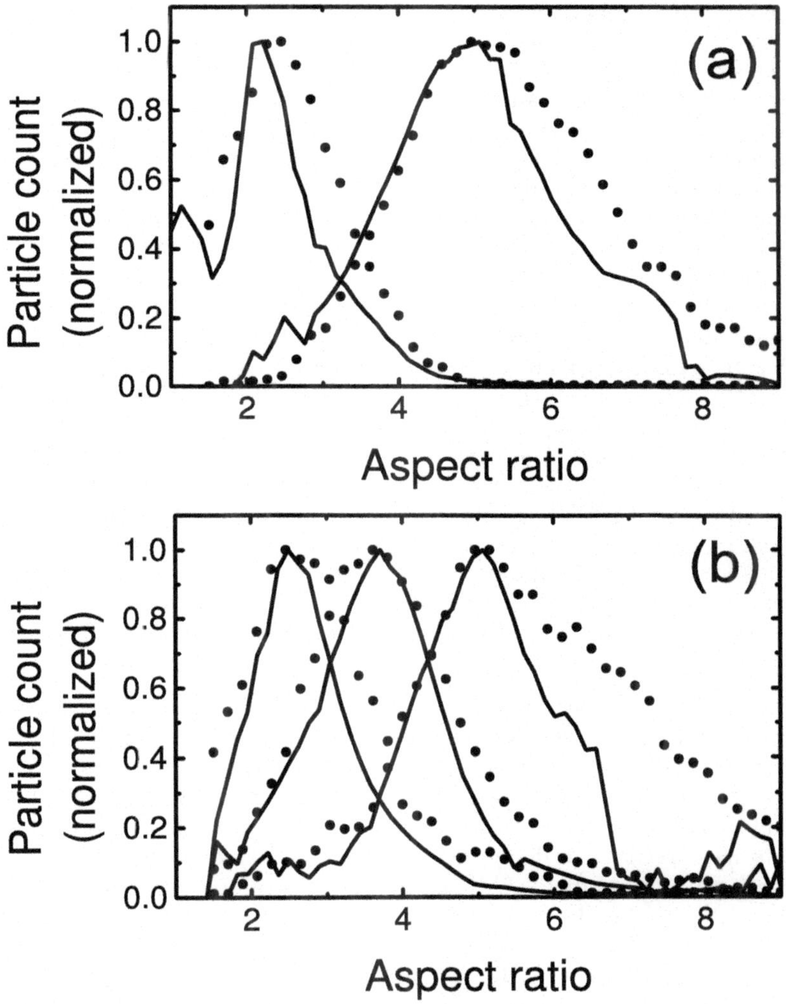

Fig. 12. Aspect-ratio distributions from TEM images by Eustis and El-Sayed (Figure 5b in Eustis & El-Sayed 2006) (squares) and from Gans fitting (solid lines) using our method with (a) $\varepsilon_m = 1.93$ and (b) $\varepsilon_m = 2.36$. The calculated absorption spectra using our fitting method are given in figure 10.

A comparison of our calculations with the absorption spectra of Eustis and El-Sayed (Figure 5c in Eustis & El-Sayed 2006) is given in figure 10. We find excellent agreement between our results and the measured absorption spectra.

6. Discussion

Using the dielectric constant of the medium as a fitting parameter, it is possible for one to obtain the aspect-ratio distribution of nanowires in solution from the absorption spectrum by fitting with Gans theory. To obtain a good match between the calculated aspect-ratio distributions and TEM measurements, we are required to set the dielectric constant of the medium to a significantly higher value than the dielectric constant of the pure solvent. For our gold nanowires with aspect ratios $R \leq 2.5$, we find that a value of $\varepsilon_m = 2.14 \pm 0.05$ gives a reasonable fit. However, the fitting has limited accuracy for very-low aspect ratios and thus it is hard to quantify the presence of spherical particles accurately. This could be the result of a limitation of Gans theory, but it can also be caused by errors in the size measurements from TEM images. We believe that the optical properties of the local nanowire environment are strongly affected by the presence of the surfactant as well as possible residue from the fabrication process. In analyzing the aspect ratios of nanowire-solutions, it has proven to be of great importance to incorporate a dependence on the dielectric properties of the local environment. We have shown that with Gans theory it is also possible to do this by optimizing the fit of the absorption spectrum for ε_m. In this way, one can not only determine the aspect-ratio distribution but also obtain insight into the local environment of the nanowires. Inversely, we have shown that Gans theory can be used to reproduce an absorption spectrum from a known particle size distribution. Here too, we find an optimal fit for $\varepsilon_m = 2.1 \pm 0.1$, which is significantly higher than the dielectric constant of water ($\varepsilon_m = 1.77$). This value for $\varepsilon_m = 2.1 \pm 0.1$ lies between the two values of $\varepsilon_m = 1.93$ and $\varepsilon_m = 2.36$ that we derived from the experimental results of Eustis and El-Sayed (Figure 5b and c in Eustis & El-Sayed 2006). The difference between the dielectric constants that we find for our particles and the experimental results of Pérez-Juste et al. could possibly be explained by differences in the nanowire synthesis procedures (see Pérez-Juste et al. 2005 and chapter 3).

We have analyzed particle sizes for a very small range of aspect ratios. Most contemporary research is done using particles with $R > 2$, whereas we have used particles with $R \leq 2.5$. The nanowires of Eustis and El-Sayed (Eustis & El-Sayed 2006) span a much larger range of aspect ratios ($1.5 \leq R \leq 9$), and when we apply our method to their results we also find a higher dielectric constant than expected for the solvent, as described in chapter 5. In fact, we need to define two different ε_m values in order to fit all of their experimental results. From their TEM measurements, it can be concluded that there is an additional variable besides the aspect ratio that influences the absorption spectrum. Three possible factors have been mentioned: end-cap geometry, particle size, and the dielectric constant of the medium. We have shown that Gans theory fits the results of Eustis and El-Sayed if we assume that not all samples have the same dielectric constant. A different dielectric constant could be the result of impurities in the solvent or different surfactant coverage densities of the nanowires. This could be the result of variations in the fabrication and growth process.

The results presented in this paper can probably be duplicated by using DDA calculations with varying particle geometries. We deem it likely that in reality a combination of these two effects occurs. In any case, the possibility that the dielectric constant of the medium in the vicinity of the nanowires is different from that of the pure solvent should not be easily dismissed. It appears that adapting the particle geometry in DDA calculations does improve the agreement between theory and experiment but still does not fully explain the difference between Gans theory and experimental results (Prescott & Mulvaney 2006). A different dielectric environment around the nanowires could explain this discrepancy. Therefore, we believe that our results show that Gans theory is a useful tool in the determination of the aspect ratios of Au nanowires and offers a good alternative to the DDA method.

7. Conclusions

Our implementation of Gans theory provides a quick, easy, and nondestructive method to accurately obtain the aspect-ratio distributions of nanowire solutions from UV-vis absorption spectra. By optimizing the dielectric constant of the medium, ε_m, in our calculations, we obtain a good agreement with TEM measurements for low-aspect-ratio gold nanowires. We find that the dielectric constant of the medium is an important factor in the characterization of gold nanowires. A comparison of Gans theory with other experimental results also shows that a higher dielectric constant of the local environment of the nanowires improves the absorption spectrum calculations for a larger range of aspect ratios. The earlier work by Eustis and El-Sayed reported mixed results in applying Gans theory to inhomogeneously broadened absorption spectra (Eustis & El-Sayed 2006). We have improved on their methods and shown that their results can be explained better under the assumption that the dielectric properties of the local environment around the nanowires have a profound effect on the gold nanowire absorption characteristics (Sprünken et al. 2007).

8. Acknowledgment

We thank Mr. Daan Sprünken, Dr. Kazuaki Furukawa, Dr. Hiroshi Nakashima, Dr. Ilya Sychugov, Prof. Yoshihiro Kobayashi, Dr. Keiichi Torimitsu, Dr. Koji Sumitomo and Dr. Katsuhiro Ajito, Mr. Seiichiro Mizuno, Mrs. Hiroko Takahata for their comments, insightful discussions, and technical assistances. D.S. is supported through a Vulcanus in Japan scholarship by the EU-Japan Centre for Industrial Cooperation.

9. References

Agarwal, A.; Huang, S.-W.; Day, K.; O'Donnell, M.; Day, M.; Kotov, N. & Ashkenazi, S. (2007). Targeted Gold Nanorod Contrast Agent for Prostate Cancer Detection by Photoacoustic Imaging, *Journal of Applied Physics,* Vol. 102, No. 6, (September 2007), pp. 064701-1-3.

Amendola, V & Meneghetti (2009), Size Evaluation of Gold Nanoparticles by UV-vis Spectroscopy, *Journal of Physical Chemistry C* Vol. 113, No. 11, (February 2009), pp.4277-4285.

Brioude, A.; Jiang, X.; Pileni, M. (2005). Optical Properties of Gold Nanorods: DDA Simulations Supported by Experiments. *Journal of Physical Chemistry B* Vol. 109, No. 27, (June 2005), pp. 13138-13142.

Chen, C.-D.; Cheng, S.-F.; Chau, L.-K. & Wang, C. C. (2007). Sensing Capability of the Localized Surface Plasmon Resonance of Gold Nanorods. *Biosensers and Bioelectronics* Vol. 22, No. 6, (January 2007), pp. 926-932.

Eustis, S. & El-Sayed, M. A. (2006). Determination of the Aspect Ratio Statistical Distribution of Gold Nanorods in Solution from a Theoretical Fit of the Observed Inhomogeneously Broadened Longitudinal Plasmon Resonance Absorption Spectrum. *Journal of Applied Physics.* Vol. 100, No.4, (August 2006), pp. 044324-1-7.

Gans, R. (1912). Über die Form ultramikroskopischer Goldteilchen, *Annalen der Physik.* Vol. 342, No. 5, 881-900.

Gou, L. & Murphy, C. (2005). Fine-Tuning the Shape of Gold Nanorods, *Chemistry of Materials* Vol. 17, No. 14, (June 2005), pp. 3668-3672.

Gulati, A.; Liao, H. & Hafner J. (2006). Monitoring Gold Nanorod Synthesis by Localized Surface Plasmon Resonance, *Journal of Physical Chemistry B,* Vol. 110, No. 45, (October 2006), pp. 22323-22327.

Hu, M.; Wang, X.; Hartland, G.; Mulvaney, P.; Pérez-Juste, J. & Sader, J. (2003). Vibrational Response of Nanorods to Ultrafast Laser Induced Heating: Theoretical and Experimental Analysis, *Journal of American Chemical Society,* Vol. 125, No. 48, (November 2003), pp. 14925-14933.

INRIA & ENPC. *Scilab 4.1,* (2006) available from www.scilab.org.

Johnson, P. B. & Christy, R. W. (1972). Optical Constants of the Noble Metals, *Physical Review B,* Vol. 6, No. 12, (December 1972), pp. 4370-4379.

Link, S.; Mohamed, M. & El-Sayed, M. (1999). Simulation of the Optical Absorption Spectra of Gold Nanorods as a Function of Their Aspect Ratio and the Effect of the Medium Dielectric Constant, *Journal of Physical Chemistry B,* Vol. 103, No. 16, (April 1999), pp. 3073-3077.

Link, S. & El-Sayed, M. (2005). Simulation of the Optical Absorption Spectra of Gold Nanorods as a Function of Their Aspect Ratio and the Effect of the Medium Dielectric Constant, *Journal of Physical Chemistry B,* Vol. 109, No. 20, (May 2005), pp. 10531-10532.

Mie, G. (1908). Beiträge zur Optik trüber Medien, speziell kolloidaler Metallösungen, *Annalen der Physik,* Vol. 330, No. 3, pp. 377-445.

Miranda, O. & Ahmadi, T. (2005). Effects of Intensity and Energy of CW UV Light on the Growth of Gold Nanorods, *Journal of Physical Chemistry B,* Vol. 109, No. 33, (July 2005) pp. 15724-15734.

Murphy, C. J.; San, T. K.; Gole, A. M.; Orendorff, C. J.; Gao, J. X.; Gou, L.; Hunyadi, S. E. & Li, T. (2005). Anisotropic Metal Nanoparticles: Synthesis, Assembly, and Optical Applications, *Journal of Physical Chemistry B,* Vol. 109, No. 29, (June 2005), pp. 13857-13870.

Murphy, C. J.; Gole, A. M.; Hunyadi, S. E. & Orendorff, C. (2006). One-Dimensional Colloidal Gold and Silver Nanostructures, *Journal of Inorganic Chemistry,* Vol. 45, No. 19, (September 2006), pp. 7544-7554.

Nehl, C. L. & Hafner, J. H. (2008), Shape-dependent Plasmon Resonances of Gold Nanoparticles, *Journal of Materials Chemistry* Vol. 18, (February 2008), pp. 2415-2419.

Nikoobakht, B. & El-Sayed, M. (2003). Preparation and Growth Mechanism of Gold Nanorods (NRs) Using Seed-Mediated Growth Method, *Chemistry of Materials*, Vol. 15, No. 10, (April 2003), pp. 1957-1962.

Noguez, C. (2007). Surface Plasmons on Metal Nanoparticles: The Influence of Shape and Physical Environment, *Journal of Physical Chemistry C*, Vol. 111, No. 10, (February 2007), 3806-3809.

Orendorff, C.; Gearheart, L.; Jand, N. & Murphy, C. (2006). Aspect Ratio Dependence on Surface Enhanced Raman Scattering using Silver and Gold Nanorod Substrates, *Physical Chemistry Chemical Physics*, Vol. 8, No. 1, (October 2006), pp. 165-170.

Pérez-Juste, J.; Pastoriza-Santos, I.; Liz-Marzan, L. M. & Mulvaney, P. (2005). Gold Nanorods: Synthesis, Characterization and Applications, *Coordination Chemical Review*, Vol. 249, No. 17-18, (September 2005), pp. 1870-1901.

Prescott, S. W. & Mulvaney, P. (2006). Gold Nanorod Extinction Spectra, *Journal of Applied Physics*, Vol. 99, No. 12, (June 2006), pp. 123504-1-8.

Sönnichsen, C.; Franzl, T.; Wilk, T.; von Plessen, G.; Feldmann, J.; Wilson, O. & Mulvaney, P. (2002). Drastic Reduction of Plasmon Damping in Gold Nanorods, *Physical Review Letters* Vol. 88, No. 7, (January 2002), pp. 077402-1-4.

Stoller, P.; Jacobsen, V. & Sandoghdar, V. (2006). Measurement of the Complex Dielectric Constant of a Single Gold Nanoparticle, *Optics Letters*, Vol. 31, No. 16, (July 2006), pp. 2474-2476.

Sprünken, D; Omi, H; Furukawa K.; Nakashima H.; Sychugov I.; Kobayashi Y. & Torimistu K. (2007). Influence of the Local Environment on Determining Aspect-Ratio Distributions of Gold Nanorods in Solution Using Gans Theory, *Journal of Physical Chemistry C*, Vol. 111, No. 39, (September 2007), pp. 14299-14306.

Suzuki, M.; Niidome, Y.; Kuwahara, Y.; Terasaki, N.; Inoue, K. & Yamada, S. (2004). Surface-Enhanced Nonresonance Raman Scattering from Size- and Morphology- Controlled Gold Nanoparticle Films, *Journal Physical Chemistry B*, Vol. 108, No. 31, (July 2004), pp. 11660-11665.

Suzuki, M.; Niidome, Y.; Terasaki, N.; Inoue, K.; Kuwahara, Y. & Yamada, S. (2004). Surface-Enhanced Nonresonance Raman Scattering of Rhodamine 6G Molecules Adsorbed on Gold Nanorod Films, *Japanese Journal of Applied Physics*, Vol. 43, (April 2004), pp. 554-556.

Sztainbuch, I. W. (2006). The Effects of Au Aggregate Morphology on Surface-Enhanced Raman Scattering Enhancement, *Journal of Chemical Physics*, Vol. 125, No. 12, (September 2006), pp. 124707-1-12.

Xu, X.; Gibbons, T. & Cortie, M. (2003). Spectrally-selective Gold Nanorod Coatings for Window Glass, *Gold Bullutin*, Vol. 39, No. 4, pp. 156-165.

Yan, B.; Yang, Y. & Wang, Y. (2003). Comment on "Simulation of the Optical Absorption Spectra of Gold Nanorods as a Function of Their Aspect Ratio and the Effect of the Medium Dielectric Constant". *Journal of Physical Chemistry B*, Vol. 107, No. 34, (July, 2003), pp. 9159-9159.

Yin, G.; Wang, S.-Y.; Xu, M.; Chen, L. Y. (2006). Theoretical Calculation of the Optical Properties of Gold Nanoparticles, *Journal of the Korean Physical Society*, Vol. 49, No. 5, pp. 2108-2111.

Yu, C. & Irudayaraj, J. (2007). Multiplex Biosensor Using Gold Nanorods, *Journal of Analytical Chemistry,* Vol. 79, No. 2, (December 2007), pp. 572-579.

Zweifel, D. & Wei, A. (2005). Sulfide-Arrested Growth of Gold Nanorods, *Chemistry of Materials,* Vol. 17, No. 16, (June 2005), pp. 4256-4261.

9

Influence of Pulse-Impact on Microstructure of Welded Joints at Various Temperatures in Liquid-Phase-Pulse-Impact Diffusion Welding Particle Reinforcement Aluminum Matrix Composites

Kelvii Wei Guo
MBE, City University of Hong Kong, Hong Kong

1. Introduction

The high specific strength, good wear-ability and corrosion resistance of Aluminum Matrix Composites (AMCs) attract substantial industrial applications. Typically, AMCs are currently used widely in automobile and aerospace industries, structural components, and heat resistant-wearable parts in engines, *etc.* (Go´mez de Salazar JM et al., 2003; Loyd DJ, 1994; Nair SV et al., 1995; Pirondi A et al., 2009; Rotundo F et al., 2010). The particles of reinforcement elements in AMCs may be either in form of particulates or as short fibers, whiskers and so forth (Loyd DJ, 1994; Maity J et al., 2009). These discontinuous natures create several problems to their joining techniques for acquiring their high strength and good quality weld-joints. Typical quality problems of those welding techniques currently available for joining AMCs (American Welding Society, 1996; Arik H et al., 2005; Feng AH et al., 2008; Fernandez GJ et al., 2004; Hsu CJ et al., 2005; Marzoli LM et al., 2006; Schell JSU et al., 2009; Shanmuga Sundaram N et al., 2010; Wert JA, 2003) are as elaborated below.

1. The distribution of particulate reinforcements in the weld

As properties of welded joints are usually influenced directly by the distribution of particulate reinforcements in the weld, their uniform distribution in the weld is likely to give tensile strength higher than 70~80 % of the parent AMCs. Conglomeration distribution or absence (viz. no-reinforcements-zone) of the particulate reinforcements in the weld generally degrades markedly the joint properties and subsequently resulted in the failure of welding.

2. The interface between the particulate reinforcements and aluminum matrix

High welding temperature in the fusion welding methods (typically: TIG, laser welding, electron beam etc.) is likely to yield pernicious Al_4C_3 phase in the interface. Long welding time (e.g. several days in certain occasions) in the solid-state welding methods (such as diffusion welding) normally leads to (i) low efficiency and (ii) formation of harmful and brittle intermetallic compounds in the interface.

To alleviate these problems incurred by the available welding processes for welding AMCs, a liquid-phase-pulse-impact diffusion welding (LPPIDW) technique has been developed (Guo W et al., 2007; Guo W et al., 2008; Guo W et al., 2008). This paper aims at providing some specifically studies the influence of pulse-impact on the microstructures of welded joints. Analysis by means of scanning electron microscope (SEM), transmission electron microscope (TEM) and X-Ray Diffraction (XRD) allows the micro-viewpoint of the effect of pulse-impact on LPPIDW to be explored in more detail.

2. Experimental material and process

2.1 Specimens

Stir-cast SiC_p/A356, P/M SiC_p/6061Al and Al_2O_{3p}/6061Al aluminum matrix composite, reinforced with 20 %, 15 % volume fraction SiC, Al_2O_3 particulate of 12 μm, 5 μm mean size, are illustrated in Figs. 1~ 3.

2.2 Experiment

Fig. 1. Microstructure of aluminum matrix composite SiC_p/ A356

The quench-hardened layer and oxides, as induced by wire-cut process, on the surfaces of aluminum matrix composite specimens were removed by careful polishing using 400 # grinding paper. The polished specimens were then properly cleaned by acetone and pure ethyl alcohol so as to remove any contaminants off its surfaces. A DSI Gleeble®-1500D thermal/mechanical simulator with a 4×10^{-1} Pa vacuum chamber was subsequently used to perform the welding.

The microstructures and the interface between the reinforcement particle and the matrix of the welded joints were analyzed by SEM and TEM.

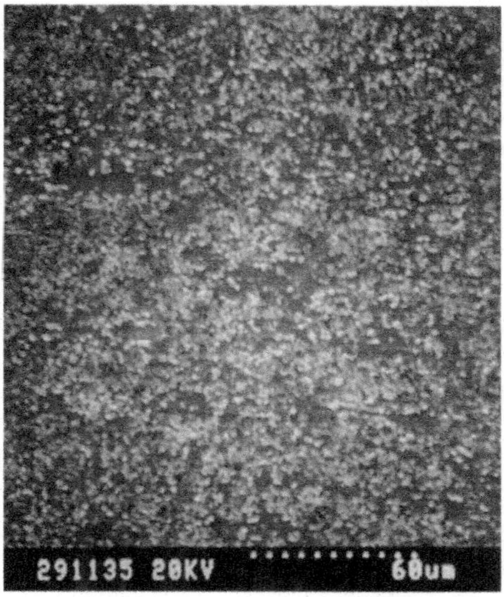

Fig. 2. Microstructure of aluminum matrix composite SiC_p/6061Al

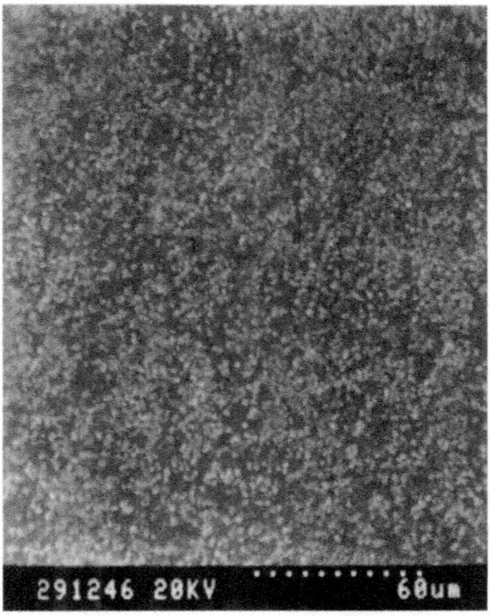

Fig. 3. Microstructure of aluminum matrix composite Al_2O_{3p}/6061Al

2.3 Operation of LPPIDW

Figure 4 illustrates a typical temperature and welding time cycle of a LPPIDW. It basically involved with: (i) an initially rapid increase of weld specimens, within a time of t_a, to an optimal temperature T_a at which heat was preserved constantly at T_a for a period of (t_b - t_a), (ii) at time t_c, a quick application of pulse impact to compress the welding specimens so as to accomplish an anticipated deformation δ within a glimpse of 10^{-4}~10^{-2} s, whilst the heat preservation was still maintained at the operational temperature T_a; and (iii) a period of natural cooling to room temperature after time t_b.

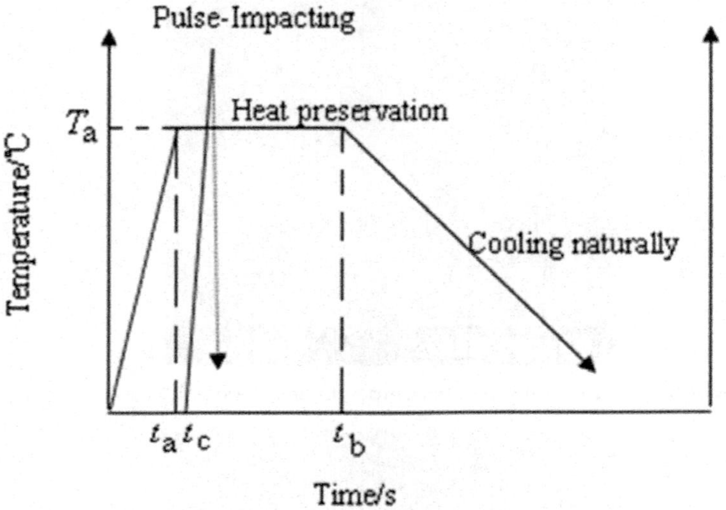

Fig. 4. Schematic diagram of liquid-phase-pulse-impact diffusion welding

3. Results and discussion

3.1 Microstructure of welded joint

Figure 5 shows the microstructures of welded joints of SiC_p/A356 at various temperatures with V_I=560 mm/s, t_I=10^{-2}~10^{-4} s, t=30 s, P_0=5 MPa, δ=1 mm, where V_I was velocity of pulse-impact, t_I was the impacting time, t was holding time for heat preservation, P_0 was holding pressure during the welding, δ was the horizontal deformation. It elucidated that when the welding temperature was 563 °C, under the effect of pulse-impact, the liquid phase matrix alloy wasn't formed enough to wet the particle reinforcements. In addition, at this temperature, the diffusion capability of the atoms within the matrix was relatively low. As a result, the welding interface between two specimens could be observed obviously as shown in Fig. 5(a) and followed by the unsuitable strength (about 118 MPa). Moreover, because of lower welding temperature, the area of the formed solid-liquid phase was smaller, which led to some streamlines scattered in the matrix (Fig. 5(a)) after the pulse-impact acting on the substrates. When the temperature reached 565 °C, the rate of the atom diffusion in the joint region within the matrix was accelerated (Fig. 5(b)). At the same time, more liquid phase matrix alloy was formed to wet reinforcements (SiC). Therefore, the interface state of reinforcement and reinforcement was improved and the reinforcements were distributed uniformly to some extent. Also, the streamlines scattered in the matrix were disappeared, and the tensile strength of welded joints was about 134 MPa higher than that of 563 °C.

When the temperature is up to 570 °C, the formed liquid phase matrix alloy was enough and suitable for wetting reinforcements effectively, and the rate of the atom diffusion was more active. As a result, for reinforcements the welding mode in the joint region changed from reinforcement - reinforcement to reinforcement - matrix - reinforcement. Consequently, the joint was welded successfully (Fig. 5(c)). The average strength of 179 MPa for the welded joints produced at welding temperature of 570 °C was about 74.6 % of the 240 MPa for the strength of parent aluminum matrix composite.

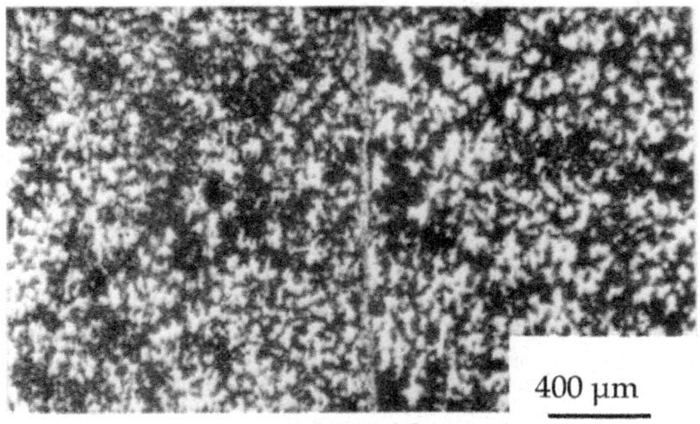

(a) 563 °C

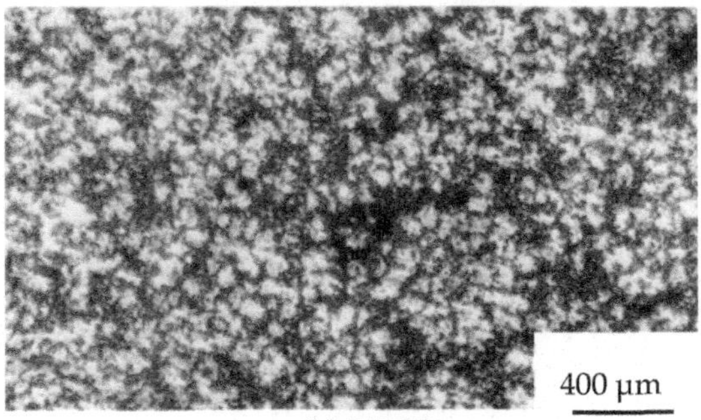

(b) 565 °C

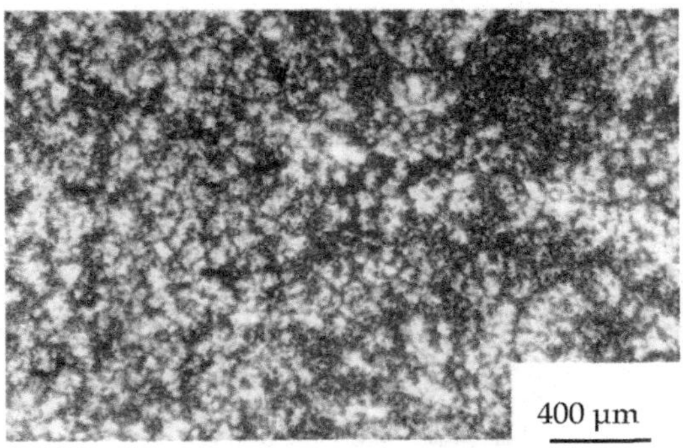

(c) 570 °C

Fig. 5. Microstructures of welded joints of SiC_p/A356 at various temperatures by LPPIDW

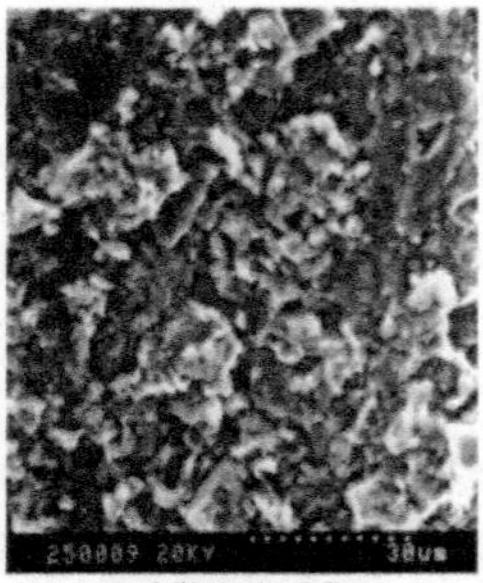

(a) 563 °C

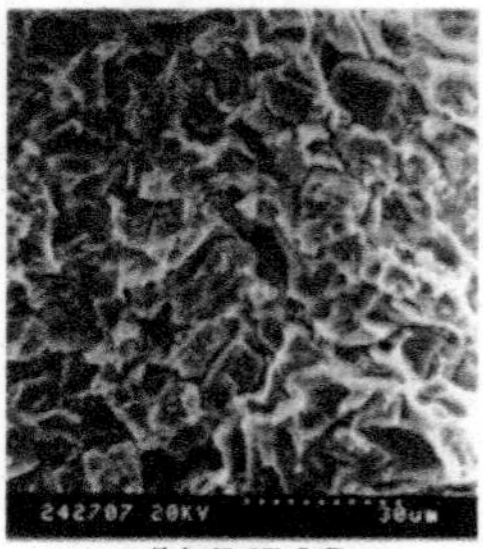

(b) 565 °C

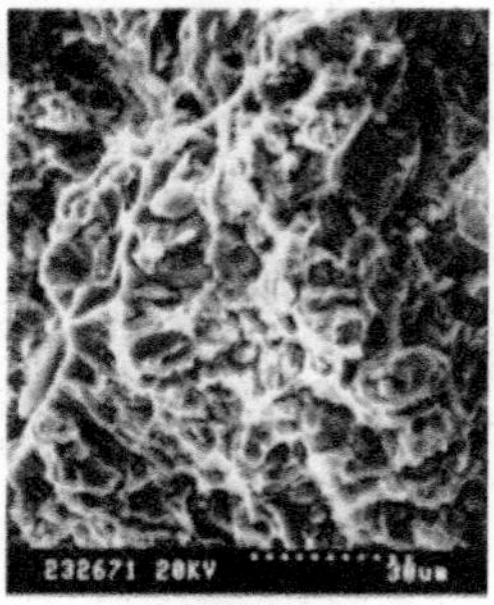

(c) 570 °C

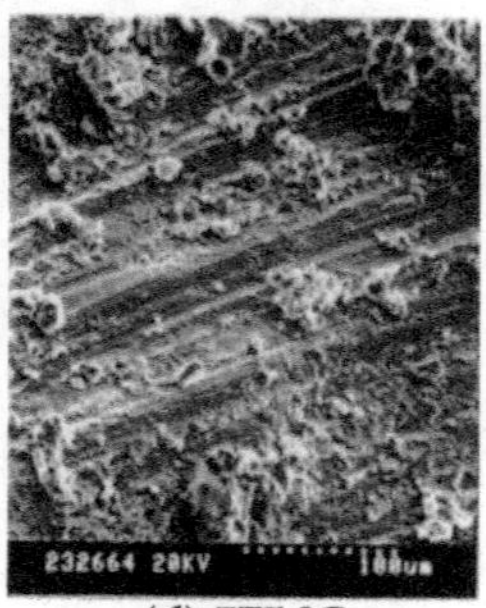

(d) 575 °C

Fig. 6. Fractographs of SiC_p/ A356 at various temperatures

The relevant fractographs are shown in Fig. 6. It illustrated that when the welding temperature was 563 °C, the initial morphology of substrate could be detected obviously, and some sporadic welded locations appeared together with some rather densely scattering bare reinforcement particles as shown in Fig. 6(a). With the welding temperature was increased to 565 °C, more liquid phase was formed. Under the effect of pulse-impact, some wet locations in the joint had been excellently welded and the aggregated solid reinforcement particles were improved. However, the bare reinforcement particles were still distributed on the fractographic surface. It indicated that substrates did not weld ideally the pieces together and it consequently resulted in a low strength joint (Fig. 6(b)). Figure 6(c) shows the fractograph of welded joint at 570 °C. It illustrated that the fracture was dimple fracture. Moreover, SEM of the fracture surface showed some reinforcement particles (SiC) in the dimple. In order to confirm the state of these reinforcement particles, particles itself and matrix neighboring to these particles were analyzed by energy dispersive X-ray analysis (EDX) respectively. It indicated that reinforcement particles (SiC) were wet by matrix alloy successfully suggesting that the reinforcement particles had been perfectly wet and the composite structure of reinforcement/reinforcement had been changed to the state of reinforcement/ matrix /reinforcement.

As welding temperature increasing to 575 °C, it led to more and more liquid phase matrix alloy distributing in the welded interface, meanwhile, more liquid phase matrix alloy reduced the effect of impact on the interface of the welded joints, subsequently the application of transient pulse-impacting would cause the relative sliding of the weldpieces that jeopardized ultimately the formation of proper joint as shown in Fig. 6(d). It demonstrated that results of fractographs were agreed with the corresponding microstructures well.

The relevant results of SiC_p/6061Al and Al_2O_{3p}/6061Al at various welding temperatures are shown in Fig. 7 to Fig. 10.

It showed that the microstructure evolutions and its corresponding fracture surfaces under the effect of pulse-impact are similar to that of SiC_p/A356.

Figures 7(a) and 9(a) show when the welding temperature was too low to form enough liquid phase matrix alloy to wet the reinforcement particles, and the diffusion capability of the atoms within the matrix was relatively low. Therefore, the indistinct welding interface between two specimens could be observed resulted in low tensile strength (about 240 MPa for SiC_p/6061Al and 270MPa for Al_2O_{3p}/6061Al). When the temperature is higher (623 °C for SiC_p/6061Al and 644 °C for Al_2O_{3p}/6061Al), the liquid phase matrix alloy was formed enough to wet the reinforcement particles (SiC), together with higher rate of the atom diffusion in the joint region (Figs. 7(b) and 9(b)). Consequently, the joints could be welded successfully with the average strength of 260 MPa for SiC_p/6061Al (about 72.2 % of the 360 MPa for the strength of parent aluminum matrix composite) and 282 MPa for Al_2O_{3p}/6061Al (about 70.5 % of the 400 MPa for the strength of parent aluminum matrix composite). As welding temperature increasing further (such as 625 °C for SiC_p/6061Al and 647 °C for Al_2O_{3p}/6061Al), more and more liquid phase matrix alloy would be distributed in the welded interface, at the same time, more liquid phase matrix alloy reduced the effect of impact on the interface of the welded joints, subsequently prompted for the descending of the joint strength (Figs. 7(c) and 9(c)).

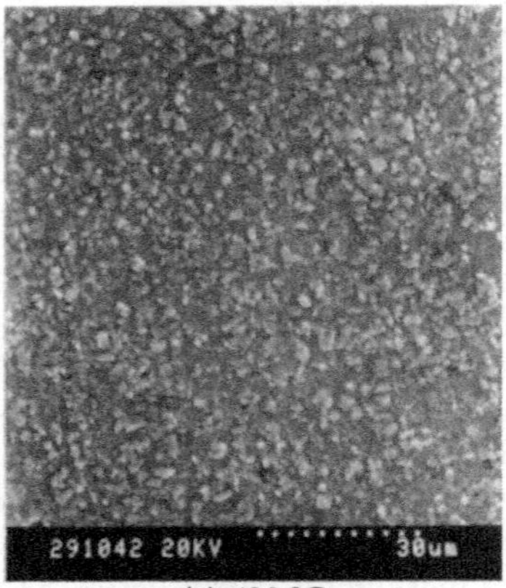

(a) 620 °C

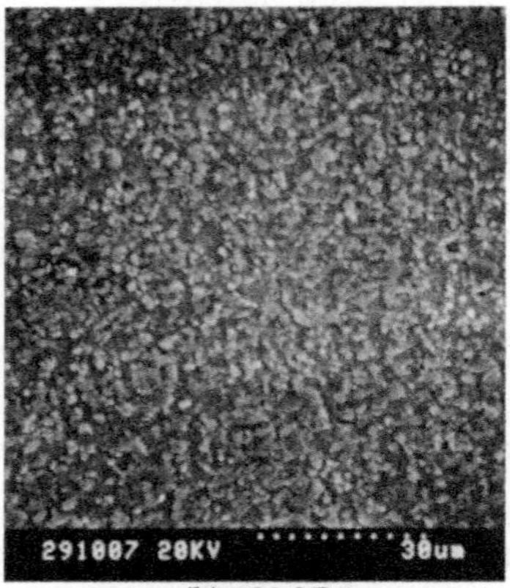

(b) 623 °C

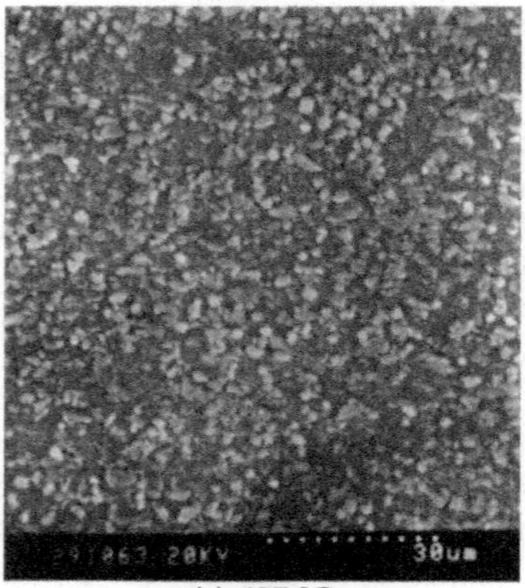

(c) 625 °C

Fig. 7. SEM micrographs of $SiC_p/6061Al$ welded joints at various welding temperatures

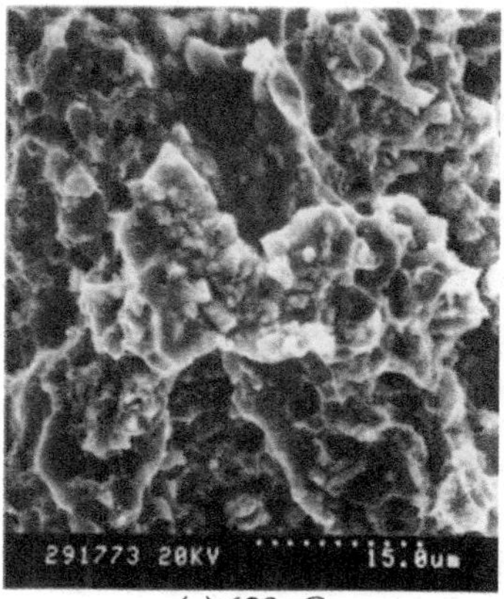

(a) 620 °C

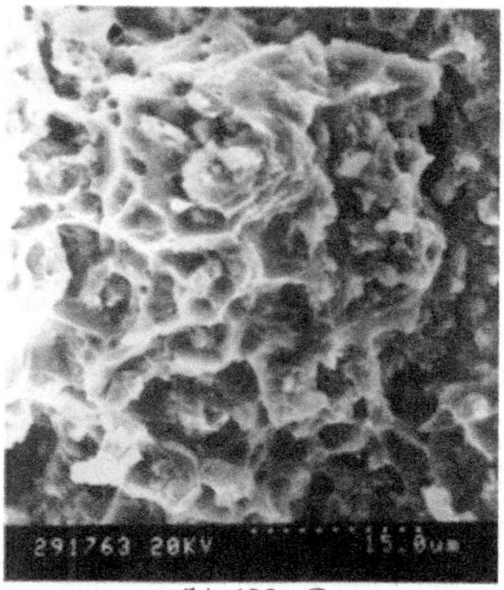

(b) 623 °C

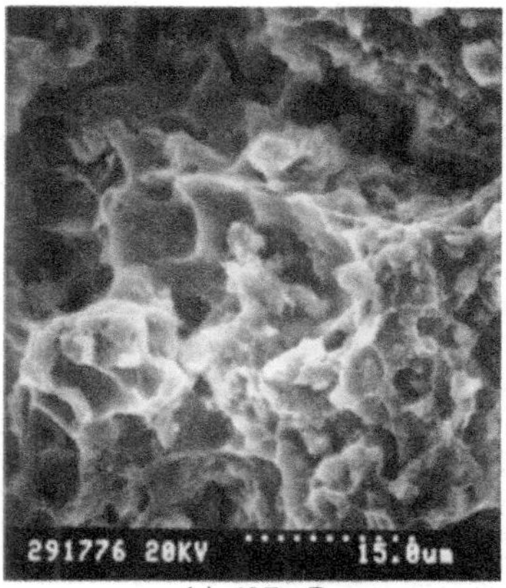

(c) 625 °C

Fig. 8. Fractographs of SiC_p/6061Al at various temperatures

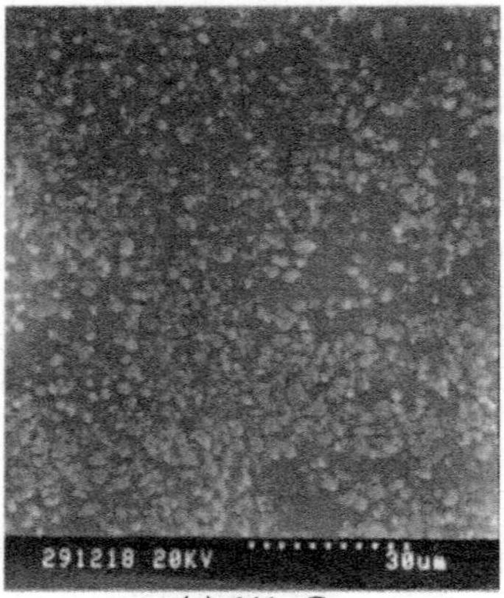

(a) 641 °C

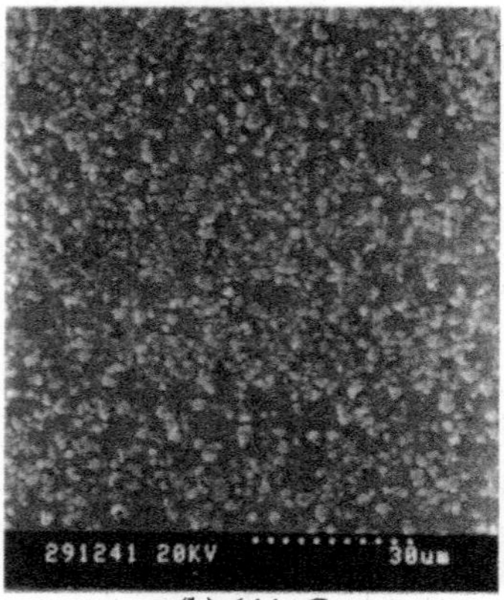

(b) 644 °C

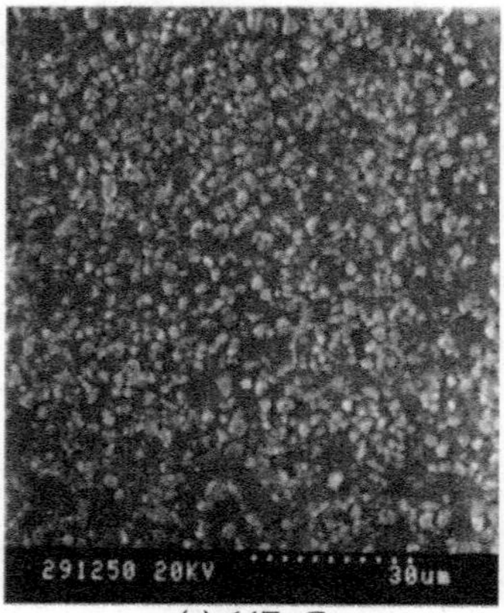

(c) 647 °C

Fig. 9. SEM micrographs of Al_2O_{3p}/6061Al welded joints at various welding temperatures

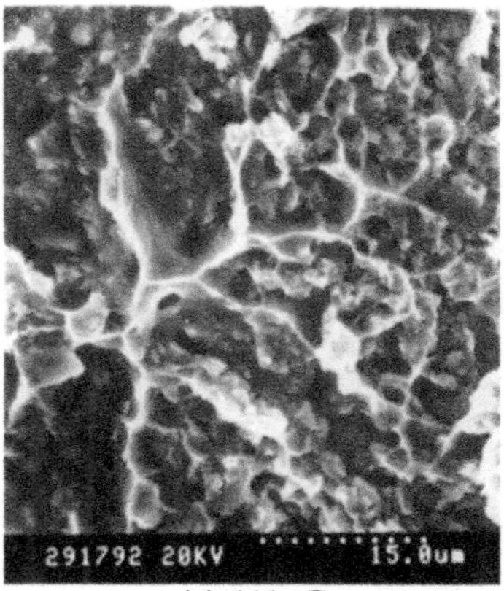

(a) 641 °C

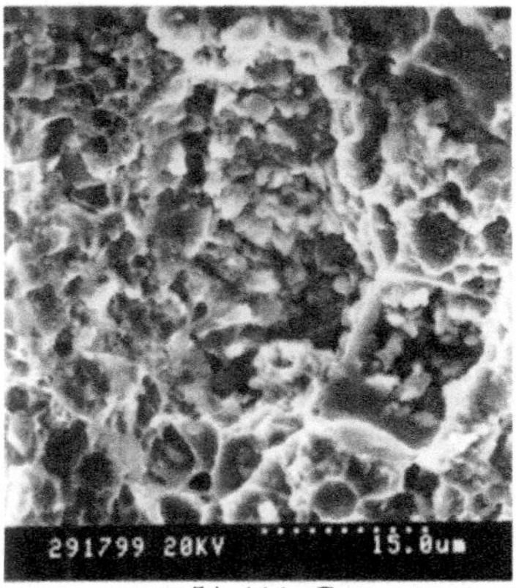

(b) 644 °C

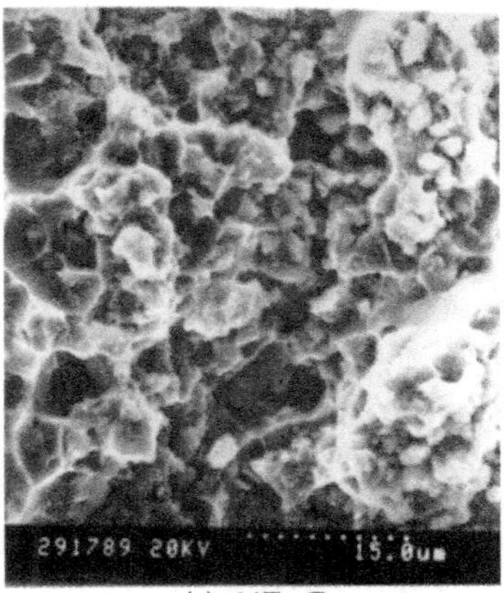

(c) 647 °C

Fig. 10. Fractographs of Al_2O_{3p}/6061Al at various temperatures

Moreover, according to the fractures of welded joints at various temperatures shown in Figs. 8 and 10, it showed that it agreed with Figs. 7 and 9 very well, and the fractures were all dimple fractures with some reinforcement particles (SiC, Al_2O_3) in the dimple. Also, the results of SiCp/6061Al were better than that of Al2O3p/6061Al due to a mild interfacial reaction between the reinforcement and matrix, which released the thermal mismatch stress to an acceptable extent between the reinforcement and matrix to allow load transfer from the matrix to reinforcement successfully. As a result, it had advantageous effect of improving the strength of welded joints further (Guo W et al., 2008).

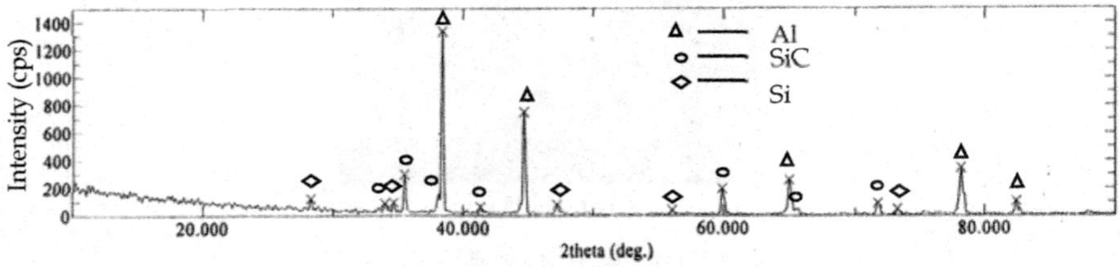

(a) SiC_p/A356

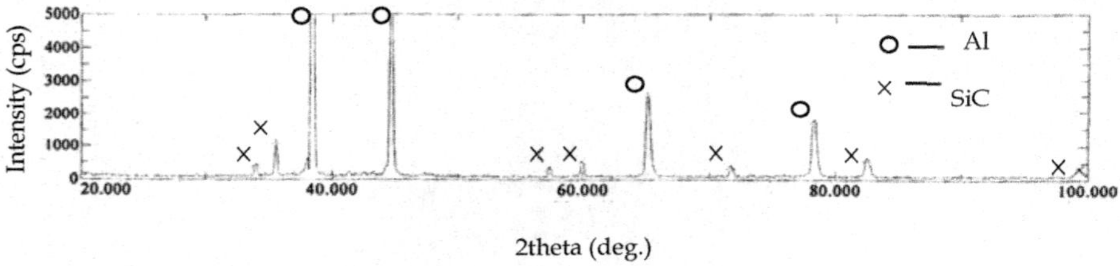

(b) SiC_p/6061Al

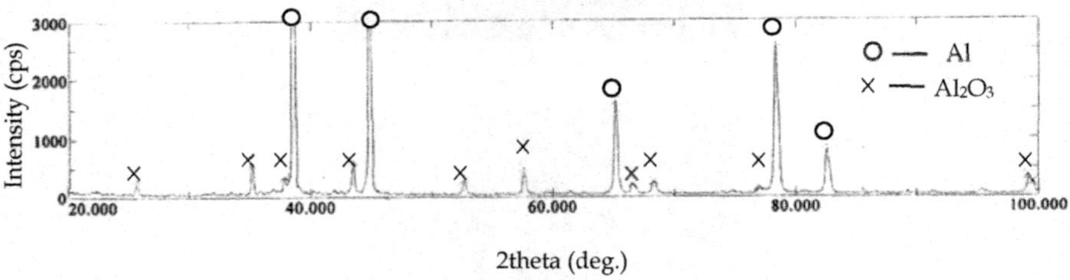

(c) Al_2O_{3p}/6061Al

Fig. 11. XRD pattern of the fracture surfaces

Based on microstructures of the welded joints with the optimal parameters (i.e., $T_{SiCp/A356}$=570 °C, $T_{SiCp/6061Al}$=623 °C, T_{Al2O3p}=644 °C, V_I=560 mm/s, t_I=10^{-2}~10^{-4} s, δ=1 mm, t=30 s, P_0=5 MPa) and its corresponding fracture surfaces as shown in Figs. 5, 6, 7-10, the welded joint displayed with uniformly distributing reinforcement particles and microstructure almost similar to that of its parent composite (Figs. 1, 2 and 3). SEM of the fracture surface showed that the reinforcement particles had been perfectly wet and the composite structure of reinforcement/reinforcement had been changed to the state of reinforcement/matrix /reinforcement. XRD pattern of the fracture surfaces (Fig. 11) did not illustrate the existence of any harmful phase or brittle phase of Al_4C_3. This suggested the effective interface transfers between reinforcement particles and matrix in the welded joint that subsequently provided favorable welding strength (Guo W et al., 2007; Guo W et al., 2008; Guo W et al., 2008).

3.2 Distribution of dislocation in the welded joint

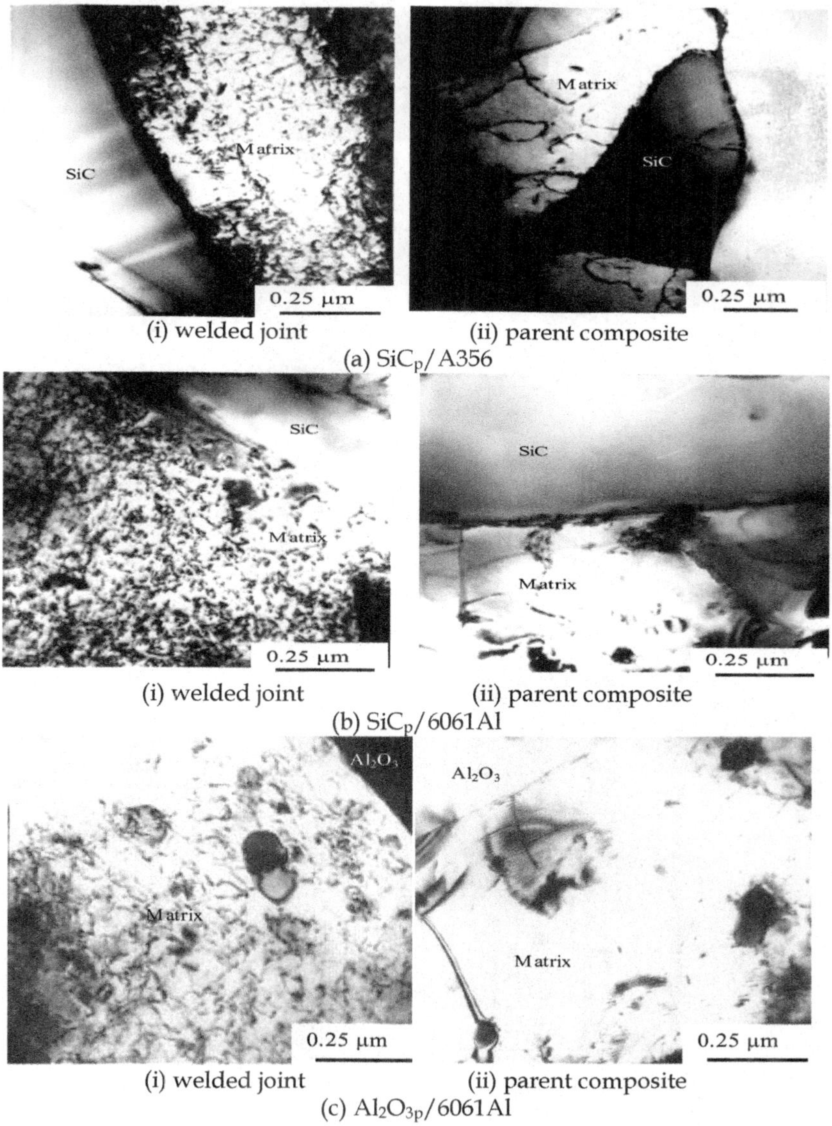

(i) welded joint (ii) parent composite
(a) SiC_p/A356
(i) welded joint (ii) parent composite
(b) SiC_p/6061Al
(i) welded joint (ii) parent composite
(c) Al_2O_{3p}/6061Al

Fig. 12. Distribution of dislocation in the matrix neighboring to the interface of the welded joint and parent composite respectively

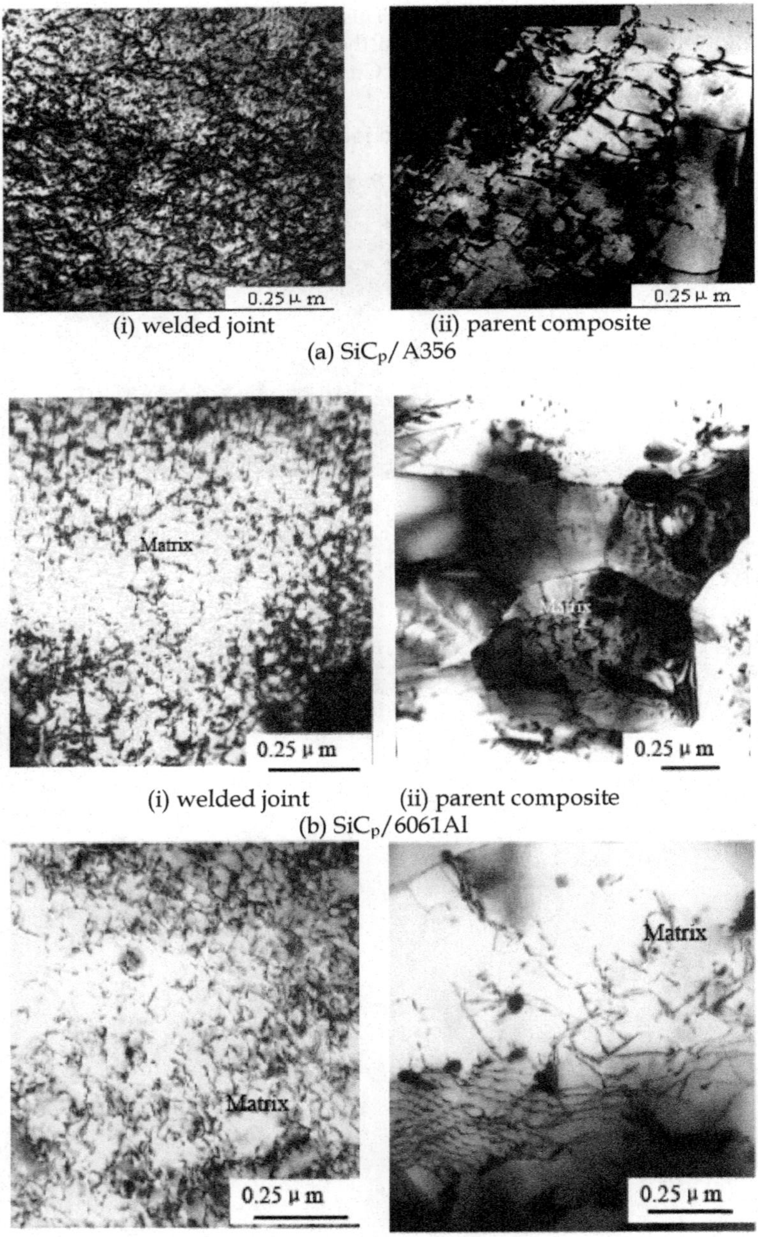

(i) welded joint (ii) parent composite
(a) SiC_p/ A356

(i) welded joint (ii) parent composite
(b) SiC_p/6061Al

(i) welded joint (ii) parent composite
(c) Al_2O_{3p}/6061Al

Fig. 13. Distribution of dislocation in the matrix away from the interface of the welded joint and parent composite respectively

The distribution of dislocation in the matrix neighboring to the interface of the welded joint by LPPIDW in comparison with its parent composite is shown in Fig. 12. The clearly distinctive interface between reinforcement particle and matrix indicated that the integration between the reinforcement particle and matrix was prominent. The effect of pulse-impact subsequently led to dislocation in the matrix lattices and showed sign of mutually entwisting to give higher welded strength. Comparatively, its dislocation distribution in the matrix neighboring to the interface was relatively denser than that in its parent composite (cf. Figs. 12(i) and 12(ii)). Similarly, the density of dislocation and dislocation entwisting in the matrix away from the welded interface was also higher than that of its parent composite (cf. Figs. 13(i) and 13(ii)). Such favorable characteristics ultimately gave relatively superior strength of the welded joint to that of conventional diffusion welding (Guo W et al., 2007; Guo W et al., 2008; Guo W et al., 2008).

3.3 Formation of nano-grains in the weld

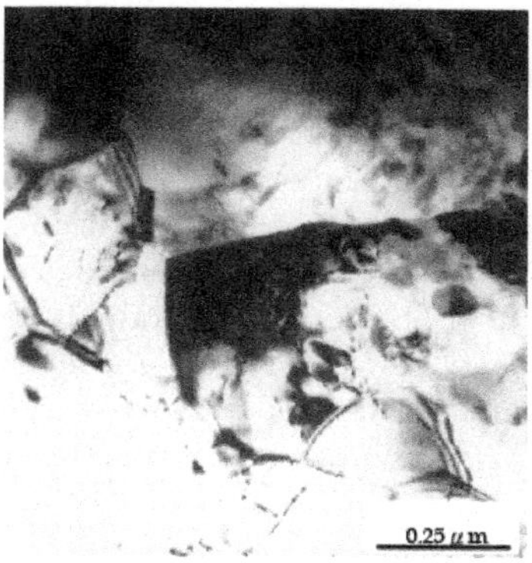

(a) SiC_p/A356

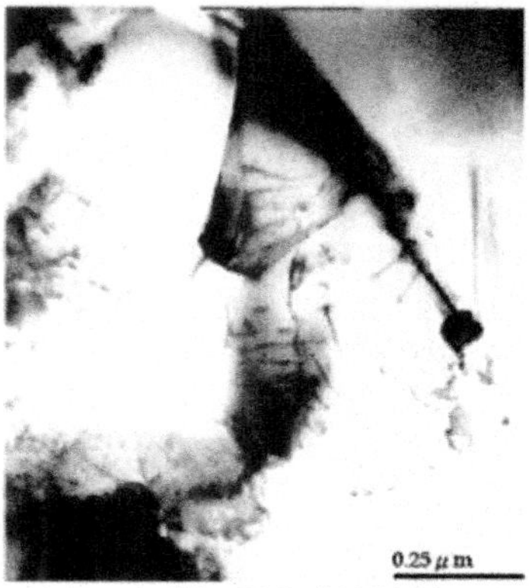

(b) SiC_p/6061Al

Fig. 14. (a,b) Nano-grains formed in the weld of particle reinforcement aluminum matrix composites during the LPPIDW.

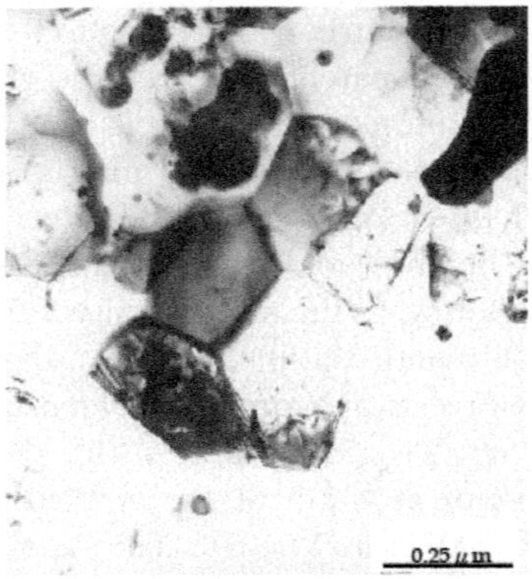

(c) $Al_2O_{3p}/6061Al$

Fig. 14. (c) Nano-grains formed in the weld of particle reinforcement aluminum matrix composites during the LPPIDW. (Continuation)

TEM micrograph (Fig. 14) of a weld by LPPIDW displayed some newly-formed nano-grains in the lattices of the joint. These nano-grains would seat in the interstices of crystal lattices and create new grain boundary in hindering the movement of neighbouring grains and subsequently improved obviously the properties of the welded joints. The formation of new nano-grains was the advantageous effect of pulse-impact in LPPIDW. In addition, XRD pattern of the fracture surface (Fig. 11) did not illustrate the existence of any harmful phase or brittle phase of Al_4C_3. This suggested the effective interface transfers between reinforcement particles and matrix in the welded joint that subsequently provided favorable welding strength (Guo W et al., 2007; Guo W et al., 2008; Guo W et al., 2008).

4. Conclusions

Results of this study on the microstructures of welded joints of particle reinforcement aluminum matrix composites ($SiC_p/A356$, $SiC_p/6061Al$, $Al_2O_{3p}/6061Al$) using liquid-phase-pulse-impact diffusion welding process show that:

1. Pulse-impact in liquid-phase-pulse-impact diffusion welding in joining particle reinforcement aluminum matrix composites ($SiC_p/A356$, $SiC_p/6061Al$, $Al_2O_{3p}/6061Al$) resulted in higher density of dislocation in the matrix neighboring to and away from the interface than their parent composite. Simultaneously, the dislocation entwisted mutually and intensively in the welded joint propitious to improve the strength of welded joints.
2. There was distinctly clear interface between reinforcement particle and matrix. It overcame some diffusion problems normally encountered in conventional diffusion welding, and prevented the formation of harmful microstructure or brittle phase in the welded joint.
3. The joint by LPPIDW process would form nano-grains. The newly-formed nano-grains would improve the properties of welded joints resulted in higher tensile strength.

5. Acknowledgement

This work is supported by City University of Hong Kong Strategic Research Grant (SRG) No. 7002582.

6. References

American Welding Society (1996). *Welding Handbook*. Miami.

Arik H, Aydin M, Kurt A, Turker M (2005). *Weldability of Al_4C_3-Al composites via diffusion welding technique*. Materials & Design, Vol. 26, No. 6, pp. 555-560.

Feng AH, Xiao BL, Ma ZY (2008). *Effect of microstructural evolution on mechanical properties of friction stir welded AA2009/SiCp composite*. Composites Science and Technology, Vol. 68, No. 9, pp. 2141-2148.

Fernandez GJ & Murr LE (2004). *Characterization of Tool Wear and Weld Optimization in the Friction-Stir Welding of Cast Aluminum 359+20% SiC Metal-Matrix Composite*. Materials Characterization, Vol. 52, No. 1, pp. 65-75.

Go´mez de Salazar JM, Barrena MI (2003). *Dissimilar Fusion Welding of AA7020/MMC Reinforced with Al_2O_3 Particles. Microstructure and mechanical properties*. Materials Science and Engineering A, Vol. 352, No. 1-2, pp. 162-168.

Guo W, Hua M & Ho JKL (2007). *Study on liquid-phase-impact diffusion welding SiCp/ZL101*. Composites Science and Technology, Vol. 67, No. 6, pp. 1041-1046.

Guo W, Hua M, Ho JKL, Law HW (2008). *Mechanism and Influence of Pulse-Impact on Properties of Liquid Phase Pulse-Impact Diffusion Welded SiCp/A356*. The International Journal of Advanced Manufacturing Technology. DOI:10.1007/s00170-008-1411-y.

Guo W, Hua M, Law HW, Ho JKL (2008). *Liquid Phase Impact Diffusion Welding of SiCp/6061Al and Its Mechanism*. Materials Science and Engineering: A. Vol. 490, No. 1-2, pp. 427-437.

Hsu CJ, Kao PW & Ho NJ (2005). *Ultrafine-Grained Al–Al_2Cu Composite Produced In-Situ by Friction Stir Processing*. Scripta Materialia, Vol. 53, No. 3, pp. 341-345.

Loyd DJ (1994). *Particle Reinforced Aluminum Magnesium Composites*. International Materials Reviews, Vol. 39, No. 1, pp. 1-22.

Maity J, Pal TK, Maiti R (2009). *Transient liquid phase diffusion bonding of 6061-15 wt% SiCp in argon environment*. Journal of Materials Processing Technology, Vol. 209, No. 7, pp. 3568-3580.

Marzoli LM, Strombeck AV, Dos Santos JF, Gambaro C, Volpone LM (2006). *Friction Stir Welding of an AA6061/Al_2O_3/20p Reinforced Alloy*. Composites Science and Technology, Vol. 66, No. 2, pp. 363-371.

Nair SV, Tien JK & Bates RC (1995). *SiC-Reinforced Aluminium Metal Matrix Composites*. International Metals Reviews, Vol. 30, No. 6, pp. 275-288.

Pirondi A, Collini L (2009). *Analysis of crack propagation resistance of Al-Al_2O_3 particulate-reinforced composite friction stir welded butt joints*. International Journal of Fatigue, Vol. 31, No. 1, pp. 111-121.

Rotundo F, Ceschini L, Morri A, Jun TS & Korsunsky AM (2010). *Mechanical and microstructural characterization of 2124Al/25vol.%SiCp joints obtained by linear friction welding (LFW)*. Composite Part A: Applied Science and Manufacturing, Vol. 41, No. 9, pp. 1028-1037.

Schell JSU, Guilleminot J, Binetruy C, Krawczak P (2009). *Computational and experimental analysis of fusion bonding in thermoplastic composites: Influence of process parameters.* Journal of Materials Processing Technology, Vol. 209, No. 11, pp. 5211-5219.

Shanmuga Sundaram N, Murugan N (2010). *Tensile behavior of dissimilar friction stir welded joints of aluminum alloys*. Materials & Design, Vol. 31, No. 9, pp. 4184-4193.

Wert JA (2003). *Microstructures of Friction Stir Weld Joints between an Aluminium-Base Metal Matrix Composite and a Monolithic Aluminium Alloy*. Scripta Materialia, Vol. 49, No. 6, pp. 607-612.

10

The Cell Ultrastructure of Diatoms - Implications for Phylogeny?

Yekaterina D. Bedoshvili and Yelena V. Likhoshway
Limnological Institute of the Siberian Branch of the Russian Academy of Sciences, Russia

1. Introduction

Diatoms (BACILLARIOPHYTA) are unicell eucaryotes with species-specific silica external skeleton forming intracellularly. The main part of ultrastructural studies are devoted to siliceous valve morphogenesis (Reimann, 1964; Drum & Pankratz, 1964; Schmid & Schultz, 1979; Pickett-Heaps et al., 1990; Pickett-Heaps, 1998; Van de Meene & Pickett-Heaps, 2002) and involvement of cytoskeleton elements in the control of nuclear division and formation large structures on the silica frustule (Pickett-Heaps & Kowalski, 1981; Pickett-Heaps et al., 1988; Tippit & Pickett-Heaps, 1977; Tippit et al., 1980; Pickett-Heaps, 1983).

According to Round et al. (1990), BACILLARIOPHYTA consists of three classes: centric diatoms with radial and bipolar symmetry of frustules (class Coscinodiscophyceae Round & Crawford) and two classes of pennate diatoms differing in presence (Bacillariophyceae Haeckel) or absence of a raphe (Fragilariophyceae Round).

For the last decade systematics of diatoms based on the morphological data (Round et al., 1990) has been subject to revision (http:\\www.algaebase.org) due to the appearance of molecular phylogeny and understanding of its data (Medlin et al., 2000; Medlin, 2009).

Molecular-phylogenetic analysis revealed that BACILLARIOPHYTA are divided into three clades. Clade 1 includes centric diatoms with radial symmetry of frustules, clade 2a consists of centric diatoms with bi- and multipolar symmetry of frustules and the order Thalassiosirales Glezer & Makarova, and clade 2b includes all pennate diatoms (Medlin, 1997).

Despite the limited number of papers describing diatom cell structure and lack of all major taxa, researchers attempted to detect characteristic features of cell organelle structure for different taxonomic groups of diatoms several times. For example, Mann (1996) and Schmid (2001) considered pyrenoid structure to be as a source of taxonomic and phylogenetic information. They discovered that pyrenoids of diatoms of phylogenetic clade 2 were more diverse than pyrenoids of clade 1. Lately, data on thin structure of chloroplasts and pyrenoids were enlarged and character sets were specified for different diatom phylogenetic clades (Bedoshvili et al., 2009).

Medlin and Kaczmarska (2004) systematized data on arrangement of Golgi bodies in diatom cells and defined three types. Type 1 is typical of centric diatoms with radial symmetry (except the order Aulacoseirales Crawford). In this case, dictyosomes are associated with ER cistern and mitochondrion and form the so-called G-ER-M unit (*Coscinodiscus* Ehrenberg - Schmid,

1987; *Stephanopyxis* (Ehrenberg) Ehrenberg - Medlin et al., 2000; *Ellerbeckia* Crawford - Schmid & Crawford, 2001). Thalassiosirales and Aulacoseirales, most of pennates and bipolar centrics have Golgi arrangement of type 2. Dictyosomes encircle the nucleus and form a perinuclear shell around it (Medlin & Kaczmarska, 2004). *Stephanodiscus niagare* Ehrenberg is a pattern (Drum et al., 1966). Modifications of this type of Golgi arrangement were revealed in cells of *Synedra ulna* (Nitzsch) Ehrenberg (Schmid, 1989) and *Pinnularia nobilis* (Ehrenberg) Ehrenberg (Drum, 1966) where dictyosomes are located along the long nuclear extension towards the poles or periphery of cell. In the cell of *Pinnularia nobilis*, dictyosomes are paired. Type 3 was found in *Biddulphiopsis titiana* (Grunow) von Stosch & Simonsen (Coscinodiscophycidae Round *et* Crawford, clade 1). In this species only outer nuclear envelope forms filose tentacles along which Golgi bodies are paired (Medlin & Kaczmarska, 2004).

This work was aimed at studying cell ultrastructure of seven diatom species from different orders and phylogenetic clades. Ultrathin cell sections were analysed paying attention to a nucleus form, Golgi arrangement and it number, form and number of mitochondria. An attempt was made to reveal specific features of diatom cell ultrastructure of different taxa.

2. Materials and methods

2.1 Cultures

The cell culture of *Synedra acus* subsp. *radians* (Kützing) Skabitschevsky (Fragilariophyceae Round, Fragilariales Round, clade 2b) was isolated from Listvennichnyi Bay of Lake Baikal and was grown by T.A. Safonova on the medium DM1 (Thompson et al., 1988). *Cymbella ventricosa* Agardh (Bacillariophyceae Haeckel, Cymbellales Mann, clade 2b) was isolated from Irkutsk Reservoir and was grown by T.N. Basharina on the same medium. Cultures of *Thalassiosira proschkinae* Makarova (Coscinodiscophyceae, Thalassiosirales Glezer *et* Makarova, clade 2a), *Attheya ussurensis* Stonik, Orlova *et* Crawford (Coscinodiscophyceae, Chaetocerotales Round *et* Crawford, clade 2a), *Ditylum brightwellii* (West) Grunow (Coscinodiscophyceae, clade 2a), and *Chaetoceros muelleri* Lemmermann (Coscinodiscophyceae, Chaetocerotales Round *et* Crawford, clade 2a) were kindly provided by N.A. Aizdaicher from A.V. Zhirmunsky Institute of Marine Biology FEB RAS (Vladivostok). Sampling of *Aulacoseira baicalensis* (K. Meyer) Simonsen (Coscinodiscophyceae, Aulacoseirales Crawford, clade 1) was carried out in Listvennichnyi Bay of Lake Baikal during its vegetation in spring.

2.2 Transmission electron microscopy

For transmission microscopy cells in their logarithmic phase were fixed by 2,5 % glutar aldehyde (Sigma-Aldrich, Germany) on PBS (pH 7.0) for 12 h and then by 1% osmium tetroxide (Merck KGaA, Germany) on the same buffer (2 h). Then the cells were placed in 1 % agarose (Helicon, Moskow, Russia). Dehydration was made in ethanol solution series (Reahimkomplekt, Angarsk, Russia) and then in ethanol and acetone dehydrated with copper sulphate (Reahimkomplekt, Angarsk, Russia) for 5 min in each solution. Dehydrated material was soaked in three mixtures with resin Araldite 502 Kit (SPI, USA) and acetone and in pure resin for 12 h. After this, the material was placed in resin with catalyst DMP-30 (SPI, USA) and polymerised in a thermostat at 60°C (for 3 days). Ultrathin sections were cut with a diamond knife ULTRA 35° (Diatom, Switzerland) on the ultramicrotome Ultracut R (Leica, Austria). Sections were stained with Reinolds's lead citrate (Reynolds, 1963). Samples were analysed under a transmission electron microscope Leo 906 E (Zeiss, Germany) at accelerating voltage 80 kV. Microphotographs were made by CCD camera MegaView II (Zeiss, Germany) and processed with the MegaVision programme.

3. Results

All organelles in the cell of *Thalassiosira proschkinae* are tightly pressed to one another. Large nucleus is located in the cell centre between the lobes of the only chloroplast (fig. 1A). Poorly-expressed nucleolus is displaced on the nucleus periphery and pressed to the nucleus envelope (fig. 1B). Nuclear pores are turned to the cytoplasm free of chloroplast (fig. 1B). Golgi arrangement is perinuclear (fig. 1C). Large mitochondria (one or two per cell) are of diverse shape on the sections: ring-shaped (fig. 1D), complicated (fig. 1E) or oblong (fig. 1F).

A big vacuole occupies the main volume of *Aulacoseira baicalensis* cell and the nucleus and cell organelles are pressed to the cell wall (fig. 2A). There is a well-defined nucleolus in the centre of the nucleus (fig. 2B); Golgi bodies (it can be seen no more than 4; fig. 2B) are situated on the side of the nucleus directed to the cell periphery. Nuclear pores are directed to Golgi bodies (fig. 2C). There are vesicles budding from the outer nuclear membrane (fig. 2C). The analysis of serial sections of mitochondria revealed that they have either an intricate branchy form (fig. 2D) or a length up to 4.5 µm (fig. 2E). Mitochondria are frequently situated near the forming silica valves (fig. 2F) or lipid drops tightly pressed to them (fig. 2G).

The nucleus of *Chaetoceros muelleri* is shifted to girdle bands and the Golgi bodies are pressed to nuclear membrane (fig. 3A, B). Nuclear pores are directed to either Golgi bodies (fig. 3A) or cell periphery (fig. 3C). Nucleolus is poorly expressed pressing to the nuclear envelope (fig. 3C). The analysis of serial sections showed that the only mitochondrion in the cell can be of diverse shape: complicated (fig. 3D), elongate, sometimes branchy or roundish (fig. 3E). Invaginations were frequently observed (fig. 3F).

A nucleus with a well-defined nucleolus is seen in the cell centre of *Attheya ussurensis* (fig. 4A). Nuclear pores are directed to the Golgi bodies on most of the sections (fig. 4B). Up to four Golgi bodies are located near the nucleus (fig. 4C). A great number of small vesicles around Golgi bodies attest to their activity (fig. 4D). Mitochondria on some sections (fig. 4E) have a commissure of outer and inner membrane looking like a pore.

The nucleus of *Synedra acus* subsp. *radians* of a complicated form with lobes and invaginations (fig. 5A) is located diagonally, ER tanks being inside the invaginations (fig. 5B). There is a roundish or oval nucleolus in the nucleus centre (fig. 5B, C). Multiple Golgi vesicles are often observed budding from the outer nuclear membrane (fig. 5D). Six Golgi bodies are located along the nucleus; nuclear pores turn to them (fig. 5E). Multiple mitochondria of up to 2.7 µm long and to 0.4 µm wide are on the cell periphery (fig. 5F).

The nucleus of *Cymbella ventricosa* is located in the wide part of the cell and pressed to girdle bands (fig. 6A). There is a nucleolus in the nucleus centre (fig. 6B). Up to five Golgi bodies are around the nucleus (fig. 6A, C). A great number of vesicles and ER tanks are detected in the cytoplasm (fig. 6D). Long and narrow mitochondria reach 5 µm on the cell periphery (fig. 6E).

Ditylum brightwellii has a roundish nucleus in the cell centre (fig. 7A). There is a nucleolus in the nucleus centre (fig. 7B), multiple nuclear pores are aligned regularly on the nuclear envelope (fig. 7C). Multiple Golgi bodies are near the nucleus (fig. 7C). The G-ER-M units are also observed (fig. 7D).

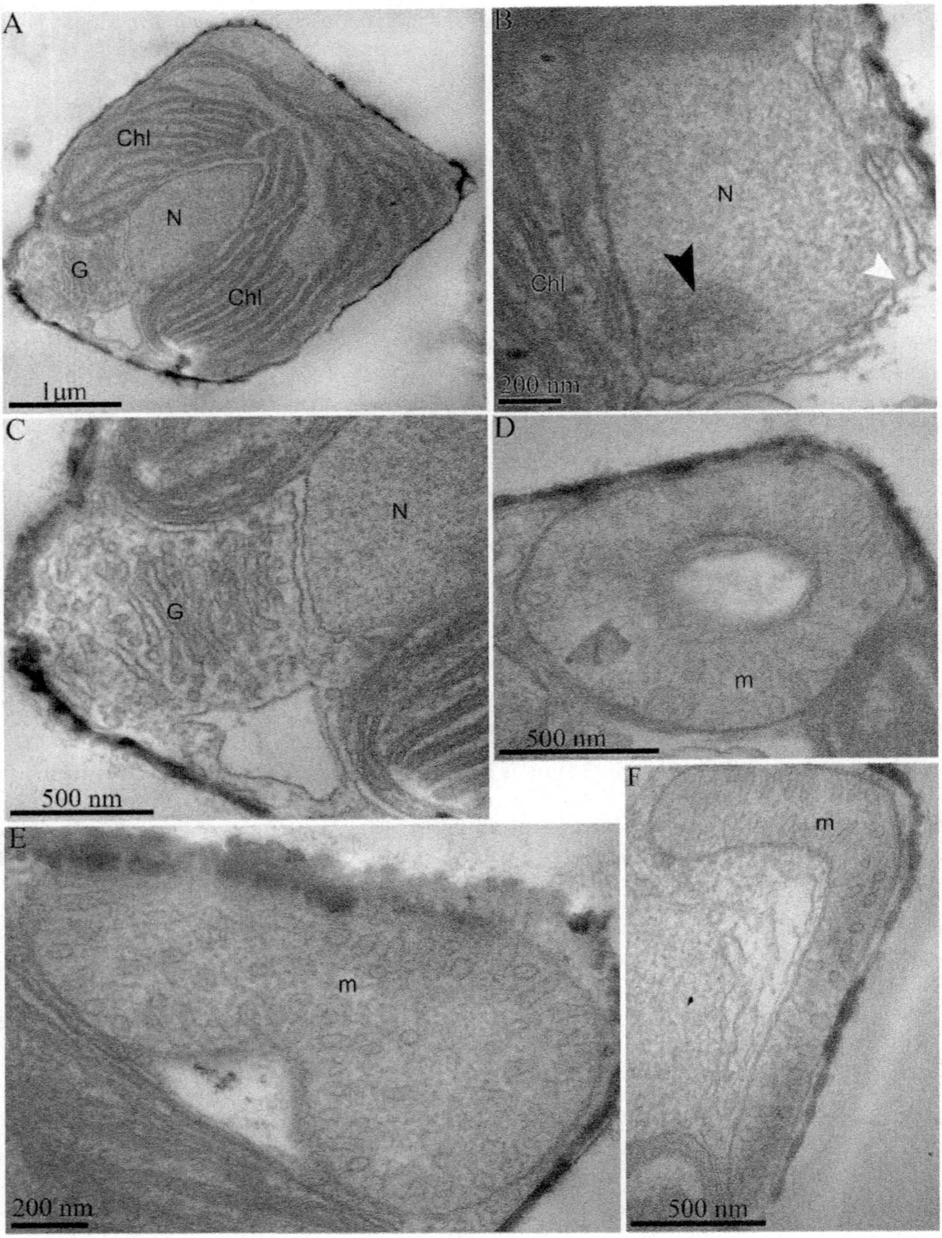

Fig. 1. The cell structure of *Thalassiosira proschkinae*. TEM. A - cell section; B - nucleus section, the nucleus is indicated by black arrowhead, nuclear pore - white arrowhead; C - Golgi bodies section; D-F - sections of mitochondria of different shape. Legend: Chl - chloroplast; G - Golgi bodies; m - mitochondrion; N - nucleus.

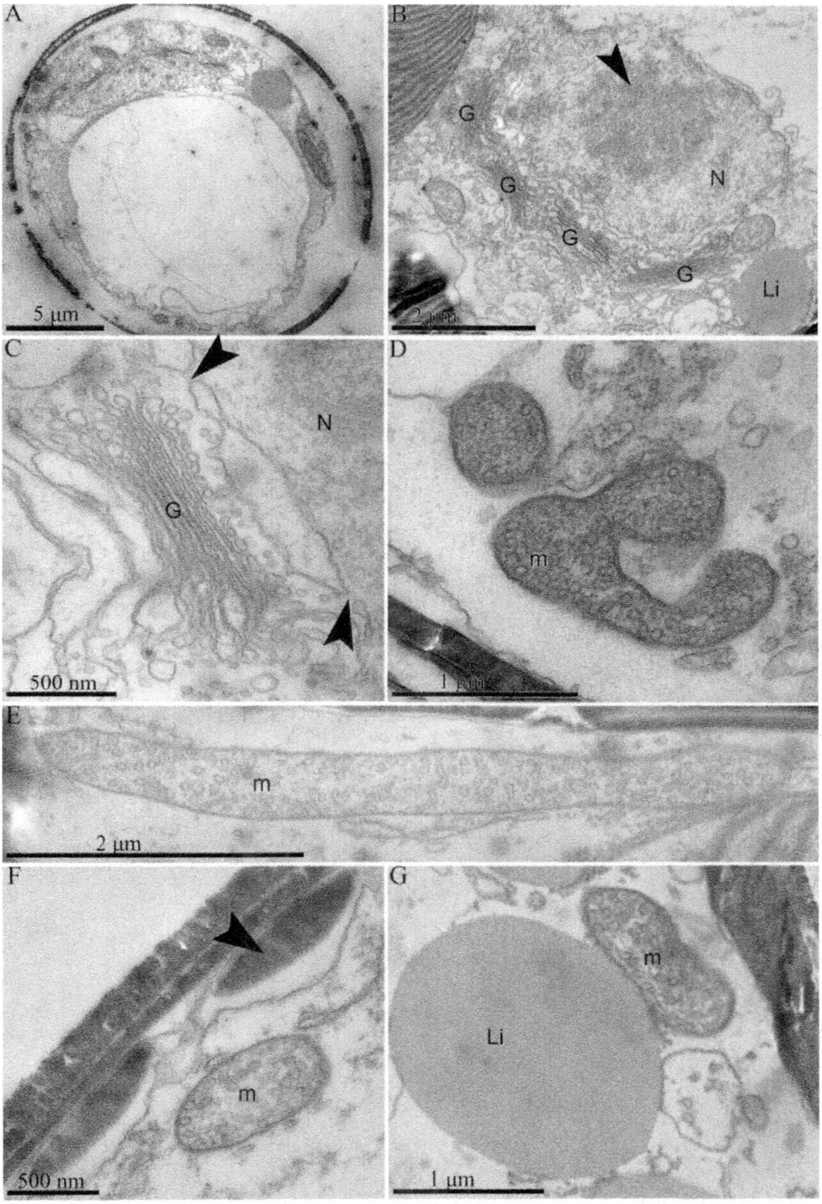

Fig. 2. The cell structure of *Aulacoseira baicalensis*. TEM. A - cell cross section; B - nucleus and Golgi bodies section; C -Golgi bodies section; D-G - mitochondria sections; D - mitochondrion longitudinal section; F - section of the mitochondrion near the forming valve (indicated by the black arrowhead); G - section of the mitochondrion near the lipid drop. Legend: G - Golgi bodies; Li -lipid drop; m - mitochondrion; N - nucleus.

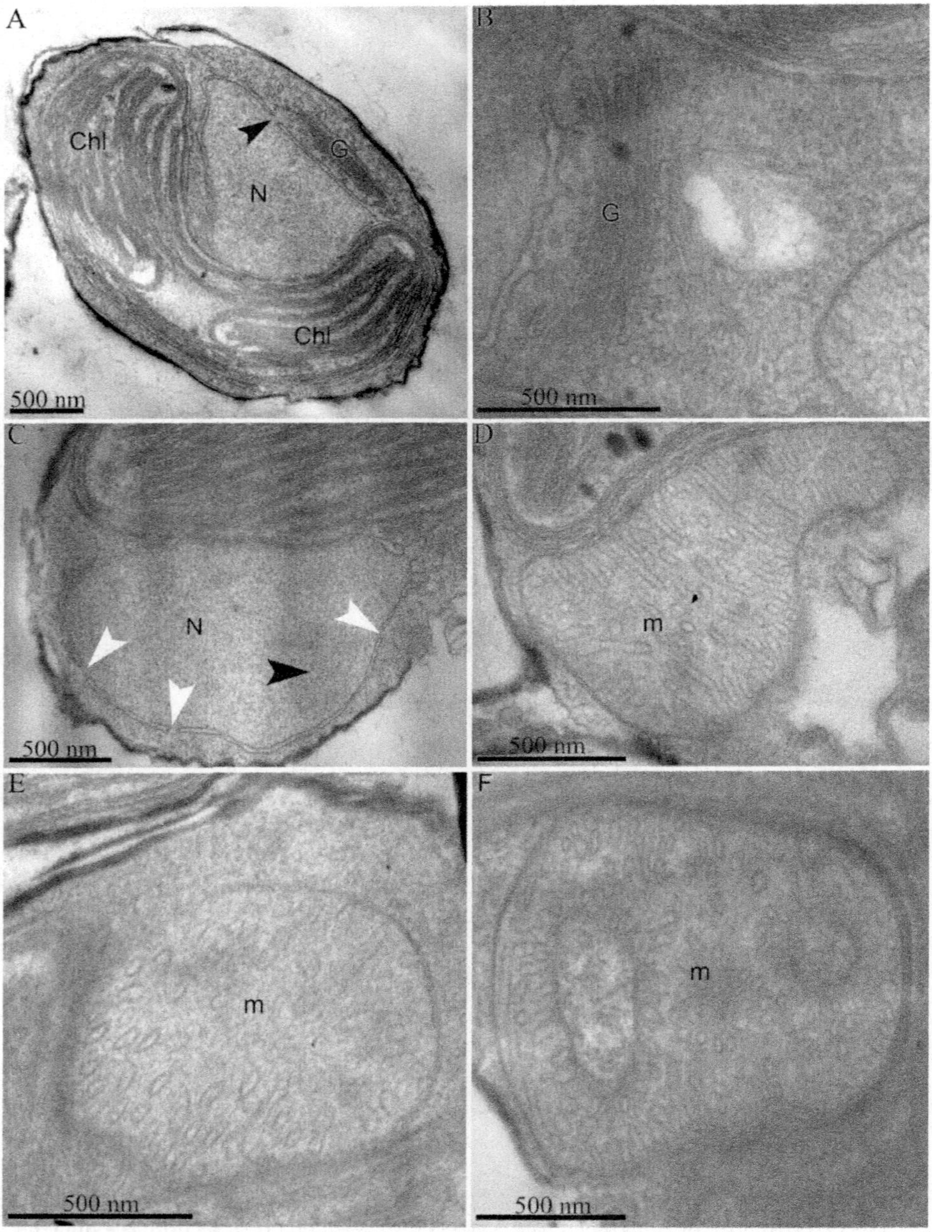

Fig. 3. The cell structure of *Chaetoceros muelleri*. TEM. A – cell section (the nuclear pore is showed by the arrowhead); B –Golgi bodies section; C – nucleus section (black arrowhead – the nucleolus; white arrowhead – the nuclear pores); D-F –sections of mitochondria of different shape. Legend: Chl – chloroplast; G – Golgi bodies; m – mitochondrion; N – nucleus.

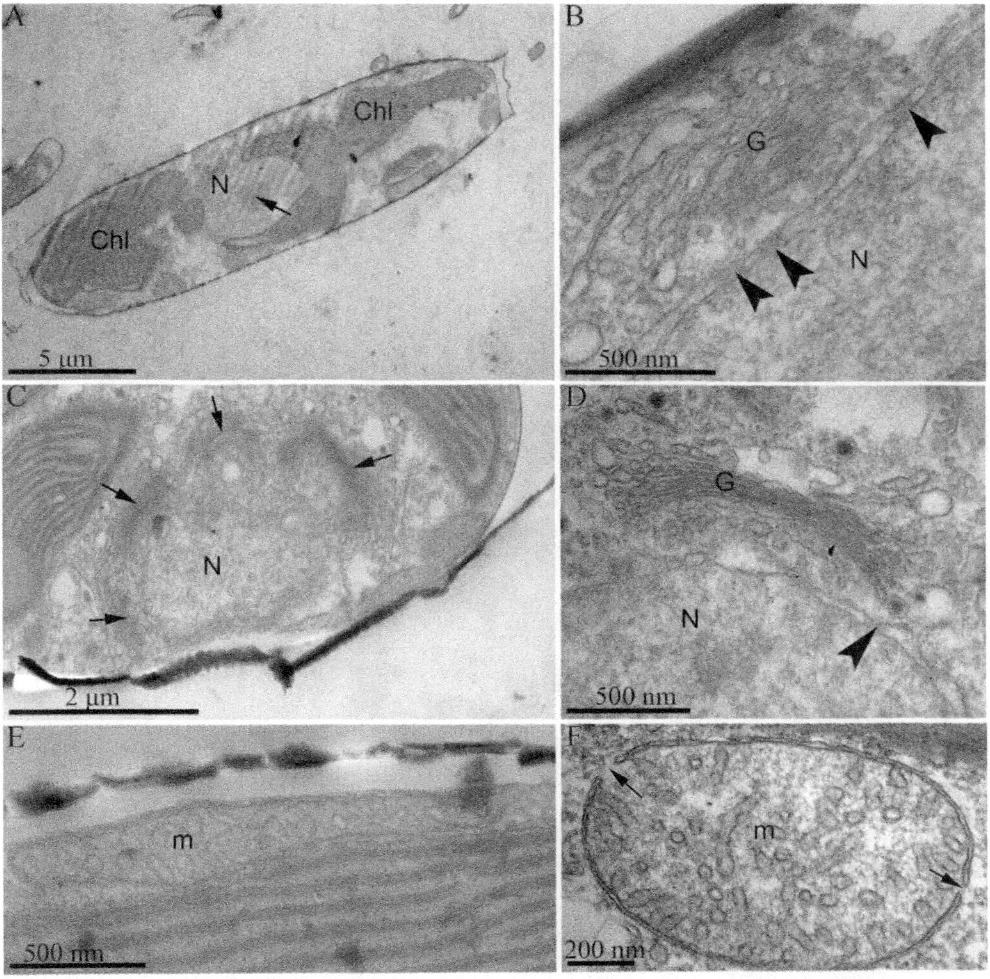

Fig. 4. The cell structure of *Attheya ussurensis*. TEM. A - cell cross section (the nucleolus is indicated by arrow); B -nucleus section (the arrows show the nuclear pores); C - nucleus and Golgi bodies section (arrows show the Golgi bodies); D - Golgi bodies section (black arrowheads - the nuclear pores); E - mitochondrion longitudinal section; F - mitochondrion cross section (arrows indicate the comissures of outer and inner mitochondrion membranes). Legend: Chl - chloroplast; G - Golgi bodies; m - mitochondrion; N - nucleus.

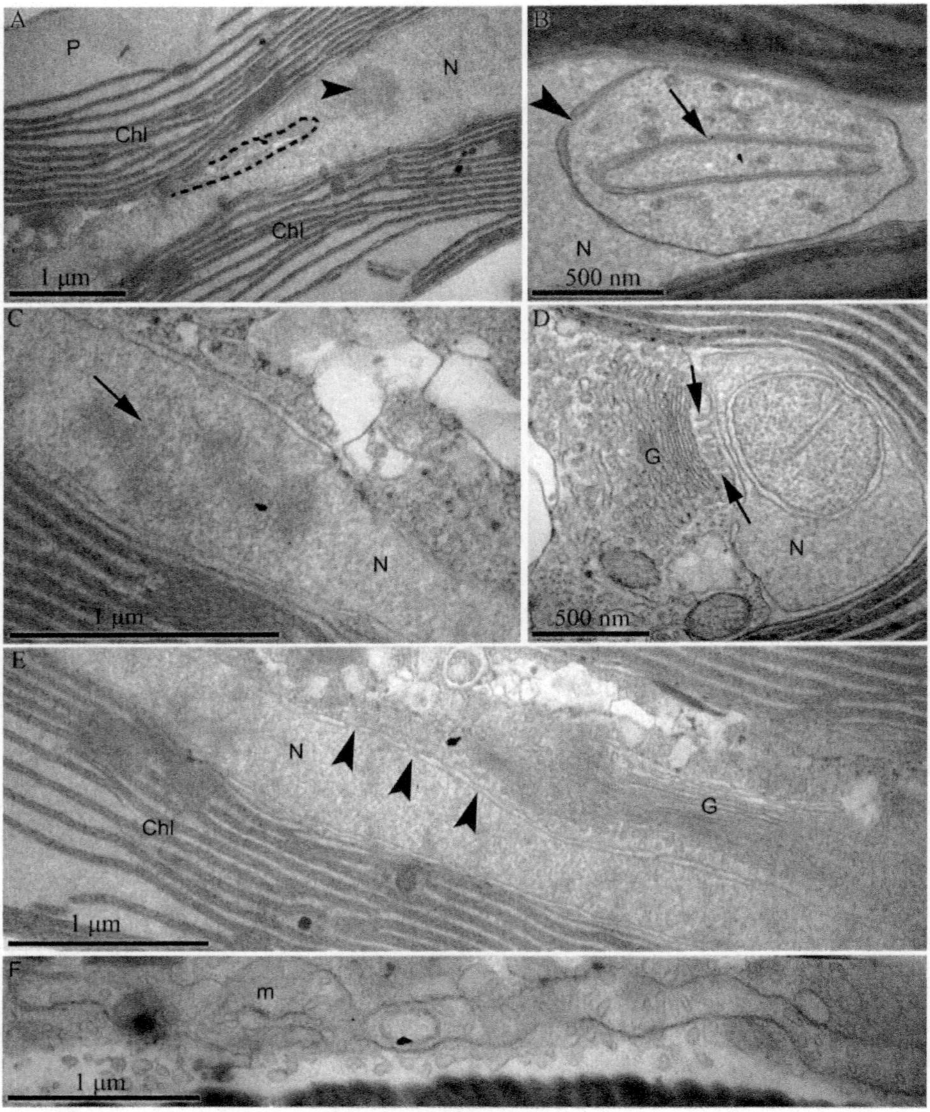

Fig. 5. The cell structure of *Synedra acus* subsp. *radians*. TEM. A – longitudinal section of the nucleus with the invagination (dotted line), the nucleolus is shown by the arrowhead; B – cross section of the nucleus invagination with the smooth ER tank inside (arrow); the arrowhead indicates the nuclear pore; C – the nucleus longitudinal section (the nucleolus is shown by the arrow); D – cross section of nucleus and Golgi bodies, the vesicles budded from the nuclear outer membrane are indicated by the arrows; E – nucleus and Golgi bodies longitudinal section, the arrowheads indicate the nuclear pores; F – mitochondrion longitudinal section. Legend: Chl – chloroplast; G – Golgi bodies; m – mitochondrion; N – nucleus.

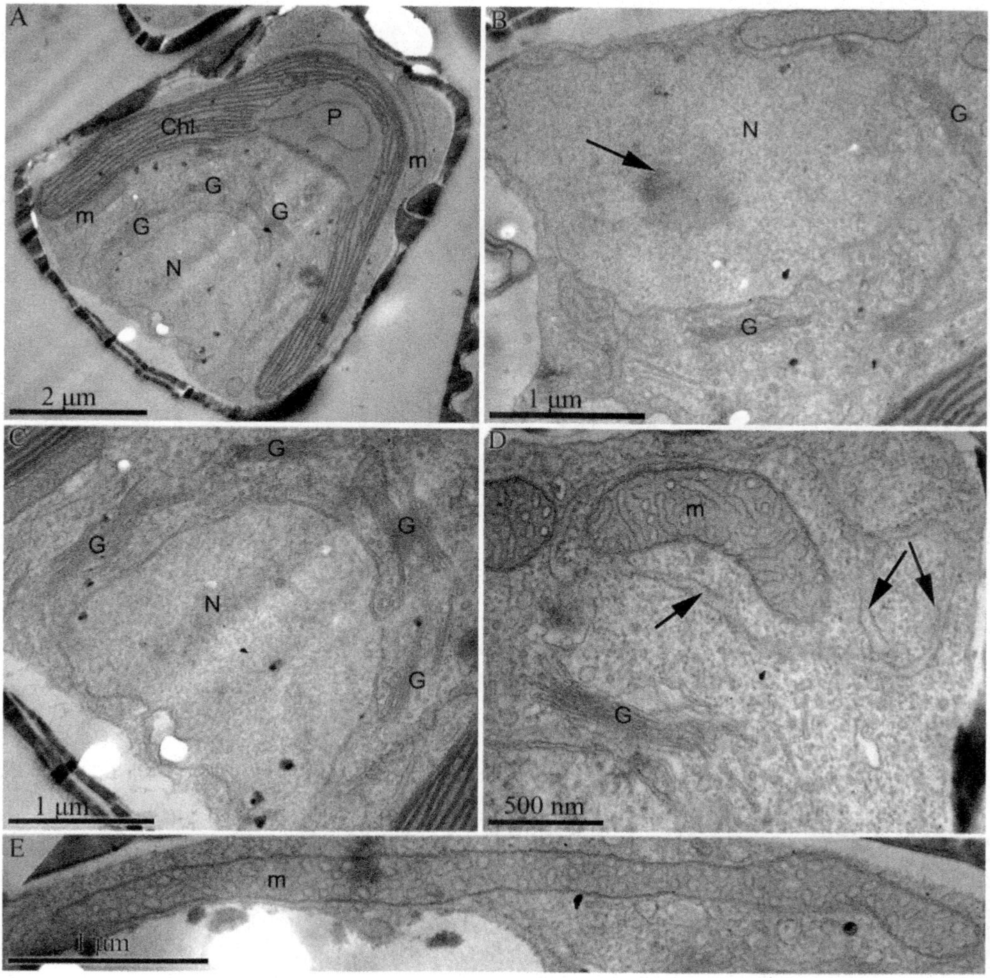

Fig. 6. The cell structure of *Cymbella ventricosa*. TEM. A - cell cross section; B - nucleus cross section (the nucleolus is shown by the arrow); C - cross section of nucleus and Golgi bodies; D - cytoplasm section (the ER tanks are shown by the arrows); E - mitochondrion longitudinal section. Legend: Chl - chloroplast; G - Golgi bodies; m - mitochondrion; N - nucleus; P - pyrenoid.

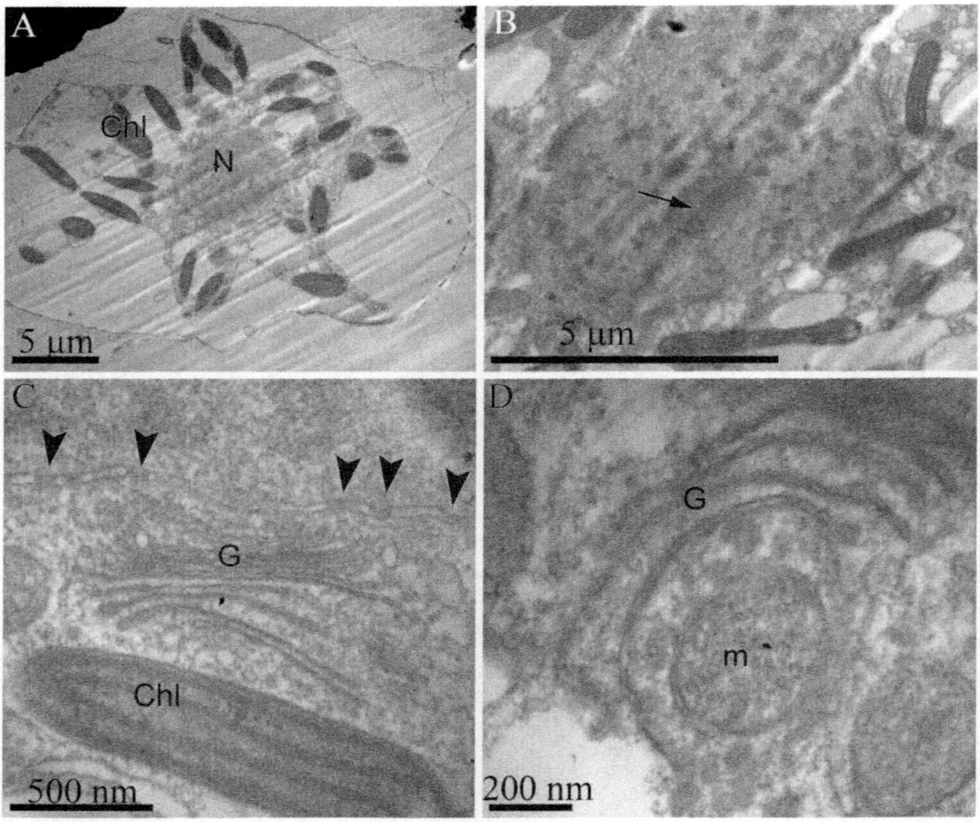

Fig. 7. The cell structure of *Ditylum brightwellii*. TEM. A - cell cross section; B - nucleus cross section (the nucleolus is indicated by the arrow); C - cross section of nucleus and Golgi bodies (the nuclear pores are shown by the arrowheads); D - cross section of G-ER-M unit. Legend: Chl - chloroplast; G - Golgi bodies; m - mitochondrion; N - nucleus.

4. Discussion

4.1 The Golgi bodies number and their arrangement relative to the nucleus

Among the examined species the relationship between the cell size and the number of Golgi bodies is observed. The only Golgi complex is recorded in small cells of *C. muelleri* and *T. proschkinae* from different orders and phylogenetic clades. In this study, *A. baicalensis* and *S. acus* subsp. *radians* have the largest cell size (140 and 290 µm long, respectively). Six Golgi complexes are located in the cells of *S. acus* subsp. *radians*. *Synedra ulna*, a larger cell species of the same genus, possesses a higher number of Golgi complexes (7) located along the filamentous extensions of the nucleus (Schmid, 1989, fig. 17). There are four Golgi bodies in *A. baicalensis* as well as in small-celled (up to 6 µm) centric *A. ussurensis* and pennate *C. ventricosa*. It is attributed to small cytoplasm volume in the *A. baicalensis* cells and to a large vacuole occupying the major part of the cell.

Golgi bodies of all examined species are located perinuclearly, with some exception. For example, we found two types of Golgi arrangement (G-ER-M unit and perinuclear dictyosomes) in the same cell of *D. brightwellii* (a member of clade 2a; fig. 7).

Golgi complexes in *S. acus.* subsp. *radians* are located along the nucleus according to type 2 contrary to *S. ulna* whose Golgi arrangement is referred to modification of this type (Medlin & Kacmarska 2004). The nucleus shape differs in these two species: long tubular extension is absent in the nucleus of *S. acus.* subsp. *radians,* whereas in *S. ulna* it is present. Our data support the conclusion of Medlin and Kacmarska (2004) - this is the same perinuclear arrangement of Golgi bodies (type 2) as well as type 3 and it is not worth marking out individual modifications in it.

Golgi bodies in *A. baicalensis* are located only on one side of the nucleus which turns to the cell periphery. The same arrangement of Golgi bodies is described for *Melosira varians* Agardh (Crawford, 1973) from the closely-related order of the same class Coscinodiscophyceae.

4.2 The nucleus shape

According to the literature data, most of diatoms have a pillow-like or spherical nucleus. However, there are some exceptions. For example, the nucleus of *Lauderia annulata* Cleve consists of two lobes situated on different valves and connected in the cell centre with a thin filament (Holmes, 1977). The nucleus in some species of *Surirella* Turpin can be of H-like shape (Drum & Pankratz, 1964). We revealed that the nucleus shape can differ in cells of species of the same genus. Both *S. ulna* and *S. acus* subsp. *radians* have an oblong nucleus but the latter can form invaginations instead of long extensions. Such nucleus shape increases its surface providing better contact with other cell organelles through the nuclear pores. All examined species have polarization of the nuclear pores which turn to Golgi bodies (figs 1C, 2C, 3A, 4B, D, 5E, 6C). This polarization is clearly expresses in *A. baicalensis* (fig. 2C). The increase of the nucleus surface and the polarization of nuclear pores make it possible to accelerate metabolism of nucleus and cytoplasm. They are not characters of any taxon or phylogenetic clade but reflect physiological state of the cell.

4.3 The number and structure of mitochondrion

Numerous studies of eukaryotes revealed that mitochondrion morphology is complicated and plastic (Hoffmann & Avers, 1973; Bereiter-Hahn & Voth, 1994; Nunnari et al., 1997; Boldogh et al., 2001). According to the literature data, long mitochondria with tubular cristae localized in the peripheral cytoplasm layer are characteristic of diatoms (Round et al., 1990, p. 60). Comparative studies of mitochondrion structure have not been performed before.

The shape of diatom mitochondrion appeared to be extremely changeable even in the cells of the same species. Small-celled *C. muelleri* and *T. proschkinae,* belonging to clade 2a, possess one large mitochondrion of various shapes in different cells (figs 1D-F, 3D-F). Lager-celled *A. ussurensis* from the same clade have many long mitochondria. The longest mitochondria (up to 5 μm) are detected in the cells (approximately 6 μm) of the pennate *C. ventricosa.* Multiple long and often branched mitochondria (figs 2 D-E, 5F) are recorded in the cells of two species from different clades having the largest cells - *A. baicalensis* (clade 2a) and *S.*

acus subsp. *radians* (clade 2b). Projections, invaginations and intricate branched shape of mitochondria are similar to complex mitochondrial networks which were detected in other eukaryotes, particularly in yeast (Rizzuto et al., 1998; Mannella, 2000; Rutter & Rizzuto, 2000).

4.4 ER and vesicles

ER is poorly developed in diatoms like in other heterokont algae (Gibbs, 1981; Round et al., 1990; Van den Hoek et al., 1997) (fig. 6D). It is of an intricate shape when it is located near the nucleus (fig. 5B, D). The outer nuclear membrane performs the functions of ER in synthesis and substance transport (fig. 5D). We believe that vesicles near the outer nuclear membrane are budded from it (instead of coalescence). Synthesized nutrients in the vesicles and transported to the Golgi complex (fig. 5D). Perinuclear location of Golgi bodies promotes rapid transport of vesicles and is observed in all cases especially in the cells with oblong nuclei (fig. 5E).

5. Conclusion

Based on the results of analysis of some diatom species, we tried to apply specific features of cell ultrastructure to systematics of diatoms. Golgi arrangement relative to the nucleus is consistent with characters of phylogenetic clades proposed by Medlin & Kacmarska (2004), i.e., perinuclear Golgi arrangement (type 2) is characteristic of the members of phylogenetic clades 2a and 2b, except *D. brightwellii* with G-ER-M units present in the cell. Both *Aulacoseira* and *Melosira* have perinuclear Golgi arrangement.

Other characteristics of cell ultrastructure are not associated with the systematic position of species. The number of Golgi bodies in species from the same clade and order can differ and depends on the cell size and/or cytoplasm volume. The number and size of mitochondria can be similar in species from different orders and also depend on cell size, whereas the shape of mitochondria can differ even in representatives of the same species.

Thus, the application of cell ultrastructure characteristics in phylogenetic reconstructions is limited except pyrenoids: their structure defines certain taxa and phylogenetic clades (Schmid, 2001; Bedoshvili et al., 2009).

6. Acknowledgements

The work was supported by the RAS Presidium Programme "Molecular and Cell Biology", Grant 6.3.

7. References

Bedoshvili, Ye.; Popkova, T. & Likhoshway, Ye. (2009). Chloroplast structure of diatoms of different classes. *Cell and Tissue Biology*, Vol. 3, № 3, pp. 297-310, ISSN *1990-519X*

Bereiter-Hahn, J., & Voth, M. (1994). Dynamics of mitochondria in living cells: shape changes, dislocations, fusion, and fission of mitochondria. *Microsc. Res. Tech.*, Vol. 27, № 3, pp. 198-219

Boldogh, I.; Yang, H. & Pon, L. (2001). Mitochondrial inheritance in budding yeast. *Traffic*, Vol. 2, № 6, pp. 368-374, ISSN 1398-9219

Crawford, R. (1973). The protoplasmic ultrastructure of the vegetative cell of *Melosira varians* C.A. Agardh. *J. Phycol.*, Vol. 9, № 1, pp. 50-61

Drum, R. (1966). Electron microscopy of paired Golgi structure in the diatom *Pinnularia nobilis*. *Journal of Ultrastructure Research*, Vol. 15, pp. 100-107

Drum, R.W. & Pankratz S. (1964). Pyrenoids, raphes, and other fine structure in diatoms. *Am. J. Bot.* Vol. 51, pp. 401-418

Drum, R.W.; Pankratz, S. & Stoermer E. *Electronmicroscopy of diatom cells. Diatomeenschalen im electronenmikroschen* Bild. Lehre, J. Cramer Verlag, 1966, Plates 514-613

Hoffmann, H. & Avers, C. (1973). Mitochondrion of yeast: ultrastructural evidence for one giant, branched organelle per cell. *Science*, Vol. 181 (24 August 1973), pp. 749-751

Holmes, R.W. (1977). *Lauderia annulata* – a marine centric diatom with an elongate bilobed nucleus. *J. Phycol.*, Vol. 13, pp. 180-183

Mann, D. (1996). Chloroplast morphology, movements and inheritance in diatoms. In: *Cytology, genetics and molecular biology of algae*, B.R. Chaundhary & S.B. Agrawal, (Eds), 249–274, SPB Academic Publishing, Amsterdam

Mannella, C. (2000). Our changing views of mitochondria. *J. Bioenerg. Biomembr.*, Vol. 32, № 1, pp. 1-4

Medlin, L. & Kaczmarska, I. (2004). Evolution of the diatoms: V. Morphological and cytological support of the major clades and taxonomic revision. *Phicologia*, Vol. 43, № 3, pp. 245-273

Medlin, L.; Kooistra, W. & Schmid, A.-M. (2000). A review of the evolution of the diatoms – a total approach using molecules, morphology and geology. In: *The origin and early evolution of the diatoms: fossil, molecular and biogeographical approaches*, A. Witkowski & W. Sieminska (Eds.), 13-35, Szafer Institute of Botany, Polish Academy Sciences, Crakow, Poland

Nunnari, J.; Marshall, W.; Straight, A.; Murray, A.; Sedat, J. & Walter, P. (1997). Mitochondrial transmission during mating in *Saccharomyces cerevisiae* is determined by mitochondrial fusion and fission and the intramitochondrial segregation of mitochondrial DNA. *Mol. Biol. Cell.*, Vol. 8, № 7, pp. 1233-1242

Pickett-Heaps, J. & Kowalski, S.E. (1981). Valve morphogenesis and the microtubule center of the diatom *Hantzschia amphioxis*. *Eur. J. Cell Biol.*, Vol. 25, pp. 150-170

Pickett-Heaps, J. (1998). Cell division and morphogenesis of the centric diatom *Chaetoceros decipiens* (Bacillariophiceae) II. Electron microscopy and a new paradigm for tip growth. *J. Phycol.*, Vol. 34, pp. 995-1004

Pickett-Heaps, J.; Cohn, S.; Schmid, A.-M. & Tippit, D. (1988). Valve morphogenesis in *Surirella* (Bacillariophiceae). *J. Phycol.*, Vol. 24, pp. 35-49

Pickett-Heaps, J. (1983). Valve morphogenesis and the microtubule center in tree species of the diatom *Nitzschia*. *J. Phycol.*, Vol. 19, pp. 269-281

Pickett-Heaps, J.D.; Schmid, A.-M. & Edgar, L.A. (1990). *The cell biology of diatom valve formation. Biopress, Bristol.*

Reimann, B. (1964). Deposition of silica inside a diatom cell. *Exp. Cell Res.*, Vol. 34, pp. 605-608

Reynolds, E. (1963). The use of lead citrate at high pH as an electronopaque stain in electron microscopy. *J. Cell Biol.*, Vol. 17, № 1, pp. 208-212

Rizzuto, R.; Pinton, P.; Carrington, W.; Fay, F.; Kevin, E.; Lifshitz, L.; Tuft & Pozzan, T. (1998). Close contacts with the endoplasmic reticulum as determinants of mitochondrial Ca21 responses. *Science*, Vol. 280, № 5370, pp. 1763-1766, ISSN 0036-8075

Round, F.; Crawford, R. & Mann, D. (1990). *The diatoms*, Cambridge University Press, Bristol

Rutter, G. & Rizzuto, R. (2000). Regulation of mitochondrial metabolism by ER Ca21 release: an intimate connection. *Trends Biochem. Sci.*, Vol. 25, № 5, pp. 215-221, ISSN 0968-0004

Schmid, A.-M. & Schultz, D. (1979). Wall morphogenesis in diatoms: deposition of silica by cytoplasmic vesicles. *Protoplasma*, Vol. 100, pp. 267-288

Schmid, A.-M. & Crawford, R. (2001). *Ellerbeckia arenaria* (Bacillariophyceae): formation of auxospores and initial cells. *Eur. J. Phycol.*, Vol. 36, pp. 307-320

Schmid, A.-M. (1989). Geitler's Plattenband in *Synedra ulna* in the light of TEM investigations. *Pl. Syst. Evol.*, Vol. 164, pp. 239-252

Schmid, A.-M. (2001). Value of pyrenoids in the systematics of the diatoms: their morphology and ultrastructure, *Proceedings of the 16th International Diatom Symposium*, Athens, August, 25 - September, 1, 2000

Schmid, A.-M. (1987). Morphogenetic forces in diatom cell wall formation, In: *Cytomechanics*, J. Bereiter-Hahn, O. Anderson & W. Reif (Eds), 1983-199, Springer, Berlin

Thompson, A., Rhodes J., Pettman, I. (Eds). (1988). *Culture collection of algae and protozoa: Catalogue of strains*, Titus: Wilson and Son, Kendal

Tippit, D. & Pickett-Heaps, J.D. (1977). Mitosis in the pennate diatom *Surirella ovalis*. *J. Cell Biol.*, Vol. 73, pp. 705-727

Tippit, D., Pickett-Heaps, J.D. & Leslie R. (1980). Cell division in two large pennate diatoms *Hantzschia* and *Nitzschia*. III. A new proposal for kinetochore function during prometaphase. *J. Cell Biol.*, Vol. 86, pp. 402-416

Van de Meene, A.M.L. & Pickett-Heaps, J.D. (2002). Valve morphogenesis in the centric diatom *Proboscia alata* Sundstrom. *J. Phycol.*, Vol. 38, pp. 351-363

Medlin, L. (1997). Evolution of the diatoms - a total approach using morphology, molecules and geology. Report of the Workshop June 1-2, 1997, *Diatom Res.*, pp. 371-379

Medlin, L. (2009). Diatoms (Bacillariophyta) In: *The timetree of life*, S.B. Hedges & S. Kumar, (Eds), 127-130, Oxford University Press

http:\\www.algaebase.org

11

TEM Investigations of Wear Mechanisms of Single and Multilayer Coatings

Łukasz Major[1], Jürgen M. Lackner[2] and Jerzy Morgiel[1]
[1]*Institute of Metallurgy and Materials Science Polish Academy of Sciences,*
[2]*Joanneum Research Forschungsges. m. b. H.,*
Institute for Surface Technologies and Photonics,
[1]*Poland*
[2]*Austria*

1. Introduction

Wear resistant ceramic coatings deposited by physical vapour deposition (PVD) techniques, like magnetron sputtering, ion beam assisted deposition, arc evaporation and pulsed laser deposition (PLD), are successfully used to protect surfaces of mechanical components working under high wear loads for nearly half century.

The titanium nitride (TiN) is of special interest due to its corrosion resistance and high hardness (R. F. Bunshah, 2001, D. S. Rickerby & A. Matthews, 1991). The other promising material for wear resistant applications is amorphous, hydrogenated carbon (a-C:H). The a-C:H coatings are characterised by very low friction and biological inertness (V. Kumar et al., 2011). The tribology- related engineering applications for highly- stressed components require the development of new multifunctional thin films materials providing superior mechanical, tribological, chemical and high- temperature performance. It could be achieved by connecting the properties of different type of materials in multilayer coatings (Li Chen et al., 2008, M. Stueber et al., 2009, E. Martinez et al., 2003, J. Smolik et al., 1999, Y. L. Su et al., 1998, N. Dück et al., 2001, M. Nordin et al., 1999).

The multilayer properties could be controlled by appropriate chemical composition, microstructure and thickness of individual layers. In many tribological applications, coatings with a stack of soft and hard layers can offer much improved mechanical properties compared to single- layered hard coatings (Ł. Major et al., 2010, Ł. Major & J. Morgiel, 2009, M. Bromark et al., 1997, A. Leyland & A. Matthews, 1994, A. Matthews & S. S. Eskilden, 1994, K. Holmberg et al., 1998). The multilayered titanium/titanium nitride (Ti/TiN) coatings display much improved fracture resistance as compared to single- layered TiN. The presence of plastically deformable metallic layers play an important role in cancelling cracks propagation mechanisms.

The coatings for medical application and especially those remaining in contact with human body fluid should both pre- sent high mechanical properties and contain as limited metallic material as possible. Otherwise such coatings may start the metalosis (the metal ions adverse interaction with human organism). The carbon, being part of most tissue of plants and animals is definitely the best material for such coatings. Carbon provides the framework for all tissues of plants and animals. These tissues are built of elements grouped around in chains and rings made of carbon atoms (J. Grabczyk et al., 2007). This is why in bio- application a development of multilayered coatings with carbon spacers are also of great importance.

The aim of the present paper was to describe the microstructure changes of single (ceramic) layered coating, multilayer ceramic/metallic (with different ratio) coating, and multilayer ceramic/ceramic coating after static mechanical tests by application of transmission electron microscopy techniques. The coatings, were subjected to static mechanical test because it simulates the uploading in their real application (as a coatings for pumps elements which support artificial heart chambers).

2. Experimental

Thin and ultra- thin coatings are preferably produced by deposition from the gaseous phase (PVD-Physical Vapour Deposition and CVD- Chemical Vapour Deposition). PVD coating requires as the first step the vaporization of the basic material- the target. CVD coating starts from mixtures of reactive gases. In contrast to CVD techniques, the PVD processes are generally characterized by lower coating temperatures, a wide range of possible coating and substrate materials, and the higher purity of the deposited coatings. Besides an optimized microstructure, the adhesion of a coating on a substrate is the most crucial property for a coating defining its durability in industrial applications. One of the techniques combining classical evaporation with plasma activation is the Pulsed Laser Deposition (PLD) technique. The target evaporation (ablation) and vapor activation take place by means of a high-energetic focused pulsed laser beam. The high- energetic vapour (plasma) flux allows deposition of highly adhesive coatings even at room temperature. In addition the industrially- scaled room- temperature PLD deposition process was linked with other vacuum coating technique- magnetron sputtering and ion- assisted deposition to combine the advantages of these techniques.

The hybrid PLD with titanium and carbon targets (99, 9at. %) was used for coatings deposition also in presented work. Single layered TiN coatings were deposited using Ti target in nitrogen atmosphere. Multilayer Ti/TiN coatings were deposited using Ti target and sequential gas atmosphere change in between argon for Ti phase deposition to nitrogen for TiN phase deposition. Multilayer TiN/Ti/a-C:H coatings were deposited using both sequential atmosphere change (nitrogen for TiN and argon for a-C:H) and sequential target change (Ti for TiN and C for a-C:H). The details of deposition process is described elsewhere (J. M. Lackner, 2006).

Analyzed coatings were deformed by pushing a diamond indenter with 20μm tip radius under 1N load into it. The test simulates the mechanical interaction by which elements are uploaded in their real application (as a coatings for pumps elements which support artificial heart chambers). The microstructure of as deposited coatings as well as coatings after mechanical tests was characterized using TECNAI G^2 F20 FEG (200kV) transmission electron microscope (TEM). Phase analysis was performed by electron diffraction pattern and confirmed by identification of high resolution images (HRTEM). Energy Dispersive X-Ray technique (EDS) was done for chemical analysis. Thin foils for TEM analysis have been prepared from a section of mechanically deformed place by the Focused Ion Beam technique (FIB). The FIB technique together with in-situ OmniProbe micro- manipulator allowed to receive foil directly from the place of interest, in this particular case from deformed part of coating after mechanical test.

3. Results and discussion

3.1 TEM analysis of microstructure of single layered coatings

3.1.1 As deposited coatings

The as deposited single layered a-C:H coatings was characterized by totally amorphous structure (Fig. 1). The atom arrangements in amorphous materials are still relatively poorly understood. An amorphous material is one where the locations of the neighboring atoms are defined by a probability function, so they average inter- atomic distance vary significantly.

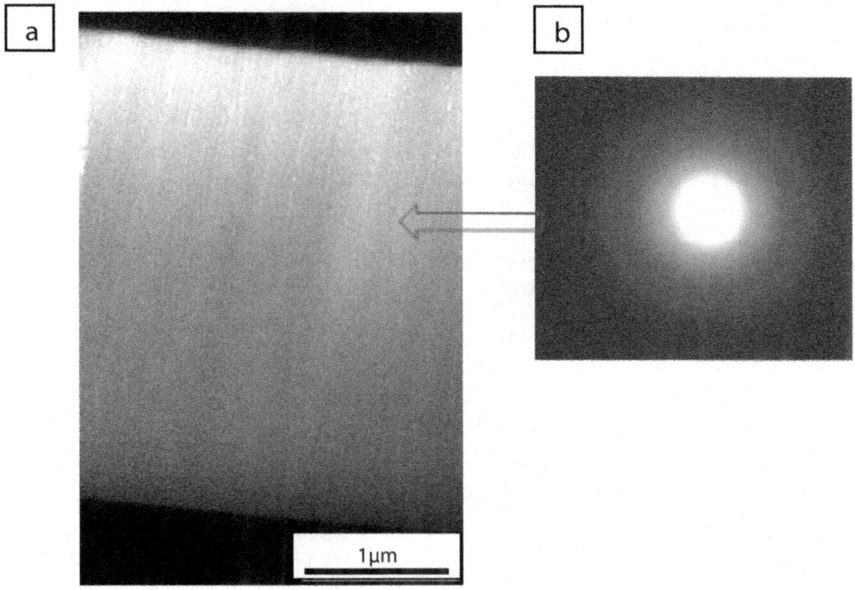

Fig. 1. Microstructure analysis of single layered amorphous carbon coating done on a cross-section by TEM technique; a). TEM Bright Field; b). electron diffraction pattern.

As a result a diffraction pattern from an amorphous material looks similar to that from polycrystalline material, but rings are broad and diffused. Aside from first ring representing distribution of "nearest neighbours" usually also similar but fainter second ring representing "next nearest neighbours" is present.

In the case of crystalline materials, like TiN, thin ceramic PVD coatings deposited at low temperatures are generally characterized by columnar structure with microcracks, pinholes and transient grain boundaries (Fig. 2a). Due to high compressive residual stress in the single layer TiN coating, microcrack-like features are forming along the column boundaries. (Fig. 2b) They consist of accumulation and pile-up of dislocations between the columns, as proved by high resolution imaging. According to literature, microcracks are nucleated in crystalline materials by three different mechanisms: dislocation pile-ups, twin intersections, and strain incompatibility (Gwidon & W. Stachowiak, 2005). In present material all these mechanisms are probably operating simultaneously.

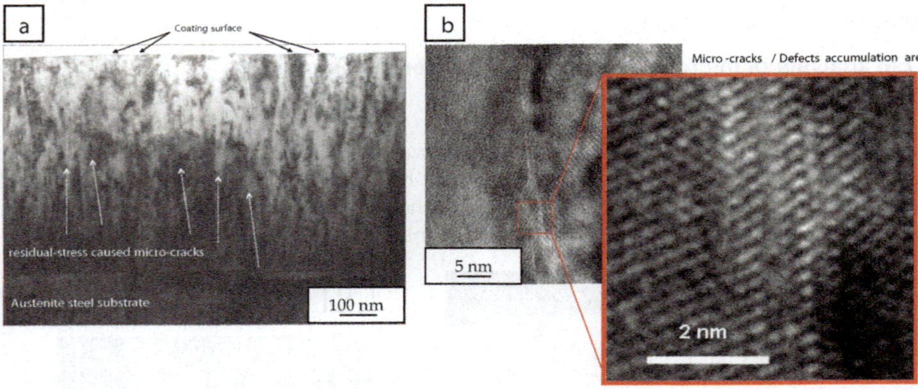

Fig. 2. Microstructure analysis of single layered TiN coating done on a cross- section by TEM technique; a). TEM Bright Field; b). high resolution image (HRTEM) of microcrack at columnar boundary.

The increase TiN single layered coating/substrate adhesion is achieved by inserting, a thin metallic Ti buffer layer (~150nm) between substrate and the coatig (Fig. 3).

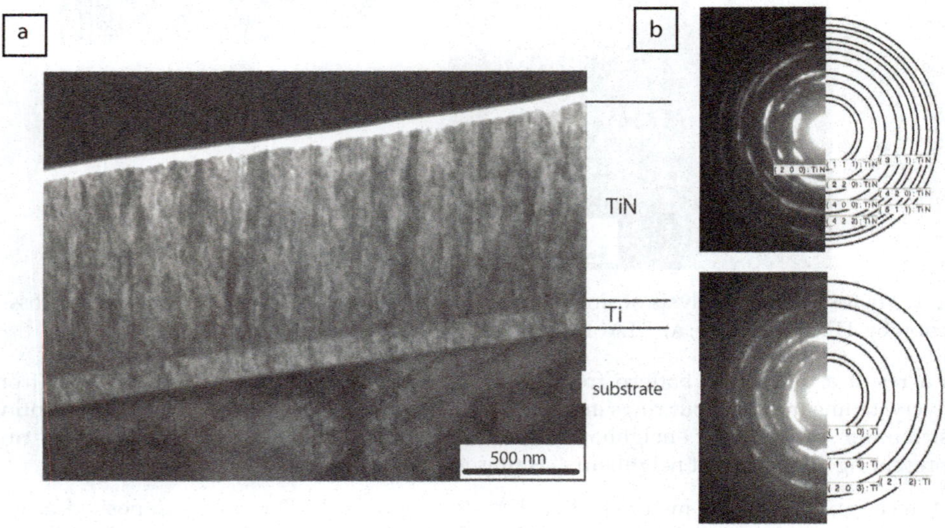

Fig. 3. Microstructure analysis of single layered TiN coating with Ti additional buffer layer done on a cross- section by TEM technique; a). TEM Bright Field; b). phase analysis done by electron diffraction pattern (SAEDP).

The ring patterns of selected area electron diffraction patterns confirmed that it belong to cubic (TiN) and hexagonal (Ti) systems. The presence of large number of fine spots in these rings indicate that layers are formed from extremely fine crystallites. Metallic buffer layer may prevent contact of substrate with environment by sealing cracks. Energy of brittle cracking in ceramic layer may be compensated by energy of plastic deformation of buffer.

3.1.2 Coatings after mechanical test

Above described coatings were subjected to mechanical test. The test simulated the condition by which coatings are uploaded in their real application. The test was static and based on pushing diamond indenter into the coating with 1N of the applied load (Fig. 4).

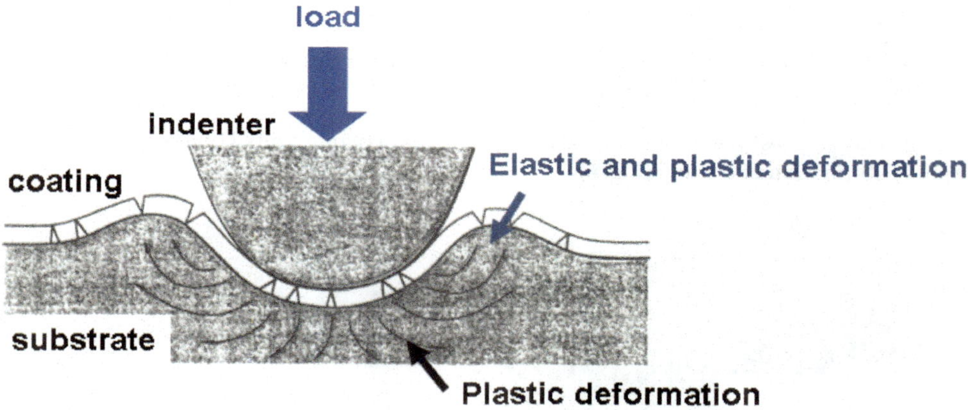

Fig. 4. Theoretical scheme of mechanical test by which analyzed coatings were subjected (R. F. Bunshah, 2001).

Reaction of coatings to the mechanical load during application is as follows: at the beginning elastic deformation dominates and next plastic deformation takes over. Because of ionic or covalent bonding, which characterize ceramic materials, plastic deformation is limited. As the dislocations are locked, brittle cracking is dominating.

First described material was amorphous carbon single layer coating after static mechanical test (Fig. 5).

Carbon, single layered coating turned out very brittle, as expected. During static mechanical test, substrate deformed plastically, while coating deformation was limited. It deformed elastically and just after that brittle cracks occured. The test caused coating delamination from the substrate. The adhesion was relatively weak.

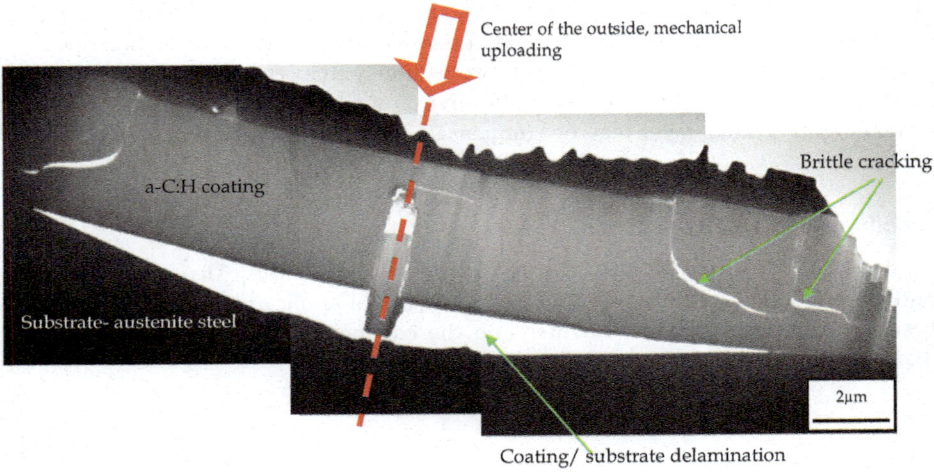

Fig. 5. Microstructure analysis of single layered a-C:H coating on a cross- section, after mechanical test, done by TEM technique.

Static mechanical test performed on TiN single layered coating with Ti buffer layer, caused serious deformation of coating (Fig. 6)

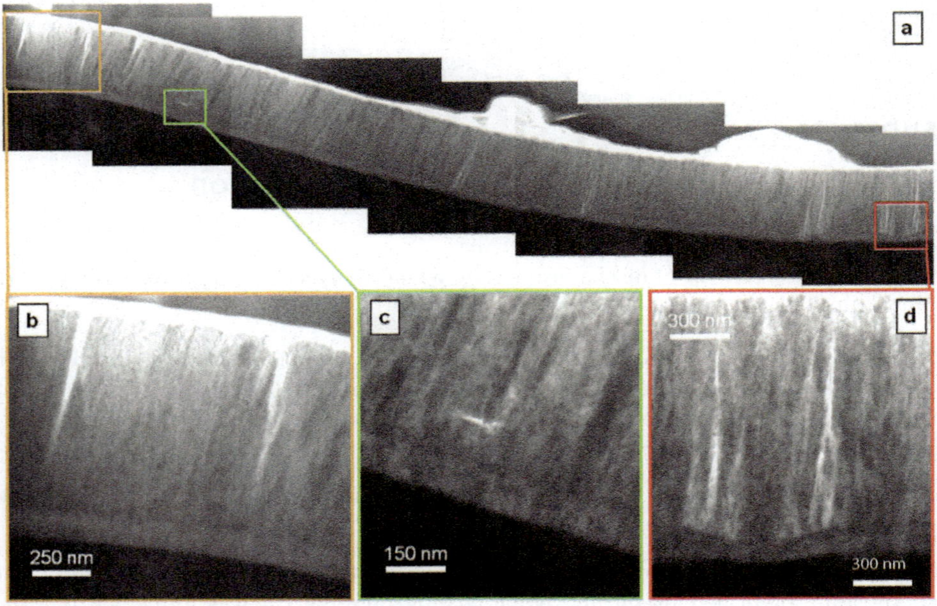

Fig. 6. Microstructure analysis of single layered TiN coating with Ti buffer layer on a cross-section, after mechanical test, by TEM technique.

TiN layer brittle cracked (Fig. 6). Cracking in ceramic crystalline materials mainly depends on the crystallographic plane which is the most responsible for cracking propagation as presents on the scheme (Fig. 7).

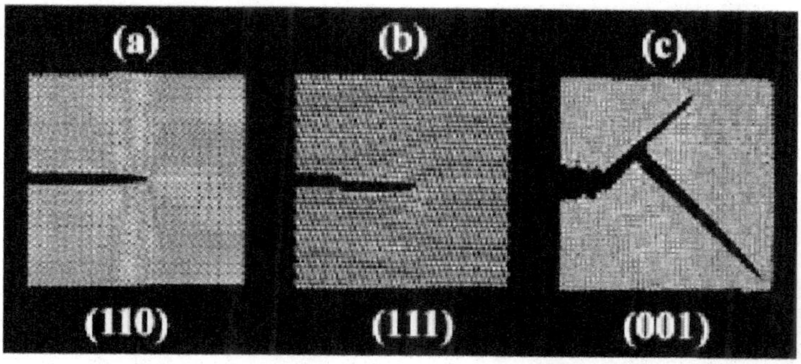

Fig. 7. Schemes of cracking character in dependence on the crystallographic plane on which it may be realized (C. L. Rountree et al. 2002).

Normally the process occurs on the most packed planes. If cracking is propagating along {110} planes the cleavage like cracking is visible. In case of cracks on {111} planes, steps like line is formed. Finally on {001} planes, branches like pattern is visible. Titanium nitride is characterized by face center cubic (fcc) structure. The most packed plane for fcc is {111} on which cracking is the most probable.

High resolution analysis of crack path performed through the TiN layer revealed that cracking was propagating along {111} plane (Fig. 8).

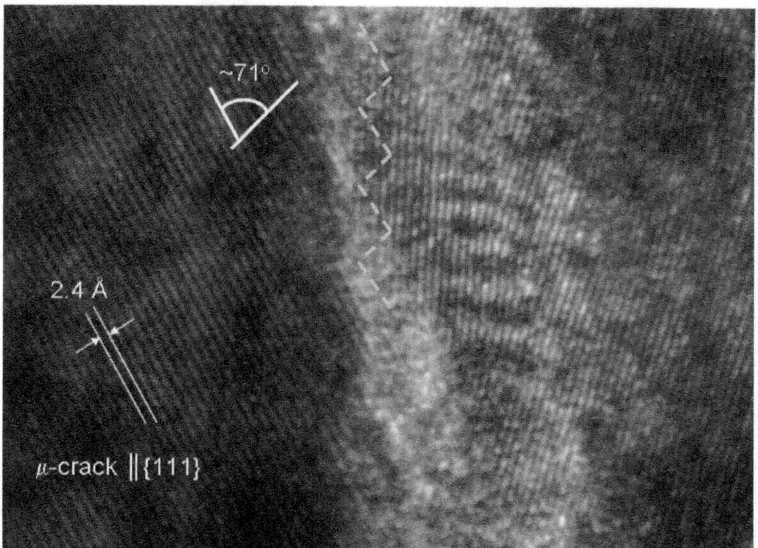

Fig. 8. HRTEM image of crack propagating through the TiN layer.

The steps like line was formed (Fig. 7b). The angle in between steps was ~71^0. Calculation of the distance in between lattice fringes and the angle consideration informed about the most plausible plane for cracking.

The presence of Ti thin buffer layer allowed to stop crack propagation through coating. Energy of brittle cracking in TiN layer transformed into the energy of plastic deformation in Ti at the interface (Fig. 9).

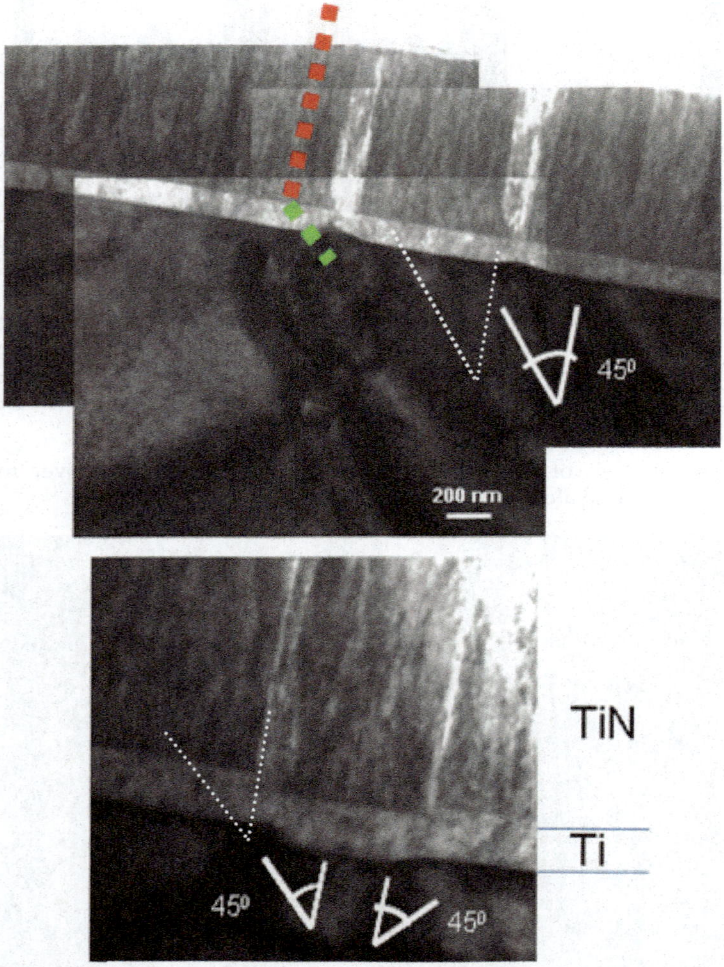

Fig. 9. Bright field image of crack propagating through the TiN layer and plastic deformation in Ti layer obtained by TEM.

The 45^0 it is a typical angle of metallic, polycrystalline materials plastic deformation. The presence of Ti phase (buffer layer) did not allow the substrate to contact with outside environment. It prevents also against catastrophic coating delamination.

3.2 TEM analysis of microstructure change of multilayer coatings

3.2.1 Coatings before mechanical test

As- deposited multilayered Ti/TiN coatings were characterized by strongly defected columnar crystallites (Fig. 10).

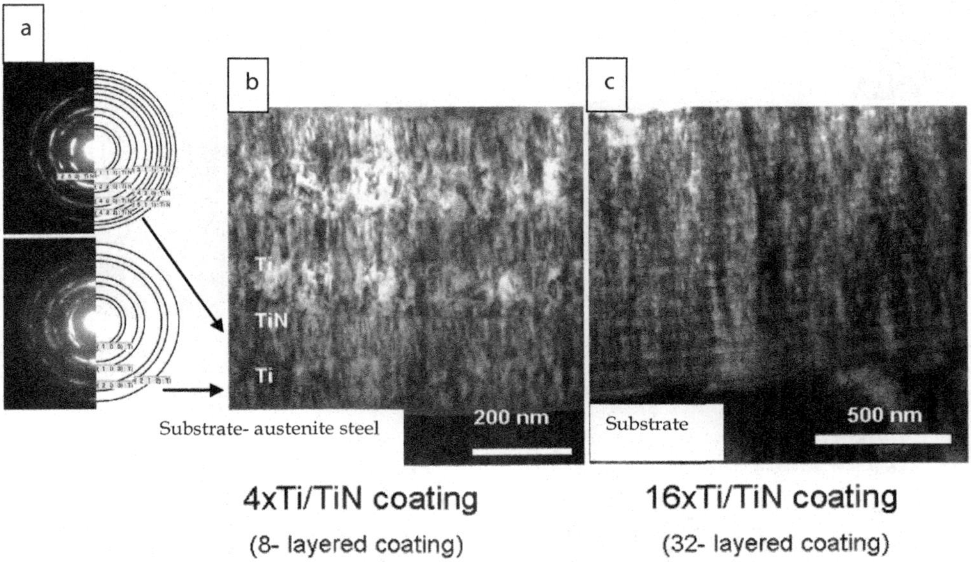

Fig. 10. Microstructure analysis of multilayered Ti/TiN coatings done on a cross- section by TEM technique; a). phase analysis done by diffraction patterns; b). 8- layered coating; c). 32- layered coating.

The same diffraction contrast, which goes through several interfaces, indicates that there are crystallographic relations in between individual layers. The first metallic buffer layer (in connection to the substrate) was characterized by much smaller crystallites then following sub- layers of the same phase.

The second group of multilayers was based on two ceramic phases: titanium nitride and amorphous carbon. To control the damage process of TiN/a-C:H multilayer coatings, small amount of metallic phase was inserted into the coating at each interface. To increase the adhesion properties of coating, metallic Ti buffer layer was deposited as a first layer on the substrate as well (Fig. 11).

The presence of carbon layers placed in a sequence with TiN ones is evident also from the line scan (Fig. 11b). Line scan confirmed presence of very thin metallic Ti layers at each TiN/a-C:H interface. It is with a good agreement with the deposition process. Energetic lines of Ti and N are very close to each other. Theoretical peak deconvolution is the only way to separate these two elements in EDS analysis. It was possible to do by confirmation phase analysis by electron diffraction pattern, and by HRTEM images. It will be presented in the 3. 2. 2 section -"Coatings after mechanical test".

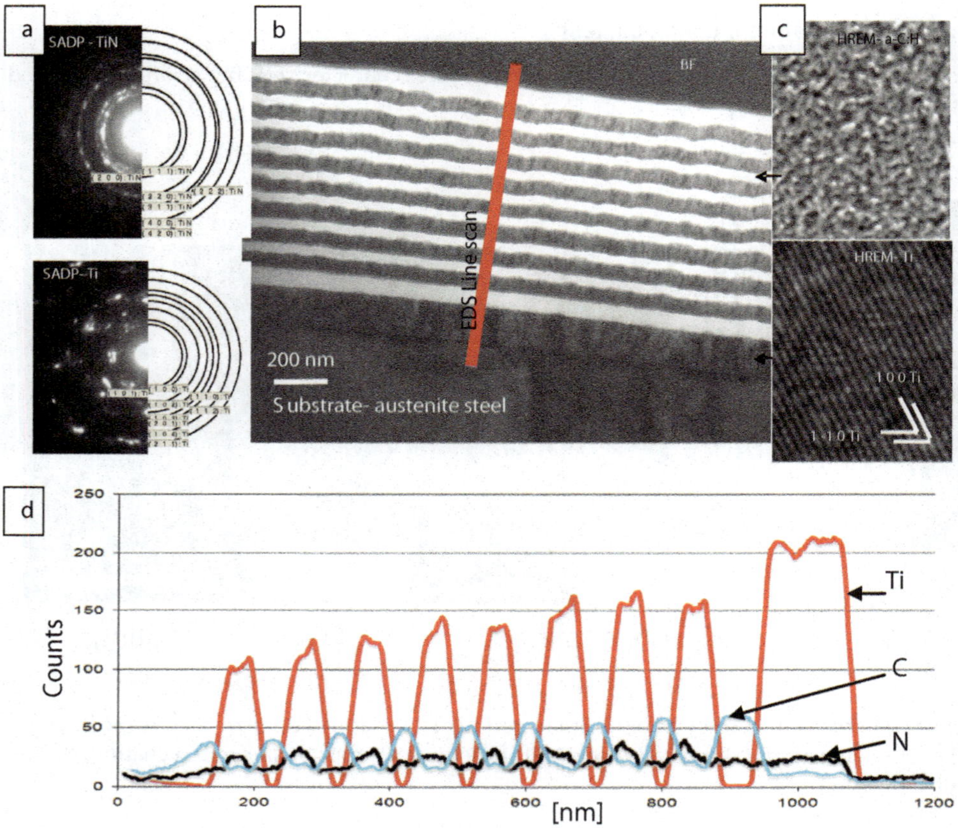

Fig. 11. Image of the TiN/Ti/a-C:H multilayer coating done by transmission electron microscopy; a). phase analysis done by SAEDP; b). bright field image and phase analysis of 8xTiN/Ti/a-C:H coating done by electron diffraction pattern analysis and high resolution; c). HRTEM images of a-C:H and Ti buffer; d). qualitative chemical analysis done by EDS (Energy Dispersive X-ray Spectroscopy) (line- scan along the line marked in the Fig. 11b).

3.2.2 Coatings after mechanical test

Both types of multilayer coatings were subjected to static mechanical tests. The tests were based on pushing diamond spherical indenter into the coatings with 1N of the applied load. Ti/TiN multilayer coatings after mechanical test were taken first under consideration focusing on microstructure change after mechanical test (Fig. 12).

The multilayer Ti/TiN coating was strongly deformed. Deformation lines appeared at an constant angle to the crystallites growth. Deformation lines were formed at 45^0 to the crystallites growth, which is a typical angle of plastic deformation for metallic polycrystalline materials (Fig. 13).

Fig. 12. Microstructure analysis of multilayered Ti/TiN coatings after static mechanical test done on a cross- section by TEM technique.

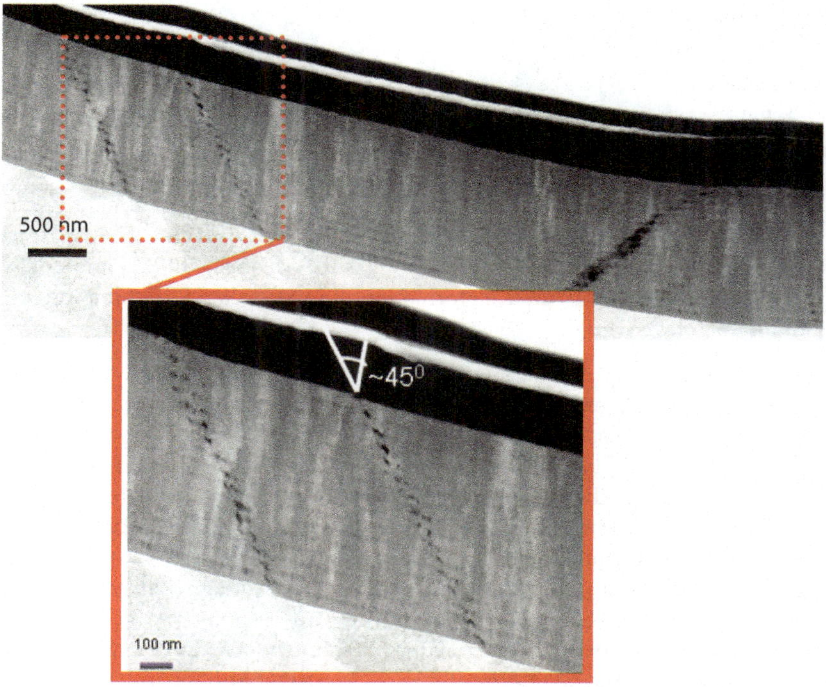

Fig. 13. Microstructure analysis of multilayered Ti/TiN coatings after static mechanical test done on a cross- section by TEM technique (higher magnification of the Fig. 12).

Detail analysis of deformation lines exhibited plastic deformation in metallic (Ti) layers and brittle cracking in ceramic (TiN) phase (Fig. 14)

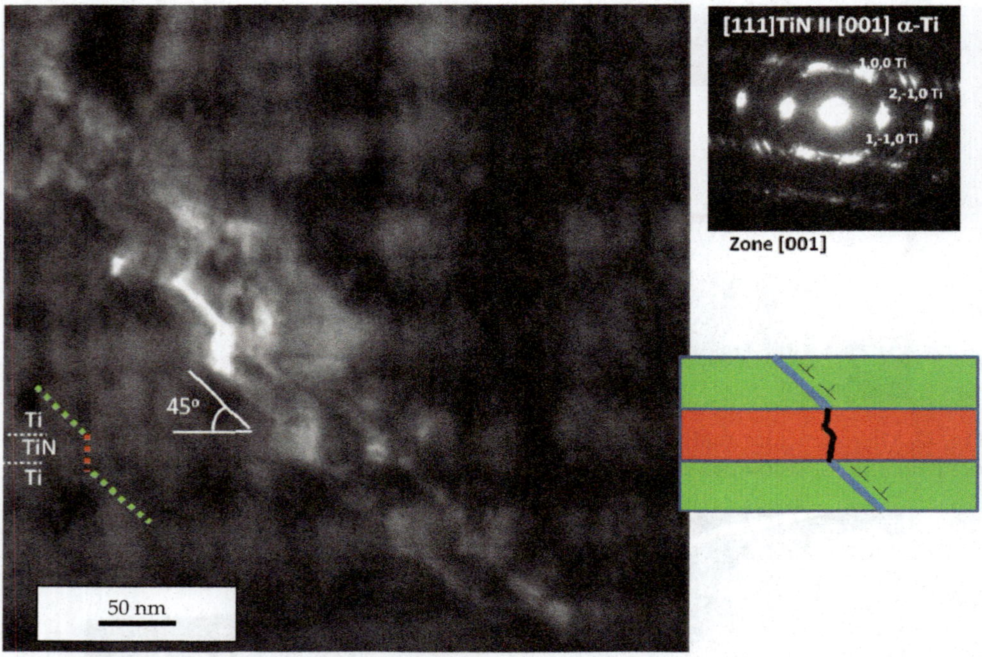

Fig. 14. Microstructure analysis of multilayered Ti/TiN coatings after static mechanical test done on a cross- section by TEM technique (detail analysis of deformation line).

Titanium metallic layers deformed plastically by slip systems while ceramic TiN brittle cracked. The character of the total deformation line suggested domination of plastic deformation in the total coating.

On one side, the presence of metallic sub- layers is crucial; they may stop crack propagation through the coating or decrease the energy of cracking; on the other side, the decrease of their amount and the increase of the hard phase, keeping constant the total thickness of coating. It may have an influence on the increase of mechanical properties, like hardness increase. One could ask, if the metallic phase reduction allow to keep control over wear. Two types of multilayer coatings have been suggested. The first with 1:2 ratio (the amount of ceramic phase two times higher than metallic one) and the second 1:4 ratio (the amount of ceramic phase four times higher than metallic ones). The advantage of multilayer coatings is fact that they contain metallic inter- layers. They are responsible for stopping cracks propagation. Their presence reduces the risk that substrate would have sudden contact with outside environment. It may suggest that these coatings would have much better corrosion properties. The characterization of the coatings with different ratio has been done in the use of transmission electron microscopy, on the cross- section (Fig. 15).

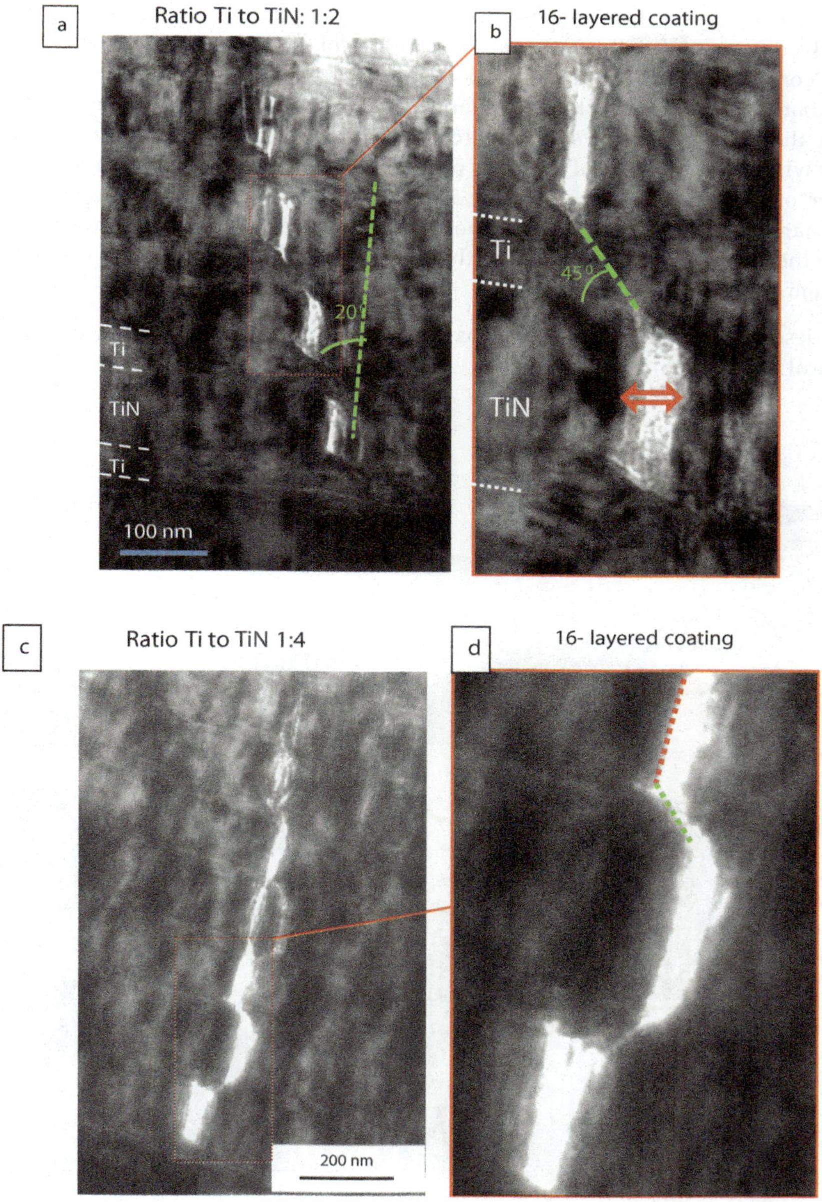

Fig. 15. Microstructure analysis of multilayered Ti/TiN coatings with different Ti to TiN ratio after static mechanical test done on a cross- section by TEM technique; a). ratio 1:2; b). higher magnification of the crack propagation through the coating with the 1:2 ratio; c). ratio 1:4; d). higher magnification of the crack propagation through the coating with the 1:4 ratio.

Ceramic inter- layers brittle cracked under the applied load, while metallic ones plastically deformed in slip systems. Plastic deformation of metallic Ti layers did not allow to contact both brittle cracked areas. The character of the total deformation line was intermediate in between plastic deformation and brittle cracking domination. The angle between the direction of crystallites growth was about 20^0. In the case of multilayer coating, where the amount of ceramic phase was four times higher than metallic one, character of the total deformation line in general was perpendicular to the substrate. Higher magnification showed that the metallic inter- layers played still its own role, not to allow to the contact of the substrate with outside environment by plastic deformation in slip system.

Second described multilayer system was TiN/Ti/a-C:H coating which was subjected to mechanical test (Fig. 16).

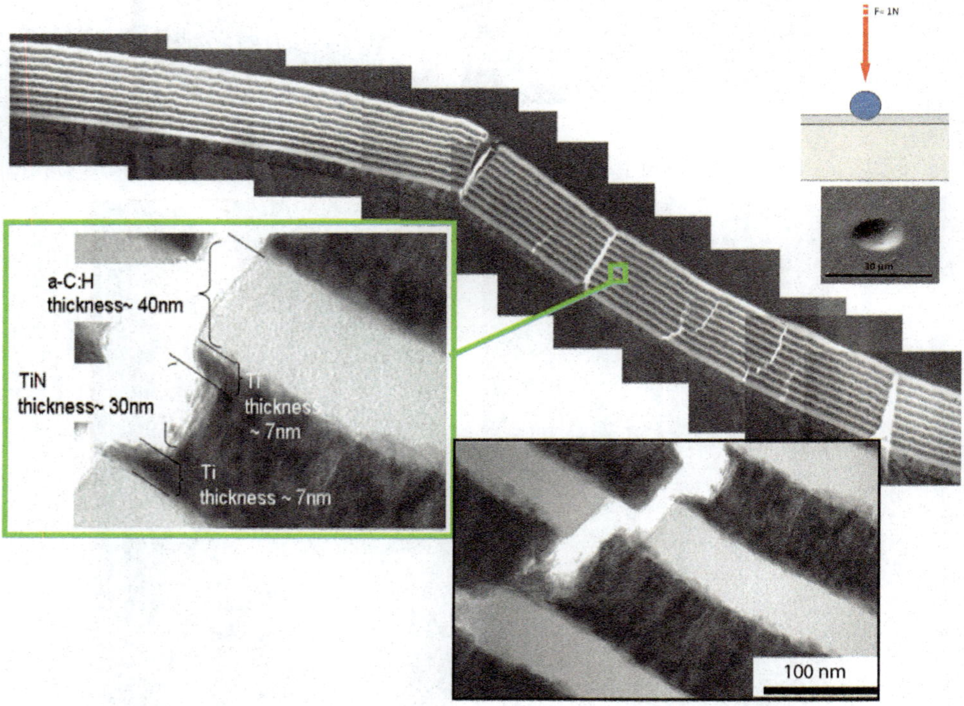

Fig. 16. Microstructure analysis of deformed a-C:H/TiN multilayer coating; a). topography of the coating after the done by SEM technique; b). bright filed analysis of deformed coating at the cross-section by TEM technique.

The presence of thin metallic layers at interfaces and their important role in damage process of described multilayer system was confirmed by bright field and high resolution TEM analysis (Fig. 17).

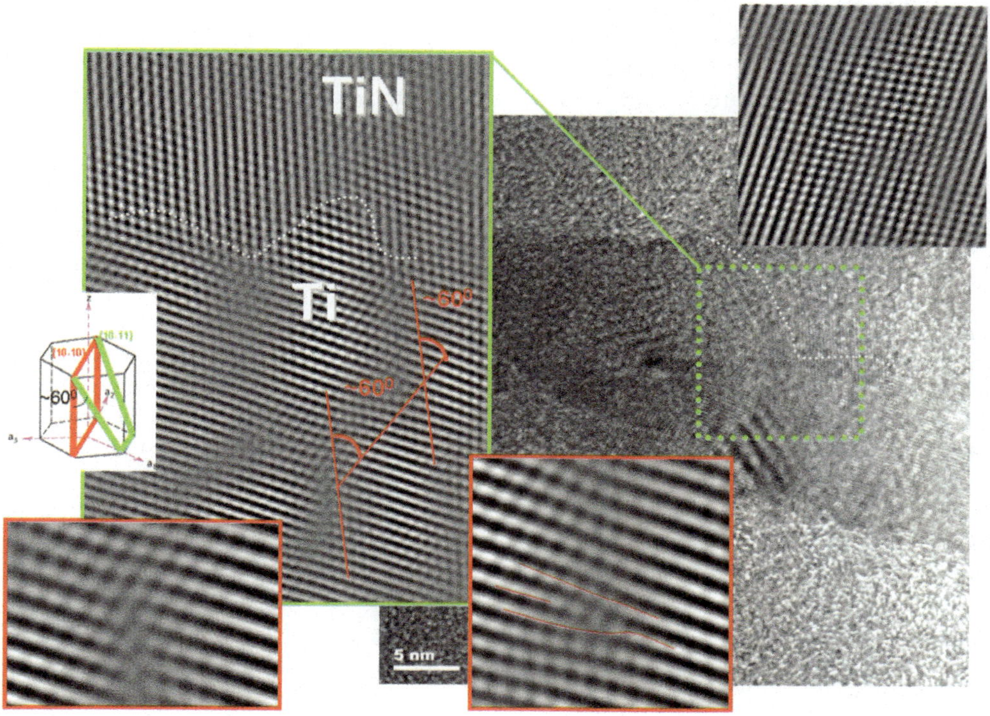

Fig. 17. Microstructure analysis of deformed TiN/Ti/a-C:H multilayer coating by high resolution technique.

Ceramic TiN, as well as a-C:H layers brittle cracked while very thin metallic layers plastically deformed. Plastic deformation propagated at 45^0 to crystals growth. The presence of plastically deformed Ti layers at interfaces as well as presence of Ti buffer layer (first layer from the substrate) play an important role in a damage process control. In zones where the impact of the external force was big, cracks were formed perpendicular to the substrate (normal behavior of brittle coatings), while in areas where the load was lower the deformation lines formed at 45^0 (Fig. 16). The multilayer TiN/Ti/a-C:H type coating may help diverting perpendicular cracks to tilted cracks without losing much of coating hardness and get some control over the coating damage. HRTEM analysis of the individual Ti crystallites visualized the presence of crystallographic planes on which deformation was realized (Fig. 17).

Titanium is characterized by hexagonal structure. The most probable planes for plastic deformation for hexagonal metals is: {10-10}, {10-11} and {0001}. The angle in between {10-10} and {10-11} is 60^0. Filtrated HRTEM image (after twin-oval-mask application on the fast fourier transform) exhibited blurred lines in the area of individual Ti crystallite. The lines formed angle of ~60^0. It informed which crystallographic planes were the most responsible for plastic deformation of titanium. Fourier masking can remove unwanted noise or enhance periodic elements of an image. The typical sequence for using masking is to perform a Fourier transform on a real- space image, mask off the desired frequencies in frequency space, and finally perform an inverse Fourier transform on the masked image (DigitalMicrograph, 1996-2003).

4. Summary

The as deposited single layered TiN and multilayered Ti/TiN coatings were characterized by columnar microstructure and high dislocation density. Carbon phase as single layered coating as well as in multilayer TiN/Ti/a-C:H system was amorphous. Coatings after mechanical tests were strongly deformed i. e. showed presence of cracks. In Ti/TiN multilayer system under the applied load characteristic deformation lines appeared. Deformation line consisted of plastic deformation in metallic layers and brittle cracks in ceramic layers. Deformation occurred at 45^0 to crystals growth. It is a typical angle for plastic deformation of metallic, multi- crystalline materials. Cracking in crystalline layers (TiN) were propagating along the most packed {111} planes. In the case of TiN/Ti/a-C:H multilayer system ceramic TiN and a-C:H layers were cracked, metallic Ti layers, presented at each interface, deformed plastically. The presence of metallic phase lead to deviation of direction of small cracks resulting in improving an overall coating cracks resistance.

5. Acknowledgement

Financial support of this work by the:

- Research project of National Science Centre (Polish – Narodowe Centrum Nauki, abbr. NCN) No: 3066/B/T02/2011/40
- Polish- Austrian exchange project PL 12/2010

are highly acknowledged.

6. References

[1] R. F. Bunshah: Handbook of Hard Coatings. ISBN 0-8155-1438-7, NP New Jersey USA (2001)

[2] D. S. Rickerby, A. Matthews. Advanced Surface Coatings; Handbook of Surface Engineering. Chapman and Hall N. Y. USA (1991)

[3] V. Kumar, A. A. Bergman, A. A. Gorokhovsky, A. M. Zaitsev; Formation of carbon nanofilms on diamond for all-carbon based temperature and chemical sensor application; Carbon 49(2011)1385-1394

[4] Li Chen, S. Q. Wang, S. Z. Zhou, Jia Li, Y. Z. Zhang: Microstructure and mechanical properties of Ti(C, N) and TiN/Ti(C, N) multilayer PVD coatings; International Journal of Refractory Metals and Hard Materials, 26 (2008) 456–460

[5] M. Stueber, H. Holleck, H. Leiste, K. Seemann, S. Ulrich and C. Ziebert: Concept for the design of advanced nanoscale PVD multilayer protective thin films; Journal of Alloys and Compounds 483 (2009) 321-333.

[6] E. Martinez, J. Romero, A. Lousa and J. Esteve: Nanoindentation stress–strain curves as a method for thin-film complete mechanical characterization: application to nanometric CrN/Cr multilayer coatings; Appl. Phys., A 77 (2003) 419-426

[7] J. Smolik and K. Zdunek: Effect of interlayer composition on the tribological properties of TiC/Ti(Cx, N1-x)/TiN anti-abrasive multi-layer coatings Vacuum 55 (1999) 147-151

[8] Y. L. Su and W. H. Kao: Optimum multilayer TiN-TiCN coatings for wear resistance and actual application; Wear 223 (1998) 119- 130

[9] N. Dück, W. Gamer, M. Gesatzke, W. Griepentrog, M. Österle, I. Sahre and Ž. Urban: Ti/TiN multilayer coatings: deposition technique, characterization and mechanical properties; Surf. Coat. Technol. 142–144 (2001) 579- 584

[10] M. Nordin, M. Larsson and S. Hogmark: Mechanical and tribological properties of multilayered PVD TiNrCrN; Wear 232 (1999) 221- 225

[11] Ł. Major, M. Kot, J. M. Lackner; The effect of metallic inter-layers on multilayer ceramic/ metal coatings properties; Materials Science, 3(175) (2010) 445-448

[12] Ł. Major, J. Morgiel: TEM analysis of wear of Ti/TiN multi- layer coatings in ball- on-disc test. Key Engineering Materials 409(2009)123-127

[13] M. Bromark, M. Larsson, P. Hedenqvist, S. Hogmark: Wear of PVD Ti/TiN multilayer coatings; Surf. Coat. Technol. 90 (1997) 217- 223

[14] A. Leyland and A. Matthews: Thick Ti/TiN multilayered coatings for abrasive and erosive wear resistance; Surf. Coat. Technol. 70 (1994) 19- 25

[15] A. Matthews and S. S. Eskilden: Engineering applications for diamond-like carbon; Diamond Relat. Mater. 3 (1994) 902- 911

[16] K. Holmberg, A. Matthews and H. Ronkainen: Coatings tribology—contact mechanics and surface design; Tribol. Int. 31 1-3 (1998) 107- 120

[17] J. Grabczyk, D. Batory, P. Louda, P. Couvrat, I. Kotela, K. Bakowicz-Mitura; Carbon coatings for medical implants; Journal of Achievements in Materials and Manufacturing Engineering, 20(2007)107-110

[18] J. M. Lackner, Industrially-scaled hybrid Pulsed Laser Deposition at Room Temperature, OREKOP, ISBN 83-921845-1-3 Krakow (2006)

[19] Gwidon W. Stachowiak, Wear - Materials, Mechanisms and Practise, Wiley, Chichester, (2005)

[20] C. L. Rountree, R. K. Kalia, E. Lidorikis, A. Nakano, L. Van Brutzel, P. Vashishta: ATOMISTIC ASPECTS OF CRACK PROPAGATION IN BRITTLE MATERIALS: Multimillion Atom Molecular Dynamics Simulations, Annual Review of Materials Research 32(2002) 377- 400

[21] DigitalMicrograph™ 3. 10. 0 for GMS 1. 5. 0 by the Gatan Software Team, Copyright © 1996-2003

12

Deposition and Characterization of Platinum and Palladium Nanoparticles on Highly Oriented Pyrolytic Graphite

Nora Elizondo et al.*,
Facultad de Ciencias Físico-Matemáticas, México

1. Introduction

Nanostructured transition metal nanoparticles are of great interest from both fundamental and practical view points because of the quantum size effect, which is derived from the dramatic reduction of the number of free electrons. [Halperin, 1986; Schmid, 1994] Volokitin et al. have found that the quantum size effect strongly affected the thermodynamics of the metal nanoparticles. [Volokitin et al., 1996; Colvin et al., 1994; Andres et al., 1992]

Platinum is considered one of the best electrocatalyst for low temperature reactions in a H_2/O_2 fuel cell. Palladium is an element of the platinum group metals and it has similar chemical properties. Nanoparticles(Nps) are of great interest because of the modification of properties observed due to size effects, modifying the catalytic, electronic, and optical properties of the monometallic Nps. Interest in platinum Nps derives mostly from the importance of highly dispersed platinum and palladium in catalysis. [Colvin et al., 1994; Andres et al., 1992] They have been concretly applied to catalysts[Lewis, 1993] for hydrogenation of olefins and dienes,[Hirai et al., 1986; Teranishi, 1996] hydration of acrylonitrile to acrylamide,[Toshima & Wang 1994] photogeneration of hydrogen from water,[Bard, 1980] and reduction of carbon dioxide,[Willner et al., 1987; Toshima et al.,1995]

Donald H. Galván[3], Lorena Álvarez-Contreras[4], Ran Tel-Vered[5], Arquímedes Cruz-López[2], Ricardo Obregón[1], Sergio Belmares-Perales[1], Manuel García-Méndez[1], Odilón Vázquez-Cuchillo [2] and Antonio A. Zaldívar[2]

[1]*Facultad de Ciencias Físico-Matemáticas, México*
[2]*Facultad de Ingeniería Civil,*
Universidad Autónoma de Nuevo León, San Nicolás de los Garza, N. L., México
[3]*Centro de Nanociencias y Nanotecnología,*
Universidad Nacional Autónoma de México, Ensenada, México
[4]*Departamento de Química de Materiales,*
Centro de Investigación en Materiales Avanzados, S. C., Chihuahua, México
[5]*Institute of Chemistry, The Hebrew University of Jerusalem, Jerusalem, Israel*

the catalytic activity and selectivity being strongly affected by the particle size. To investigate the physical and chemical properties of metal nanoparticles, especially the size-dependent properties, precise control of the particle size is essentially required. Moreover, the precise control of particle size is also required for the organization of metal nanoparticles.[Teranishi et al., 1997; Reetz et al.,1997]

It has been found from mass spectral studies that the binding energies of metal nanoparticles consisting fewer than ca. 1000 atoms vary periodically due to the quantum size effect.[Sugano, 1991; Katakuse, 1994] This phenomenon was first found by Knight et al.[Knight et al., 1984] through experiments involving Na nanoparticles. This suggests the discontinuous existence of nanoparticles having certain stable structures. Therefore, a synthetic technique is required to produce the nanoparticles with any sizes within a few angstroms in standard deviation.

The usual synthetic technique for making such nanoparticles involves chemical or electrochemical reduction of metal ions in the presence of a stabilizer such as linear polymers, [Teranishi, 1996; Toshima et al., 1991, 1995; Hirai et al., 1985; Henglein, 1993; Bradley et al., 1991; Hirai, 1979] ligands, [Schmid, 1992; Poulin et al., 1995; Amiens, 1993] surfactants, [Toshima & Takahashi, 1992; Yonezawa et al., 1995; Esumi et al., 1995; Leff et al., 1995]

An "ideal" model system for investigating a particle size effect in electrochemical reactions-such as the methanol oxidation and oxygen reduction reactions- would possess all of the following characteristics: (1) Platinum nanocrystals should be size and shape monodisperse. (2) Nanocrystals should be dispersed on, and electrically connected to, a technologically relevant support surface that facilities spectroscopic characterization of the particles and of adsorbed intermediates. For many electrocataysis reactions the preferred support material is graphite. (3) Individual platinum particles on this support should be well-separated from one another. (4) The structure of the platinum nanocrystals on the support surface should be accessible both before and following the involvement of these particles in the catalytic process of interest. (5) Supported and platinum Nps should be stable for days [Zoval et al., 1996; Thomas et al.,1996; Bronstein et al., 2000; Han et al., 1998; Freelink et al. 1995].

Control of particle size by using the ligands has been extensively studied, [Schmid et al., 1992] many shell-structured nanoparticles being synthesized, such as 2-shell Au,[Schmid et al., 1984] 4-shell Pt,[Schmid et al., 1989] and 5, 7, and 8-shell Pd[Vargaftik et al.,1991; Schmid, 1988; Schmid et al.,1993] nanoparticles. When a linear polymer is used as a protective agent, modification of the functional groups can offer a specific reaction field around the metal nanoparticles that promotes the catalytic activity of the nanoparticles and may change the electronic structures of the metal nanoparticles. Ahmadi et al. succeeded in controlling the shape of Pt nanoparticles by using sodium polyacrylate as a capping polymer.[Ahmadi et al., 1996] We have controlled the size of Pt and Pd nanoparticles by using poly (vinylpyrrolidone) (PVP) and succeeded in their two dimensional organization. Thus, linear polymers have great potential as protective agents for nanoparticles and as stabilizers and ethylenglycol as a reductor.[Park et al., 2008; Cao, 2004] Such procedure yield different morphologies and sizes of metal Nps (including platinum and palladium).[Teranishi & Miyake, 1998] The interaction between PVP and metal precursors has effect on the formation of the colloidal metal nanoparticles. Strength of the interaction between PVP and metal nanoparticles has direct influence on the stability and the size of the PVP-stabilized metal nanoparticles. Therein, species of the metal precursors and amount of the stabilizer are main factors on the strength of the interaction.[Wei-xia, 2008]

The polyol method has been reported to produce the platinum and palladium nanoparticles as the final product,[Yonezawa & Toshima, 1993],[Kan et al., 2003; Sanchez-Ramirez, 2001; Viau et al., 2001] easily changing surface modifiers. This technique does not require an additional reducing agent since the solvent by itself reduces the metallic species.[Tekaia-Elhsissen et al., 1999; Bonet et al., 1999] However, besides the stoichiometry and order of addition of reagents in the synthesis process, one of the most important[Turkevich et al., 1951] parameters in the preparation is the temperature. Modifications in temperature influence the reaction by changing the stabilization of the complexes formed between Pt and Pd respectively with the surface modifiers, e.g., PVP, and the nucleation rate of the reduced metallic atoms.[Garcia-Gutierrez et al., 2008]

In this chapter, a colloidal method of synthesis has been proposed to obtain metallic Nps; the polyol method has been reported to produce small Nps as the final product. [Alvarez et al., 1997; Henglein et al., 2000; Grabar, 1995; Baker, 1996]

The gas-phase method of platinum and palladium salt particles disposed on a graphite surface by H_2 was used in this work. This method has the potential to yield nanocrystals that are disposed in direct contact with a substrate surface. [Bartolomew & Boudart, 1972]

In this work we describe an electrochemical method for preparing dispersions of platinum and palladium nanocrystals on a graphite basal plane surface involving the pulsed potentiostatic deposition of platinum from dilute $PtCl_6^{2-}$ using large overpotentials ($E_{overvoltage} \approx 500$ mV) and by overpotential deposition also with higher Pd^{2+} concentrations in solution. [Reetz et al., 1992; Allongue & Souteyrand,1992 ; Cachet et al., 1992; Diculescu et al., 1992]

A novel approach to characterize this kind of particles is based on the use of HAADF technique, in a high resolution transmission electron microscope (HRTEM), which allowed us the observation of the elements due to atomic number, densities, or the presence of strain fields due to differences in lattice parameters, structure, the presence of surfactants or any other surface modifier besides the size of the particle. [Williams & Carter, 1996]

The interaction between metal nanoparticles and substrate surfaces has long attracted attention. This is because of underlying interests in understanding (1) the behavioral transition from atomic to bulklike properties, as a function of nanoparticle size, (2) the effect of size-dependent electronic structures in heterogeneous catalysis, (3) the dimensionally controlled fabrication of supported nanoparticles, (4) thin film deposition onto heterogeneous surfaces and (5) the adhesion associated with interfacial interactions. Studies have indicated that the evaporation, sputter, and electrochemical depositions of Pt and Pd all lead to particle formation on such substrates, due to both a lack of substrate wetting and relatively weak interfacial interactions. [Yang et al., 2006]

The HOPG is described as consisting of a lamellar structure the freshly cleaved surface consists of atomic steps and steps of several or dozens of atomic layers. The crystallographic planes do have a definite structure and the height of a single step is 0.34 nm. [Pauling,

1960]The Moiré patterns and rotations between the first and second layers of HOPG in the direction perpendicular to the basal plane by effect of heating Pt Nps on HOPG (in H_2 flow) were observed. [Williams & Carter, 1996; Pauling, 1960; Beyer et al., 1999]

2. Experimental section

The polyol method was followed to obtain platinum and palladium metallic Nps passivated with poly(vinylpyrrolidone) (PVP). Hexachloroplatinic (IV) acid (H_2PtCl_6) hydrate (99.99%), Palladium Chloride(II) and poly (N-vinyl-2-pyrrolidone) (PVP-K30, MW = 40000) were purchased from Sigma Aldrich, and 1,2-ethylenediol (99.95%) was purchased from Fischer Chemicals; all the materials were used without any further treatment.

A 0.4 g sample of Poly (N-vinyl-2-pyrrolidone) (PVP) was dissolved in 50 mL of 1,2-ethylenediol (EG) under vigorous stirring, heating in reflux, until the desired temperature was reached (working temperatures ranged from 100 °C to 190 °C in increments of 10 °C). For the Pt and Pd metallic Nps, a 0.1 mM EG-solution of the metal precursor was added to the EG-PVP solution, with continuous agitation for 3 h in reflux. When preparing the Pt and Pd metallic Nps, the following criterion was used: after complete dissolution of PVP in EG, 2 mL of an EG solution of H_2PtCl_6 (0.05 M) and $PdCl_2$ was added to the EG-PVP solution in a period of 1 h. The reaction was carried out for 3 h at constant temperature. For this work the Pt and Pd Nps presented the smaller average size for a synthesis performed at 130 °C. These Pt and Pd Nps in a solution of ethanol were impregnated on HOPG and dried in a oven at 80°C.

The Pt and Pd Nps preparation on HOPG by direct reduction of Platinum and Palladium salts in H_2 flow on HPOG (the gas phase method) consisted in the partial oxidation of the support in a muffle furnace at 600°C for 24 hours. Then the impregnation of preoxidized HOPG with chloroplatinic acid and palladium chloride solutions respectively in a four to one mixture of benzene to ethanol (absolute). The metal concentrations were adjusted to produce the desired total metal loading (10 Wt % Pt and 10 Wt % Pd). The amount of solvent was fixed using 50 mL/g of HOPG. A mixture of salt solution and HOPG was shaken while nitrogen was bubbled through the suspension at flow rate of 200-500 mL/min until to solution evaporated to dryness, i.e., after 40-60 h for a 10 g sample in 500 mL of solution.

The samples prepared by these methods (Pt and Pd Nps on HOPG) were heated in a H_2 flow at a temperature range from 450 °C to 1000 °C during time intervals from 2h to 5 h. Samples were then exfoliated with a scotch tape for TEM observation.

Pt and Pd Nps were electrochemically deposited on a highly oriented pyrolytic graphite (HOPG, grade-1) which was obtained from SPI Supplies (West Chester, PA). In the case of Platinum depositions were carried out by immersing the potential in which electroless platinum deposition was not observed), followed by stepping the potential of the graphite surface to a deposition potential of -0.6 V, for 100 ms. Following the application of the deposition pulse, the electrode potential was stepped back to 0.2 V, and the working electrode was removed from the plating solution. All electrochemical experiments were performed using a CH Instruments potentiostat model CHI 900B (CH Instruments, Austin, TX). A platinum coil (d=0.5 mm) and a Hg/Hg_2SO_4 were used as the counter and reference electrodes, respectively.

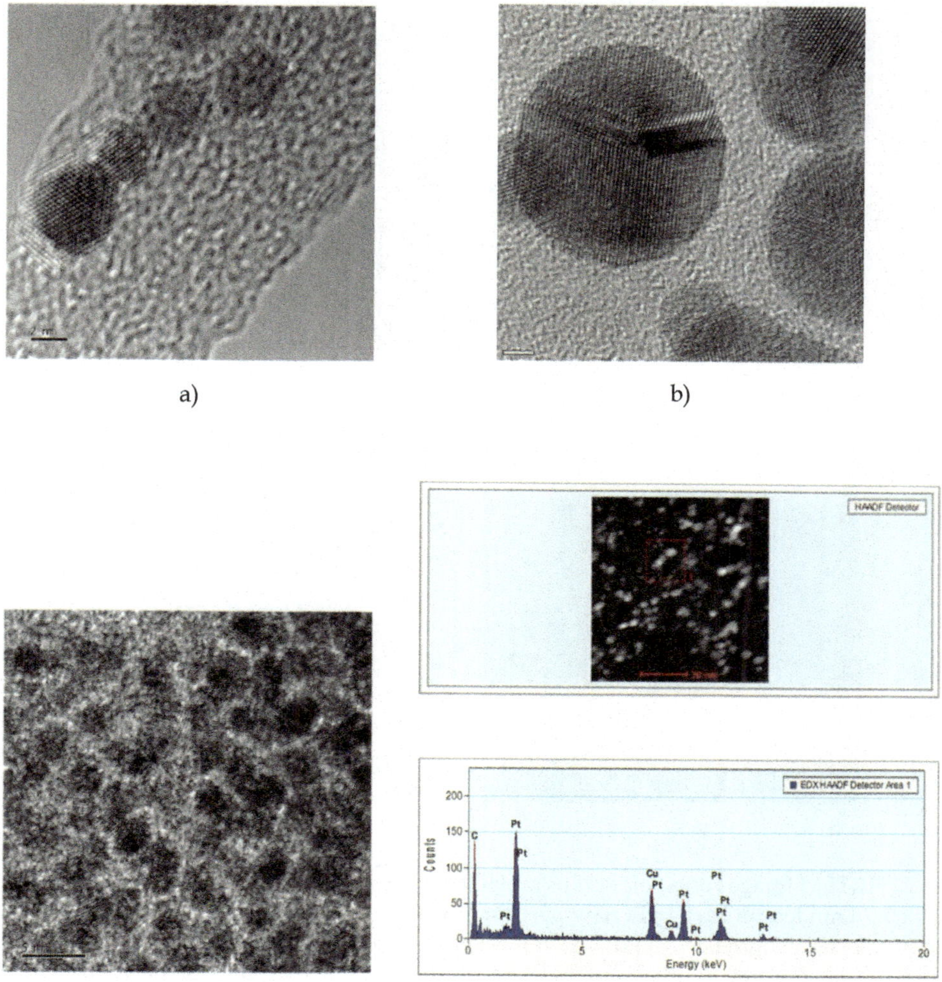

Fig. 1. A microscopy images by HRTEM, show metallic Nps synthesized by polyol method at: a) platinum(2 nm length bar), b) palladium(2 nm length bar), c) platinum well dispersed synthesized at 145 °C(5 nm length bar), and d) HAADF shows of Pt nanoparticles and EDX HAADF analyze of Pt nanoparticles(20 nm length bar), peaks for Cu and C are from the grid used.

The Pt and Pd Nps on HOPG for the electron microscopy analysis were prepared over lacey carbon TEM grids. HRTEM images were taken with a JEOL 2010F and a Titan FEI microscopes. HAADF images were taken with a JEOL 2010F microscope in the STEM mode, with the use of a HAADF detector with collection angles from 50 mrad to 110 mrad.

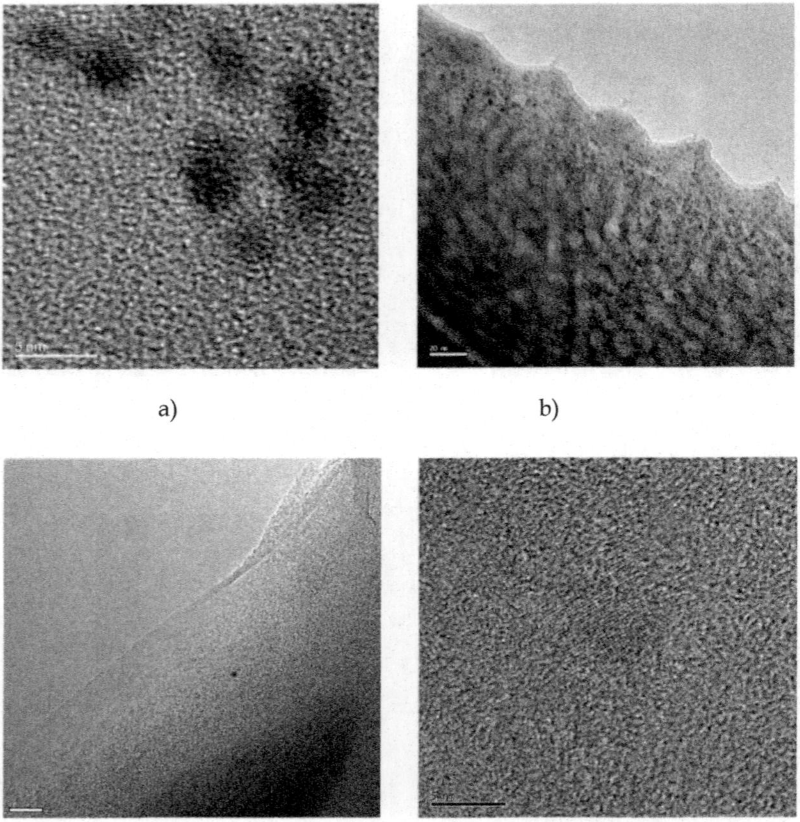

a) b)

c) d)

Fig. 2. HRTEM images of: a) of Pt Nps on HOPG synthesized by polyol heated in H_2 flow at 450°C(5 nm length bar), b) Pt nanoparticle synthesized on HOPG by the electrochemical method(20 nm length bar), c) Pd Np synthesized by polyol on HOPG(20 nm length bar), and d) Pt nanoparticle synthesized on HOPG graphite by direct reduction of chloroplatinc acid impregnated on HOPG and heated in H_2 flow at 950 °C(5 nm length bar).

3. Electronic properties of bilayered graphene with platinum and bilayered graphene with palladium atoms on one of the layers

High resolution transmission electron microscopy analysis of two-layered graphene yielded Moiré patterns induced by Pt atoms/clusters located at the top of one of the layers, which induce rotation between planes. The rotations measured vary between 3 and 5 degrees. Theoretical analysis was performed using an extended Hückel tight-binding scheme on two-layered graphene with either containing two carbon atom vacancies or with two Pt atoms, with a cluster of 6 Pt atoms or with a cluster of 13 Pt atoms located at the top of one of the graphene layers. In most of the cases, the system remains semi metallic, except when a cluster of 6 Pt atoms was located on the surface of one of the graphene layers. For this case, the system behaves as a semi conductor with an energy gap $E_g \sim 0.05$ eV.

We have also investigated the scenario of graphene when irradiated with high energetic protons and latter decorated with Pd atoms on one of the layers. Theoretical analysis were performed on graphene 2L (two layers) with vacancies (carbon 3 and 13), graphene 2L with vacancies and intercalated in between the two carbon layers, graphene 2L with the vacancies intercalated and subsequently with two Pd atoms on one of the layers (called surface), and last but not least, graphene 2L with vacancies intercalated and decorated with six Pd atoms on the surface. For the three cases enunciated formerly, energy bands were performed and provided information about the metallic behavior, showing more metallic character for the first case, while less metallic behavior for the second one. Moreover, for graphene 2L with vacancies and intercalated with six Pd atoms on the surface showed a mini gap (between the Conduction and Valence Bands) of the order of 0.02 eV and manifesting semiconductor behavior.

4. Theoretical analysis

The calculations reported in this work, have been carried out by means of the tight-binding method [Whangbo & Hoffmann, 1978], within the extended Hückel [Hoffmann, 1963].framework using YAeHMOP (Yet Another extended Huckel Molecular Orbital Program) computer package with f-orbitals.[Glassey et al., 1999] It is good to stress that the extended Hückel method is a semi empirical approach for solving Schrödinger equation for a system of electrons, based on the variational theorem. In this approach, explicit electron correlation is not considered except for the intrinsic contributions included in the parameter set. Furthermore, neither the relativistic contributions form the Dirac electrons were considered, except that they indirectly appear in the parameter set which were obtained from S. Alvarez et al.[Alvarez, 1993] or from the most accurate *ab initio* calculations. More details about the mathematical formulation of this method have been described elsewhere.[Galván, 1998]

Calculations were performed on a system selected as a repeated cluster originated from a super cell. The super cell was generated from an infinite single sheet of carbon atoms, as depicted in Fig. 3, which aroused from crystalline graphite using the following primitive vectors: a = 2.456 Å, c = 6.696 Å, space group 186.[Hull, 1917; Hoffmann & Wilm, 1936]

The structure for graphene was made of two hexagonal layers located one at the top of the other layer separated by a distance of 3.348 Å apart. In order to make the calculations easier, a new cluster smaller than the original could be constructed. This new cluster is made-up of 40 carbon atoms in single layer, and then duplicates the second layer with a similar number of atoms separated by a distance of 3.348 Å. apart. For the graphene with platinum atoms at the top of the layer, considered as the surface, two Pt atoms were located at the center of the hexagonal rings, while the second Pt atom was located in an adjacent hexagonal ring. We also investigated the existence of vacancies created when two carbon atoms were removed from the original structure. In order to extend our investigation further whenever the number of Pt atoms was increased to a 6 Pt atoms cluster which was located it on the top one of the graphene layers. Furthermore, in order to complete to a more *realistic* problem, a cluster made of 13 Pt atoms previously optimized, was located at top the graphene layer with vacancies, and latter one of the graphene layers was rotated by 3 and 5 degrees with respect to the other layer.

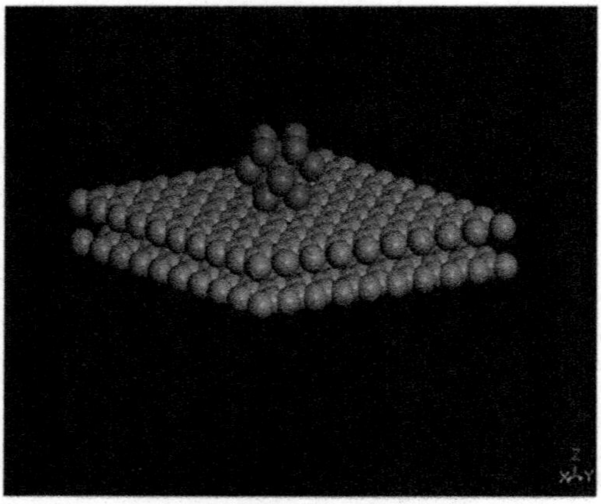

Fig. 3. Inifinite hexagonal arrays of carbon atoms used in order to generate the honeycomb structure for graphene. Grey balls are carbon atoms while blue balls correspond to Pt atoms.

Atomic parameters for carbon, platinum and palladium atoms used through the calculations were obtained from Alvarez et al.[Alvarez, 1993] and provided in Table 1.

Atom	Orbital	H_{ii}	ς_{i1}	C_1	ς_{i2}	C_2
C	2s	-21.40	1.62			
	2p	-11.40	1.62	0.0000	0.0000	0.0000
Pt	6s	-9.07	2.55			
	6p	-5.47	2.55			0.5513
	5d	-12.59	6.01	0.6334	2.6960	
Pd	5s	-7.32	2.19			
	5p	-3.75	2.15			
	4d	-12.02	5.98	0.5535	2.613	0.6701

Table 1. Atomic parameters used in the Extended Hückel tight-binding calculations, H_{ii} (eV) and ς (Valence orbital ionization potential and exponent of Slater type orbitals). The d-orbitals for Pt and Pd are given as a linear combination of two Slater type orbitals. Each exponent is followed by a weighting coefficient in parentheses. A modified Wolfsberg-Helmholtz formula was used to calculate H_{ij}.[Wolfsberg & Helmholtz, 1952]

Experimental lattice parameters instead of optimized values were used searching for a best matching of our theoretical results with the available experimental information. In order to explain the diffraction patterns in 2-L rotated graphene with vacancies with two Pt atoms and latter on with 6 Pt cluster located on one of the layers, we shall proceed with the same kind of reasoning as we proceeded in different systems like crystalline graphite and dichalcogenides, when these systems were irradiated with different sources like, electrons γ-rays, etc. as reported by our group.

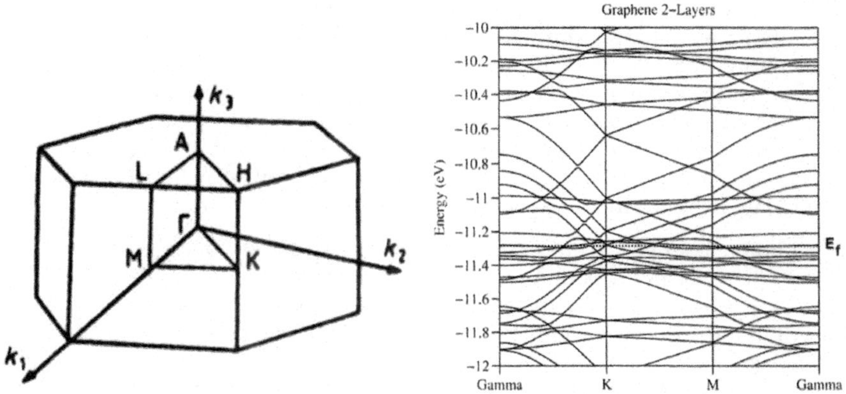

Fig. 4. Energy bands for graphene 2-L. The inset depicts the Wigner-Seitz cell for a hexagonal configuration.

Band structure calculations for 2-L graphene, graphene with two carbon vacancies, graphene with vacancies and two Pt atoms located on the top of one of the layers (from now on we shall call it surface), graphene with vacancies and a cluster of 6 Pt atoms located on the surface rotated by 3 degrees then rotated by 5 degrees were calculated using 51 k-points for each case, and sampling the First Brillouin zone (FBZ) as depicted in figure 4, figures 5 (a) to (b) and 6 (a) to (b), respectively.

First, notice that for graphene 2-L, the average energy is -5622.2462 eV and looking at Fig. 4, the system has a metallic behavior due that many degenerate bands overlap at the E_f.

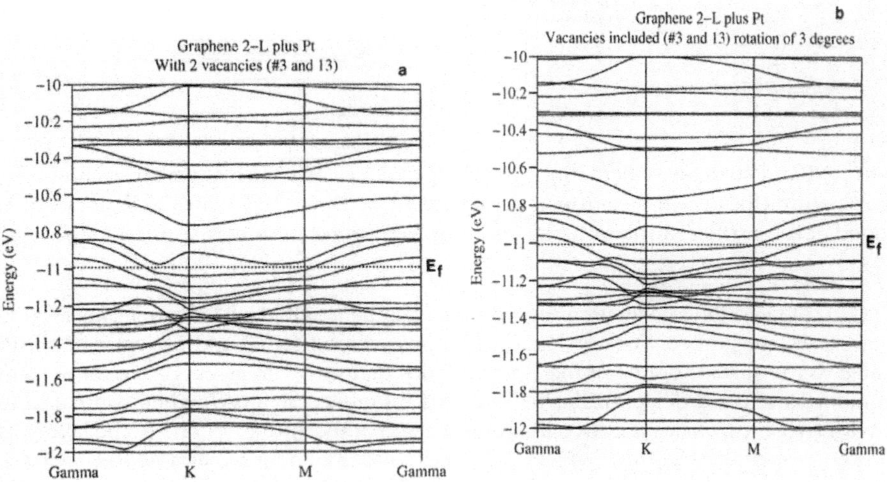

Fig. 5. (a) to (b) Band structure calculations for graphene 2-L plus two Pt atoms at the surface with two vacancies, graphene 2-L with 2 vacancies and two Pt atoms located at the surface when a rotation of 3 degrees was applied.

In order to answer the question regarding to what will it happen to the graphene when two platinum atoms were located on the surface of one of the platinum layers? Notice that the average energy for the new system changes from -5622.2462 eV to -5888.8251 eV respectively. At the same time, the E_f slightly changes due to the extra electrons provided by each Pt atom, Fig. 5 (a). Moreover, E_f is shifted to more negative values and due that Pt d-orbitals, which are located in the vicinity of the E_f, the system tends to behave as a semi metal. Then a rotation of 3 degrees is applied to one of the layers (the surface layer), as depicted in Fig. 5 (b), the average energy changes from -5888.8251 eV to -5888.8727 eV.

Notice a similar energy, except a small difference is noted after the decimal point. This means that the small perturbation was not enough to change the original system to another state and the space group remains the same as the unrotated one. On the other hand, when rotations of 4 and 5 degrees were applied to one of the planes, the average energy changes from -5888.8251 eV to -5891.6542 eV and -5889.3285 eV respectively, the perturbations applied were too severe that the systems obtained were different because the original symmetry was broken hence we have an unfavorable state.

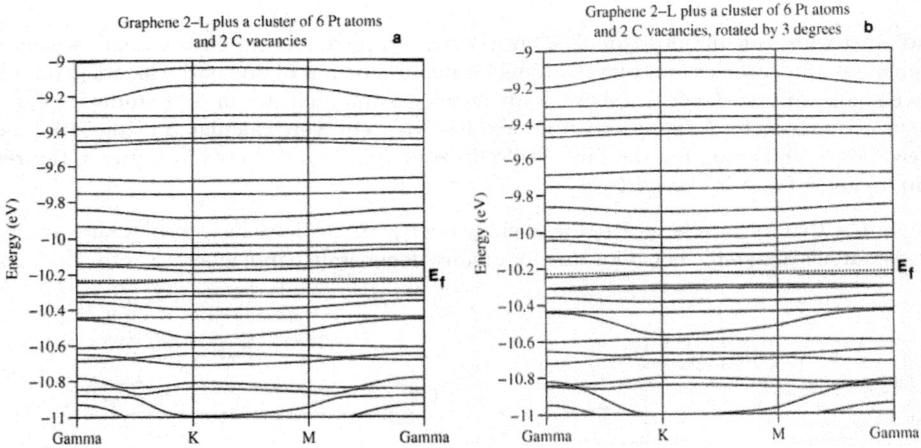

Fig. 6. (a) to (b) Band structure calculations for graphene 2-L , two carbon vacancies and a cluster of 6 Pt atoms located at the surface, graphene 2-L with vacancies and a cluster of Pt atoms located at the surface, a rotation of 3 and five degrees rotation was applied to the surface.

Figures 6 (a) to (b) yield the information regarding to graphene with two carbon vacancies (the order of vacancies considered through the calculations were of 2.5%) and a cluster of six Pt atoms located at the surface of one of the layers. Notice that the average energy of the system without rotation is of the order of -6246.7627 eV. The difference obtained, when compared to graphene with two C vacancies and two Pt atoms located at the surface of one of the layers, is attributed to the extra Pt atoms, which promotes that the new system tends to be a semiconductor with a forbidden energy gap $E_g \sim 0.05$ eV between the VB and CB, as depicted in Fig. 6 (a). When rotation in between layers of three degrees is applied, the average energy obtained was -6246.9018 eV similar to the unrotated case. The space group remains the same as the unrotated case, meaning that the rotation by three degrees is favored.

On the other hand, when a rotation of 5 degrees is applied to the surface layer, an average energy of -6247.2204 eV was obtained. Indicating that the perturbation applied was enough to place the system into another different state than the original one; we observe that the energy bands reflect a semi conductor behavior for the system with an $E_g \sim 0.05$ eV.

Moreover, when the 13 Pt atom cluster was located at the top (surface) of one of the layers a different scenario was observed. The total average energy for the new system change to -7119.7938 eV and the energy bands provide indication of semi metallic behavior. Furthermore, when a 3 degree rotation was applied to the surface, the average energy changes to -7120.0174 eV. The fact that the average energy changes (when compared to the un rotated case), provide an indication that the rotation applied was severe enough as to break the original symmetric group ending into another group, although the semi metallic behavior is still present. This behavior could be attributed to the stiffness provided to the material created when the cluster was located at the top surface layer.

We have been able to explain the original diffraction patterns observed in graphene with Pt atoms decorated on the surface, these are the direct result of a rotation of layers of the carbon honeycomb network. The rotations of 3 to 5 degrees are favored.

If the system is going to be a good promoter of the catalytic activity, the graphene (2-L) with vacancies and two Pt atoms on the surface, present metallic behavior, being the 3 to 5 degrees rotation the states which are favored.

On the other hand, when graphene and Pd case was analyzed, we proceeded in the following way: To prepare our theoretical *sample*, an infinite hexagonal honeycomb arrangement of carbon atoms were considered as depicted in Fig. 7 (a). To make the calculations simpler, a graphite single sheet made of 40 carbon atoms as depicted in Fig. 7 (b) were considered and in which 2 carbon atoms (3 and 13) were removed, as it would be a likely case of the effect of the proton irradiated beam on the sheet of carbon atoms. Moreover, these two carbon atoms were located at a distance of 1.67 Å (half way between the two layers) from the top layer and intercalated in between the second sheet, following a duplication of the second sheet and constructed of 40 carbon atoms in the AA configuration in order to create the two layers of graphene. The former configuration was made with the purpose of simulating a *real* scenario if -graphene 2Lwere subjected to proton irradiation.

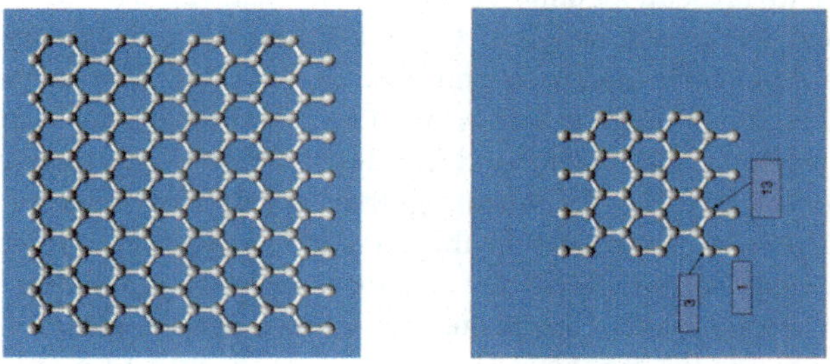

a) b)

Fig. 7. (a) Super cell generated from an infinite single sheet of carbon atoms. (b) A cell made out of 40 carbon atoms per layer was constructed.

A natural candidate to generate magnetism in coated graphene is Pd, due to its very polarizable bands. Furthermore, Pd is not intrinsically magnetic due to its close shell configuration ($[Kr]^{36}4d^{10}$) of the d orbitals. Although, in bulk, the s and d bands of Pd hybridize and exhibit Pauli paramagnetism with a high density of states in the vicinity of the Fermi level.

With these considerations in mind, our goal is to consider graphene two layers (2L) with two vacancies (carbon 3 and 13) intercalated in between the two layers and decorated with Pd atoms on one of the layers, in order to inquire about the changes occurred in the electronic and magnetic properties.

Band structure calculations for graphene two layers with vacancies (carbons 3 and 13), from now on indentified as graphene 2L (a), graphene two layers with two vacancies and intercalated in between the two layers, identified as graphene 2L (b), graphene two layers with vacancies and intercalated between the two layers with two Pd atoms located at the top of one of the layers (called surface), identified as graphene 2L (c) and last but not least, graphene two layers with vacancies intercalated between the two layers with 6 Pd atoms located in the center of six hexagons of the carbon layer (surface), identified as graphene 2L (d) are depicted on figures 8 (a) through (d) respectively.

For each figure the vertical axis is the total energy in eV *vs* k points in reciprocal space (horizontal axis). Figure 8 (a) provides information regarding energy bands for graphene 2L (a). Notice that a set of multiple degenerate bands cross the Fermi level, which is located at -11.24 eV. Obvious to say, that the system presents a metallic behavior. Figure 8 (b) yields energy bands for graphene 2 L (b). The two carbon atoms were located at 1.64 Å from the lower layer. Notice some differences were encountered when compared with figure 8 (a). The Fermi level was shifted to -11.08 eV with respect to the lower level. Moreover, the system behaves as less metallic due that only two degenerate bands interlaced and cross the Fermi level. The difference could be attributed to the two carbon atoms displaced form the network and intercalated in between the two layers.

In addition, Figure 8 (c) yields information regarding graphene 2L (c). Separation in between two Pd atoms is of the order of 2.84 Å and these two atoms were located at 0.98 Å from the surface. Note that the Fermi level has been displaced to -10.98 eV. Furthermore, the system continues to present metallic behavior like case (b). The difference encountered could be attributed to the interaction of the two Pd atoms with one of the carbon layers. Moreover, figure 8 (d) corresponds to graphene 2L (d). The differences encountered with respect to graphene 2L (a) are the following: The Fermi level was displaced to -10.25 eV while a minigap (between the Valence and Conduction bands) was manifested. The minigap was of the order of ~ 0.02 eV, indicating that the system behaves like a small gap semiconductor. This result implies that the 6 Pd atoms strongly influence the system for changing from metallic to semiconductor behavior. From the calculations described herein, the following conclusions about graphene decorated with Pd atoms can be drawn. The calculated energy bands for graphene 2L(a), graphene 2L (b) and graphene 2L (c) showed metallic behavior, while for graphene 2L (d) yielded a small gap semiconductor.

For each figure the vertical axis is the total energy in eV *vs* k points in reciprocal space (horizontal axis). Figure 8 (a) provides information regarding energy bands for graphene 2L (a). Notice that a set of multiple degenerate bands cross the Fermi level, which is located at -11.24 eV. Obvious to say, that the system presents a metallic behavior. Figure 8 (b) yields energy bands for graphene 2L (b). The two carbon atoms were located at 1.64 Å from the lower layer. Notice some differences were encountered when compared with figure 8 (a). The Fermi level was shifted to -11.08 eV with respect to the lower level. Moreover, the system behaves as less metallic due that only two degenerate bands interlaced and cross the Fermi level. The difference could be attributed to the two carbon atoms displaced form the network and intercalated in between the two layers.

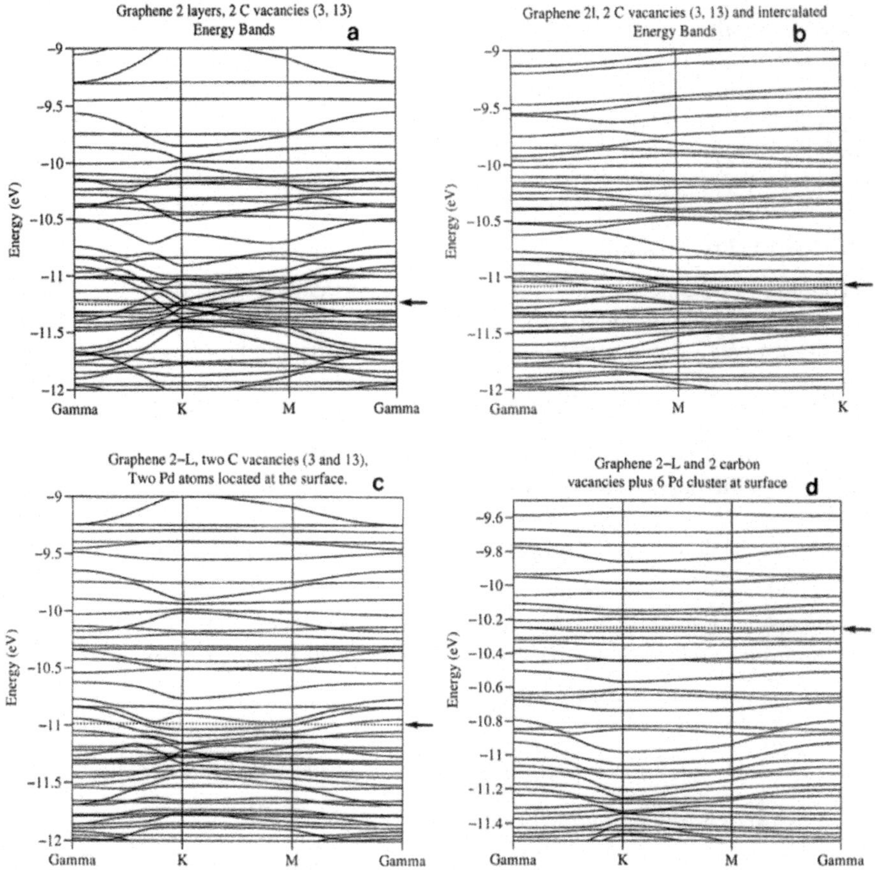

Fig. 8. (a) Graphene two layers with two Carbon vacancies (3 and 13) energy bands. The inset depicts the Wiegner-Seitz cell for a hexagonal configuration. (b) Graphene two layers with two Carbon vacancies (3 and 13) and intercalated in between the two Carbon layers. (c) Graphene two layers with two Carbon vacancies and intercalated inbetween the two Carbon layers with two Pd atoms on the surface layer. (d) Graphene two layers with two Carbon vacancies and intercalated in between the two Carbon layers with 6 Pd atoms on the surface layer.

In addition, Figure 8 (c) yields information regarding graphene 2L (c). Separation in between two Pd atoms is of the order of 2.84 Å and these two atoms were located at 0.98 Å from the surface. Note that the Fermi level has been displaced to -10.98 eV. Furthermore, the system continues to present metallic behavior like case (b). The difference encountered could be attributed to the interaction of the two Pd atoms with one of the carbon layers. Moreover, figure 8 (d) corresponds to graphene 2L (d). The differences encountered with respect to graphene 2L (a) are the following: The Fermi level was displaced to -10.25 eV while a minigap (between the Valence and Conduction bands) was manifested. The minigap was of the order of ~ 0.02 eV, indicating that the system behaves like a small gap semiconductor. This result implies that the 6 Pd atoms strongly influence the system for changing from metallic to semiconductor behavior. From the calculations described herein, the following conclusions about graphene decorated with Pd atoms can be drawn. The calculated energy bands for graphene 2L(a), graphene 2L (b) and graphene 2L (c) showed metallic behavior, while for graphene 2L (d) yielded a small gap semiconductor.

5. Results and discussion

The morphology and size distribution of metallic particles produced by the reduction of metallic salts in solution depends on various reaction conditions such as temperature, time, concentration, molar ratio of metallic salt/reducing agent, mode and order of addition of reagents, presence and type of protective agents, degree and type of agitation, and whether nucleation is homogeneous or heterogeneous [Sanguesa, C. D., et al., 1992].

Following the polyol method with ethylene glycol as solvent reductor, it was possible to obtain monometallic nanoparticles with narrow size distributions in systems and different structures depending on the temperature of reaction. The monometallic synthesis of nanoparticles by itself showed distinctive morphologies of the nanoparticles depending on the temperature of reaction.

Reaction proceeds in general as an oxidation of the ethylene glycol reducing the metallic precursor to its zero-valence state. [Carotenuto, G., et al. 2000; Sun, Y. et al., 2002]

$$OH\text{-}CH_2\text{-}CH_2\text{-}OH \rightarrow CH_3\text{-}CHO + H_2O \quad (1)$$

$$4(CH3 - CHO) + H_2PtCl_6 \rightarrow 2(CH_3 - CO - CO - CH_3) + Pt^0 + 6HCl \quad (2)$$

This reaction describes the reduction of Pt^{+2} to Pt^0.

$$OH - CH_2 - CH_2 - OH \rightarrow CH_3 - CHO + H_2O \quad (3)$$

$$2(CH_3 - CHO) + PdCl_2 \rightarrow (CH_3 - CO - CO - CH_3) + Pd^0 + 2HCl \quad (4)$$

This reaction describes the reduction of Pd^{+2} to Pd^0.

In the presence of a surface modifier, the reaction changes depending on the ability of the metal to coordinate with it, as in the case of PVP where the metallic precursor could coordinate with the oxygen of the pyrrolidone group, when the particles are in the nanometer size range, while when they are in the micrometer size range the coordination is mainly with the nitrogen, as reported by F. Bonet et al.[Sun, Y. et al., 2002; Bonet, F. et al., 2000]

By the polyol method with ethylene glycol as solvent-reductor, was possible to obtain monometallic platinum and palladium Nps with narrow size distributions in systems with small particles (2-4 nm) and different structures depending on the temperature of reaction. The structure of platinum and palladium Nps are cubic, face centered. The monometallic synthesis of Pt and Pd Nps by itself showed a distinctive morphology of quasi hemispherical small Pt and Pd Nps, which does not depend on the temperature of reaction has can be seen from Figure 1 (a) and (b).

The Pt and Pd Nps synthesized by polyol method, were deposited successfully on the HOPG in figure 2 can be seen Pt Nps on HOPG.

In the case when the Pt Nps were synthesized by direct reduction of platinum salt in H_2 flow on HOPG we obtained Nps with a considerable size distributions in systems of Pt Nps and different structures depending on the temperature of reaction. [Tomita &Tamai,1974; Baker et al., 1982]

By immersing the graphite electrode into a platinum plating solution were obtained Pt Nps in a narrow particle size distribution also for mean crystallite diameters smaller than 4 nm.

The formation of small Pd Nps with a uniform distribution over the electrode is achieved fast also by overpotential deposition as in the case of platinum with higher Pd^{2+} concentrations in solution.

It is important to observe that several patterns of diffraction of the samples Pt Nps on HOPG presented rotations of some degrees between the layers of HOPG in the direction perpendicular to the basal plane by effect of heating these samples in H_2. Also Moire patterns were observed in some of these samples. Honeycomb structures were observed on the HOPG surface.

For preparing supported platinum and palladium Nps on HOPG graphite for investigations of electrocatalysis, the advantage of the polyol method is the small size of the particles and the narrow distribution sizes of them. The second method have the potential to yield nanocrystals that are disposed in direct contact with HOPG; however, in neither case has it been possible to achieve good particle size monodispersity for platinum and palladium across a wide range of particle sizes.

Electrochemical deposition resulted an effective method to obtain directly nanoscale platinum and palladium particles on HOPG with a narrow size distribution.

The system remains semi metallic for Pt Nps calculations, except when a cluster of 6 Pt atoms was located on the surface of one of the graphene layers. For this case, the system behaves as a semi conductor with an energy gap $E_g \sim 0.05$ Ev, In most of the cases.

Energy bands were performed and provided information about the metallic behavior in the case of Pd Nps, for the three cases enunciated formerly, showing more metallic character for the first case, while less metallic behavior for the second one. Moreover, for graphene 2L with vacancies and intercalated with six Pd atoms on the surface showed a mini gap (between the Conduction and Valence Bands) of the order of 0.02 eV and manifesting semiconductor behavior.

6. Conclusions

In all cases, the deposition is sufficiently extensive to bind the Pt and Pd nanoparticles with HOPG and to permit the observation of a crystalline reorganization during the dissipation of the heat into the surroundings. We have been able to explain the original diffraction patterns observed in graphene with Pt atoms decorated on the surface, these are the direct result of a rotation of layers of the carbon honeycomb network.

From the theoretical analysis described herein, the following conclusions about graphene decorated with Pd atoms can be drawn that the graphene shown metallic behavior in some specific cases, which is very important to improve the physicochemical properties of graphene. These conclusions might apply as well to graphite, which also shows a two-dimensional spectrum, carbon nanotubes, which have a similar properties, and other materials with similar electronic structure.

7. Acknowledgment

Authors would like to acknowledge Professor Allen J. Bard, Facultad de Ciencias Físico Matemáticas, PAICYT Project and Microscopy Laboratory of CIIDIT de la Universidad Autónoma de Nuevo León for their support.

8. References

Ahmadi, T. S., Wang, Z. L., Green, T. C., Henglein, A., El-Sayed, M. A. (1996). Shape-Controlled Synthesis of Colloidal Platinum Nanoparticles. *Science,* 272, 1924.

Allongue, P., Souteyrand, E. (1993).*J. Electroanal. Chem.*362, 79.

Alvarez, M. M., Khoury, J. T., Schaaff, G., Shafigullin, M. N., Vezmar, I., Whetten, R. L. (1997). Optical Absorption Spectra of Nanocrystal Gold Molecules.*J. Phys. Chem. B.* 101, 3706.

Alvarez, S. (1993). *Tables of parameters for extended Hückel calculations.* (March 1993), Universitat de Barcelona, Spain.

Amiens, C., de Caro, D., Chaudret, B., Bradley, J. S., Mazel, R.; Roucau, C. (1993). *J. Am. Chem. Soc.* 115, 11638.

Andres, R. P., et al. (1992). Nanostructured Materials Promise to Advance Range of Technologies. *Chem. Eng. News,* (Nov 18, 1992).

Baker, B. E., Kline, N. J., Treado, P. J., Natan, M. J.(1996).*J. American Chemical Society,* 118, 8721.

Baker, R. T., Sherwood, R.S., Derouane, E. O. (1982).*J. Cat.* 75, 382.

Bard, A. J. Photochemistry.*Science.* (1980). 207, 139.

Bartolomew, C. H., Boudart, M. (1972).*J. Catalysis,* 25, 173.

Beyer, H., Müller, M., Schimmel, T. (1999).*Applied Physics A*, Science and Prosessing, Vol. 68, No. 2, pp.0947-8396.

Bonet, F., Delmas, V., Grugeon, S., Herrera-Urbina, R., Silvert, P. Y., Tekaia-Elhsissen, K. (1999). *NanoStruct. Materials*. 11, 1277.

Bonet, F., Tekaia-Elhsissen, K., Sarathy, K. V. (2000). *Bull. Mater. Sci. 23*, 165.

Bradley, J. S., Millar, J. M., Hill, E. W., Behal, S., Chaudret, B., Duteil, A. (1991). Surface Chemistry on Colloidal Metals: Spectroscopic Study of Adsorption of Small Molecules *Faraday Discuss. 92*, 255.

Bradley, J. S.; Hill, E. W.; Behal, S.; Klein, C. (1992). Preparation and characterization of organosols of monodispersed nanoscale palladium. Particle size effects in the binding geometry of adsorbed carbon monoxide.*Chem. Mater.* 4, 1234.

Bronstein, L. M., Chernyshov, D. M., Volkov, I. O., Ezernitskaya, M. G., Valetsky, P. M., Matveeva, V. G., Sulman, E. M. (2000). Structure and Properties of Bimetallic Colloids Formed in Polystyrene-*block*-Poly-4-vinylpyridine Micelles: Catalytic Behavior in Selective Hydrogenation of Dehydrolinalool. *J. Catal.* 196, 302.

Cachet, H., Froment, M., Souteyrand, E., Denning, C.(1992) *J. Elecrochem. Soc.* 139, 2920.

Cao, G. (2004). *Nanostructures and nanomaterials, synthesis, properties and applications*(First Edition). Imperial College Press, England.

Colvin, V. L., Schlamp, M. C., Alivisatos, A. P. (1994). Light-Emitting Diodes Made From Cadmium Selenide Nanocrystals and a Semiconducting Polymer.*Nature,* 370, 354.

Diculescu, V.C., Chiorcea-Paquim, A.M., Corduneanu, O., Oliveira Brett, A.M. (2007). Palladium nanoparticles and nanowires deposited electrochemically: AFM and electrochemical characterization, *J.Solid State Electrochem.* 11, 887-898.

Esumi, K., Matsuhisa, K., Torigoe, K.(1995). Preparation of Rodlike Gold Particles by UV Irradiation Using Cationic Micelles as a Template. *Langmuir.*11, 3285.

Freelink, T., Visscher, W., Van Veen, J. A. R.(1995). *J. Electroanal. Chem.* 382, 65.

Galván, D. H.(1998) Extended Hückel calculations on cubic boron nitride and diamond. *J. Mat. Sci. Lett.*17, 10, 805-810.

Galván, D. H., Posada-Amarillas, A., Elizondo, N., Mejia, S., Pérez-Tijerina E., José-Yacamán, M. (2009). Diffraction Patterns Observed in Two-layered Graphene and Their Theoretical Explanation. *Fullerenes, Nanotubes and Carbon Nanostructures.*17, 258-272.

Galván, D. H., Posada-Amarillas, A., Núñez-González, R., Mejía S., José-Yacamán M. (2010). Study of Vacancies and Pd Atom Decoration on the Electronic Properties of Bilayer Graphene. *Journal of Superconductivity and Novel Magnetism,* 23, 1543-1550.

García Gutiérrez D.I., Gutiérrez-Wing, C. E., Giovanetti, L., Ramallo-López, J. M., Requejo, F. G., José-Yacamán, M.(2005). Temperature Effect on the Synthesis of Au-Pt Bimetallic Nanoparticles. *J. Phys. Chem. B* , 109, 3813-3821.

Glassey, V. W., Papoian, G. A., Hoffmann, R., (1999). The energy partition within one-electron formalism: A Hamiltonean population of study surface-CO interaction in the c(2x2)-CO/Ni (100) chemisorption system. *J. Chem. Phys.* 111, 3, 893-910.

Grabar, K. C., Freeman, R. G., Hommer, M.B., Natan, M. *J.* (1995). Preparation and Characterization of Au Colloid Monolayers. *Anal. Chem.* Vol. 67, No. 4, (February 1995), pp.735-743.

Halperin, W. P. (1986). Quantum Size Effects in Metallic Powders*Rev. Mod. Phys.* 58, 533.

Han, S. W., Kim, Y. Kim, K.,(1998). Dodecanethiol-Derivatized Au/Ag Bimetallic Nanoparticles: TEM,. UV/VIS, XPS, and FTIR Analysis. *J. Colloid Interface Sci.*, 208, 272.

Henglein, A. (1993). Physicochemical properties of small metal particles in solution: "microelectrode" reactions, chemisorption, composite metal particles, and the atom-to-metal transition.*J. Phys. Chem. 97*, 5457.

Henglein, A. (2000). Preparation and Optical Aborption Spectra of $Au_{core}Pt_{shell}$ and $Pt_{core}Au_{shell}$ Colloidal Nanoparticles in Aqueous Solution. *J. Phys. Chem. B*.104, 2201.

Hirai, H. (1979). Formation and catalytic functionality of synthetic polymer noble metal colloid.*J. Macromol. Sci.-Chem. A*, 13, 633.

Hirai, H., Chawanya, H., Toshima, N. (1985). Colloidal palladium protected with poly(*N*-vinyl-2-pyrrolidone) for selective hydrogenation of cyclopentadiene. *React. Polym. 3*, 127.

Hirai, H.; Toshima, (1986) N. *Tailored Metal Catalysts*; Iwasawa, Y., Ed., D. Reidel, Dordrecht, pp 87-140.

Hoffmann, R. (1963).An Extended Hückel theory. I. hydrocarbons. *J. Chem. Phys.* 39, 6, 1397-1412.

Hoffmann, U., Wilm, D. Z. (1936). *Elektrochem.*, 42, 42.

Hull, A. W., (1917). A new method of X-ray crystal analysis, *Phys. Rev.* 10, 6, 661-696.

Kan, C.; Cai, W., Li, C.; Zhang, L., Hofmesiter, H. (2003) *J. Phys. D: Appl. Phys. 36*, 1609.

Katakuse, I. (1994). *J. Mass Spectrom. Soc. Jpn. 42*, 67.

Knight, W. D., Clemenger, K., deHeer, W. A., Saunders, W. A., Chou, M. Y., Cohen, M. (1984). *Phys. Rev. Lett. 52*, 2141.

Leff, D. V., Ohara, P. C., Heath, J. R., Gelbart, W. M. (1995). *J. Phys. Chem.* 99, 7036.

Lewis, L. N. (1993). *Chem. Rev.* (Washington, D.C.) 93, 2693.

Moiseev, I. I. (1995). *J. Mol. Catal. 95*, 109.

Park, H. K.; Lim, Y. T.; Kim, J. K.; Park, H. G.; Chung, B. H. (2008). *Ultramicroscopy* 108,10, 1115.

Pauling, L. (1960). The Nature of the Chemical Bond, p. 235, 3rd. Edition.

Poulin, J. C., Kagan, H. B., Vargaftik, M. N., Stolarov, I. P., Moiseev, I. I. *J. Mol. Catal.* (1995), 95, 109-113.

Reetz, M. T., Helbig, W.(1992).*J. Am. Chem. Society*, 116, 7401.

Reetz, M. T., Winter, M., Tesche, B. (1997). *Chem. Commun.* 147.

Sanchez-Ramirez, J. F., Pal, U. (2001). *Superficies Vac. 13*, 114.

Schmid, G. (1988). *Polyhedron*, 7, 2321.

Schmid, G. (1992). *Chem. Rev.* (Washington, D.C.)*92*, 1709.

Schmid, G. (1994).*Clusters and Colloids* (First Edition), Verlagsgesellschaft, *Weinheim*, ISBN 527-29043-5, Germany.

Schmid, G., Giebel, U., Huster, W., Schwenk, A. (1984). *Inorg. Chim. Acta 85*, 97.

Schmid, G., Harms, M., Malm, J.O., Bovin, J.O., van Ruitenbeck, J., Zandbergen, H. W., Fu, W. T.(1993), *J. Am. Chem. Soc.* 115, 2046.

Schmid, G., Morun, B., Malm, J. O. (1989). *Angew. Chem., Int.* Ed. Engl. *28*, 778.

Snow, D., Major, M., Green, L. (1996). *Microelectronic Engr.* 30, 969.

Sugano, S. (1991). *Microcluster Physics*; Springer-Verlag: Berlin.

Sun, Y., Yin, Y., Mayers, B. T., Herricks, T., Xia, Y. (2002). *Chem. Mater. 14*, 4736.

Tekaia-Elhsissen, K., Bonet, F., Silvert, P. T., Herrera-Urbina. R. (1999). *J. Alloys Compounds 292*, 96.

Teranishi, T., Miyake, M. (1998). *Chem. Mater.* 10, 594-600.

Teranishi, T., Hosoe, M., Miyake, M. (1997). *Adv. Mater. 9*, 65.

Teranishi, T., Nakata, K., Miyake, M., Toshima, N. (1996). *Chem. Lett.* 277.

Thomas, J. M., Raja, R., Johnson, B. F. G., Hermans, S., Jones, M. D., Khimyak, T.(2003). *Ind. Eng. Chem. Res.* 42, 1563.

Tomita A.,Tamai, Y. (1974).*J. Phys. Chem.*, 78, 2254.

Toshima, N., Harada, M., Yonezawa, T., Kushihashi, K., Asakura, K. (1991). *J. Phys. Chem. 95*, 7448.

Toshima, N., Kuriyama, M., Yamada, Y., Hirai, H. (1981). *Chem. Lett.* 793.

Toshima, N., Takahashi, T. (1992). *Bull. Chem. Soc. Jpn.* 65, 400.

Toshima, N., Wang, Y. (1994). *Adv. Mater.* 6, 245.

Toshima, N., Yamaji, Y., Teranishi, T., Yonezawa, T. Z. (1995). *Naturforsch. 50a*, 283.

Turkevich, J.; Stevenson, P.; Hillier, J. (1951). *Discuss. Faraday Soc.* 11, 55.

Vargaftik, N. M.; Moiseev, I. I.; Kochubey, D. I.; Zamaraev, K. I. (1991). *Faraday Discuss. Chem. Soc.* 92, 13.

Viau, G., Toneguzzo, P., Pierrard, A., Acher, O., Fievet-Vincent, F., Fievet, F. (2001). *Scripta Mater.* 44, 2263.

Volokitin, Y., Sinzig, J., de Jong, L. J., Schmid, G., Vargaftik, M. N., Moiseev, I. I. (1996). *Nature,* 384, 621.

Wei-xia, T. *(2008). Chinese Journal of Polymer Science.* 26, 1, 23−29.

Whangbo, M-H., Hoffmann, R. (1978). The band structure of the tetracyanoplatinate chain. J. *Am. Chem. Soc.,* 100 ,19, 6093-6098.

Williams, D. B., Carter, C. B.(1996).*Transmission Electron Microscopy,* Plenum Press, New York.

Willner, I., Maidan, R., Mandler, D., Dürr, H., Dörr, G., Zengerle, K. (1987). *J. Am. Chem. Soc.* 109, 6080.

Wolfsberg, M. W.; Helmholtz, L. The spectra and electronic structure of the tetrahedral ions MnO_4^{-1}, CrO_4^{-2}, and $ClO4^{-1}$. (1952). *Chem. Phys.* 20, 5, 837-843.

Yang, D-Q., Zhang, G.-X., Sacher, E. , José-Yacamán, M., Elizondo N. (2006). Evidence of the Interaction of Evaporated Pt Nanoparticles with Variously Treated Surfaces of Highly Oriented Pyrolytic Graphite.*J. Phys. Chem. B.* 110, 8348-8356.

Yonezawa, T., Tominaga, T., Toshima, N. (1995). Novel Characterization of the Structure of Surfactants on Nanoscopic Metal Clusters by a Physicochemical Method. *Langmuir.* 11, 4601.

Yonezawa, T., Toshima, N. (1993). Polymer- and micelle-protected gold/platinum bimetallic systems. Preparation, application to catalysis for visible-light-induced hydrogen evolution, and analysis of formation process with optical methods. *J. Mol. Catal. 83*, 167.

Zoval, J. V., Lee, J., Gorer, S., Penner, R. M.(1998). Electrochemical preparation of platinum nanocrystallites with size selectivity on basal plane oriented graphite surfaces.*J. Phys. Chem. B*, 102, 1166-1175.

13

Ultrastructural Mechanisms of Aposporous Embryo Sac Initial Cell Appearance and Its Developmental Process in Gametophytic Apomicts of Guinea Grass (*Panicum maximum*)

Lanzhuang Chen[1*] and Liming Guan[2]
[1]*Faculty of Environmental and Horticultural Science, Minamikyushu University, Tateno, Miyakonojo city, Miyazaki,*
[2]*Faculty of Education and Culture, University of Miyazaki, Gakuen Kibanadai Nishi, Miyazaki, Japan*

1. Introduction

The ultrastructural mechanisms of aposporous embryo sac initial cell (AIC) appearance and its developmental process have been studied in gametophytic apomicts of guinea grass (*Panicum maximum* Jacq.), using ovary clearing treatment and Nomarski differential interference-c The ultrastructural mechanisms of aposporous embryo sac initial cell (AIC) appearance and its developmental process have been studied in gametophytic apomicts of guinea grass (*Panicum maximum* Jacq.), using ovary clearing treatment and Nomarski differential interference-contrast microscopy (DIC), and ultra-thin section and the transmission electron microscopy (TEM). For the observation of AIC appearance by DIC, AIC appears while megaspore degenerated in apomictic accessions, and most of the ovules contain several AICs, and the number of AICs increased as the ovary grew before anthesis. That is, several AICs in the same ovule did not differentiate synchronously, but instead, they seemed following a continuous course and appeared one by one during the period from after megasporogenesis to the first AIC-derived embryo sac maturity. It was also found that the higher the frequency of apospory was, the greater the number of AICs was, and the longer the duration of AICs appearance should be. For the mechanism of seed-forming embryo development in polyembryonic ovules in apomictic accessions observed by DIC, the first AIC is located dominantly in micropylar end, and the percentage of mature embryo sacs in micropylar end was higher than that in the other ends. The rates of the ovules contained developed embryo sacs in micropylar end at 4 days after anthesis were 33~91% comparable to 0~2% of that in the other ends. The embryo of the developed sac in micropylar end, in final, became a seed-forming embryo, and in contrast, the other sacs,

* Corresponding Author

were crowed out to chalazal end and degenerated at 10 or more days after anthesis. Form the results described here, in can be concluded that the AIC-derived embryo sac in micropylar end, in most of cases, has temporal dominant in embryo sac mature, and has the positional dominant in fertilization and subsequent development of embryo sac when compared with the cases in the other end in polyembryonic ovules.

For the AIC appearance and its development in apomictic accessions observed by TEM, the first sign observed is the degeneration in dyad stage, and accompanying it, the cell adjacent to the dyad in chalazal end, derived from nucellus began to increase its size of cell and change its organelles and shape, and thus, the cell become to the first AIC usually located in micropylar end. The AICs appear in order continently adjacent to the first AIC in chalazal end up to the stage of anthesis. For the AICs further development, they, as the substitute of megaspore, form four-nucleate embryo sac, containing one egg, one polar and two synergids at anthesis. The characteristics of reproductive cells formed in embryo sac derived from AIC are observed in detail, and the interdependent patterns of the nucellar cell and AIC, and their relationship are also discussed at subcellular level. These ultrastructural evidence obtained here provides important information to isolate AIC-specific genes (=apomixis-specific gene) and to understand the mechanism how the somatic nucellar cell changes into reproductive AIC by using single cell manipulation.

Apomixis is a reproduction mode that bypasses female meiosis and syngamy to produce embryos genetically identical to the maternal parent. In general, three major mechanisms of apomixis have been identified based on the origin and development of cells from which the embryo derives, and those are considered as important reproduction modes in plant breeding (Asker 1979, Asker and Jerling 1992, Bashaw and Hanna 1990). Apomixis phenomenon involves in many kind of plants in different types. Apomixis not only represents an interesting reproduction mode that can result in the reproduction of 4- and 8-nucleate embryo sacs simultaneously, i.e. in *Panicum maximum* (guinea grass), but also provides a method for cloning plants through seeds. From the unique reproduction mode, the significance of apomixis in two standpoints, i.e. macro and micro, can be considered. In the macro, apomixis can promise to bring the economic benefits to human being more than that of the "Green Revolution", by means of fixation of F1 varieties to cost down F1 production expense, fixation of hopeful breeding materials to shorten the times of breeding generation, and transformation of apomixis-specific gene into vegetative plants, like sweet potato and potato, to produce seeds to avoid the usage of seed tubers. And in micro, to understand the mechanism how the somatic nucellar cell changes into reproductive aposporous embryo sac initial cell (AIC) will take place another "Revolution of Reproduction" in plant. For example, somatic cell can be changed into sexual one like apomixis, and in the reverse, sexual cell can be the somatic one, by means of artificial biotechnology. In this meaning, the understanding and resolution of sexual reproduction can be expected not only in plant but also in human being.

In this chapter, we will focus on the appearance of AIC and its further developmental process in transmission electron microscopy (TEM) using the guinea grass as a model, based on the same observation in Nomarski differential interference-contrast microscopy (DIC). We will structure the wider discussion around the knowledge of apomixis we have accumulated from our study of *P. maximum*, which we established a model system of apospory.

2. Cytological mechanisms of AIC appearance and its developmental process using DIC

Up to now, three major mechanisms of apomixis, that is, apospory, diplospory and adventitious embryony, have been identified based on the origin and development of cells from which the embryo derives (Hanna and Bashaw 1987). Warmke (1954) firstly described that in guinea grass aposporous apomixis with pseudogamy is the mechanism of apomictic reproduction. Following that, researches on the reproduction mode in guinea grass have been performed to reveal the mode inheritance of apomixis in sexual (Smith 1972, Hanna et al. 1973) and in apomictic accessions (Warmke 1954, Savidan 1975, Savidan and Pernes 1982, Nakajima and Mochizuki 1983, Nakagawa 1990, Chen et al. 2000, 2001). And in the other species *Pennisetum*, co-inheritance of apomictic reproduction and two molecular markers has been demonstrated by Ozias-Akins et al (1993). Among the researches above, an important and key point is that by how many genes the apomixis is controlled, and the point has attracted many scientists working on it day and night for over a half century. They had attempted to clarify the genetic mode of apomixis using the traditional method of crossing hybridization (Daniel et al. 1998a, b). Unfortunately, the plant materials which appear apomictic figures are usually belonging to tetraploids, so that it is more difficult to find out the solution in tetraploids than that in diploids. However, the researches made by the scientists could be concluded with that apomixis may be mainly controlled by a single dominant or a few tightly linked genes (Savidan 1975, 1989, Nakajima and Mochizuki 1983, Chen and Kozono 1994a).

Although some researches have been reported on cytology and inheritance of apomixis in guinea grass, the period of apomictic gene expression, the key to cloning the apomixis genes has not been identified yet. When observing the differences between sexual and apomicts, the appearance of AIC, from which aposporous embryo sac is derived, should be considered the most relevant stage for the expression of apomixis genes, as these cells appear only in apomictic but not in sexual plants. Here, the problem is that the mechanism of AIC development is not well understood. And as a hypothesis that if an observable or measurable index could be found out through cytological evidence, it can be used to estimate the range of period of AIC appearance based on it the apomixis genes may be cloned. This study mainly describes cytological observation of AIC appearance and its development using DIC, and based on it provides information on the best timing for sampling materials in the program for apomixis gene isolation.

The application of DIC was carried out to study cytologically the mechanism of AIC appearance and its developmental process in guinea grass (*P. maximum*) using an improved method (Herr 1982). For obtaining the reliable data, seven facultively apomictic accessions and three obligately sexual accessions collected from Tanzania, Zambia and Japan were used in our study, and one hundred to 300 bus or flowers staged before and at thesis were collected per accessions for embryo sac analysis. The important point is that the pre-treatment of the samples be performed with FPA50 (formalin : propionic acid : 50% ethanol = 5:5:90) (Chen and Kozono 1994a) for one week at 4 °C. After the treatment of ethanol series, the samples are cleared in Herr (1982) fluid of benzyl-benzoate-four-and-half for over 2h at 0-4 °C. The observation of the samples was conducted suing DIC.

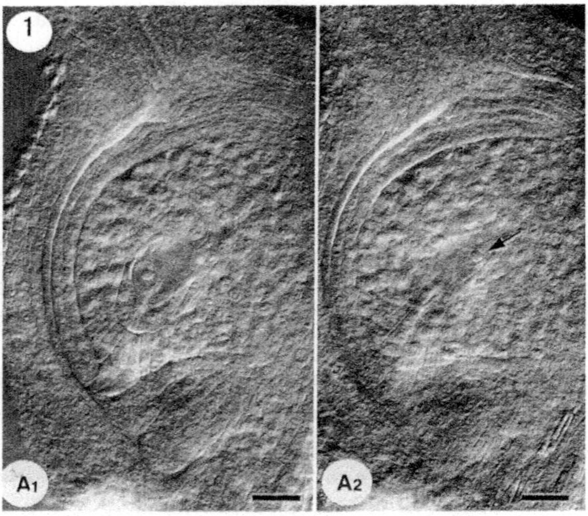

Fig. 1. Appearance of aposporous embryo sac initial cell (AIC) in facultative apomictic *Panicum maximum*. (A1) appearance of AIC in micropylar end and (A2) remaining of functional megaspore without nuclear membrane in chalazal end (arrow) in one ovule. Bar= 50μm.

Until megasporogenesis, differences between obligate sexual and aposporous accessions were not observed in ovule development. After that, the megaspore of sexual ovules showed a manner typical of the *Gramineae* family forming 8-nucleated embryo sacs as described by Hanna et al. (1973). In contrast, the same of apomictic ovules stopped its continuous development or degenerated. Usually at the same times that the megaspore lost its function, the unreduced (2n) nucellar cells around the degenerated megaspore appeared and enlarged their sizes, moved to the empty space and developed into a functional AIC (Fig. 1). The earliest AIC usually appears in ovule always located in where would turn toward and finally become micropylar end as the ovary grows. And then, the AIC divided twice and directly into 4-nucleated embryo sacs, containing one egg, two synergids and one polar nucleus. The AIC usually exists with megaspore, which often degenerated in the same ovule. Usually, several AICs appear in one ovule and their number increase as the ovary grows until anthesis. To clarify the mechanism of appearance of AIC, ovary length was selected as an index and measured when they were observed in different AIC appearance and its development. From the ovary length measuring result, it is understood that the AICs do not appear together in same time, instead, they seemed following a continuous course and appeared one by one during the period from megasporegenesis even to first embryo sac maturity. And accessions with higher frequency of apomixis have plural embryo sacs and also showed wider range of period of AIC appearance, when ovary length was measured. As another meaning, that is, their developmental stages of the ovary can be estimated using the ovary length as an index. The AIC appearance is a unique event in apospory different from the sexual, and may correspond with the time of gene expression of apomixis. Here, based on a hypothesis that the time of apomixis gene expression is just before the time of AIC appearance, it could be considered that the more the number of embryo sacs per ovule are, the longer the expression period will be. Therefore, as the materials for apomixis gene cloning, the accessions containing higher number of embryo sacs / higher frequency of apomixis, should be advantageous for their having wider duration of AIC appearance.

The ovary length of stages of first AIC was longer than that of functional megaspore in most accessions also indicates that apospory is initiated after megasporogenesis. And more, the ovary length staged in functional megaspore was wide and close to that staged in degeneration of embryo sac. These results also support that the development of sexual embryo sac is often terminated in many aposporous apomicts at the megaspore mother or megaspore stage, and the products of sexual process degenerate (Nogler 1984, Asker and Jerling 1992). However, as the limitation of DIC observation, the sexual termination in stage of megaspore mother cell was not observed in this study.

About the sexual embryo sac formation in ovules with plural embryo sacs in facultative apomictic plants, Nakajima and Mochizuki (1983) have reported that a few ovules had two sexual embryo sacs, and that in polyembryonic ovules having both sexual and apomictic embryo sacs, number of sexual embryo sacs was limited to be one. In this study, five types of embryo sac formation were recognized. That is, SS: only two 8- or 5-nucleate mature sexual embryo sacs in an ovule; S: only one 8-nucleate embryo sac in an ovule; An: one or more 4-nucleate mature apomictic embryo sacs; S+An: one 8- and one or more 4-nucleate embryo sacs in an ovule; SS+An: two 8- and one or more 4-nucleate embryo sacs in an ovule. Fore the case of ovules with one or two 5(8)-nucleate embryo sacs appeared with or without 4-nucleate ones in one ovule, two pathways could be considered that 1) the sexual embryo sac formation results from the direct division of one or two megaspore(s) though the AIC(s) appeared (or not) in the same ovules; and that 2) it is derived from AIC(s). In particular, as the ovules with two megaspores in chalazal end were not observed in this study while AIC(s) appeared in the micropylar end, the former pathway could be hardly considered as putative one. If try to explain the appearance of ovules containing one or two sexual embryo sacs with or without 4-nucleate apomictic ones, the later, however, seems reasonable based on that AICs develop into not only 4-nucleate but also, at a low frequency, 5-nucleate embryo sacs in *Panicum* (Nakajima and Mochizuki 1983), or 8-nucleate ones in *Hieracium* (Nogler 1984). For the 5-nucleate embryo sac formation, it could be imaged that after megaspore or AIC divided into two nuclei, only the micropylar nucleus continued to divide twice and to form 4-nuclei, the chalazal nucleus and one of the four micropylar nuclei pair with each other to form two polar nuclei. On the other hand, a rare case of 4-nucleate embryo sac divided from AIC, with an egg cell, one synergid cell and two polar nuclei was also depicted for *Panicum* by Bashaw and Hanna (1990). From the above observation of embryo sac formation reported in this study, it could be concluded that facultative apomictic guinea grass forms not only common and rare *Panicum*-type (4-nucleate), but also its typical type (5-nucleate), and in addition, *Polygonum*-type (8-nucleate), within the same one ovule.

Based on the observation of this study that the period of AIC appearance is the key to catch the apomixis genes, and ovary length can be used as an index to do sampling of the key period of AIC appearance, apomixis-specific gene-1 (*ASG*-1) and its family genes have been cloned from the key period of AIC appearance in facultative apomictic guinea grass using differential screening method (Chen et al 1999).

3. Cytological mechanisms of AIC-deriving egg cell and seed-forming embryo development

From the above observation of DIC, we have known that facultative apomicts of guinea grass usually form one more embryo sacs derived from AICs in the same ovule. While doing germination test of seeds, only one germinated from the facultative apomicts. There is an important problem in facultative apomicts that which one will be the seed-forming embryo in progeny, and it is important for breeder to know the frequency of sexual or apomixis when doing crossing experiments. Therefore, an effective and repaid method for estimating the degrees of sexuality or apomixis is needed to be developed for their utilization in various plant breeding programs (Chen and Kozono 1994b). Up to now, two methods to estimate the degree of sexuality, i.e. embryo sac analysis and progeny test were often combined and used by many breeders (Warmke 1954, Smith 1972, Hanna et al. 1973, Savidan 1975, Savidan and Pernes 1982, Nakajima and Mochizuki 1983, Nakagawa 1990). Progeny test is impractical to provide direct estimation of frequencies of sexual and apomictic embryo sacs, and to observe entire variability in progeny (Nakajima and Mochizuki 1983). Embryo sac analysis permits it possible to identify the morphology of embryo sacs (Nakagawa 1990). Nakajima and Mochizuki (1983) had compared the two methods and indicated that two degrees of sexuality expressed as embryo sac type and frequency of off-type plants agreed mostly well in both of highly and lowly sexual accessions, but disagreed in the facultative apomicts. As two types of sexual and apomictic embryo sacs coexist in a same ovule, it remains unknown in facultative apomicts which embryo sac will become the seed-forming one. This study mainly focuses to demonstrate cytologically the events at the stages of embryo sac formation and seed-forming embryo development in polyembryonic ovules of facultative apomicts, to find out the relations between the stages and then, to provide reliable information for the estimation of the degree of apomixis or sexuality without progeny test.

To clarify the cytological mechanism of seed-forming embryo development in polyembryonic ovules in facultative apomictic guinea grass, the samples staged after anthesis of seven facultative apomicts were collected every day up to 10th day after open pollination (DAP), and observed using DIC and improved clearing method (Chen and Kozono 1994b). The continuous observation of the ovules indicate that, 1) the first AIC is located dominantly in micropylar end, and the percentage of mature embryo sacs in micropylar end was higher than that in the other end; 2) the rates of ovules containing embryo at 4 DAP was 90% in micropylar end higher than 2% in the other end; 3) the embryo in micropylar end, in final, became a seed-forming embryo, and in contrast, the others were crowded out to chalazal end and degenerated at 10 or days. The above results suggest that the degrees of sexuality or apomixis can be estimated based on the frequency in present generation, even without progeny test.

For the AIC appearance, Chen and Kozono (1994a) indicated that during the period from after megasporogenesis to anthesis, AIC appears one by one in facultative apomictic plants of *P. maximum*, according to a continuous course, and the first AIC appeared in the ovule always located dominantly in the micropylar end, forming mature 4-nucleate embryo sac . On the other hand, Nakajima and Mochizuki (1983) described that the most vigorous embryo sac, sexual or apomictic one, is considered to be representative of the polyembryonic ovule. In fact, however, it is difficult to decide which is the most vigorous one, especially, for the beginner while observation. To find out a solution of deciding which embryo sac will develop into seed-forming embryo, is very important not only in establishment of determination standard for all of observers but also in evaluation of sexuality or apomixis degree (Chen and Kozono 1994b). This study attempted to compare the percentage of mature sacs and the sizes of nucleoli of the embryo sacs positioned in micropylar end with those in the other ends, to determine the reproduction mode of the polyembryonic ovules. The result that the percentage of mature sacs and sizes of nucleoli of embryo sac in micropylar end were more advantageous than those in the other end, means that the AIC formed in micropylar end has the temporal dominant in formation and maturity of the embryo sac when compared with the other sacs (Fig. 2). In other words, the dominantly matured sac in micropylar end has been ready for fertilization. This result was also supported by that if numerous embryo sacs exist in an ovule, the embryo sac in the favorable position, i. e., closest to the micropylar end of the ovule, is usually the one that the pollen tube enters (Koltunow 1993).

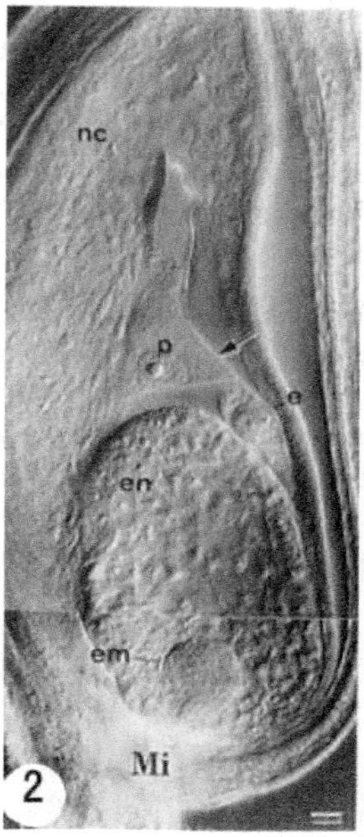

Fig. 2. Dominant development of AIC-derived embryo sac in micropylar end with pre-globular-stage embryo and well-developed endosperm at 3 DAP in polyembryonic ovule of facultative apomictic *Panicum maximum*. e = egg cell; en = free-nuclear endosperm; p = polar nucleus; em = embryo; nc = nucellar cell; Mi = micropylar end; bar = 50μm.

Accession	No. ovules observed	No. embryo sacs	E.s. in micropylar end				E.s. in the other end			
			Embryo and endosperm		Embryo only	Endo-sperm only	Embryo and endosperm		Embryo only	Endo-sperm only
	A		B	(B/A)			C	(C/A)		
T-75	46	66	42	(91%)	2	2	0	(0%)	0	0
T-41	47	144	34 (1)[1]	(68%)	5 (1)[2]	2	0 (1)[1]	(0%)	2 (1)[2]	0
N68/84-1-s-6	40	115	20	(43%)	0	0	0	(0%)	0	0
Natsukaze	54	63	18	(33%)	0	0	0	(0%)	0	0
N68/96-8-o-10	40	55	30	(75%)	0	0	0	(0%)	0	0
N68/96-8-o-11	102	235	78 (6)[1]	(76%)	3	0	2 (6)[1]	(2%)[3]	0	0

[1] Twin embryo sacs containing both of embryo and endosperm, respectively, in an ovule. [2] Twin embryo sac containing one embryo but no endosperm, respectively, in an ovule. [3] The embryo sac in the other end, instead o in the micropylar end, containing developed embryo and endosperm.

Table 1. Development of embryo sac (e.s.) in the flowers 4 DAP in facultative apomictic accessions of *Panicum maximum*.

Next, the fertilization degrees of different embryo sacs in an ovule were examined to certify that if the position of embryo sac located has the advantage in fertilization. The results were obtained that 33~98% of the ovule contained the developed sac in micropylar end in all of the 6 accessions tested, but in the other ends, 0% in 5 accessions and 2% in N68/96-8-o-11, at 4 DAP, respectively (Table 1). It strongly supports the above proposition. When observing the process of the embryo and endosperm developments, it is found out that the sac in micropylar end is usually the one that will be fertilized (or pseudogamy) dominantly (here, the fertilization means that when endosperm formation it is needed) and, just the egg develops automatically to embryo in the sac without fertilization, in final, became to the seed-forming embryo. On the other hand, the other sacs have no chance to receive entrance of pollen tube for fertilization, and be crowded out by the developed sac to chalazal end, in final, be completely degenerated after 10 DAP.

This study provides reliable data for effective embryo sac analysis. 1) If the matured embryo sac in micropylar end is apomictic one, apomictic seed-forming embryo should be formed, which can be evaluated at anthesis; 2) If the sac is sexual one, sexual seed-forming embryo should be formed, which can also be evaluated at anthesis. Therefore, it can be concluded that with this method mentioned here, it should be possible that using the frequency of apomictic and sexual embryo sacs in micropylar end at anthesis estimate the degree of sexual or apomixis, even without progeny test. It is interesting as a further project to compare the two degrees of sexuality or apomixis estimated by embryo sac analysis as described here and by progeny test. Although the same project had been conducted (Nakajima and Mochizuki 1983), a new conclusion will be expected using the method described here.

4. Ultrastructural mechanism of AIC appearance

Chen et al (1999) has reported that the *ASG*-1 gene was cloned using differential screening method based on ovary length as an index to sample the different developmental stages of facultative apomictic guinea grass. And then, the *ASG*-1 was used as a probe to do in situ hybridization in apomictic guinea grass, indicating that the signals of the gene expression were detected not only in the expected AIC and AIC-derived embryo sac but also in unexpected anther (Chen et al. 2005). The above result indicates that AIC appearance is most different event from sexual, so further characterization of real timing of AIC appearance means importance not only in clarification of the mechanism but also in cloning of aposporous gene.

For the observation of AIC in guinea grass, many efforts have been turned to this work for a long time in section method (Warmke 1954) and in embryo sac analysis method (Savidan, 1975, Nakajima and Mochizuki 1983, Nogler 1984, Nakagawa 1990, Chen and Kozono 1994a, b). They have made it clear that 1) the first difference between sexual and apomicts was degeneration of megaspore in facultative apomictic guinea grass; 2) after megasporogenesis, it is almost at the same time that megaspore degenerates and AIC appears; 3) when examining the degree of sexuality or apomixis in facultative or obligate apomictic ovary with polyembryonic ovules, the embryo sac located in micropylar end represents the fate of the ovary as it develops into the only one of seed-forming embryo (Chen and Kozono 1994b). Naumova and Willems (1995) firstly reported an ultrastructural characterization of aposporous megagametophytes in *P. maximum*, indicating that AIC usually differentiates adjacent to degenerating tetrad, and megaspore or 1-nucleated sexual embryo sacs. And they also indicated that little is known about the processes that regulate cellularization and differentiation of aposporous megagametophytes.

In the section, we focus the ultrastructural features of the developing aposporous megagametophytes in guinea grass (*P. maximum*), in particular, on the timing of degeneration in process of megasporogenesis, AIC differentiation and AIC-derived embryo sac formation. And the interdependent patterns of the nucellar cell and AIC, and their relationship are also discussed at subcellular level.

The facultative apomicts of guinea grass, "Petrie" and "Gatton" of 2 varieties were used and the samples were collected from 3 classes of the stages 1) before and completion of megasporogenesis, 2) the AIC appearance and 3) formation of the AIC-derived embryo sac based on ovary length as an index (Chen and Kozono 1994a). The ovaries were fixed and the fixed samples were dehydrated by a graded ethanol series and embedded in Spurr's resin (Guan and Adachi 1997). Ultrathin sections were cut on an ultramicrotome using a glass knife and double stained with uranyl acetate and lead citrate. The sections were viewed with a HITACHI H-800 MV transmission electron microscopy (TEM) at 75 kV.

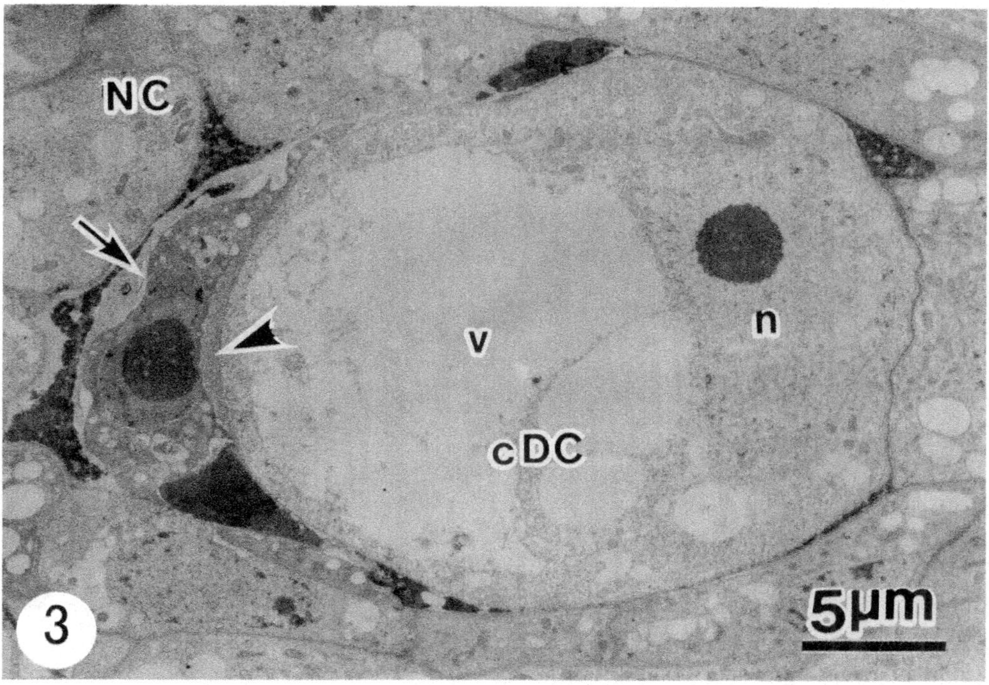

Fig. 3. A dyad stage of the megasporocyte displaying the degenerated dyad cell (arrow) in micropylar end and well-developed chalazal dyad cell (cDC) with the electron-dense material in the transverse wall (arrowhead). cDC = chalazal dyad cell; fn = free nuclear; n = nucleus; NC = nucellar cell; v = vacuole.

This study mainly focused on the timing and processes in degeneration of sexual cells and the appearance of AIC at ultrastructural level. It is found that the degeneration occurred as early as the stage of dyad. As first event, micropylar dyad was degenerated and chalazal dyad showed vacuoles in high degree during megasporocyte division (Fig. 3). The occurrence of vacuoles is normal phenomenon in megasporegenesis process (Guan and Adachi 1994), and it is also an auspice of cell degeneration (Guan and Adachi 1997) reported in *Fagopyrum esculentum*. The ultrastructure of dyad cells of *P. maximum* is comparable to that observed in other plants (Russel 1979, 1985). The vacuole appearance may be one of auspice of sexual cell degeneration in the case of *P. maximum*. The thickness of chalazal dyad cell wall between micropylar end and chalazal end is different, and there usually appeared thick layer of callose in micropylar cell wall. That some electron dense materials are present in the transverse wall of the dyad and tetrad was also observed in this study. It is bright comparison between that micropylar dyad degenerated completely and that there appeared electron-dense wall in chalazal dyad toward micropylar end. As Naumova and Willems (1995) indicated, the incomplete callose wall is probably an early sign of low activities, which will be followed by degeneration of the megaspore. As the appearance of both complete and incomplete callose in different ovules, the incomplete callose wall is probably not the reason for the onset of apospory, but a sign of apospory.

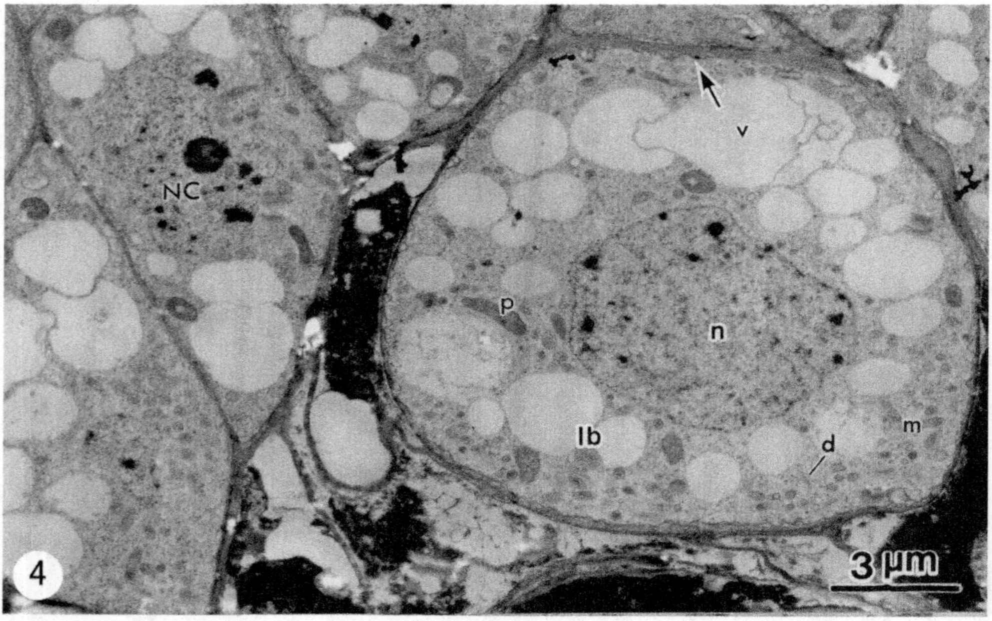

Fig. 4. A near round shape nucellar cell (NC) occurred adjacent to the degenerated cell located in lower side and to the other NC in other directions. The NC formed its original cell wall with two cell layers between the round NC and normal NC (arrow), and contains abundant contents with vacuole (v), lipid body (lb), dictyosomes (d), plastid (p), mitochondria (m) and a big nucleus (n) located in the center of the NC, that differs from the normal NC. Here, the newly formed round NC was named as AIC.

For the AIC appearance, it is found that the dyad generated, at the same time, the cell of nucellus showed subtle change, meaning appearance of AIC (Fig. 4). The stage of AIC appearance is earlier than tetrad stage reported by Naumova and Williams (1995). At the chalazal end of degenerated sexual cells, it is usually observed that some nucellar cells appear with bigger size than normal nucellar cells and with round shape. These cells have their independent cell walls. As ovule develops in early stage, the cells continue to enlarge their size, and show vacuole, a host of RES and high dense of nucleotides. These subtle changes described here mean the cell will develop into AIC. There are no differences observed between the nucellar cells around megasporocyte in ultrastructure. The nucellar cells have capability to form aposporous embryo sacs, and they depend upon and inhibit each other by the communication between thin cell wall and cytoplasm, with which the internal balance is kept. However, the capability to become AIC is very different according to the position of nucellar cells located in (Chen and Kozono 1994a, b, Chen et al. 2000, 2001, Naumova and Willems 1995). Although every nucellar cell has the chance to develop into AIC, only one or some cells break down the balance and inhibition among the nucellar cells, and after extreme competition, become the AIC to develop into AIC-derived aposporous embryo sac, in final.

And then, how the AIC develops into embryo sac will be described. The AIC will continues to differentiate and forms ellipse initial cell, and the size of which is 8.5 times of the round nucellar cells in volume. The AIC has uniformly thick and complete wall, and lack communication substance between cells. The thickness of cell wall increases, and plasmodesmata connections diminish leading to the end of symplastic transport and separating the cell from the other.

Chen et al. (1999) indicated that the *ASG*-1 gene isolated from stage of AIC appearance in facultative apomictic guinea grass, containing 1177 bp and 305 amino acids, shows the 45-48% positive similarity to polygalacturonase 1 beta chain precursor (Polyg 1) of *Lycopersicon esculentum* (Zhang et al 1992) that has a bifunctional plant proteins that interact with both structural components of the cell wall and catalytic proteins to localize and/or regulate metabolic activities within the cell wall. *ASG*-1 existing in AIC and AIC-derived embryo sac is also reported by in situ hybridization (Chen et al. 2005). *ASG*-1 with cell wall growing function is also supported by the ultrastructural analysis as described here. The structure of cell wall thickness means that AIC has already had the differentiation capability of self-complete and genetic system. The AIC, finally, results in formation of aposporous embryo sac.

In early stage of ovule development, around the AIC many degenerating cells at different stages of megasporogenesis with high density and black degenerated cytoplasm in center of the ovule. These degenerated cells could be considered as that 1) they are dyad, tetrad, megaspores, or degenerated sexual cells staged in sexual embryo sac formation; 2) some nucellar cells degenerated. As AIC appears, it competes with nucellar cells round it, and absorbs nutrients from them for its own differentiation. At the same time, as the volume of the AIC increases, it pushes and affects the other cell around it. Therefore, it is considered that when the AIC divides into embryo sac, it absorbs part nutrients from nucellar cells around it, and from degenerated sexual cells or sexual embryo sac.

Concluding the above mentions in ultrastructural analysis, it can be found that from the subtle change of organelles, volume and sizes, and shape of nucellar cells, aposporous development has already began while sexual cell, e. g. dyad, degenerated. Generally, sexual cells and nucellar cells are localized in balance. However, it is considered that nucellar cells located in chalazal end inherit and develop dominantly when they sense some signals from sexual ones which will be the fate of abortion and degeneration, and/or lose capacity of division. At the same time, one of nucellar cell-derived AIC cells appears, sexual cells and other nucellar cells provide nutrients for AIC differentiation (Guan et al. 2006, 2007, 2008). In some meanings, the life of the ovule (or ovary) will be kept by whatever either sexual or asexual (nucellar) cell becomes to embryo sac. In this way, that over 90% of ovules contain aposporous embryo sac, and lower 10% of megaspores in small ratio developed dominantly into sexual embryo sac, was observed in both of facultative apomictic guinea grass, 'Petrie' and 'Gatton' (Chen and Kozono 1994a, b).

5. Ultrastructural mechanism of AIC-derived embryo sac formation

For the AIC differentiation in facultative apomictic guinea grass, the above section has already given a clear description in ultrastructural level. However, the mechanism of AIC-derived embryo sac formation has not yet been described in ultrastructural level, although DIC observation of it had been reported by Chen and Kozono (1994a, b). In a practical meaning of breeding, when examining the degree of sexual or apomixis in facultative apomictic ovule with AIC-derived embryo sac in *P. maximum*, the method of progeny test, i. e. percentage of apomictic origin, was used, in general. After the ovule-clearing technique (so called embryo sac analysis) used, the most vigorous embryo sac is considered to be representative of multiple embryo sacs in ovule for estimation of apomixis degrees (Nakajima and Mochizuki 1983). However, the standard of estimation of the most vigorous embryo sac in the ovule should be different according to the observers. Using the improved ovule clearing technique, Chen and Kozono (1994b) have defined that, the embryo sac located in micropylar end represents the fate of the ovule as the sac develops into the only one of seed-forming embryo. And a new challenge for isolation of reproductive cells of aposporous embryo sac has been reported by Chen et al. (2005), indicating the differences of size, morphology and numbers between egg, synergid and polar in the sac. Little is still known about ultrastructural characteristics of reproductive cells in aposporous embryo sac in *P. maximum*. This study wants to make it clear why the AIC located in micropylar end develops dominantly, and what the functions of the reproductive cells are, as the characteristics of AIC in ultrastructural analysis may provide information to understand the overall functions of gametophytic apomict.

For the sample preparation, the facultative apomicts of guinea grass, “Petrie” and “Gatton” of 2 varieties were used and the samples were collected. The ovaries were fixed and the fixed samples were dehydrated by a graded ethanol series and embedded in Spurr’s resin (Guan and Adachi 1997). Ultrathin sections were cut on an ultramicrotome using a glass knife and double stained with uranyl acetate and lead citrate. The sections were viewed with a HITACHI H-800 MV transmission electron microscopy (TEM) at 75 kV.

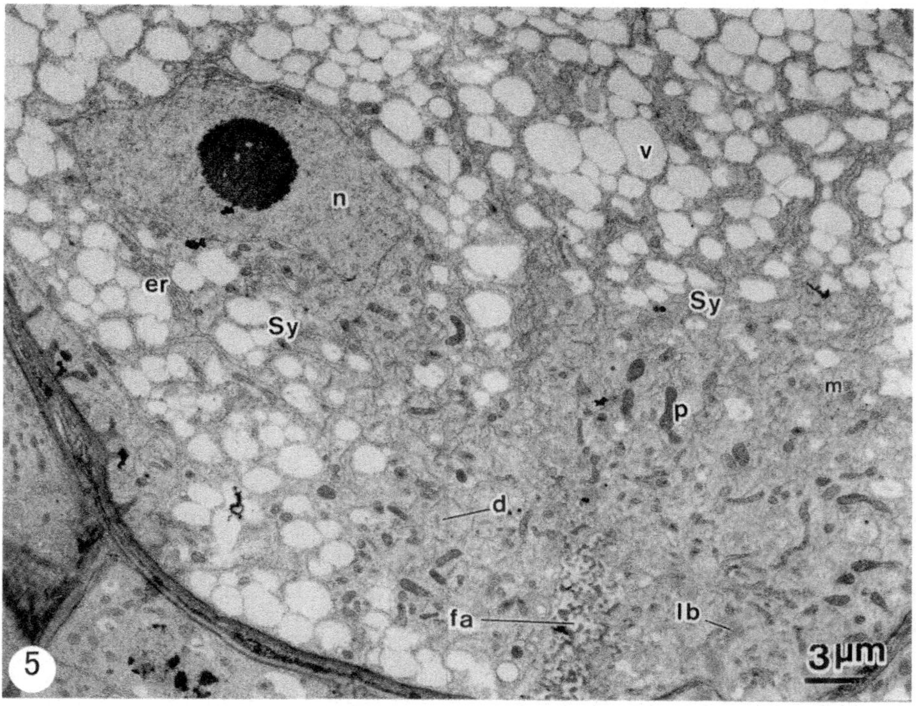

Fig. 5. Two synergids (Sy) with filiform apparatus (fa) in the micropylar end of the cell. v = vacuole; m = mitochondria; p = plastid; lb = lipid body; d = distyosome; er = endoplasmic reticulum.

For cloning of apomixis-specific gene, two strategies are mainly tried out, i. e. using genomic DNA (Ozais-Akins et al 1998), and using differential screening of mRNA of AIC (Chen et al 1999). In particular, when using mRNA to do gene cloning the examination of special developmental stage, which is considered as the target gene expression stage, will be important (Chen et al 2005). Moreover, for the developmental stage determination, ultrastructural data is stricter than that of observation of normal microscopy. However, ultrastructural data about apospory and other types of apomixis are scarce and not discussed in recent publication (Koltunow 1993). And thereafter, ultrastructural analysis of AIC appearance in *P. maximum* have been reported in KK-15 by Naumova and Willemse (1995), and in two varieties of 'Petrie' and 'Gatton' by Chen and Guan (2006), respectively. This study shows ultrastructural characteristics of the reproductive cells in AIC-derived embryo sac.

In sexual and aposporous *P. maximum*, the developmental process of gametophytes is generally reported as shown in Fig. 6, according to the previous reports (Warmke 1954, Savidan 1975, Chen and Kozono 1994a). And Chen and Kozono (1994a, b) has indicated that while the megaspore failed to divide, some nucellar cells enlarge and become the AIC to replace the megaspore to form 4-nucleate sac, based on the observation of embryo sac analysis by using DIC.

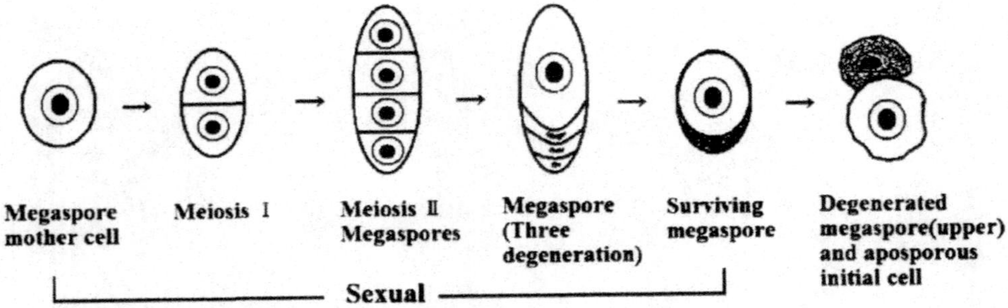

Fig. 6. The illustration of megasporogenesis in sexual and aposporous *P. maximum*.

In multiple aposporous embryo sac ovule of *P. maximum*, each sac displays difference from the other in size, reproductive cell number, cell location and cell developmental stage. In the same ovule, embryo sacs usually are located in the center of ovule, and every ovule has different number of sac, i. e. usually from two to four. As the number of the sac increases the difference in its sizes among the sacs becomes bigger. This result agrees with that AIC appears as ovary length grows (Chen and Kozono 1994a). The micropylar end of the sac displays half-sphere shape, and the chalazal end is irregular.

Here, different number of reproductive cells between the embryo sacs located in micropylar and the other end was found out by DIC and TEM. The sac located in micropylar end usually consists of four cells. And the sac develops dominantly from micropylar end so that the numbers of reproductive cells in the sac and the size of the sac are more and bigger than that located in chalazal end. At anthesis, the sacs in different developmental stages coexist, and later formed AICs develop to immature sacs containing 1-nucleate, coenocytes or incomplete developed cells. All of the sac have monopolarity and lack antipodals. This constitution of sac in aposporous *P. maximum* is different from another aposporous *Hieracium aurantiacum*, which have seven cells with eight nuclei same to sexual embryo sac (Bicknell 1997, Koltunow et al. 1998).

For the first aposporous embryo sac development, which is usually located in micropylar end of the ovule, there are the egg apparatus with two synergids and one egg cell arranged triangularly in the sac. The egg apparatus differentiates ultrastructurally with morphological characteristics similar to those found in sexually functional synergids and egg cell. The triangular organization of the egg apparatus is not necessarily conversed as the unreduced egg cell may appear in more lateral position with respect to one of synergids (Naumova and Vielle-Calzada 2006). The first aposporous embryo sac with 4 cells usually develops dominantly and becomes the only seed-forming embryo (Chen and Kozono 1994b). It can be considered that as the first aposporous embryo sac contains four cells and is located in micropylar end from where pollen tubes go through the filiform apparatus and enter into the synergid (Guan and Adachi 1997).

For the other sacs, they showed containing less than four cells, and feathers of degeneration. During development of the sacs, there are competition in space and nutrition among the sacs, so that some sacs may fail to develop. The degenerated embryo sac, in some meaning, will provide nutrition for aposporous embryo sac development. This degeneration often takes place in multiple aposporous embryo sac ovule, indicating that the more the number of the sac are, the rates of degeneration are higher. From the points of structure and nutrition, the results that the sac located in micropylar end gives 76% of rates of embryo formation, but the other only 2% (Chen and Kozono 1994b), can be understood well. This result also indicates that the sac in micropylar end decides the fate of the seed formation in multiple sac ovule. This also supports the estimation method reported by Chen and Kozono (1994b) that the embryo sac in micropylar end represents sexuality of the ovary.

There showed clear deference between the both ends of the egg apparatus. That is, the micropylar end is separated by many normal cell walls of synergid, egg-synergid, synergid-polar, and egg-polar, and in the chalazal end, it is only surrounded by the plasma membrane. For the cell wall ingrowths, Chapman and Busri (1994) indicated that in facultative apomictic genotype in *Pennisetum*, many wall projections were observed in the micropylar region, but few ingrowths were found in the chalazal region of both sexual and apomictic megagametophytes. The results obtained in this study are also agreement with their observations.

For the development of synergids in the first sac, the cell wall between two synergids is expanded to form conveniently orientated filiform apparatus in the micropylar end. However, variational filiform apparatus presents in micropylar end of the synergid located in the other sac, where only one synergid formed. The content feature of the single synergid was not different from that of the two synergids. The synergids had characteristics of highly vacuolated cells and high localized cytoplasm containing a large and irregular shaped nucleus. The filiform apparatus formed only in first sac has conformity with that only one seed-forming embryo formed in multiple embryo sacs in ovule of guinea grass (Chen and Kozono 1994b). This fact also confirms indirectly the estimation to be correct, that is, in ovule with multiple embryo sacs, the first sac located in micropylar end represents the reproductive mode of the ovule (Chen and Kozono 1994b). According to the hypothesis of Savidan (1989), which postulates through a complete egg cell wall, the sperm-cell penetration cannot be taken place. However, such a complete wall in this study was not observed in the mature aposporous embryo sac.

For the ultrastructures of the aposporous embryo sac and the reproductive cell in the sac, they have own characteristics, respectively. Egg cell containing highly dense cytoplasm, and around its nucleus plastids and mitochondria are remarkable. Lipid bodies and rER and dictyosomes are few, and big vacuoles are distributed in cytoplasm. The polar cell is occupied almost by one big vacuole, the lower dense cytoplasm than that of egg cell wraps one to two nuclei and is distributed around egg apparatus. That only in the sac in micropylar end the cell wall of the polar exists with ingrowth of cell wall shows that this structure transports the nutrition for the embryo development. Synergid usually contains lowest dense of cytoplasm, and lipid bodies, plastids and mitochondria are distributed along to the filiform apparatus in micropylar end. Extreme rER and vacuoles distributed in chalazal end of the synergid might contain abundant inorganic substance and they are absorbed as a part of nutrition for the developing embryo (Fig. 5). The ultrastructural analysis of AIC and AIC-derived embryo sac formation described in this study will give information not only for gene cloning but also for single AIC manipulation in next step.

6. References

Asker SE. 1979. Progress in apomixis research. *Hereditas* 91: 231-240.

Asker SE, Jerling L. 1992. Apomixis in plants. *CRC Press*, Boca Raton.

Bashaw EC, Hanna WW. 1990. Apomictic reproduction. In Reproduction Versatility in the Grasses, GP Chapman, ed. (Cambridge: *Cambridge University Press*), pp 100-130.

Bicknell RA. 1997. Isolation of a diploid, apomictic plant of *Hieracium aurantiacum*. *Sex. Plant Rerpod.* 10:168-172.

Chapman GP, Busri N. 1994. Apomixis in *Pennisetum*. An ultrastructural study. *Int. J. plant Sci.* 155:492-497.

Chen LZ, Amano S, Guan LM, Adachi T. 2005. Preliminary isolation of viable protoplasts from egg cells in facultative apomictic guineagrass (*Panicum maximum*). *Cytologia* 70:171-179.

Chen LZ, Guan LM, Kojima A, Adachi T. 2000. The mechanism of appearance of aposporous initial cell and apomictic embryo sac formation in *Paspalum notatum*. *Cytologia* 65:333-341.

Chen LZ, Guan LM, Kojima A, Adachi T. 2001. The mechanism of polyembryonic seed set in *Paspalum notatum*. *Cytologia* 66:157-165.

Chen LZ, Guan LM, Seo M, Hoffmann F, Adachi T. 2005. Developmental expression of ASG-1 during gametogenesis in apomictic guinea grass (*Panicum maximum*). *J. Plant Physiol.* 162: 1141-1148.

Chen LZ, Kozono T. 1994a. Cytology and quantitative analysis of aposporous embryo sac development in guinea grass (*Panicum maximum* Jacq.). *Cytologia* 59: 253–260.

Chen LZ, Kozono T. 1994b. Cytological evidence of seed-forming embryo development in polyembryonic ovules of facultatively apomictic guinea grass (*Panicum maximum* Jacq.). *Cytologia* 59: 351–359.

Chen LZ, Miyazaki C, Kojima A, Saito A, Adachi T. 1999. Isolation and characterization of a gene expressed during early embryo sac development in apomictic guinea grass (*Panicum maximum*). *J. Plant Physiol.* 154: 55–62.

Daniel G., Leblanc O., Espinosa E., Perotti E., Leon D., Savidan Y. 1998a. Mapping diplosporous apomixis in tetraploid *Tripsacum*: one gene or several genes? *Heredity* 80: 33-39.

Daniel G., Leblanc O., Espinosa E., Perotti E., Leon D., Savidan Y. 1998b. Non-mendelian transmission of apomixis in maize-Tripsacum hybrids caused by a transmission ratio distortion. *Heredity* 80: 40-47.

Guan LM, Adachi T. 1994. Ultrastructural changes of the mature embryo sac in buckwheat (*Fagopyrum esculentum*) as a result of high temperature exposure. *Cytologia* 59:281-292.

Guan LM, Adachi T. 1997. Ultrastructural studies of embryogenesis in *Fagopyrum esculentum*. *Int. J. Plant Sci.* 158:110-120.

Guan LM, Chen LZ, Terao H. 2006. Ultrastructural studies of gametophytic apomicts in guinea grass (*Panicum maximum*). I. Differentiation of aposporous initial cell. *Cytologia* 71:379-389.

Guan LM, Chen LZ, Terao H. 2007. Ultrastructural studies of gametophytic apomicts in guinea grass (*Panicum maximum*). II. Characteristics of aposporous initial cell-derived embryo sac. *Cytologia* 72:145-153.

Guan LM, Chen LZ, Adachi T. 2008. Ultrastructural studies of embryo abortion in buckwheat (*Fagopyrum esculentum*) as a heat-stress. *Cytologia* 73:371-379.

Hanna WW, Bashaw EC. 1987. Apomixis: Its identification and use in plant breeding. *Crop Sci.* 27:1136-1139.

Hanna WW, Powell J, Millot JC, Burton GW. 1973. Cytology of obligate sexual plants in *Panicum maximum* Jacq. And their use in controlled hybrids. *Crop Sci.* 13:695-697.

Herr JMJr. 1982. An analysis of methods for permanently mounting ovules cleared in four-and-a-half type clearing fluids. *Stain. Technol.* 57:161-169.

Koltunow AM. 1993. Apomixis: embryo sac and embryos formed without meiosis or fertilization in ovules. *Plant cell* 5: 1425-1437.

Koltunow AM, Johnson S, Bicknell RA. 1998. Sexual and apomictic development in *Hieracium*. *Sex. Plant Reprod.* 11:213-230.

Nakagawa H. 1990. Embryo sac analysis and crossing procedure for breeding apomictic guineagrass (*Panicum maximum* Jacq.). *JARQ* 24:163-168.

Nakajima K, Mochizuki N. 1983. Degrees of sexuality in sexual plants of guinea grass (*Panicum maximum* Jacq.). *Japan J. Breed.* 33: 45-54.

Naumova TN, Vielle-Calzada JP. 2006. Ultrastructural analysis of apomictic development. In: Savidan Y, Carman JG, Dresselhaus T. (eds.). The flowering of apomixis: from mechanism to genetic engineering. *CIMMYT*, European Union, and Institut de recherché pour le developpement. pp 44-63.

Naumova TN, Willemse MTM. 1995. Ultrastructural characterization of apospory in *Panicum maximum*. *Sex. Plant Reprod.* 8:197-204.

Nogler GA. 1984. Gametophytic apomixis. In: Johri BM (ed.): *Embryology of Angiosperm. Springer-Verlag* Berlin pp 475-518.

Ozias-Akins P, Lubbers EL, Hanna WW, McNay JW. 1993. Transmission of the apomictic mode of reproduction in *Pennisetum*: co-inheritance of the trait and molecular markers. *Theor. Appl. Genet.* 85:632-638.

Ozias-Akins P, Roche D, Hanna WW. 1998. Tight clustering and hemizygosity of apomixis-linked molecular markers in *Pennisetum squamulatum* implies genetic control of apospory by a divergent locus that may have no allelic form in sexual genotypes. *Proc. Natl. Acad. Sci. USA* 95: 5127-5132.

Russell SD. 1979. Fine structure of megagametophytes development in *Zea mays*. *Canad. J. Bot.* 57:1093-1110.

Russell SD. 1985. Preferential fertilization in *Plumbago*: Ultrastructural evidence for gamete-level recognition in an angiosperm. *Proc. Natl. Acad. Sci.* (USA) 82:6129-6132.

Savidan YH. 1975. Heredite de l'apomixie, contribution a l'etude de l'heredite de l'apomixie sue *panicum maximum* Jacq. (analyse des sacs embryonnaires). *Cah. ORSTOM ser Biol* 10: 91-95.

Savidan Y. 1989. Apomixis in plant breeding: transfer *vs.* synthesis. *Apomixis Newsletter* 1:22-24.

Savidan Y, Pernes J. 1982. Diplod-tetraploid-dihaploid cycles and the evolution of *Panicum maximum* Jacq. *Evolution* 36:596-600.

Smith RL. 1972. Sexual reproduction in *Panicum maximum* Jacq. *Crop Sci.* 12:624-627.

Warmke HE. 1954. Apomixis in *Panicum maximum*. *Amer. J. Bot.* 41:5-11.

Zhang L, Heupel RC, DellaPenna D. 1992. The beta subunit of tomato fruit polygalacturonase isoenzyme 1: isolation, characterization, and identification of unique structural features. *Plant Cell* 4:1147-1156.

14

Ultrastructure and Cell Wall Thickness Modification and Its Detection After Chemical Treatments in Huanglongbing Infected Citrus Plants

Hajivand Shokrollah[1], Thohirah Lee Abdullah[2] and Kamaruzaman Sijam[3]

[1]*Department of Plant and Seed Improvement, Agriculture and Natural Resources Research Center of Qazvin, Qazvin,*
[2]*Departments of Crop Science, Faculty of Agriculture, University Putra Malaysia, Serdang,*
[3]*Department of Plant Protection, Faculty of Agriculture, University Putra Malaysia, Serdang,*
[1]*Iran*
[2,3]*Malaysia*

1. Introduction

Citrus is the leading fruit crop grown in the world. Many factors are known to limit the production and processing of citrus. Unfortunately many diseases such as Huanglongbing (HLB) have been threatening citrus industry in the world. The disorder of citrus described as "greening" disease has caused great damage in the citrus industry and is known as Huanglongbing (HLB) disease (Bove, 2006). It was detected in most citriculture areas of Asia, Africa, USA and most *Citrus* species are susceptible to this disease.

The HLB disease pathogens are fastidious phloem-limited bacteria in the genus *Candidatus* Liberibacter. The isolate from South Africa has been named *Candidatus* Liberibacter africanus, and the isolate from Asia has been named *Candidatus* Liberibacter asiaticus (Jagoueix et al., 1994). Symptoms of HLB were observed in sweet orange trees near the city of Araraquara in São Paulo State. This was the first reported case of HLB from the American Continent (Coletta-Filho et al., 2005; Teixeira et al., 2008). A new type of HLB (*Candidatus* Liberibacter) species is genetically characterized, and the bacterium is designated as "*Candidatus* Liberibacter psyllaurous". This bacterium infects the psyllid *Bactericerca cockerelli* and its solanaceous host plants potato and tomato, potentially resulting in "psyllid yellowing" (Hansen et al., 2008).

Virus and bacteria such as HLB pathogen can be detected using transmission electron microscope (TEM). This method has been used to detect the HLB pathogen since 1993. The shape and length of bacteria can be determined by TEM but the disadvantage of TEM methods is that it required special equipment (Bove and Garnier, 2002). Different protocols have been developed to detect the bacteria disease in plant by Roistacher (1991). A ''mycoplasma-like organism'' was observed in citrus phloem tissue infected with citrus greening disease using a transmission electron microscope (TEM) (Bove and Garnier, 2002). Electron microscopy measurements on thin sections showed that the filamentous forms of the liberibacteria have a diameter of 0.2-0.3 micrometer (µm). Variations in diameter occurred between organisms and sometimes within a single organism. Round forms were larger (0.5 µm) with a less dense cytoplasm and often showed plasmolysis. The *Candidatus* Liberibacter spp., envelope was a membranous wall characteristic of Gram-negative bacteria, but the peptidoglycan layer was hardly visible (Garnier et al., 1984). There is no evidence of flagella or pili. *Candidatus* L. asiaticus and *Candidatus* L. africanus cannot be distinguished on the basis of morphology. As far as it is known, none of the HLB strains have been obtained in a media culture (Bove and Garnier, 2002).

However, the HLB organisms were about 2000 nanometer (nm) long and 100-200 nm in diameter. Similar bodies were observed in both vectors of the citrus greening disease, *Ttioza erytreae* (Moll and Martin, 1973) and *Diaphorina citri* Kuwayama (Chen *et al.*, 1973). The greening organism was compared with citrus stubborn organism (spiroplasma) by Saglio *et al.*, (1971), the study showed that the outer membrane of the greening organism was much thicker (25 nm) than that of the spiroplasma (10 nm). HLB bacterium was found to be Gram-negative bacteria type (Garnier *et al.*, 1984).

Molecular detection methods have been difficult to develop since the greening organism hasn't been cultured (Halbert and Manjunath, 2004). Recently, molecular methods such as polymerase chain reaction (PCR) have become available to detect HLB disease and provided the credible detection of HLB disease. *Candidatus* Liberibacter spp., in infected plants and psyllid vector has been detected using DNA technology. Halbert and Manjunath (2004) reported that Jagoueix *et al.* (1996) used universal primers for general amplification of prokaryotic, 16S ribosomal DNA (rDNA).

Based on sequence information, primers have been developed to amplify 1,160 baise pair (bp) region of ribosomal DNA for detection of greening by PCR. Ribosomal DNA primers have been used widely for detection of both forms of HLB. These primers have do not amplify 16S ribosomal sequences of other citrus pathogens.

Detection of citrus HLB pathogens from asymptomatic tissue is inconsistent by any known method. Similarly, the molecular assays sometimes are complicated to run, and results are not always reliable. Clearly, more accurate, timely, and robust detection methodologies are needed. The outputs of the study may improve fruit quantity and quality, and reduce the risk of HLB in Malaysian citrus industry and citrus industries around the world. Therefore, the present research was carried out to detect *Candidatus* Liberibacter asiaticus (*Ca.* L. a) pathogen causing HLB disease in infected *Citrus reticulata,* and to determine the cell wall thickness modification in infected citrus after chemical treatments using TEM technique.

2. Materials and methods

2.1 Research area description

The research was conducted in Kuala Berang, Terengganu, Malaysia, between Jan 2008 and October 2009. The state of Terengganu overlooks the South China Sea on the east coast of Malaysia. It has a strong tropical monsoon climate, with a relatively uniform temperature ranging between 21 degree centigrade (°C) and 32 °C. Between January and April, the weather is dry and warm, with consistent high humidity in the lowlands ranging between 82 and 86 percent annually. Terengganu's average rainfall ranges from 2,032 millimetre (mm) to 2,540 mm per year, with the most rain falling between November and January. The orchard of *Citrus reticulata* cv. Limau madu was cultivated in 2000 and it has been unproductive since that time because of HLB (Fig.1).

A total of 72 trees of *Citrus reticulata* were selected for this experiment. A randomized complete block design (RCBD) with three replications was used. Three trees were selected as a treatment unit for each replication for each treatment. Gibberellic acid (GA_3), antibiotics, foliar fertilizer and a combination of those chemicals were tested (Table 1). Aqueous solutions of the treatments were prepared and sprayed on whole trees. Moreover, 2 grams per litre (g/L) of antibiotic was injected into the stem at 15 to 25 centimetre (cm) above the soil (Fig.2). The treatments were applied before flowering and during fruit set formation (Fig.3). The following parameters were evaluated to measure the severity of the HLB disease and the effects of the treatment on plant rehabilitation. HLB detection using conventional PCR, ultrastructure of HLB and cell wall thickness modification using transmission electron microscopy (TEM) were carried out.

Fig. 1. Overview of HLB-infected *Citrus reticulata* in Hulu Paka, Terengganu, Malaysia.

Treatments	
T1	*Antibiotic (Oxi-tetracycline) 2 g/L (as injection)
T2	Gibberellic acid (GA_3) 15 mg/L (as foliar application)
T3	Foliar fertilizer (Vita-Grow) 5 ml / L (as foliar application)
T4	Antibiotic (2 g/L) + GA_3 (15 mg/L)
T5	Antibiotic (2 g/L) + foliar fertilizer (20 ml /4 L)
T6	GA_3 (15 mg/L) + foliar fertilizer (20 ml /4 L)
T7	Antibiotic (2 g/L)+ GA_3 (15 mg/L)+ foliar fertilizer (20 ml /4 L)
T8	Control (non treated)

*Antibiotic was injected in the trunk of the tree 15 cm above the soil surface.
A total of 72 trees of *Citrus reticulata* on a RCBD with 3 replications were used.

Table 1. List of chemical treatments to reduce the HLB disease severity and improve fruit quantity and quality.

Fig. 2. Injection of antibiotics; (A) making a hole on the trunk of the tree, (B) placing the tube, (C) preparing the antibiotic container, (D) an overall view of the whole injection set up.

Fig. 3. Chemical treatments (GA_3 and foliar fertilizer) application before flowering (A) and after flowering (B) using tractor mounted sprayer.

2.2 Transmission Electron Microscopy (TEM) examination

A TEM technique described by Childers and Achor (1991) was used as TEM examination. New terminal shoots of the infected citrus trees were collected and washed. Midribs were removed from the leaves and chopped into small pieces, which were fixed in 4% glutaraldehyde for 12 to 24 hours at 4°C and washed using a 0.1molar cacodylate buffer at three stages, each for 30 minutes (min). The specimens were first rinsed in the buffer followed by post-fixation in 1% osmium tetraoxide for 2 hours at 4°C. It was washed for a second time and dehydrated through a series of acetone from 35%, 50%, 75%, 95% and 100%, each for 30 min. Specimens were later infiltrated with acetone and resin mixtures at 1/1, 1/3 and 100 resin (two stapes) for 24, 12 and 2 hours, respectively, and embedded into beam capsules and filled with resin. It was polymerized at 60°C for 72 hours.

Thick sectioning (1micromolar) was done using a glass knife and an ultra-microtome (Leica UCT, Australia) to cut ultra-thin (60-80 nm) sections. The samples were placed onto a glass slide and stained with toluidine blue. Subsequently, these were dried on a hot plate and the stain was washed. Ultra-thin sections of the samples were cut with a diamond knife on an LKB-Huxley ultra-microtome (Cambridge, UK) and stained with 2% uranyl acetate (10 min). This was followed by filtered alcohol (50%) and washed by double distilled water prior to examination using a transmission electron microscope (TEM) at 60 kilovolts (kV).

For each treatment, six cells were randomly chosen to measure the thickness of cell walls and the middle lamella. The thickness of cell walls and middle lamella were measured using three points for each cell and the average thickness was used for comparison.

2.3 Polymerase Chain Reaction (PCR) test

The leaves of new citrus flushes were collected from the plots every two months after treatment. DNA was extracted from the midribs of the leaves using the method described by Hung et al. (2000). This was followed by grinding the tissue in a mortar and pestle using liquid nitrogen until there were no longer any large pieces of tissue remaining. DNA was extracted from the HLB-infected tissue using the hexadecyltrimethyl ammonium bromide (CTAB) method. The pellets were washed with 70% ethanol, dried and diluted with 100 micro litres (µl) of Tris-EDTA (TA) buffer (pH 8.0). The DNA quality was checked by electrophoresis on 1% agarose gel.

Conventional PCR was performed using 25µl of reaction mixture containing 20 milimolar (mM) Tris-HCl (pH 8.0), 50 mM KCl, 4 mM MgCl2, 0.2 mM of dATP, dTTP, dCTP and dGTP (the 5'-triphosphates of deoxyadenosine, deoxycytidine, deoxyguanosine and deoxythymidine respectively), 50 ng forward primer, 50 ng reverse primer, 0.75 units of Taq DNA polymerase and 200 ng genomic DNA. The thermal cycle conditions were: one cycle at 95°C for 2 min, 35 cycles at 95°C for 40 sec, 60°C for 1 min and 72°C for 1 min followed by a 72°C extension for 10 min. Specific primer pairs, composed of the forward primer of OI1 (5′-GCG CGT ATG CAA TAC GAG CGG CA-3′) {The four bases of DNA are adenine (abbreviated A), cytosine (C), guanine (G) and thymine (T)} and reverse primer of OI2c (5′-GCC TCG CGA CTT CGC AAC CCA T-3′), were used to amplify the 16S ribosomal DNA fragment. Amplification of DNA was determined by electrophoresis on 1.2% agarose gel for approximately 30-45 min and visualized by ethidium bromide staining. PCR products were purified directly on spin columns (Qiagen PCR purification kit).

2.4 Statistical analysis

A quick method was adopted to analyse the PCR product using a 1% agarose type III (Sigma brand). The gel was run in a TBE buffer (Tris base, boric acid and 2 mM EDTA, pH 8.0). The resolved DNA bands were detected by staining the gels with 0.5µg/mL of ethidium bromide, followed by destaining with distilled water. The electrophoresis was turned on for 30-40 min using a high voltage (100V) and finally photographed under UV illumination (SYNGENE; GENE Genious Bio Imaging System). A one kilo base pair (1Kbp) DNA ladder was used as a convenient marker for size estimates of the products.

Data of cell thickness modification were collected and analyzed by using analysis of variance (ANOVA), and significant differences between treatments were compared by the Duncan's multiple range tests (DMRT) method with a statistical analysis system (SAS Version 9 -TS M0) to separate the treatment means at the 0.05% level of probability. The percentage data were converted to their arcsine-square root equivalents in order to normalize data sets prior to analysis.

3. Results

3.1 TEM detection of HLB, ultrastructurre of the pathogen and cell wall thickness modification after treatments using TEM examination

Results have shown abundant bacteria cells damaging the cell wall in sieve tube cell at 30-70 kV magnification (Fig. 4 and 5) in infected samples. TEM detection of HLB has shown that the number of bacteria in sieve tubes is higher in leaves with strong mottle. According to these results, the organisms are seen surrounded by a cell wall. Electron micrographs of sieve tube elements of infected *Citrus reticulate* leaf at high magnification (30-100 kV) were observed (Fig.4). In figures 4 and 5 there were two shapes of Candidatus liberibacter asiaticus that were present; spherical and rod shape. Electron micrographs of sieve tube elements of the infected *Citrus reticulata* leaf at high magnification (30-100 kV) were observed (Fig.4 –Fig.6). The HLB bacteria organisms were found to be surrounded by a cell wall (Fig.5 and Fig.6; T8). Electron micrographs of the sieve tube of *Citrus reticulata* leaf showing the fusion confluent of middle lamella and cell wall structure are shown in (Fig.6; T8). The cell wall and middle lamella were damaged by the *Candidatus* Liberibacter asiaticus, which penetrated through the cells (Fig. 6; T8).

The obvious change in the total cell wall thickness, relating to the change of the cell wall and middle lamella thickness, were observed using TEM study (Fig.6; T1-T7) after treatments. The comparison of means of the total thickness of cell wall, cell wall thickness and middle lamella thickness are presented in table 2 at the 0.05% probability level.

Total cell wall thickness was measured and was found to be statistically significant for certain treatments (Table 2). T4 (942.72 nm), T7 (811.62 nm) and T5 (766.07 nm) had the highest total cell wall thickness compared to the control treatment (T8; 214.07 nm) while a low rate of total cell wall thickness was measured for T3 (43.31 nm), T2 (469.78 nm) and T1 (564.96 nm). There was no significant difference for T1, T2 and T3, which were used as direct treatments in this study.

However, a significant difference in the cell wall thickness was observed among the treatments compared to the control (Table 2). Thick cell walls were observed for T4 (580.82nm) followed by T7 (476.39 nm), while thin cell walls were recorded for T8 (as control; 86.33 nm) followed by T3 (209.83 nm), T2 (212.64 nm) and T1 (305.53 nm).

Significant differences between the treatments were also observed concerning the thickness of the middle lamella (Table 2). A thick middle lamella was recorded for T7 (403.15 nm) and T4 (361.90 nm), while a thin middle lamella was observed for T8 (126.98 nm), T3 (223.48 nm), T1 (254.72 nm), T2 (256.20 nm), T6 (278.7 nm) and T5 (289.68nm). These were found to be statistically similar (Table 2).

Treatment*	Cell wall	Middle lamella	Total thickness
T1	305.53 cd	254.72 b	564.96 dc
T2	212.64 de	256.20 b	469.78 d
T3	209.83 de	223.48 b	433.31 d
T4	580.82 a	361.90 a	942.72 a
T5	425.69 bc	289.68 b	766.07 b
T6	378.65 bc	278.75 b	671.66 bc
T7	476.39 ab	403.15 a	811.62 ab
T8	86.33 e	126.98 c	214.07 e

* Means within a column with different letters are significantly different and means followed by the same letters are not significantly different according to DMRT test at $P \leq 0.05$ levels.
T1: Antibiotic. T2: GA_3. T3: foliar fertilizer. T4: Antibiotic + GA_3. T5: Antibiotic + foliar fertilizer. T6: GA_3 + foliar fertilizer. T7: Antibiotic + GA_3 + foliar fertilizer. T8: Control.

Table 2. Mean Comparison of cell wall, middle lamella and total thickness of cells after treatments using Duncan multiple range test.

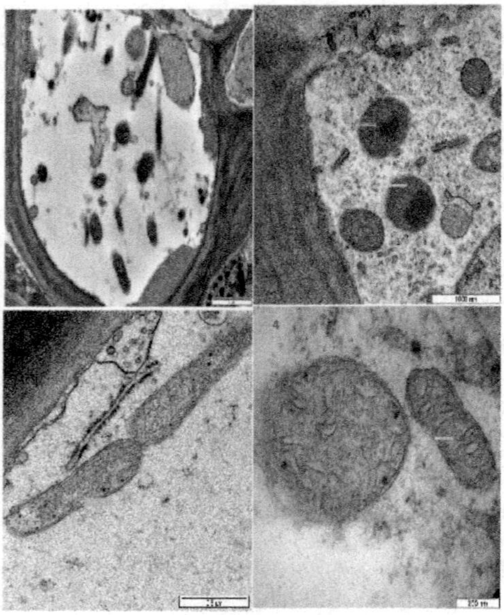

Fig. 4. Electron micrographs of the vascular system from the infected midrib of *C. reticulata* showing abundant bacteria cells damaging the cell wall (cw) and middle lamella (ml) in sieve tube cell (Fig.1-2) at magnification, 30 kV; and cross section of sieve tube cells containing spherical (s) and rod (r) shape bacteria (Fig.3-4) of HLB disease at high magnification (100 kV).

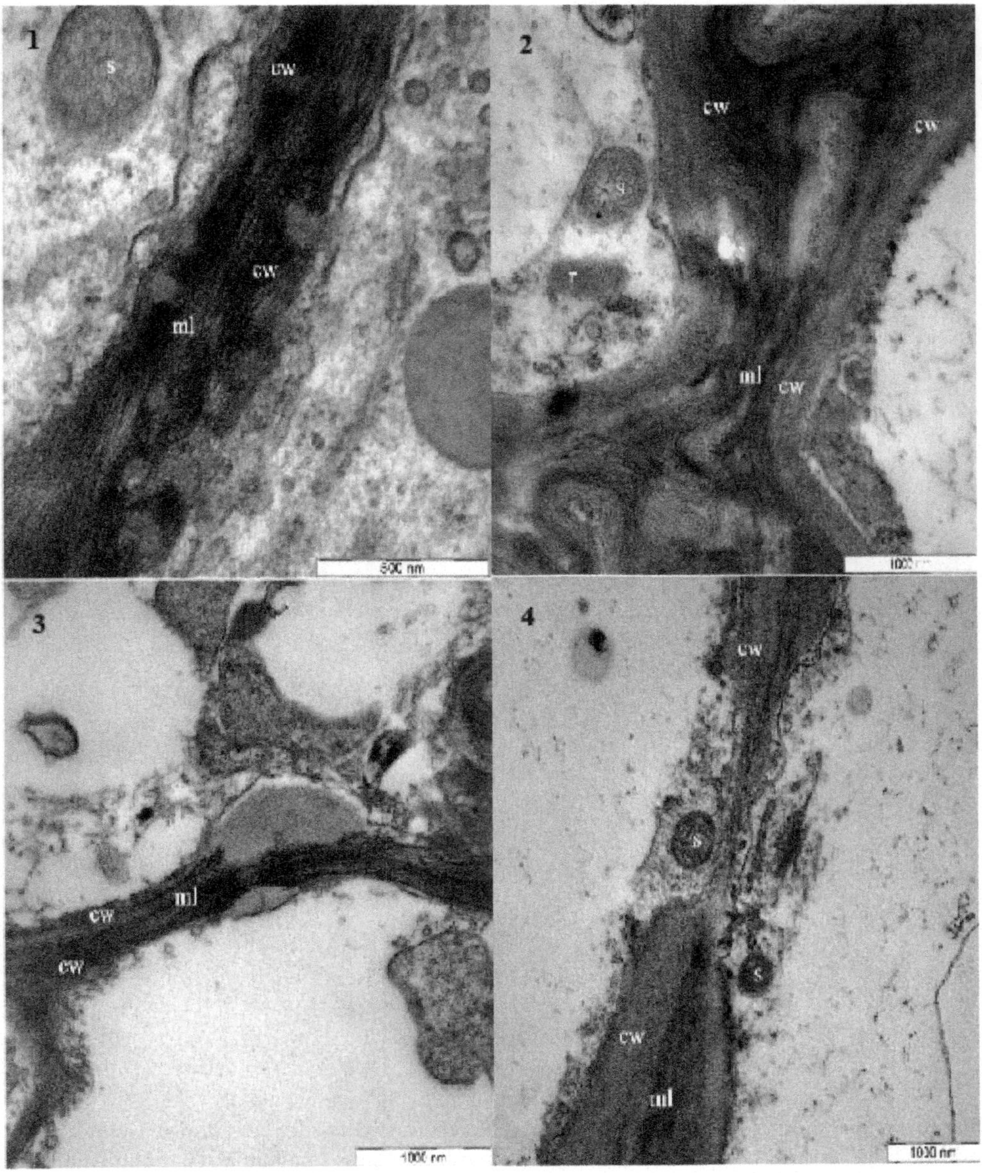

Fig. 5. Electron micrographs of the sieve tube of *Citrus reticulata* leaf; section (1 and 2) showing the fusion confluent of middle lamella (ml) and cell wall (cw) structure (1-mag. at 70 kV and 2-mag., at 30 kV magnification). Cross section of sieve tube (3 and 4) showing damaged cell wall (cw) and middle lamella (ml) caused by *Candidatus* Liberibacter asiaticus penetrating through the cells (3-mag. at 30 kV and 4-mag. at 20 kV).

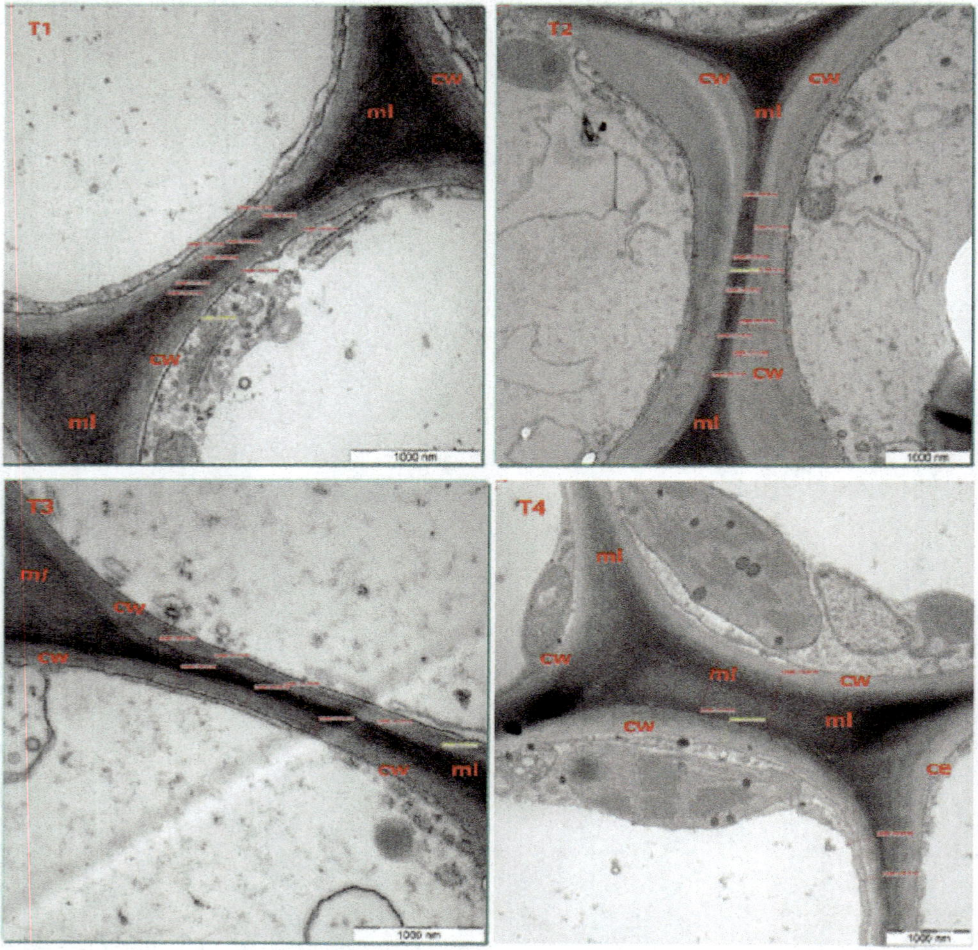

Fig. 6. Electron micrographs of the sieve tube of *Citrus reticulata* leaf: section showing the middle lamella and cell wall structure and cell wall thickness (mag at 70 kV magnifications). T1: Antibiotic. T2: GA3. T3: foliar fertilizer. T4: Antibiotic+GA3. T5: Antibiotic + foliar fertilizer. T6: GA3 + foliar fertilizer. T7: Antibiotic + GA3 + foliar fertilizer. T8: Control.

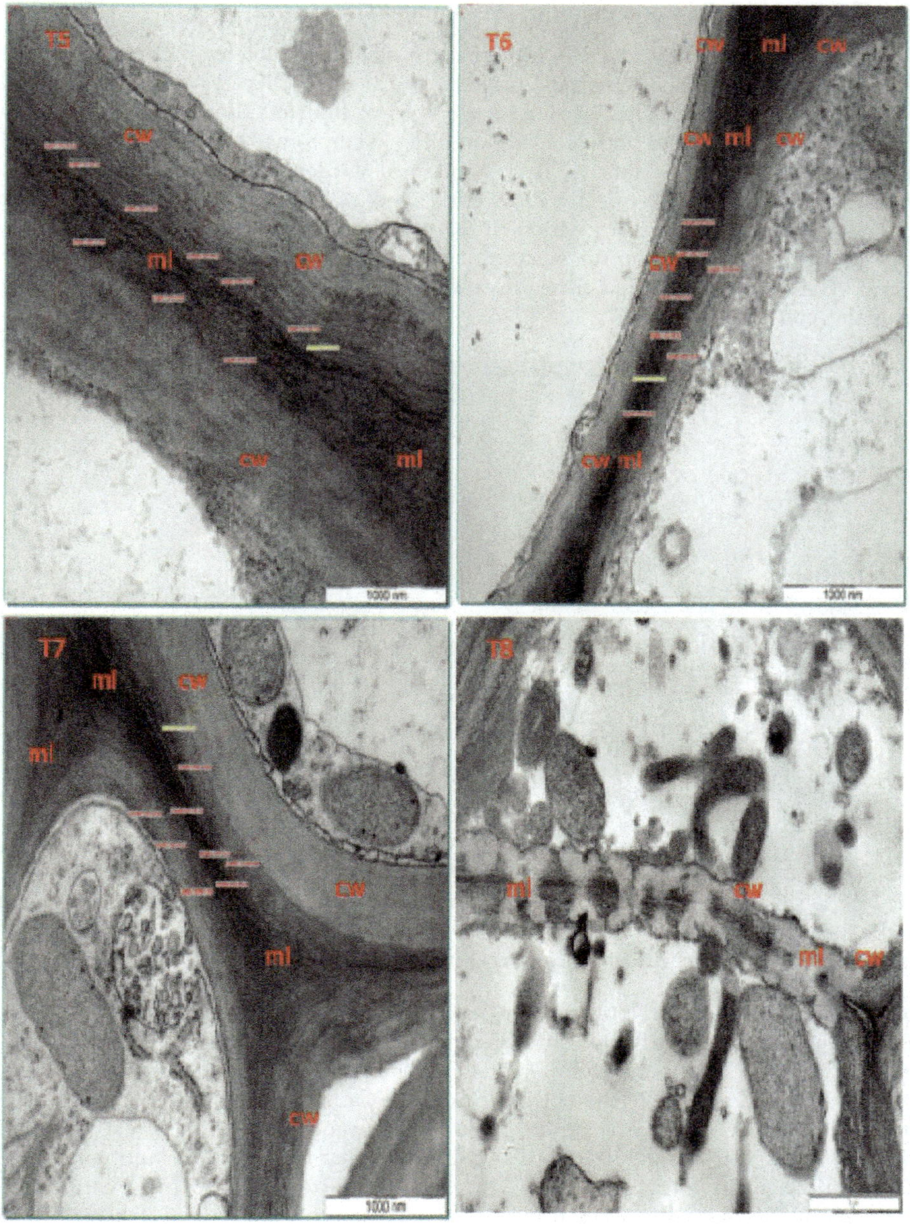

Fig. 6. (Continue): Electron micrographs of the sieve tube of *Citrus reticulata* leaf: section showing the middle lamella and cell wall structure and cell wall thickness (mag at 70 kV magnifications). T5: Antibiotic + foliar fertilizer. T6: GA3 + foliar fertilizer. T7: Antibiotic + GA3 + foliar fertilizer. T8: Control.

3.2 Detection of HLB bacteria using conventional PCR

For PCR detection of the *Candidatus* liberibacter asiaticus, blotchy mottled and midrib yellowing leaves were used. One pair primers: f-OI1 and r-OI2c for amplification of 16S rDNA was used. Line 4 of Fig.7 shows strong band of midrib yellowing symptoms while line 5 shows weak band of blotchy mottling symptoms after conventional PCR amplification. It is true to state, as it is sometimes written, that it is difficult to detect liberibacters. Therefore HLB infection was confirmed by PCR amplification in the typical symptoms. A polymerase chain reaction (PCR) with specific primers OI1/OI2c and A2/J5 were used for detection and produced specific bands of 1160 bp and 703 bp, respectively in Thailand. These were amplified from diseased leaves whereas no product from healthy citrus plants could be obtained.

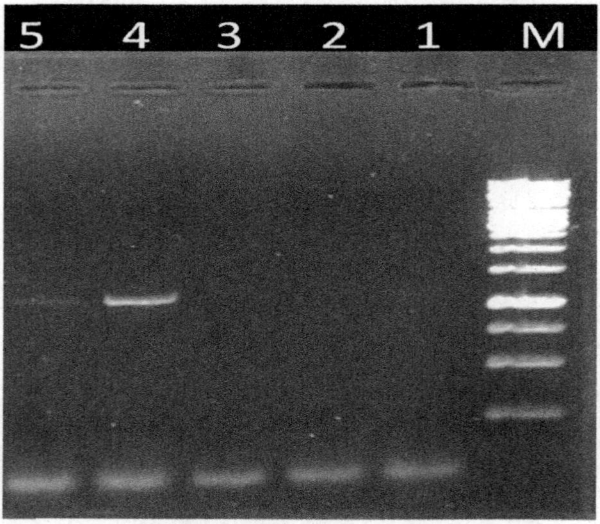

Fig. 7. 16S rDNA fragments with molecular weight of 1160 bp were successfully amplified from the infected samples; (M): Marker (100 bp invitrogen); (line 1): Water; (line 2 &3): Negative sample; (line 4): Midrib yellowing; (line 5): Blotchy mottling symptoms; (line 5-7).

A conventional PCR of 16S rDNA fragment with 1160 bp was successfully amplified and the results are presented in Figs.8 and 9 after treatments. Samples of *Citrus reticulata* leaves were collected and transferred to the pathology Lab to detect the *Candidatus* Liberibacter asiaticus (*Ca*. L. a). Conventional PCR was not able to detect Ca. L. bacteria after 35 cycles of PCR with primer pair OI1 and OI2s under T1 (antibiotic) and T7 (antibiotic + GA3 + foliar fertilizer) (Fig.8). HLB was detected on T2 (GA3), T3 (foliar fertilizer), T4 (antibiotic + GA3), T5 (antibiotic + foliar fertilizer), T6 (GA3 + foliar fertilizer) and T8 (no treatment = as control) (Fig.8). Second conventional PCR with the same primer pair (OI1 and OI2s) was also conducted to detect HLB bacteria on the leaves of *Citrus reticulata* under different treatments. However, products of about 1160bp in size were obtained from the leaves of *Citrus reticulata* under all treatments after the second conventional PCR with primer pair OI1 and OI2s. These included T1, T2, T3, T4, T5, T6, T7 and T8 (Fig.9).

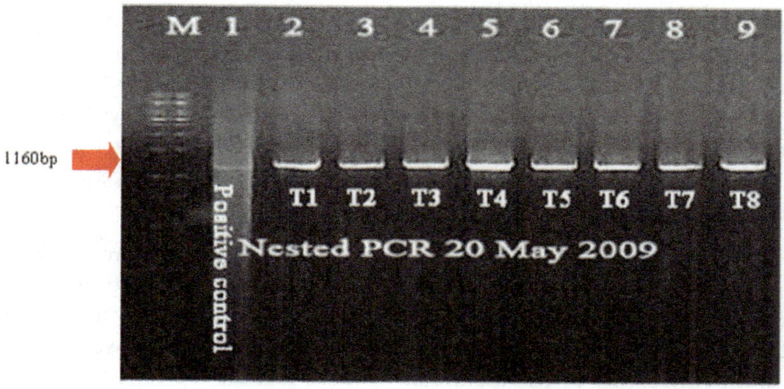

Fig. 8. Conventional PCR of 16s rDNA fragment with molecular weight of 1160 bp were successfully amplified from the leaves of *Citrus reticulata*. T1: antibiotic, T2: GA3, T3: foliar fertilizer, T4: antibiotic +GA3, T5: antibiotic +foliar fertilizer T6: GA3+ foliar fertilizer T7: antibiotic + GA3 + foliar fertilizer T8: control.

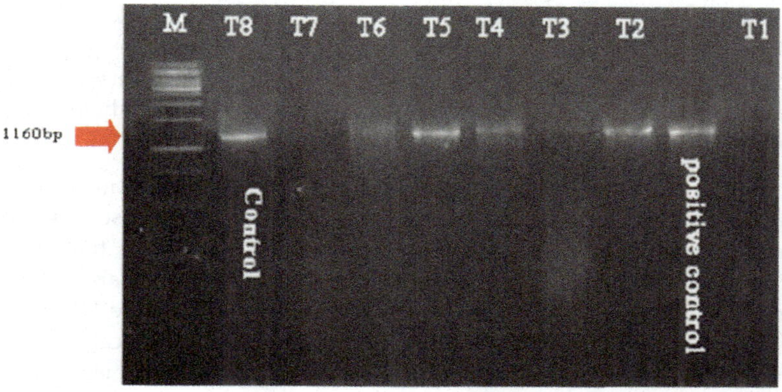

Fig. 9. Second conventional PCR of 16s rDNA fragment with molecular weight of 1160 bp were successfully amplified from the leaves of *Citrus reticulata*. T1: antibiotic, T2: GA3, T3: foliar fertilizer, T4: antibiotic +GA3 T5: antibiotic +foliar fertilizer T6: GA3+ foliar fertilizer T7: antibiotic + GA3 + foliar fertilizer T8: control.

4. Discussion

The chemical treatments had an effect on the cell wall and middle lamella thickness. The cell wall thickness was significantly increased with the T4 and T7 treatments compared to the control (T8). However, Agrios (1997) reported that after pathogen penetration the plants usually respond by forming one or more types of structure namely, cytoplasmic defence reaction, cell wall defence structure and histological defence structure which are successful in defending the plant from the next invasion.

Middle lamella and, finally, the total thickness of cell wall were significantly increased for T4 and T7 compared to the control. For cell wall defence structures, three main models have been detected in the plant. First, the plant produces amorphous, fibrillar material on the outer layer of the cell wall of parenchyma cells in contact with incompatible bacteria, which subsequently, surrounds and traps the bacteria and controls their multiplication. Second, the cell walls thicken by producing cellulosic material that infuses with a phenolic compound, which increases the cell wall resistance to pathogen penetration. Finally, callose papillae are deposited on the inner side of the cell walls within minutes after injury by microorganisms (Agrios, 1997; Lindenthal, 2005).

When the pathogen penetrates the host, the pathogens may produce a mechanical signal by the physical pressure on the plant cell. It was found that the stimulation of a mechanical signal induces the generation and accumulation of phytoalexin in cucumber and significantly enhances the resistance (Zhao et al., 2005). Some changes were observed in the structural and permeability of cell membranes. These include the release of molecules that are important in signal transduction within and around the cell after an attack by pathogens; the release and accumulation of reactive oxygen species; the activation of phenol oxidases; and the oxidation of phenolics (Agrios, 1997).

Ahmad et al. (2008) detected HLB in peninsular Malaysia using OI1 and OI2c primer. In general, HLB-infected citrus plants were stunted with twig dieback and showed open growth (Bove, 2006; Gottwald et al., 2007; Graca, 1991). Among the growth regulators, considerable work has been done with respect to the effect of pre-harvest applications of GA3 on citrus quantity and quality (Ladaniya, 2004), however, there is little or no information about the effects of GA_3 on HLB-infected fruits.

To our knowledge, this is the first report on the effects of chemical treatments in citrus HLB-infected by TEM study. Cell wall and middle lamella thickness were also increased with the application of a combination of T4 and T7. This study has shown that in non-treated samples irregularity of the cell wall increased, and the middle lamella and cell wall became disorganized compared to the treated samples. The TEM study of the HLB-infected sieve tube of *Citrus reticulata* leaf also showed the fusion confluent of the cell wall and middle lamella structure. A cross-section of the sieve tube showed the damaged cell wall and middle lamella caused by *Candidatus* Liberibacter asiaticus penetrating through the cells. The TEM investigation of the vascular system of HLB-infected *Citrus reticulata* showed spherical and rod shaped bacteria from the midrib with blotchy mottling symptoms of the disease. Finally it can be concluded that, the findings of this research are important and useful for detection the HLB and control programmes elsewhere. Choosing a strategy to control the HLB disease is imperative. Control of HLB is difficult if inoculum sources are widespread. Different chemicals were used to increase the life span of citrus mandarin in a HLB-infected orchard.

5. References

Agrios, G.N. 1997. Plant pathology. Academic press.

Ahmad Khirolmazmi, K.S., Habibuddin H, Jugah Kadir, and Syed Omar Syed Rastan. 2008. Occurrence and Spread of *Candiidatus liberibacter asiaticus*, the Causal Agent of

Huanglongbing Disease of Citrus in Malaysia. Research Journal of Agriculture and Biology Science 4:103-111.

Bove, J.M. 2006. Huanglongbing: A destructive, newly-emerging, century-old disease of citrus. Journal of Plant Pathology 88:7-37.

Bove, J.M., and M. Garnier. 2002. Phloem- and xylem-restricted plant pathogenic bacteria. Plant Science 163:1083-1098.

Chen, M.H., T. Miyakawa, and C. Matsui. 1973. Citrus likubin pathogen in thesalivary glands of Diaphorina citri. Phytopathology 63:194-195.

Garnier, M., N. Danel, and J.M. Bové. 1984. Aetiology of citrus greening disease. Annales de Institut Pasteur. Microbiologie (Inst. Pasteur) 135:169-179.

Gottwald, T.R., J.V. da Graça, and R.B. Bassanezi. 2007. Citrus Huanglongbing the pathogen and its impact. Online Plant Health Progress.

Graca, J.V. 1991. Citrus greening disease. Annual Review of Phytopathology 29:109-136.

Hajivand, S., L.A. Thohirah, S. Kamaruzaman, and N.A.A. Siti. 2009. Differential reaction of citrus species in Malaysia to Huanglongbing (HLB) disease using grafting method. American Journal of Agricultural and Biological Sciences 4:32-38.

Halbert, S.E., and K.L. Manjunath. 2004. Asian citrus psyllids (Sternorrhyncha: Psyllidae) and greening disease of citrus: a literature review and assessment of risk in Florida. The Florida Entomologist 87:330-353.

Hansen, A.K., J.T. Trumble, R. Stouthamer, and T.D. Paine. 2008. New Huanglongbing (HLB) *Candidatus* species," *Ca.* Liberibacter psyllaurous", found to infect tomato and potato is vectored by the psyllid *Bactericerca cockerelli* (Sulc). Applied and Environmental Microbiology 74(18) 5862-5865.

Hung, T.H., M.L. Wu, and H.J. Su. 2000. Identification of alternative hosts of the fastidious bacterium causing citrus greening disease. . J. Phytopathology 148:321-326.

Jagoueix, S., J.M. Bove, and M. Garnier. 1994. The phloem-limited bacterium of greening disease of citrus is a member of a subdivision of the Proteobacteria. International Journal of Systematic Bacteriology 44:379-386.

Jagoueix, S., J.M. Bové, and M. Garnier. 1996. PCR detection of the two *Candidatus* Liberobacter species associated with greening disease of citrus. Molecular and Cellular Probes 10:43-50.

Ladaniya, M.S. 2004. Reduction in post-harvest losses of fruit and vegetabale. Final report of NATP Project, Submitted to ICAR, New Delhi. .

Lindenthal, M., U. Steiner, Dehne.H.-W., and E.-C. Oerke. 2005. Effect of downy mildew development on transpiration of cucumber leaves visualized by digital infrared thermography. Phytopathology 95:233-240.

Moll, J.N., and M.N. Martin. 1973. Electron microscope evidence that citrus psylla (*Trioza erytreae*) is a vector of citrus greening in South Africa. Phytophylactica 5:41- 44.

Roistacher, C.N. 1991. Techniques for biological detection of specific citrus graft transmissible disease,.FAO Rome. pp:35-45.

Saglio, P., D. Lafleche, C. Bonnisol, and J.M. Bove. 1971. Isolement, culture et observations an microscope electronique des structures de type mycoplasme associes a la maladie du stubborn des agrumes et luer comparison avec les structures observees dans le cas de la maladie greening de argumes. Physiologie Vegetable 9:569-582.

Zhao, H.C., G.J. Li, and J.B. Wang. 2005. The accumulation of phytoalexin in cucumber plant after stress. *Colloids and surfaces B: Biointerfaces* 43:187-193.

15

Cathodoluminescence of Surface Plasmon Induced Light Emission

Naoki Yamamoto
Physics Department, Tokyo Institute of Technology, Tokyo, Japan

1. Introduction

Cathodoluminescence (CL) is light emitting phenomena induced by high energy electrons incident on materials. The CL is sometimes defined as light emission caused by the inner electronic transition between the electronic states or bands associated with excitation by the incident electron beam. The typical one is the radiative recombination of carriers in semiconductors, such as the band to band transition (Yamamoto, 2010). CL measurement has been performed by combining a light detection system with a scanning electron microscope (SEM) or scanning transmission electron microscope (STEM), in which a focused electron beam is scanned over a specimen and simultaneously a scanning image is recoarded by using an emitted light signal. This technique is called cathodoluminescence. This technique provides us the information of local electronic states from the CL spectrum, and spatial distribution of luminescence centers from the CL image, and thus has been applied to the characterization of structural defects such as dislocations and semiconductor quantum structures. On the other hand there are other types of light emission caused by a high-energy electron passing through or nearby a medium (Yamamoto et al., 1996, 2001a), such as Cherenkov radiation, transition radiation and Smith-Purcell radiation. A high-energy electron incident on a metal surface can excite a surface plasmon which propagates on a corrugated metal surface to emit light. Recently such light emission induced by surface plasmon has been intensively studied by the CL technique.

Surface plasmon (SP) is collective oscillation of surface charge on a metal surface, and is classified into two types, localized surface plasmon (LSP) and surface plasmon polariton (SPP). The LSP is localized at a narrow space of nano-structures such as particles, and SPP is a transverse magnetic (TM) mode electromagnetic wave propagating at a metal/dielectric interface, which is evanescently confined in the perpendicular direction (Raether, 1988). SPs can produce strong electromagnetic field at a localized area and are utilized for bio-sensors (Homola, 2008) and solar cells (Catchpole and Polman, 2008). SPP can be confined into a narrow area beyond the diffraction limit of light and has a potential in optical circuits and optical computer applications. This recently developed field is called "Plasmonics" (Maier, 2007; Brongsma and Kik, 2007).

The author has intensively studied light emission from metal nano-particles and nano-structures using a light detection system combined with a scanning transmission electron microscope (STEM) (Yamamoto et al., 2001a). The use of this system has an advantage for selectively examining a localized region using a converged electron beam with a probe size of 1 nm, and also observing emission intensity distribution as a monochromatic photon map. The emission angle can be selected by a light detection system using a parabolic mirror and a movable mask, which enables angle-resolved measurement. Mie and gap modes of LSP in spherical metal particles were investigated, and the size dependences of the resonant energies of the modes were found (Yamamoto et al., 2001b). The dispersion relation of SPP propagating on a flat surface was derived by the observation of the beam-scan spectral image across a surface step (Yamamoto and Suzuki, 2008). Dispersion patterns of SPP in 1D and 2D plasmonic crystals were obtained from the angle-resolved spectral patterns, and the SPP modes were identified from the photon maps, which mimicked standing SPP waves in a real space (Suzuki and Yamamoto, 2009; Takeuchi and Yamamoto, 2011).

2. Cathodoluminescence technique

2.1 System

Cathodoluminescence system is based on a scanning transmission electron microscope (STEM) or a transmission electron microscope (TEM) equipped with a beam scanning system, which is combined with a light detection system. It is needed to alter an ordinary setup of STEM to insert a light collection mirror inside a pole piece gap of the objective lens. Geometry of a light collection mirror and set-up of the light detection system to select polarization of light and emission direction are shown.

Pole piece

A standard high-resolution STEM equips a narrow gap pole piece for the objective lens. However, there is not enough space to insert a light collection mirror around a specimen between a specimen holder and an upper part of the pole piece. For example, the pole piece gap of JEM2000FX TEM (JEOL Co. Ltd.) for the analytical version is 15 mm and a height of the mirror is 5 mm. The use of wide gap pole piece reduces the resolution of STEM. In addition a probe size of the electron beam becomes 10 nm or larger, when a large beam current of the order of 1 nA is used to induce sufficient light intensity. However, recent development of the spherical aberration (Cs) corrector and field emission gun (FEG) drastically improve the resolution of TEM/STEM. We realized a 1 nm size probe at 1 nA beam current at 80 kV and 200kV by using FEG and Cs corrector in JEM2100F STEM.

Light collection mirror

An ellipsoidal mirror and parabolic mirror were used for light collection in STEM-CL, and the latter has an advantage in angle-resolved measurement. The parabolic mirror is set up above the sample. The parabolic curve of the mirror is expressed as

$$4p(p-x) = y^2 + z^2 , \tag{1}$$

using the Cartesian coordinates as shown in Fig. 1, where the origin is taken at the focal point of the parabola. The distance p between the focal point and mirror surface along the x direction is taken to be 2 mm in our system. The parabola mirror was cut into two pieces and then each mirror is 10 mm in width and 5 mm in height. In addition the bottom of the mirror is 0.2 mm higher than the parabolic axis to make a clearance between a specimen surface and the mirror. The mirror has a 0.6 mm hole located 4 mm (=2p) above the focal point, through which the electron beam illuminates the specimen. Total solid angle subtended by the mirror from the focal point is about 60% of the solid angle of hemisphere. The mirror surface is expressed by the polar coordinates as

$$\begin{cases} x = r\sin\theta\cos\varphi = 2p\dfrac{\sin\theta\cos\varphi}{1+\sin\theta\cos\varphi} \\ y = r\sin\theta\sin\varphi = 2p\dfrac{\sin\theta\sin\varphi}{1+\sin\theta\cos\varphi} \\ z = r\cos\theta \qquad = 2p\dfrac{\cos\theta}{1+\sin\theta\cos\varphi}, \end{cases} \tag{2}$$

where the radial distance r is expressed by the polar angle θ and azimuth angle φ as

$$r = \frac{2p}{1+\sin\theta\cos\varphi}. \tag{3}$$

In particular case that φ=0 and then y=0, the following relation holds between θ and z,

$$\sin\theta = \frac{4p^2 - z^2}{4p^2 + z^2} = \frac{1-\eta^2}{1+\eta^2}, \tag{4}$$

where we put $\eta = \dfrac{z}{2p}$.

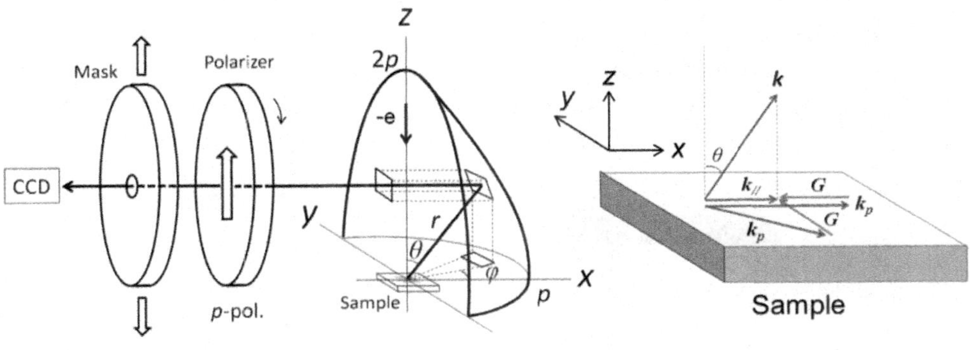

Fig. 1. (a) Geometry of a parabolic mirror and experimental set up for the angle resolved measurement. (b) Relation among the wave vectors of an emitting photon and associated SPPs.

It is needed to precisely adjust a mirror position with respect to a light emerging point on a specimen illuminated by the electron probe. The mirror is supported by a mirror holder fixed to the STEM, and the xyz positions of the mirror can be adjusted by three micrometers in the holder. The mirror can be retracted from the electron beam pass by the mirror holder.

The emitted light becomes a parallel ray after reflected by the parabolic mirror, and passes through a hollow inside the mirror holder into outside the STEM. Several parts such as an anti-contamination tip surrounding the pole piece are modified for the insertion of the mirror and mirror holder.

Light detection system

A light detection system equipped with the STEM is illustrated in Fig.2. The parallel ray from the parabolic mirror passes through a polarizer and a lens, and is switched to three paths by a flat mirror. One path leads to a photomultiplier tube for mainly photon map imaging, and another path leads to a CCD camera for imaging the parabolic mirror. The mirror position is adjusted using a luminescent specimen so as that an emission pattern in the mirror image spreads over the mirror plane observed by the CCD camera, which ensures that the parabolic mirror is focused at the light emerging point. In the straight path, the parallel ray goes through a mask with a small hole supported by an X-Y stage. The emission image on the parabolic mirror is transformed into a mask plane by the single lens. A part of the parallel ray passes through the hole, and is detected by a CCD detector to measure an angle resolved spectrum.

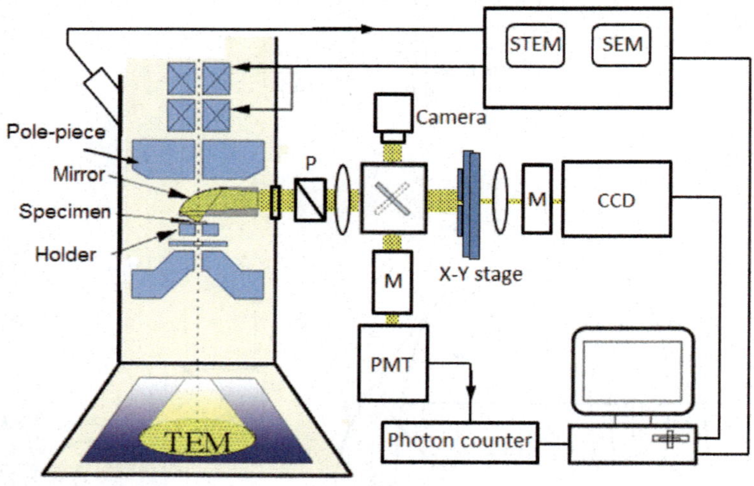

Fig. 2. Light detection system combined with a TEM/STEM

The polarization direction of light is determined by the polarizer. Here we define *p*-polarization as parallel to the emission plane (the x-z plane in Fig.1) subtended by the axis of parabola and normal to the specimen surface, and *s*-polarization is normal to the emission plane. For the angle-resolved measurement, a mask with a small hole is set on the X-Y stage. The mask position selects the emission direction of the detected light, which is specified by the angles θ and φ. The mask is moved vertically when the azimuth angle $\varphi = 0°$ and the emission spectra are successively measured to determine the θ-dependence. This procedure provides an emission intensity distribution of an angle-resolved spectrum (ARS), which shows variations with angle θ and forms an ARS pattern, $I(\theta, E)$ (Suzuki, 2009).

2.2 CL measurements

Photon map

Photon map is a 2 dimensional scanning image as shown in Fig.3 (a). The electron beam is controlled by a computer through the beam scanning system of the TEM/STEM, and is scanned on a specimen surface step by step. A photon signal is successively acquired by the PMT or CCD during the duration time at each point where the electron beam is stopped. Each point corresponds to a pixel in a scanning image of a photon map. The step interval and duration time for one pixel are adjustable, and simultaneous observation of STEM image for a thin specimen and SEM image for a bulk specimen is possible. The SEM image can be observed in the SEM mode using the beam scanning system of TEM/STEM. The simultaneous observation of photon map and STEM or SEM image enables us to directly compare the emission distribution with a specimen structure.

Polarization and wavelength of the parallel ray are selected by the polarizer and monochrometer, so we observe a polarized monochromatic photon map. A panchromatic photon map is sometimes useful for quick scan due to strong intensity, which is possible by setting a specular reflection in the monochrometer. In the CL system shown in Fig.2, the photon map observed by the PMT uses the light intensity collected over a wide solid angle subtended by the parabolic mirror. When using the CCD detector, an emission spectrum is stored for each pixel, so monochromatic photon maps over the continuous photon energy range can be obtained by one scan measurement. However, measurement of the photon map by the CCD takes more than 10 times longer than that by the PMT.

Beam-scan spectral image

Emission spectra were successively measured by scanning the electron beam along a line in the experimental arrangement of Fig. 3(b). The lower figure in Fig. 3(b) shows a Beam-scan spectral image created by aligning the observed spectra with respect to the electron beam position, where the vertical axis indicates photon energy. When combining the angle resolved measurement to select emission angle of the detected light, a beam-scan spectral image can reveal distribution of a standing SPP wave as observed in the plasmonic crystal (Suzuki and Yamamoto, 2009; Takeuchi and Yamamoto, 2011).

Angle-resolved spectral pattern

Figure 3(c) shows the experimental arrangement for the measurement of angle-resolved spectral (ARS) pattern. The mask can be moved in a plane perpendicular to the x axis by an X-Y stage. Only a part of the parallel ray can go through a small hole in the mask, and this means that the emission direction of the detected light specified by angles θ and φ is selected by the mask. We move the mask vertically under the condition of $\varphi = 0°$ and measure emission spectra successively to find the θ-dependence. Thus the direction of the detected light lies on the x-z plane, i.e., the azimuth angle $\varphi = 0°$. The mixing of the polarization by the reflection at the mirror can be neglected for the light emitted in the direction parallel to the x-z plane. This measurement gives successive emission spectra which show variation with emission angle θ. ARS pattern is created by aligning the observed spectra with respect to the emission angle as in Fig. 3(c), where the vertical axis indicates photon energy. In the original data, the horizontal axis indicates the z position of the hole instead of the emission angle. The horizontal coordinate in the ARS pattern is changed from z to θ by using eq. (4), and then the ARS pattern is expressed as a function of θ and E, i.e., $I(\theta, E)$.

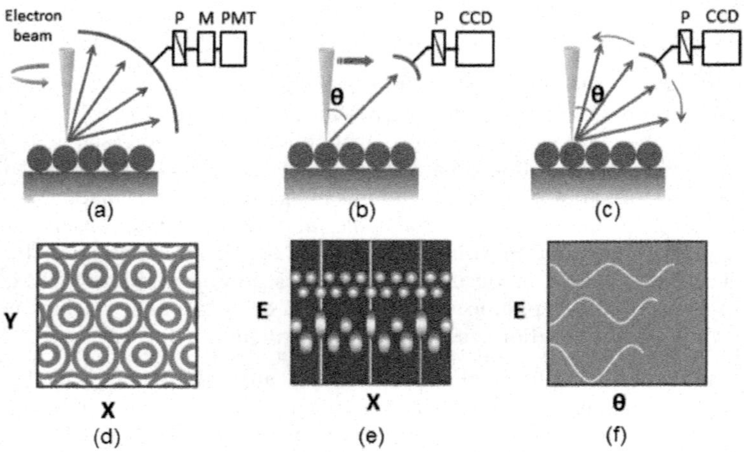

Fig. 3. Schematic diagram of experimental setup for the measurement of (a) photon map, (b) Beam-scan spectral image and (c) ARS pattern, and (d)-(f) show the observed patterns.

Dispersion pattern

The ARS pattern is transformed into the dispersion pattern, $I(k_{//}, E)$, by changing the emission angle to the wave vector component parallel to the surface, $k_{//}$, using the relation (Fig. 1)

$$|k_{//}| = |k| \sin\theta = \frac{E_{ph}}{\hbar c} \sin\theta \tag{5}$$

where $\boldsymbol{k}$ is the wave vector for the emitted light, $\hbar$ is Plank constant divided by 2π, and c is the velocity of light.

The SPP can be converted to a photon to emit light when it propagates on a periodic structure. The dispersion pattern transformed from the ARS pattern observed in the plasmonic crystal reveals the dispersion relation of SPP on the plasmonic crystal. The conversion between the photon and SPP mediated by plasmonic crystals must satisfy the following conditions in wave vector and energy,

$$k_p - k_{//} = G \tag{6a}$$

$$E_{SPP} = E_{ph} \tag{6b}$$

where k_p is the wave vector of the SPP, $k_{//}$ is the surface parallel component of the wave vector of light, and $\boldsymbol{G}$ is a reciprocal lattice vector of the surface structure. In eq.(6b) , E_{SPP} and E_{ph} are energies of the SPP and light, respectively. The SPP is converted to photon at the crossing point between the dispersion curves of the SPP and dispersion line of photon emitted to the vacuum, i.e. $\omega = ck/\sin\theta$. Consequently the light is emitted along the dispersion curves of the SPP inside the light line as shown in Fig. 4(c).

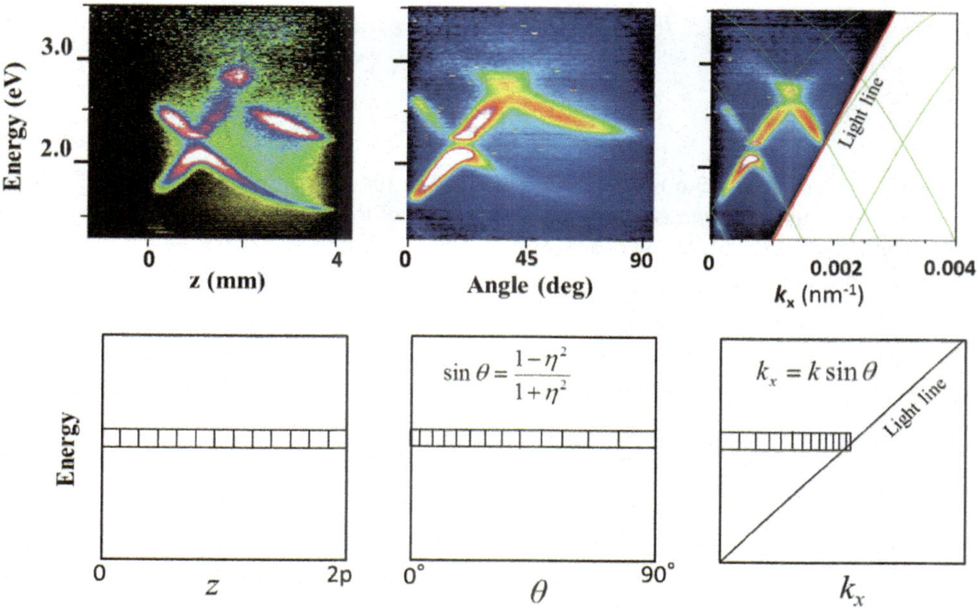

Fig. 4. (a) An original ARS pattern of an emission from a 1D-plasmonic crystal as a function of the hole position z, and (b) the ARS pattern transformed from (a) by changing the horizontal coordinate from z to θ. (c) Dispersion pattern transformed from (b) by changing to the wave vector component. (d) to (e) illustrate conversion of the intensity distribution from (a) to (c). Each square indicates an interval between neighboring data points corresponding to the original ones in (d).

Correction of spectrum and ARS pattern

1. Collection function of spectrum

The observed spectrum is modified from the true one due to the wavelength dependent efficiency of the detectors and other optical elements. The collection function of the total detection system is obtained as follows. A standard light source whose emission spectrum is known is set at the specimen position in TEM. We observe the emission spectrum by the detection system, and a collection function is obtained by dividing the true spectrum of the light source by the observed one. The transition radiation from an aluminum surface is frequently used as a standard light source, because it has a widely spreading spectrum over an optical range with a monotonously decreasing function inversely proportional to wavelength.

2. Conversion from wavelength to energy

Spectrum measured by the CCD detector consists of pixels with the same wavelength width. When the scale of the spectrum changes from wavelength to energy, the intensity should be collected by multiplying the following factor,

$$I(E)=I(\lambda)\left|\frac{d\lambda}{dE}\right|=I(\lambda)\frac{\lambda^2}{hc}=I(\lambda)\frac{hc}{E^2} \tag{7}$$

where h is Plank constant.

3. **z-θ conversion**

The mask has a hole of 0.5 mm in diameter and set in the X-Y stage. Solid angle subtended by the hole changes with emission angle, and is 0.1% of the total solid angle of hemisphere at θ=0° and 0.4% at θ=90°. In the conversion of the ARS pattern from $I(z,E)$ to $I(\theta,E)$, the intensity should be collected for the angular dependence. From the relation

$$I(\theta,\varphi)\sin\theta\, d\theta\, d\varphi = I(y,z)\, dy\, dz \tag{8}$$

and eq. (4),

$$d\varphi=\frac{1}{1-\eta^2}\frac{1}{p}dy\;\; d\theta=-\frac{2}{1+\eta^2}d\eta=-\frac{1}{1+\eta^2}\frac{1}{p}dz \tag{9}$$

,

we obtain

$$I(\theta)=I(z)p^2\left(1+\eta^2\right)^2 \tag{10}$$

The term in the right side indicates the solid angle collection factor of the ARS pattern taken along the z axis.

4. **θ-k_x conversion**

In the conversion from the ARS pattern $I(\theta,E)$ to the dispersion pattern $I(k_x,E)$, the intensity should be collected according to the relation between θ and k_x. From the equations,

$$I(\theta,\varphi)\sin\theta\, d\theta\, d\varphi = I(k_x,k_y)\, dk_x\, dk_y \tag{11}$$

$k_y = k\sin\theta\sin\varphi$

and $k_x = k\sin\theta\cos\varphi$ and, the intensity should be multiplied by the following factor.

$$I(k_x)=I(\theta)\frac{1}{k^2\cos\theta} \tag{12}$$

3. Transition radiation

Light emission caused by a high-energy electron passing through a medium are classified into several types from their characters, such as cathodoluminescence, Cherenkov radiation and transition radiation. Cherenkov radiation is generated when an electron moves in a transparent medium with a velocity higher than that of light, and transition radiation is generated when an electron passes across a boundary between two media with different dielectric constants. Characteristic features were observed in the emission spectra from thin films of mica and silicon (Yamamoto et al., 1996, 2001a). In the case of mica films, the forward emission spectra showed a characteristic shape of Cherenkov radiation, whereas the backward emission spectra showed the intensity oscillation with respect to wavelength where the oscillation periodicity depends on accelerating voltage and film thickness. Similar oscillation was also observed in the emission spectra from silicon films. Such emissions are explained well from the theory derived by Ritchie and Eldrigde (1962) and Pafomov and Frank (1967).

Here we describe the theory of transition radiation from a single boundary, because this radiation is always involved in the detected light from metal surfaces bombarded by electrons and must be considered in the analysis of the surface plasmon induced radiation. When an electron passes a boundary between two different media, transition radiation is generated from the boundary. Ginzburg and Frank (1946) first derived the formula of transition radiation for a single planar boundary. The formula for backward emission is written as

$$\frac{d^2N}{d\lambda\, d\Omega} = \frac{\alpha\beta^2}{\pi\lambda} n_1 \sin^2\theta_1 \cos^2\theta_1 \times \left| \frac{(\varepsilon_2-\varepsilon_1)\left(1-\beta^2\varepsilon_1+\beta\sqrt{\varepsilon_2-\varepsilon_1\sin^2\theta_1}\right)}{\left(1-\beta^2\varepsilon_1\cos^2\theta_1\right)\left(1+\beta\sqrt{\varepsilon_2-\varepsilon_1\sin^2\theta_1}\right)\left(\varepsilon_2\cos\theta_1+\sqrt{\varepsilon_1\varepsilon_2-\varepsilon_1^2\sin^2\theta_1}\right)} \right|^2 \quad (13)$$

where an electron passes through a planar boundary from medium 1 to medium 2 with a velocity v at normal incidence [Fig. 5(a)]. This formula gives a photon flux produced by a single electron in a wavelength range of $d\lambda$ and a solid angle of $d\Omega$, in which α is the fine structure constant ($\alpha = 1/137$), $\beta = v/c$ and $\varepsilon_1(\omega)$ and $\varepsilon_2(\omega)$ are the dielectric functions of the two media. The emission contains only the *p*-polarized component under the condition of normal incidence of the electron. Because the formula involves a factor β^2 in front, it is advantageous to use a higher accelerating voltage to observe transition radiation.

Spectra of the transition radiation is shown in Fig. 5(b) and (c) for normal incidence of an electron from a vacuum onto a silver surface calculated by using eq. (13) and the dielectric constant data of silver (Palik, 1985). Spectral shape changes with emission angle so as that the spectrum has a broad peak around 2.5-3 eV with a sharp dip at 3.8 eV (the bulk plasmon frequency of Ag), and the peak energy shifts to the lower energies with increasing emission angle.

Dependence of the transition radiation on emission angle is shown for several accelerating voltages in Fig. 5(d) for photon energy of 2.2 eV. The curve has a broad peak around 50° to 60°, and the peak position shifts to the lower angles with increasing accelerating voltage. Figures 5(e) and (f) show angular dependence for different photon energies at 200 kV. The peak energy is seen to shift to the lower angles with increasing photon energy.

In the case of a thin film surrounded by air, theoretical fomula of the light emission intensity was derived by Ritchie and Eldrigde (1962) and Pafomov and Frank (1967). Light emission from thin films of silver had been studied experimentally by many authors (Steinmann, 1961; Hattendorff, 1977), because the existence of the radiative surface plasmon mode was predicted by Ferrell (1958) (the Ferrell mode) and the wavelength of the mode is in the optical region for silver. The emission corresponding to that mode was observed as a sharp peak at the bulk plasmon frequency in the backward emission spectra from a silver film. The observed dependence of spectral shape and intensity on accelerating voltage is in good agreement with the theory, when the specimen is a single-crystal film with a flat surface (Yamamoto et al., 2001a).

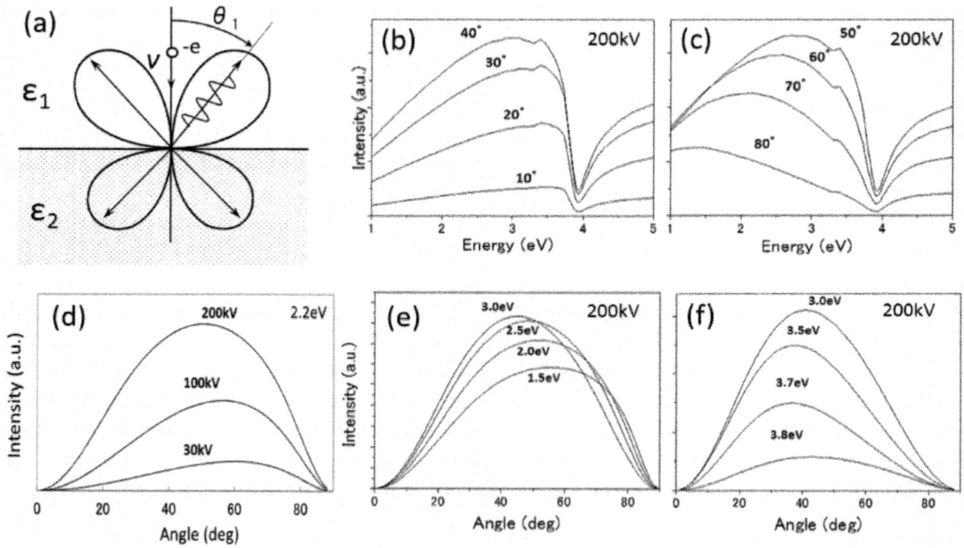

Fig. 5. (a) Diagram of angular distribution of transition radiation, (b), (c) angular dependence of TR spectrum from a silver surface, (d) accelerating voltage dependence of TR intensity, and (e), (f) photon energy dependence of angular distribution of TR.

4. Localized surface plasmon (LSP) in metal spheres

4.1 Single sphere - Mie mode

Light scattering by a metal sphere is well known and was treated by the classical electromagnetic theory (Mie, 1908). Electromagnetic oscillation modes are excited in a sphere by incidence of electrons and are transformed to photons to emit outside. In a non-retarded treatment, the energy of the sphere plasmons can be obtained by solving Poisson's equation, leading to the relation

$$\varepsilon(\omega_\ell) = -\frac{\ell+1}{\ell} \tag{14}$$

where ω_ℓ is the frequency of the localized surface plasmon modes associated with the orbital momentum number n and $\varepsilon(\omega)$ is the frequency-dependent dielectric function of the material of which the sphere is made up. A similar process is expected in the case of excitation of a metal sphere by an electron beam. Ferrell and Echenique (1985) derived a formula of electron energy loss probability for an electron passing near a dielectric sphere, and obtained the same condition for the resonance frequencies as eq. (14) calculated within a non-retarded treatment. Using the dielectric function of silver, the wavelengths obtained from this equation corresponding to the resonance frequencies are distributed in a narrow range that goes from 337 nm ($\ell = \infty$) to 354 nm ($\ell = 1$). However, the observed peaks appear at longer wavelengths, which cannot be explained by the above equation for the plasma oscillation frequencies in a sphere. Equation (14) determines the position of the loss peaks in the electron energy loss spectra calculated within a non-retarded treatment of the sphere electron interaction.

In the relativistic treatment involving the retardation effect, the probability of photon emission takes the following form (in atomic units) (Garcia de Abajo, 2002, 2010), when the electron trajectory does not pass through the sphere, as shown in Fig. 6(a),

$$\Gamma^{rad}(\omega) = \frac{1}{c\omega}\sum_{\ell=1}^{\infty}\sum_{m=-\ell}^{\ell} K_m^2\left(\frac{\omega b}{v\gamma}\right)\left[C_{\ell m}^M\left|t_\ell^M\right|^2 + C_{\ell m}^E\left|t_\ell^E\right|^2\right] \tag{15}$$

where ω is the photon frequency, K_m is a modified Bessel function, v is the electron velocity, $\gamma = 1/\sqrt{1-\beta^2}$ is the Lorentz factor accounting for the retarded reduction of the sphere oscillation frequency as seen from the frame of the incoming electron, and b is the electron impact parameter with respect to the sphere center. The coefficients $C_{\ell m}^M$ and $C_{\ell m}^E$ inside the square brackets depend exclusively on β. t_ℓ^M and t_ℓ^E are scattering *t*-matrices for magnetic and electric modes in the Mie theory (Mie 1908). The electric modes are dominant for metallic spheres and for the electron velocities under consideration (v/c=0.7). Therefore, a given peak in an emission spectrum can be associated with a maximum of the scattering matrix element t_ℓ^E for a particular ℓ th-order mode. t_ℓ^E depends both on the sphere diameter D and on the dielectric function, but not on the electron velocity. This theory explains that the resonance energy of the Mie modes changes with diameter of sphere.

We have observed some anomalous features in the light emission from silver particles of 50~500 nm in diameter supported by a carbon thin film as shown in Fig. 6(a). Their spectra are characterized by multiple peaks whose relative intensities change with the position of the electron beam, which is focused on the target (Fig. 6(b)). Those peaks correspond to the Mie modes; the lowest energy peak indicated by red arrows in Fig. 6(b) corresponds to the dipole mode, and next one indicated by green arrows corresponds to the quadrupole mode. The observed peak energy shift with diameter of sphere is in good correspondence with theoretical one shown as colored back ground plot in Fig. 6(c).

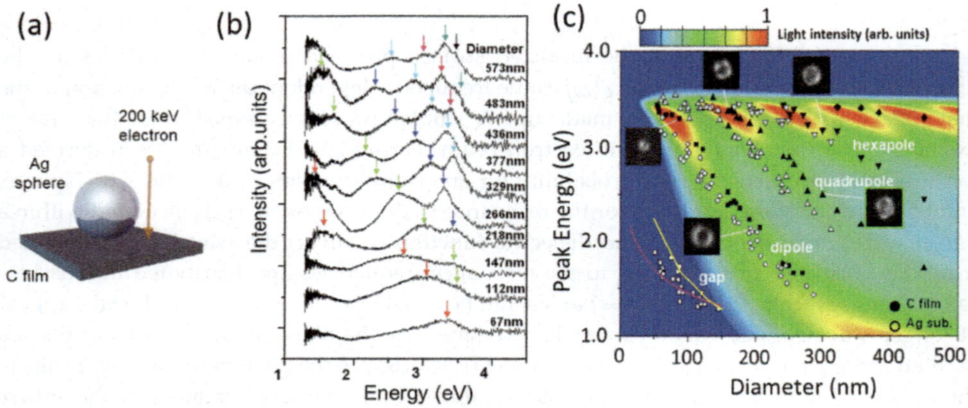

Fig. 6. (a) Excitation of Mie plasmons in silver particles excited by a grazing electron beam, and (b) emission spectra from silver particles with different diameters. (c) The size-dependent measured energies of Mie plasmons are represented and compared to the calculated cathodolulminescence intensity from self-standing particles under grazing incidence (background density plot). Open circles are observed peak energies from particles on a thick silver substrate (see ref. (Yamamoto et al., 2011)).

4.2 Metal sphere on substrate – Gap mode

Gap plasmons are observed at lower energies in the spectra from silver particles deposited on silver substrate when the beam spot is close to the center of the spheres, as shown in Fig. 7 (a). Angular dependence of the emission was measured under the arrangement of Fig. 7(a), which is shown in Fig. 7(b). Three peaks are identified from the beam-scan spectral images and photon maps taken at the emission energies; peak 1 is quadrupole Mie mode, peak 2 is dipole Mie mode and peak 3 corresponds to gap mode. The emission by the Mie mode is directed normal to the substrate surface, whereas the emission by the gap mode is directed over the higher angles. The angular distributions of those emissions are indicated in Fig. 7(c), where the emission intensities are plotted with open circles (dipole mode) and squares (gap mode).

The observed gap energies can be derived from the expression $\varepsilon(\omega) = -\varepsilon_d(1+2R/d)^{0.5}$, where ε and ε_d are the permittivities of silver and the material inside the gap, respectively, d is the gap distance, and R is the particle radius (Rendell and Scalapino, 1981). Red solid line in Fig. 6(c) is theoretical curve calculated by this expression. The yellow curve is another theoretical curve calculated for the emission from a dipole situated in the upper part of the spherical particle, opposite to the gap position, and oriented along the axis of symmetry (Yamamoto et al., 2011). The angular distribution is well explained by the emission from a dipole situated close to a planar silver surface and oriented perpendicular to it (m = 0 symmetry). The emission from the dipoles (solid curves in Fig. 7(c)) is calculated from the closed-form expressions (Novotny and Hecht, 2006).

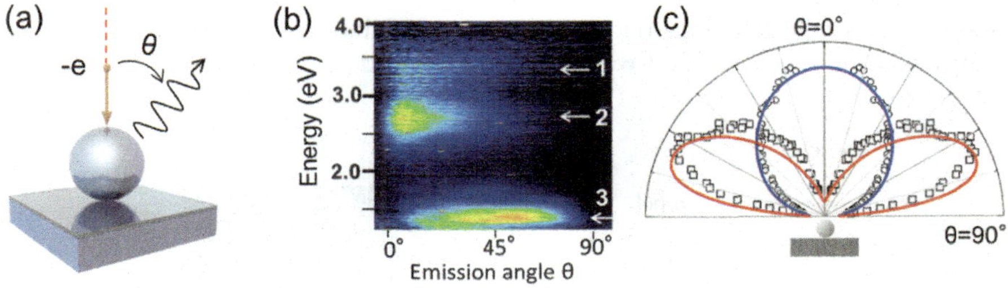

Fig. 7. (a) Arrangement for excitation of gap mode by an electron beam. (b) Angular dependence of the emission from dipole Mie (2) and gap (3) plasmons in supported spheres, and (c) schematic representation of the angular distribution. Theoretical curves are shown for gap plasmon (red) and dipole Mie plasmon (blue) (see Ref. (Yamamoto et al., 2011)).

5. SPP to light conversion at surface steps

5.1 Beam scan spectral image

A rectangular shaped terrace was fabricated on an InP substrate by electron beam lithography with a height of 100 nm and a width of 20 μm. Silver was deposited on the sample with a thickness of 200 nm by thermal evaporation in vacuum. Surface steps around the terrace were used for CL measurements and the experimental arrangement is illustrated in Fig. 8(a). Figure 8(b) shows a panchromatic photon map of the rectangular shaped terrace taken by the total photon intensity of the *p*-polarized light detected by the parabolic mirror. A fringe contrast appeared parallel to the surface steps when the monochromatic photon maps were observed at any photon energy (Yamamoto and Suzuki, 2008). Figure 8(c) clearly reveals a change in the fringe period with photon energy in the spectral image, which was taken by scanning the incident electron beam along the line A-B in Fig.8 (b). The direction of light detection was inclined by 43 °from the surface normal to the right side as depicted by the arrow in Fig. 8(b). The horizontal axis in the spectral image indicates the beam position, whereas the vertical axis indicates photon energy. The surface step was positioned at the center of the line, so the fringe patterns were discontinuous at the central vertical line in the spectral image.

In addition, the fringe contrast changed with the polarization direction. Herein *p*- (*s*-) polarization denotes the polarization direction parallel (perpendicular) to the plane determined by the surface normal and the emission direction of detected light. Figures 8(c) and (d) are the *p*- and *s*- polarized spectral images along the line A-B, respectively. A strong fringe contrast appeared in (c), but (d) had a negligible photon intensity and the contrast disappeared, indicating that the emission is strongly polarized in the p polarization direction. Figures 8(e) and (f) are *p*- and *s*- polarized spectral images along the line C-D with the same detection direction of light as that in (c) and (d), respectively. A fringe contrast, which had periods equal to those in neighboring terraces across the step, appeared in Fig. 8(e). However, the fringe contrast was asymmetrical about the step; the fringe started from the step with a bright contrast on the upper terrace [left side in Fig. 8(e)] and ended with a dark contrast on the lower terrace (right side). The phase of the fringe changed with photon energy. In Fig. 8(f), the fringe contrast disappeared on the terraces, but a bright contrast localized at the step. The bright contrast can be attributed to the emission by a localized surface plasmon at the step edge (Dobrzynski and Maradudin, 1972). The emission had two broad peaks in the spectrum, and was strongly polarized perpendicular to the step as seen from Figs. 8(d) and (f).

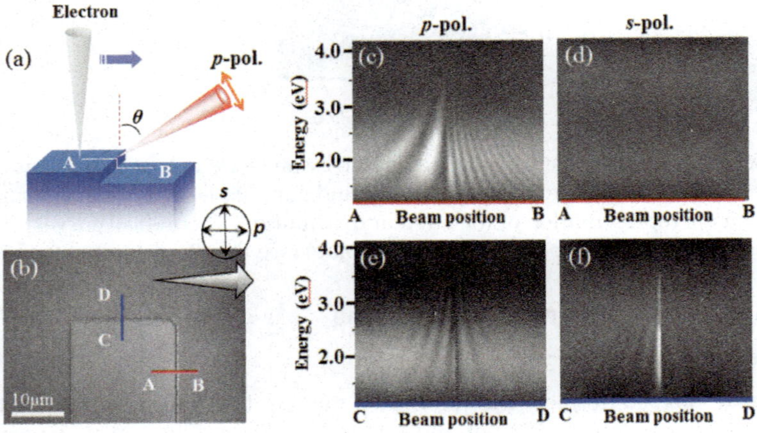

Fig. 8. (a) Panchromatic photon map of a rectangular shaped terrace. Spectral images taken by scanning the incident electron beam along the line A-B with (b) *p*-polarized and (c) *s*-polarized light emitted in the direction depicted by the arrow in (a). Spectral images taken along the line C-D with (d) *p*- and (e) *s*-polarized light. The light detection direction is tilted by 43°from the surface normal, and scanning distance along the lines is 8μm. (Yamamoto and Suzuki, 2008)

The periods of the fringes in Fig. 8(c) differed on the two terraces; the period was shorter on the right terrace than on the left terrace. This property depended on the detection direction of light, but not on the type of the surface step. The fringe contrast can be explained by the interference between transition radiation (TR) and SPP induced radiation (SPPR) at the step as illustrated in Fig. 9 (Yamamoto and Suzuki, 2008). When the incident electron passed through the silver surface, transition radiation was generated due to the induced surface charge response. The transition radiation was *p*-polarized, and had a peculiar angular distribution. Simultaneously incident electrons excited SPPs with different energies. Each SPP propagated as a two-dimensional spherical wave from the incident point. When the SPP reached the surface step, it had a given probability of being decoupled into a photon radiating into free space.

The TR and the SPPR emitting in the direction of angle θ with a frequency ω are formulated as follows:

$$\psi_{TR} = A\frac{1}{R}\exp i(kR - \omega t) \tag{16}$$

$$\psi_{SPPR} = \int_{-\infty}^{\infty} E \cdot \frac{\exp(ik_p r)}{\sqrt{r}} S(\theta)\exp(-i\boldsymbol{k}\cdot\boldsymbol{r})\frac{1}{R}\exp i(kR - \omega t)dy \quad , \tag{17}$$

where the amplitude A of the TR is expressed as a function of θ (Yamamoto and Suzuki, 2008) and the dielectric constants of silver (eq. (13)), and R is a distance from the incident point to the detector. In eq. (17), E is the excitation amplitude of SPP, S is the SPP to photon conversion efficiency expressed as a function of ω and θ, and k_p and k are wave numbers of the SPP and photon, respectively. Figure 9 depicts the Cartesian coordinates where the origin is set at the incident point. The integral in eq. (17) is taken along the step in the y direction where $\boldsymbol{r}$ is the position vector on the step, $\boldsymbol{r} = (\pm x_0, y)$, where the + and - signs correspond to the cases in Fig. 9(a) and Fig. 9(b), respectively. In these equations, the polarization of each emission is selected to be parallel to the scattering plane formed by the surface normal and the emission direction (the p-polarization). The phase in the two-dimensional spherical wave of SPP in eq. (17) adds to the phase shift of light emitted from the step because it is expressed by:

$$k_p r = \omega_p \frac{r}{v_p} - \omega \Delta t, \tag{18}$$

where v_p is the phase velocity of SPP and Δt is the time necessary for a SPP to propagate from the incident point to the step. Herein the frequency is conserved in the SPP to photon conversion process, i.e., $\omega_p = \omega$.

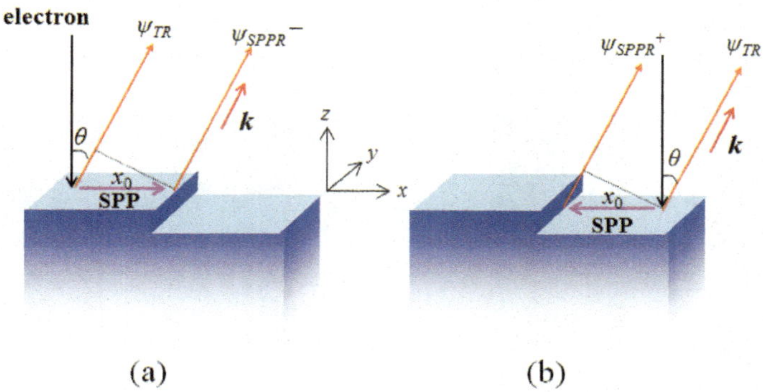

Fig. 9. Interference between transition radiation and SPP induced radiation at a surface step. (a) Electron beam is incident on the left side of the step, and (b) on the right side of the step.

Thus, the total emission intensity is given by the sum of the TR and SPP,

$$I_{Tot} = \left| \psi_{TR} + \psi_{SPPR} \right|^2 \quad . \tag{19}$$

According to the analogy with the Fresnel zone in optics, it is anticipated that the main contribution in the integral in eq. (17) is due to part of the step inside the nearest zone defined by $x_0 < r < x_0 + \lambda_p/4$. The positive and negative contributions in the integral in eq. (17) are indicated by different colors in Fig. 10(a). The integrand in eq. (17) is indicated in Fig. 10(b) for $x_0 = \lambda$. It is noticed that only the central colored region can contribute to the integral. Therefore, when the incident point is sufficiently far from the step, i.e., $x_0 >> \lambda_p$, only the SPPs propagating nearly normal to the step contribute to the SPPR.

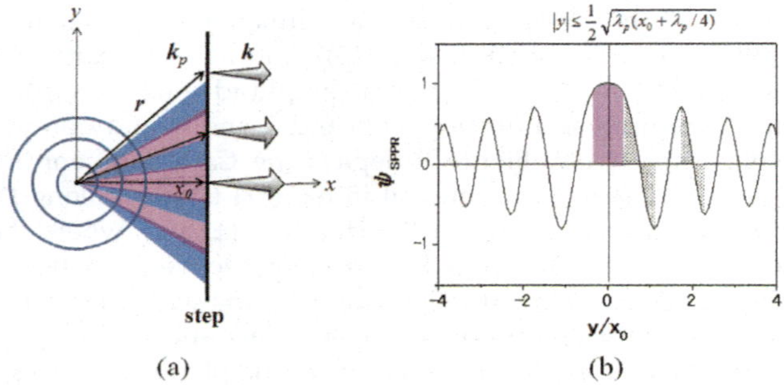

Fig. 10. (a) Geometry of the SPP to light conversion at a step. (b) The integrand in eq. (17) as a function of y for the case that $x_0 = \lambda_p$.

To confirm this, we observed beam-scan spectral images near the corner of the rectangular terrace with the detection angle of 43°as shown in Fig. 11(a). Spectral images (b) to (d) were taken along the lines in (a) far from the step by distances of (b) 1μm, (c) 2μm, and (d) 3μm, respectively. Bright fringe contrasts are seen to appear in the spectral images which are due to the interference mentioned above. Their position can be derived from the fringe positions in Fig. 8(b). It is noticed that the bright fringe contrasts disappear in the left half of the images where the horizontal step is terminated at the center of the images. This fact clearly indicates that only the SPPs propagating nearly normal to the step contribute to the SPPR.

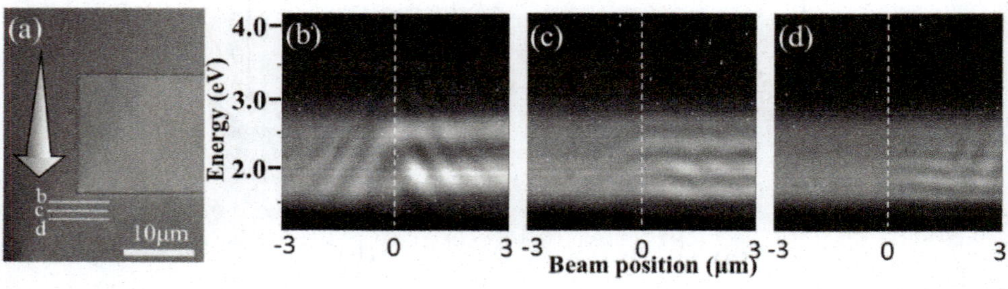

Fig. 11. (a) Panchromatic photon map of a rectangular shaped terrace. Spectral images (b) to (d) were taken by scanning the incident electron beam along the lines in (a) far from the step by distances of (b) 1μm, (c) 2μm, and (d) 3μm, respectively.

Consequently, the condition that the two radiations emitted to the angle θ are cooperatively interfered is approximately expressed by the following equation:

$$k_p r - \boldsymbol{k}\cdot\boldsymbol{r} + \delta \cong k_p x_0 \mp k x_0 \sin\theta + \delta = 2n\pi \quad , \tag{20}$$

where $\boldsymbol{k}$=(k sinθ, 0, k cosθ), and δ is a phase shift derived from the difference in phase among A, E, and S. The interference condition differs between the two cases shown in Figs. 9(a) and 9(b); the - (+) sign corresponds to the case of Fig. 9(a) (Fig. 9(b)) where the incident point is located on the left (right) side of the step. The apparent wave numbers are defined from the observed fringe periods, Λ^- and Λ^+, on the left and right side terraces, as follows:

$$K^{+} = k_{p} + k\sin\theta = \frac{2\pi}{\Lambda^{+}} \quad , \quad K^{-} = k_{p} - k\sin\theta = \frac{2\pi}{\Lambda^{-}} . \tag{21}$$

The interference condition of eq. (20) can be written using these parameters as:

$$K^{\pm}x_{0} + \delta^{\pm} = 2n\pi . \tag{22}$$

It is possible to derive the wave number of SPP and the emission angle from the measured values of Λ^{-} and Λ^{+} as:

$$k_{p} = \frac{1}{2}(K^{+} + K^{-}), \quad \sin\theta = \frac{1}{2k}(K^{+} - K^{-}). \tag{23}$$

In the case of Fig.8 (e) where $\boldsymbol{k}\cdot\boldsymbol{r}=0$, then $K^{+}=K^{-}=k_{p}$ and fringe contrast with the same period appears in the neighboring terraces. The period is equal to the wavelength of SPP.

Figure 12(a) shows the intensity profile along the section of the spectral image in Fig. 8(c) at an energy of 2.26 eV (550 nm in wavelength) where the origin of the horizontal axis is taken at the step. The bold solid lines depict the fitting curves calculated by eq. (19). Thus, the obtained fringe periods were Λ^{+}=314±6 nm and Λ^{-}=1410±30 nm for the right and left side terraces, respectively. Using these values and eq. (23), the wavelength of SPP with an energy of 2.26 eV was 514±15 nm, which is nearly equal to the expected value of 525 nm calculated for a SPP on a flat silver surface from the dielectric constants within an experimental error. The emission angle used to observe the spectral image was determined to be θ=43±2°, which coincides with the angle estimated from the mask position. The initial phase of the fringes was δ^{+}=- 0.47π in the right side terrace, and δ^{-}=1.05π for the left side terrace. This difference is due to the property of $S(\theta)$, which has a different value depending on how the SPP passes through the step, i.e., from the lower terrace to upper terrace or vice versa.

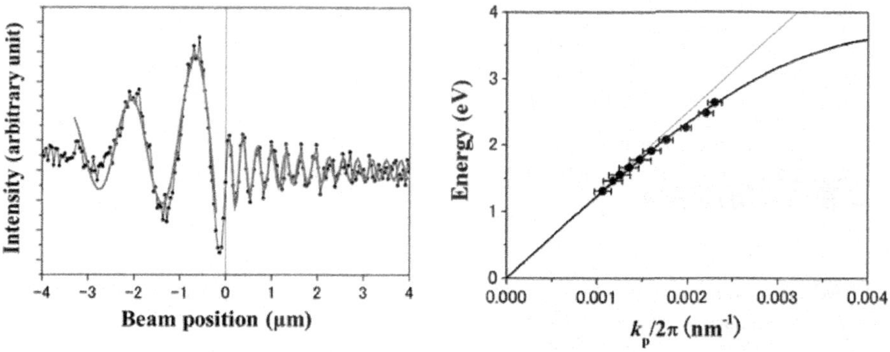

(a) (b)

Fig. 12. (a) Intensity profile along the section of the spectral image in Fig. 8(b) at an energy of 2.26 eV (550 nm in wavelength). Bold solid lines are fitting curves for the data on both sides of the step. (b) Dispersion relation of SPP on a silver surface. Solid line is a dispersion curve calculated from the optical data. (Yamamoto and Suzuki, 2008)

The dispersion relation of SPP can be derived from the energy of photon and the wave number of the SPP calculated from the fringe periods at the same energy in the spectral image. Solid circles plotted in Fig. 12(b) show the result derived from the analysis of the interference fringes in Fig. 8(c). The solid line in Fig. 12 indicates the dispersion curve of the SPP propagating on a flat silver surface calculated from the optical data of silver (Palik, 1985). The plots agreed well with the solid line within the experimental error, indicating that the oxide layer on the silver surface does not affect the dispersion of SPP in our sample because the dispersion curve shifts to a lower energy if the oxide layer is thick.

The phase shift $\delta^{\pm}$ of the interference fringes is expressed by the phases in the transition radiation, excitation function and the SPP to photon conversion efficiency in eq. (16) and (17), and is written as follows,

$$\delta^{\pm} = \delta_E + \delta_S^{\ \pm} - \delta_{TR} \,. \tag{24}$$

The phase shift depends on the emission angle and photon energy, and can be obtained from the beam scan spectral images taken at several different emission angles as shown in Figs.13 (a)-(e). The asymmetry of the fringe contrast about the step becomes strong with the emission angle.

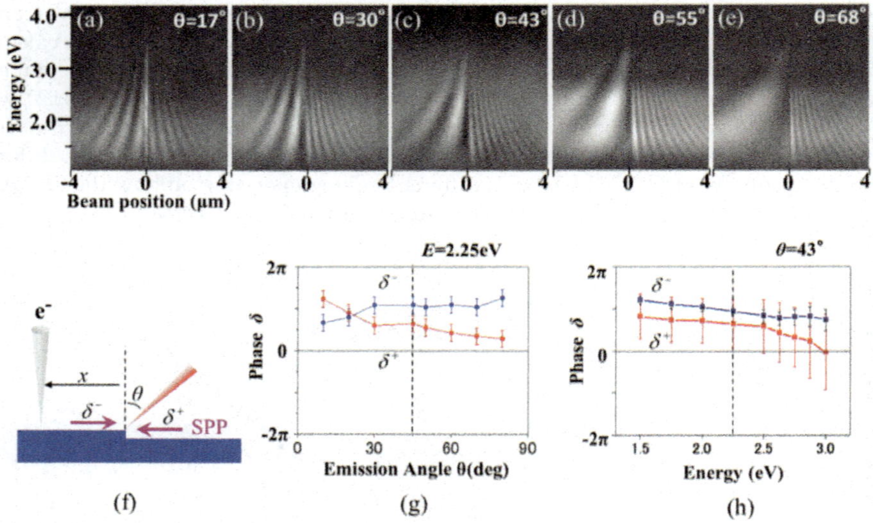

Fig. 13. The beam-scan spectral images taken at several emission angles, (a) 17°, (b) 30°, (c) 43°, (d) 55°, (e) 68°, respectively. (f) Schema of the phase shift in the SPP to light conversion at the step. (g) and (h) show dependence of the phase shift on emission angle and photon energy, respectively.

The two phase shifts δ^{+} and δ^{-} are schematically depicted in Fig.13 (f). The phase shifts can be calculated from the analysis of these images using eq. (22). Thus we can obtain the dependence of the phase shift on emission angle and photon energy as shown in Fig.13 (g) and (h), respectively. Both of the phase shifts is close to π, and nearly constant in the wide range of emission angle, except that they are crossing near θ=20°. The phase shifts are also seen to gradually decrease with photon energy (Fig. 13(h)).

Angular dependence of the SPPR was obtained by the angle-resolved measurement with an electron beam illuminated near a step. Figure 14(a) and (b) show angle-resolved spectral images taken with the electron beam fixed at a position (a) far from the step and (b) near the step. The emission pattern in Fig. 14(a) is mainly attributed to the transition radiation, whereas that in Fig. 14(b) is due to both transition radiation and SPP induced radiation at the step. Thus the intensity modulation results from interference between TR and SPPR as shown in Fig. 14(c), and can be expressed by eq. (19). Intensity profile of TR as a function of emission angle at an energy of 2.25 eV is plotted in Fig. 14(d). The theoretical curve of TR (blue line) shows good fit with the observed profile. Similarly the intensity modulation in Fig.14(b) is plotted in (e), and is fitted using eq.(19) with a proper parameter, a relative amplitude of SPPR with respect to TR. The angular distribution of the SPPR is derived as a red curve in Fig. 14(f). This distribution is rather broad compared to the theoretical one calculated by Sanchez-Gil (1998), in which the emission has sharp directionality at θ=70°. This could be attributed to the rounding of the step edge, which has a curvature of about 10 nm.

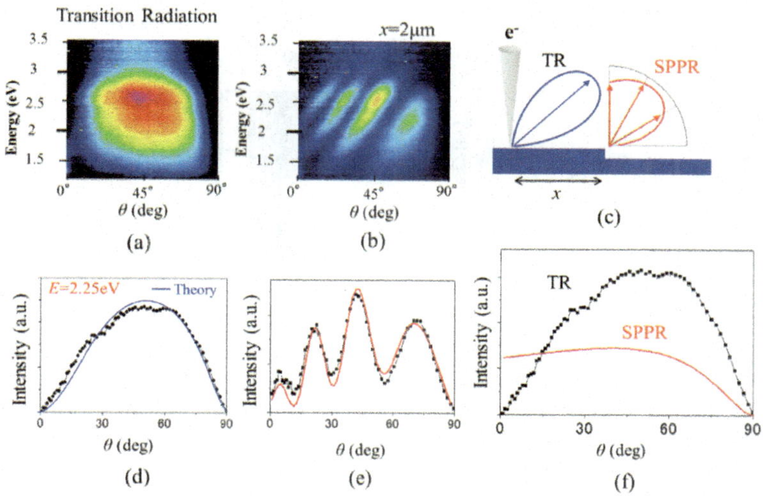

Fig. 14. Angle-resolved spectral images taken at a position (a) far from the step and (b) near the step (x=2μm). (c) shows interference between TR and SPPR. (d) , (e) show intensity profiles in (a) and (b) at an energy of 2.25 eV. (f) shows an angular distribution of SPPR (red curve).

6. Plasmonic crystals

6.1 1D-plasmonic crystal

Angle-resolved measurements were performed with the sample in two different experimental arrangements shown in Fig. 15. In the first arrangement, shown in Fig. 15(a), the sample was placed such that the grating direction was parallel to the x axis. Figure 15(b) and (c) show angle-resolved spectral (ARS) patterns of p- and s-polarized emissions, respectively. During the measurement the electron beam was scanned over an area of 5 × 5 μm². In the *p*-polarized ARS pattern of Fig. 15(b), a bright curved shape was observed, in which anti-crossing occurred at the crossing points of the curves. However, in the *s*-polarized ARS pattern of Fig. 15(c), the emission vanished and no contrast appeared. These results indicate that the detected emission is purely *p*-polarized in the setup of Fig. 15(a). Figure 15(e) and (f) were the ARS pattern measured in the setup shown in Fig. 15(d), in which the grating direction was perpendicular to the *x* axis. In the *p*-polarized ARS pattern of Fig. 15(e), spectra with a broad emission peak were observed over a large angular range. This emission is mainly due to the transition radiation (TR) emitted when an electron passes through a metal surface. The TR is also seen in Fig. 15(b) as a weak background. In the *s*-polarized ARS pattern of Fig. 15(f), several curved bright lines appear, which differ in shape from those in Fig. 15(b).

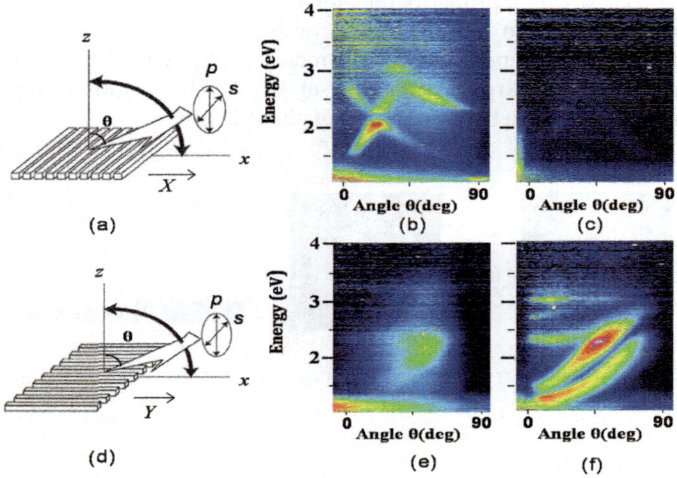

Fig. 15. (a) The arrangement for the angle-resolved measurements with the grating direction (the *x* direction) parallel to the parabolic axis of the mirror. (b, c) ARS patterns taken with the *p*- and *s*-polarized light, respectively, in the (a) arrangement. (d) The arrangement with a grating direction perpendicular to the parabolic axis. (e, f) ARS patterns taken with the *p*- and *s*-polarized light, respectively, arranged as in (d). (Suzuki and Yamamoto, 2009)

The ARS pattern is transformed into a dispersion pattern by changing the emission angle to the wave vector component parallel to the surface, $k_{//}$, using the relation of eq.(5). After the conversion of the abscissa, the ARS patterns in Fig. 15(b) and (f) are transformed into the dispersion patterns in Fig. 16(a) and (b), respectively.

A surface plasmon polariton on a periodic structure is similar to a Bloch wave. When the wave vector is far from the Brillouin zone boundary, the SPP is plane wave-like, whereas it becomes a standing wave when the wave vector is at the Brillouin zone boundary. Solid lines in Figs. 16(a) and (b) indicate dispersion curves of SPP in the 1D-plasmonic crystal along the grating direction (the X direction in Fig. 15(a)) and the perpendicular direction, respectively. These curves are made by successive shift of the dispersion curve of SPP in a flat surface by the reciprocal lattice vector ***G*** (the empty lattice approximation), though energy gaps open up at the Brillouin zone boundaries in the real plasmonic crystal. As mentioned in section 2, the dispersion curves of SPP are revealed inside the light line in the dispersion pattern. Similar result was obtained by Heitmann (1977). The SPP dispersion curves are seen to fit with the observed contrasts in Fig.16. The splitting of the bright line contrasts in Fig. 16(b) is due to dissociation of the degenerated SPP modes along the Brillouin zone boundary, corresponding to the two SPP standing wave modes (Suzuki and Yamamoto, 2009).

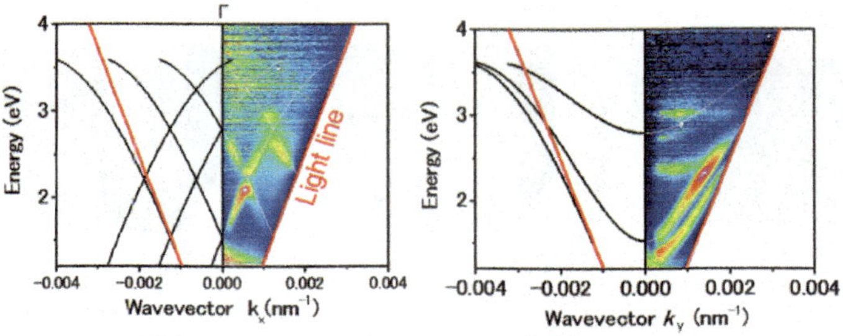

Fig. 16. (a, b) Dispersion patterns transformed from Fig.2 (b) and (f), respectively. (Suzuki and Yamamoto, 2009)

Emission spectra were successively measured by scanning the electron beam along the x direction in the experimental arrangement of Fig. 14(a), with the mask fixed at the position indicated by an arrow (θ = 20°) in Fig. 17 (a). Figure 17(b) shows a spectral image created by aligning the observed spectra with respect to the electron beam position, where the vertical axis indicates photon energy. This is *a* typical example of the beam-scan spectral image of the 1D plasmonic crystal. This image reveals spatial distribution of the SPP standing waves of the band edge at the Brillouin zone boundary. The SPP standing wave at E=2.28eV is a symmetry mode (ω+ mode) having an antinode at the center of the terrace, and that at E=2.09eV is a asymmetry mode (ω— mode) having a node at the center. The charge distributions of the two modes are illustrated in Fig. 17(c).

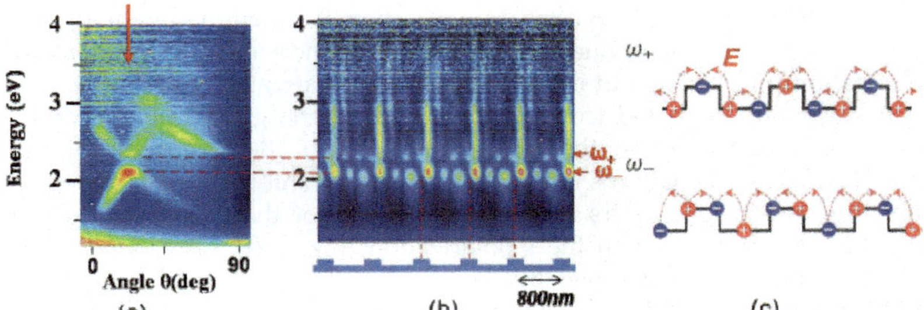

Fig. 17. (a) The ARS pattern in Fig. 14 (b). (b) A beam-scan spectral image taken with a scanning electron beam across the grating. (c) Illustration of charge distribution and the electric field of the two SPP standing waves. (Suzuki and Yamamoto, 2009)

6.2 2D-plasmonic crystal

Figure 18(a) shows the SEM image of an array of cylindrical holes on an InP substrate created by electron-beam lithography. A square lattice in the array has a periodicity of 600 nm, and the holes have a diameter and depth of 300 nm and 100 nm, respectively. A 200-nm thick silver layer is evaporated onto the substrate by thermal evaporation in a vacuum. Figure 18(b) illustrates the reciprocal lattice of this 2-D plasmonic crystal structure. Figures 18(c)-(f) show the dispersion patterns along the Γ–X direction for (c) p-polarized and (d) s-polarized emissions, and along the Γ–M direction for (e) p-polarized and (f) s-polarized emissions. These patterns were made by folding a right-half pattern to the left.

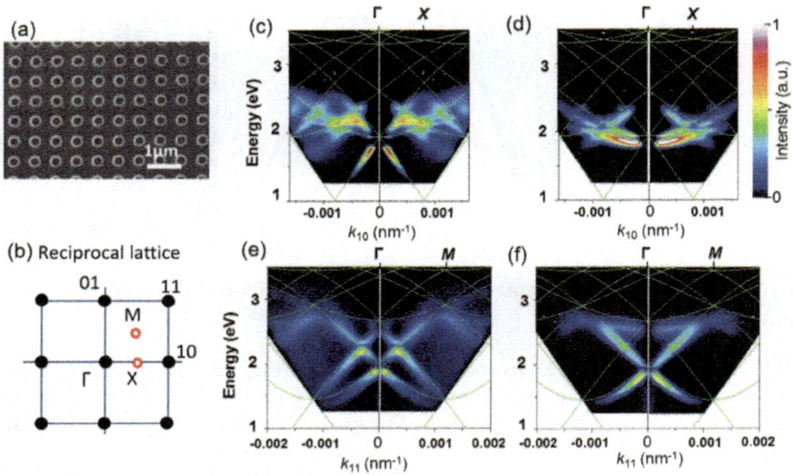

Fig. 18. Dispersion patterns of 2-D plasmonic crystal. (a) SEM image of a specimen, (b) a reciprocal lattice. (c) and (d) show dispersion patterns along the Γ–X direction for p-polarized and s-polarized emissions, and (e) and (f) are those along the Γ–M direction. (Takeuchi and Yamamoto, 2011)

In the dispersion patterns, bright curved contrasts, which appear along the SPP dispersion curves (green lines), depend on the polarization direction of the emitted light. In the dispersion pattern in the Γ–X direction, split line contrasts along the horizontal dispersion curve appear in the s-polarization (Fig. 18(d)), but disappear in the p-polarization (Fig. 18(c)). However, the other line contrasts are elongated from nearly the same energy position (~1.9 eV) at the Γ point. These contrasts appear in the p-polarization (Fig. 18(c)), but disappear in the s-polarization. The dispersion patterns in the Γ–M direction exhibit a similar dependence on the polarization direction; a gap occurs at the closing point of the dispersion curves in the p-polarization (Fig. 18(e)), but not in the s-polarization (Fig. 18(f)). The dependence of these contrasts on polarization is explained by considering the SPP modes associated with the photon emission under the condition of eq. (6) (Takeuchi and Yamamoto, 2011).

Figure 19(a) shows an ARS pattern of p-polarized emission near the surface normal direction acquired by tilting the specimen in the [1,0] direction with respect to the incident beam direction and Fig. 19(b) illustrates a schematic diagram of the dispersion curves along the Γ–X direction. The four dispersion lines gather at the Γ point. The line contrast of the highest energy mode appears only in p-polarization, although it disappears at the Γ point. Therefore, this mode should correspond to the A mode, which does not emit light, due to the symmetrical charge distribution around the hole. The E mode at the Γ point is doubly degenerated, and the lowest energy mode can correspond to this mode. Those modes can be identified from pattern in the photon maps.

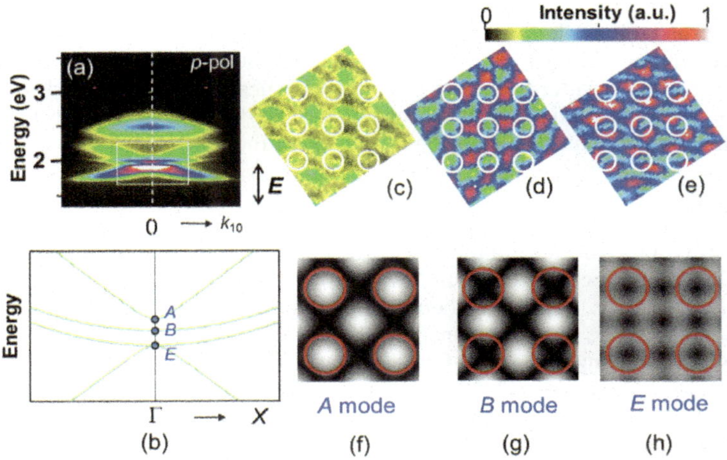

Fig. 19. ARS pattern near the Γ point taken by p-polarized emission. (b) Schematic diagram of the dispersion curves along the Γ–X direction. Photon maps taken by the p-polarized emission at peak energies of (c) 2.067 eV, (d) 2.002 eV, (e) 1.944 eV. (i)–(k) show patterns of the time-averaged field strength associated with the SPP modes at the Γ point. Red circles indicate holes. (Takeuchi and Yamamoto, 2011)

Figures 19(c)-(e) are photon maps taken with light emitted near the surface normal direction at the peak energies of the Γ point for the p-polarization where the white circles denote hole positions. It has been reported that a photon map acquired via the cathodoluminescence technique under the proper conditions mimics the field distribution of standing SPP waves or the local electromagnetic density of states (EMLDOS) (Suzuki and Yamamoto, 2009; García de Abajo and Kociak, 2008; Kuttge et al., 2009; Boudarham et al. 2010). Figures 19(f)-(h) show patterns of the time-averaged field strength associated with the SPP modes at the Γ point, i.e., the square modulus of the field amplitude. There is a clear correlation between the photon maps and calculated patterns. The SPP to photon conversion mechanism in the plasmonic crystal can be explained by the idea that a collective oscillation of an induced surface charge of the SPP forms an oscillating electric dipole around the edge of the hole to emit light (Takeuchi and Yamamoto 2011).

7. Conclusion

High energy electrons can excite surface plasmon polaritons (SPPs) in metal particles and nanostructures on metal surfaces. The excited SPPs are converted to photons via surface morphologies. The emitted light induced by incident electrons through SPP excitation inside a TEM/STEM can be detected by the TEM/STEM-CL system. The characteristic properties of CL in SPP excitation on metal surfaces are as follows: (1) it can generate an SPP point source at any position on the sample surface, regardless of the shape of the surface, (2) all the SPP modes with continuous energies can be excited, and (3) the excited SPP propagates in all directions from the electron incident point as a 2-dimensional spherical wave. The electron beam can be focused to a nanometer-sized spot, thus high spatial resolution can be achieved for CL imaging. By selecting the direction of the SPP-induced light in the angle-resolved measurement, we can investigate an individual SPP mode with a specific energy and propagation direction. Then the dispersion relation of the SPP in the specific direction was derived from the angle-resolved spectrum. The emission intensity distribution in the spectral image taken with scanning electron beam at the selected detection angle revealed the SPP standing wave patterns at the band edge energies. These results indicate that CL is a useful technique for investigating the properties of SPP in plasmonic crystals and related structures such as cavity and wave guide composed of metallic materials.

8. References

Brongsma, M. L. and Kik, P.G., ed.,2007,"Surface Plasmon Nanophotonics, Springer Series in Optical Sciences," Vol.131.

Boudarham, G., Feth, N., Myroshnychenko, V., Linden, S., Garcı´a de Abajo, F. J., Wegener, M., and Kociak, M., Phys. Rev. Lett. 105, 255501 (2010)

Catchpole, K.R. and Polman A., 2008, Opt. Express, 16, 21793.

Dobrzynski, L. and Maradudin, A.A., 1972, Phys. Rev. B 6, 3810.

Ferrell, R.A., 1958, *Phys. Rev.* 111, 1214.

Ferrell, T.L., Echenique, P.M., Phys. Rev. Lett. 1985, 55, 1526.

García de Abajo, F. J., Howie, A. *Phys. Rev. B* 2002, *65*, 115418.

García de Abajo, F. J., *Rev. Mod. Phys.* 2010, *82*, 209–275.

García de Abajo, F. J., and Kociak. M., Phy. Rev. Lett. 100, 106804 (2008).

Ginzburg V.L., Frank I.M., 1946, JETP (USSR) 16: 15; 1946, *J. Phys.* 9: 353.

Heitmann, D.,1977, J. Phys. C, 10, 397.

Hattendorff, H.D., 1977, *Phys. Status Solidi A* 42,489.

Homola, J., Chem. Rev. 108 (2008) 462-493.

Kuttge, M., Vesseur, E. J. R., Koenderink, A. F., Lezec, H. j., Atwater, H, A, Garc´ıa de Abajo, F. J., and Polman A.,2009, Phys. Rev. B 79, 113405.

Maier, S.A., 2007, "Plasmonics: Fundamentals and Applications", Springer.

Mie, G., 1908, *Ann. Phys. (Leipzig)*, *25*, 377–445.

Novotny, L. and Hecht, B., 2006, Principles of Nano-Optics; Cambridge University Press: New York.

Pafomov, V.E. and Frank I.M.., 1967, Sov. J. Nucl. Phys. 5: 448.

Palik, E.D. 1985, *Handbook of Optical Constants of Solids*, (Academic Press, New York).

Raether, H., 1988, "*Surface plasmons on smooth and rough surfaces and on gratings*" (Springer-Verlag, Berlin).

Rendell, R. W. and Scalapino, D. J., 1981, *Phys. Rev. B*, *24*, 3276–3294.

Ritchie, R.H, Eldridge, H.B., 1962, Phys. Rev. 126, 1935-1947.

Sanchez-Gil, J.A., 1998, Appl. Phys. Lett. 73, 3509 .

Steinmann, W., 1961, *Phys. Rev. Lett.* 5, 470.

Suzuki, T and Yamamoto, N., 2009, Opt. Express, 17, 23664-23671.

Takeuchi, K., and Yamamoto, N., 2011, Opt. Express, 19, 12365-12374.

Yamamoto, N., Sugiyama, H. and Toda, A. 1996, Proc.Roy.Soc.London A, 452, 2279-2301.

Yamamoto, N., Araya, K., Toda, A. and Sugiyama, H., 2001a, Surf. Interface Anal. 31, 79-86.

Yamamoto, N., Araya, K. and García de Abajo, F. J., 2001b, Phys. Rev.B 64, 205419.

Yamamoto, N., Nakano, M. and Suzuki, T., 2006, Surf. Interface Anal.38, 1725-1730.

Yamamoto, N. and Suzuki, T., 2008, Appl. Phys. Lett. 93, 093114.

Yamamoto, N., "Handbook of Nanophysics ; Nanoelectronics and Nanophotonics", (2010) Taylor&Francis Pub. (CRC Press), Ch.21, p1-25.

Yamamoto, N., S. Ohtani, S., and García de Abajo, F.J., 2011, Nano Lett., 11, 91–95.

16

Ulinastatin and Septic Cardiac Dysfunction

Jian-Dong Lin and Ming-Rui Lin
Department of Intensive Care Unit,
The First Affiliated Hospital of Fujian Medical University,
Fuzhou, Fujian,
China

1. Introduction

Sepsis is a common disease in the domain of Critical Care Medicine. It is a systemic inflammatory response syndrome (SIRS), which is response to infection. And severe sepsis may lead to septic shock and multiple organ dysfunction syndrome (MODS). The pathogenesy underlying cardiac dysfunction in sepsis is still incompletely understood and the mortality is high, although considerable effort has been applicable to understand the mechanism of inflammatory cascade response and multiple-system organ failure induced by sepsis (Wheeler and Bernard 1999). The research also found that the incidence rates of severe sepsis in the patients of hospitalization were annualized increase by 8.7 percent in the United States in the past decades years ago (Bernard et al. 1997; Martin et al. 2003; Melamed and Sorvillo 2009). However the mortality decreased in a certain, from 27.8 percent 1979 to 17.9 percent in 1995 (Martin et al. 2003; Melamed and Sorvillo 2009). But in the ICUs, incidence rate of severe sepsis occurs was still high which is up to 11% of the patients, and the mortality rates range from 18% to 55% in the developed country (Finfer et al. 2004; Karlsson et al. 2007).

2. Septic cardiac dysfunction

Defective cardiac function often occurs during sepsis, and it has been regarded as "cardiomyopathy of sepsis" (Niederbichler et al. 2006). The cardiac dysfunction greatly contributes to the the occurrence of sepsis shock and MODS in the sepsis patients. Decrease of the cardiac output can lead to hypoperfusion of the vital organs, the supply of oxygen and nutrition in tissue is reducing, the immunity is suppressed, and then the organs dysfunction happens (Kumar, Haery and Parrillo 2000).

3. The pathogenesy of septic cardiac dysfunction

Our understanding of septic cardiac dysfunction is ever more clear following the more and more studies on it. Early researches had revealed that myocardium intake 70%-75% oxygen from the coronary artery, and then saturation of blood oxygen in the coronary artery is only remains 25%-35%. In this situation when the need of the myocardium increase, ever more oxygen can't get from the blood of the coronary artery. Under this circumstances, myocardium want to intake ever more oxygen only through increasing blood flow of the coronary artery (Messer and Neill 1962). Based on this theory, early researchers assumed that myocardial depression and myocardial ischemia are interrelated during the septic shock. However, 1986 Cunnion et al (Cunnion et al. 1986) discovered that the relation between myocardial depression and decrease of the blood flow in the coronary artery is not obvious. We review almost recently researches and found out septic cardiac dysfunction may refer to following factors.

3.1 Effect of circulating causative factors such as IL-1, TNF-α, HSP-70 in the progress of myocardial depression in sepsis

Recently related review refered to the importance of pro-inflammatory cytokines and inflammatory protein in the septic cardiac dysfunction. These mediator are IL-1, TNF-α, C5a, CRP, HSP70, myocardial depressant factor (MDF), NO, epinephrine, endothelin and so on (Cohen 2002; Grocott-Mason and Shah 1998; Price et al. 1999; Sharma 2007; Tracey 2002). IL-1 and TNF-α are the most importance pro-inflammatory factors in the early stage of sepsis, and play a important in "cytokine storm" which the injury or endotoxin induces (Dinarello 1997; van der Poll and van Deventer 1999). Then the pleiotropic transcription factor NF-κB is activated and involves in the regulation of multiple biological phenomena, including stress response, cell growth, apoptosis, innate immunity, septic shock and so on (Brown and Jones 2004).

3.2 Cytopathic hypoxia and oxidative stress damage may be the causative factors of myocardial depression in sepsis

Evidence indicates that mitochondria is the energy plant of the cell metablism, it plays a critical role in oxygen utilization of cells and is responsible for supplying 90% ATP which the cell needed (Rolfe and Brown 1997). Redox-stimulated production of reactive oxygen species (ROS) may be one of the pathophysiologic mechanisms of sepsis. In that reactive oxygen may be through direct damage nucleotide of mitochondria then the DNA replication can be affected and the oxygen consumption can be reduced.

In the cecal ligation and puncture (CLP) animal models, through intraperitoneally administrating with or without polyethylene glycole-catalase: PEG-CAT (H_2O_2 and O_2^- radical scavenger) 4 hours after CLP. Hirata discovered that the level of caspase-3, OGG1 in the former is significantly higher than the later, however, 8-Oxo-2'-deoxyguanosine (8-oxo-dG), Coenzyme (Co)Q1/CoQ1H2 and Cytochrome C are significantly lowered. Thus he realized oxidative stress may make the mitochondria DNA and the respiratory chain especially the complex II damaged. And it is the major causative factor of mitochondria dysfunction (Hirata 2009). In the other hand, Ritter et al (Ritter et al. 2003) found out the ratio of superoxide dismutase (SOD) to hydrogen peroxidase was increasing. As known that SOD mainly catalyze the production of H_2O_2 and O_2^- and then make cell damaged. IL-1β and TNF-α may be the potential factors of the ratio imbalance (Schulze-Osthoff et al. 1992). All of these oxygen free radical can damage mitochondria and then damage the tissue and organs (Exline and Crouser 2008). Thus oxidative stress can damage mitochondria, decrease oxygen utilization of cell then lead to relative hypoxia.

3.3 The changes of expression of inflammation, energy metabolism and contractile related genes have been shown to be effective in the progress of septic cardiac dysfunction

In animal experiment, the differences in gene expression of cardiac tissue in septic rats was detected, which is under the CLP, by DNA microarrays. Y Liu et al (Y Liu et al. 2009) has found out Ace gene showed up-regulation may increase expression of Ang I, so that they elucidated Ace may play an important in pro-inflammatory reaction during sepsis. Available evidence suggests that Ace can activate renin-angiotensin system, increase the level of leucocyte and dose-dependently increased the poorer outcome of sepsis (Danser et al. 1995; Westendorp et al. 1997). A recent study demonstrated angiotensin II stimulation can increase the chemokine monocyte chemoattractant protein-1 (MCP-1) expression in vascular smooth muscle cells (SMCs) of rats so as to act as pro-inflammatory role (Hernandez-Presa et al. 1997). At the same time, another study verified this inference from the reverse side. The local concentrations of nitric oxide may be raised by the lower Ace enzyme activity with the II genotype, and then may make the efficiency of mitochondrial respiration better (Borutaite et al. 2001). Thus contractile function can be improved in both cardiac and skeletal muscle. In addition, angiotensin-converting enzyme inhibitor (ACEI) can reduce the level of MCP-1 induced by Angiotensin II, which in turn interfering with $CD4^+$ T cell proliferation and cytokine production mediated by MCP-1, so as to play anti-inflammatory effects (Hogaboam et al. 1998). Thus the pro-inflammatory effects of the Ace may be proved from the other side. What is more, our research team discovered that down regulation of intracellular corticosteroid receptor gene NR1D1 expression may be an important factor in septic cardiac dysfunction animal experiments. Overexpression of NF-κB induced by the pro-inflammatory cytokines such as IL-1b, Osmr, maybe the cause (Y Liu et al. 2009). Down regulation of NR1D1 expression makes the function of glucocorticoid receptor (GCR) changed, and the number decreased, so that it makes septic rats insensitive to the anti-inflammatory action of GCs (Rook et al. 2000). In addition, up-regulation of G-protein coupled receptor protein signaling transduction related gene (Gpr88) may be harmful to the heart. Because G-protein coupled receptor protein signaling pathway may along with the classic TPK signal transduction pathways to take part in signaling transduction of the heart tissue. Moreover up-regulation of transcription related gene (Cebpb) implies CCAAT/enhancer-binding protein beta (C/EBPβ) may play an important part in sepsis(Y Liu et al. 2009). In 2010, through researching in nitric oxide synthase (NOS) gene $^{-/-}$ rats, Dos Santos C C et al (dos Santos et al. 2010) discovered that rats of NOS $^{-/-}$ group failed to down-regulate the level of bioenergy and metabolism related genes expression such as PEX11b, PEX19 and PEX3 and PGC-1α-related genes. So they regarded inducible nitric oxide synthase may be the cause of sepsis-induced myocardial depression.

4. The micro-characterization of septic heart tissue

As to the micro-characterization of septic heart tissue, inner and outer membranes of mitochondria was discovered patchily disrupted by electron microscopy (Suliman et al. 2004). In this experiments, electron-lucent matrix, distorted cristae and variable swelling were found in the rat heart (Suliman et al. 2004). Also, in the rats experiments, Tatsumi et al (Tatsumi et al. 2004) has the similar findings. Compare to the sham group, there was less electron density in the mitochondria of septic hearts (Watts et al. 2004). In addition, autophagy was presented in the hearts of CLP animals by examination of electron micrographs (Watts et al. 2004).

5. The research progress of effect of the Ulinastatin (UTI) on septic cardiac dysfunction

Rcently years UTI as endogenous inhibitor has obvious effect on inhibiting various prolease. It maybe an effective therapeutic drug on sepsis cardiac dysfunction. From it was discovered to now, it has been commonly used in patients with inflammatory disorders including sepsis, surgery, disseminated intravascular coagulation, shock, pancreatitis and so on. As a reasonable and effective drug is mainly used in China, Japan, Korea. Research also suggests that UTI may suppress the expression of various inflammatory factors above mentioned.

Ulinastatin is a kind of serpin with 2 Kunitz-type domains, can inhibit the over expression of pro-inflammatory cytokines and then prevent subsequent organ failure induced by injuring or bacterial endotoxin. It was from human urine, as a urinary trypsin inhibitor, was described by Astrup T in 1964 (Astrup and Nissen 1964). It take part in regulating the physiological responses in the blood such as blood clotting response, complement activation, inflammation and so on to keep balance of internal environment. As for its protection role to septic cardiac depress maybe involve in the following aspects.

5.1 Inhibit releasing of IL-1 to act as protective action

Interleukin (IL)-1, a fever inducing substance secreted by activated leukocytes which lies at the center of the inflammatory response, was originally reported in the 1940s (Atkins 1960). It has two sub unites that are IL-1α and IL-1β. They have major roles in pro-inflammatory activities and host responses to not only exogenous but also endogenous noxious stimuli. Certainly these effects are contribute to the clearance of bacteria. However, in some cases these effects are detrimental, such as in serious septic shock and septic MODS (Miller et al. 2006). The mechanism of the pro-inflammatory response is as follows(Ninomiya-Tsuji et al. 1999). IL-1 binds to IL-1RI to form a complex protein associated with IL-1, then induces myeloid differentiation primary response gene 88 (MyD88) and the IL-1R–associated kinase (IRAK) recruitment to its receptor. And then IRAK dissociates from IRAK-receptor complex and interacts with tumour-necrosis factor receptor-associated factor 6 (TRAF6). Activated TRAF6 can make IL-1 signal transmit to the NF-kB-inducing kinase Ik B a/b (NIK-IKK-a/b) kinase pathway (Cao, Henzel and Gao 1996; Cao et al. 1996; Wesche et al. 1997). Interestingly, TNF receptor associated factor 6 (TRAF6) may activate both Jun N-terminal kinase (JNK) and NF-κB pathway (Song et al. 1997). Of course, the mechanism is different. That is NIK is the key mediators of IL-1-mediated NF-κB activation (Karin and Delhase 1998; Song et al. 1997). TAK1 can activate JNK pathway (Moriguchi et al. 1996; Shirakabe et al. 1997; Yamaguchi et al. 1995). However, TAK1 can be activated by TAB1 protein(Ninomiya-Tsuji et al. 1999). At last NIK ± IKK cascade storm happens under the regulation of IL-1 signal pathway. Moreover, evidence also showed that under administration of LPS, the level of the mRNA for leptin in adipose tissue can be increased by IL-1 or TNF-a (Grunfeld et al. 1996). It was known that leptin not only through decreasing food intake, but also increasing energy expenditure via activation of the autonomic nervous system to regulate energy balance (Pelleymounter et al. 1995). Thus IL-1 not only regulates inflammatory reaction but also energy metabolism. However protein levels of IL-1b in the liver is significantly greater in UTI $^{-/-}$ than in normal WT mice after LPS challenge under the administration of UTI (Takano et al. 2009). So UTI may play an important part in inhibiting the expression of IL-1.

5.2 Inhibit releasing of IL-6 to act as protective action

The level of IL-6 is an sensitive marker of tissue damage, and directly related to endotoxin production of the infectious diseases. It is commonly known that endotoxin is an important element of activated complement and cytokine which the inflammatory reaction related and may be an important factor of systemic inflammatory response syndrome (SIRS) (Pallua, Low and von Heimburg 2003). In addition, IL-6 as a endogenous pyrogen can stimulate body to generate neutrophil elastase and take part in pathophysiological process of sepsis and acute respiratory distress syndrome (ARDS) (Inoue et al. 2009). Pathan, N. et al (Pathan et al. 2004) found out the level of IL-6 was positively correlated to the degrees of cardiac suppression and plays a central role in cardiac suppression induced by bacterial infection. At the same time, in order to confirm this discovery, their research team verified this effect in the septic rat model with recombinant IL-6. However, Park, J. H. et al (Park et al. 2010a) found out UTI can reduce the level of stress reaction related cytokine such as IL-6 during gastrointestinal operation, also in the trauma patients with hemorrhagic shock has the similar discovery (Park et al. 2010b). This function was also verified by the septic rat model (Cao et al. 2010). Thus, UTI may regulate the level of IL-6 to act as anti-inflammatory role.

5.3 Inhibit releasing of IL-8 to act as protective action

It is well known that IL-8, a major mediator for neutrophil-mediated cell and tissue injury, is a member of C-X-C cytokine family. It is secreted by macrophages, epithelial cells and other cell types included. As a pro-inflammatory factor, it can through enhancing neutrophil infiltration, releasing of lysosomal enzyme and generating of superoxide anions, then induce tissue damage (Sekido et al. 1993). Kurt, A. N. et al (Kurt et al. 2007) found out the level of serum IL-8 in the patients with positive blood culture was found elevated significantly more than negative ones. It was also reported that UTI, as a broad-spectrum anti-inflammatory substance, may significantly down-regulate the production of IL-8 (Cao et al. 2010). Bae, H. B. et al (Bae et al. 2011) discovered the level of IL-8 in the UTI preconditioning sepsis rats is significantly lower than those in sepsis rats without treatment, and associated pathological section, they hold the opinion that UTI may through inhibiting IL-8 and so on mediators of inflammation in septic rats to act as a protective role in the septic process. Also there is evidence which shows that UTI can suppress the expression of IL-8 in vitro (Nakamura et al. 1997)

5.4 Inhibit releasing of TNF-α to act as protective action

Serum TNF-α is secreted by the activated macrophage. Its relative molecular weight is 17000. It can stimulate other inflammatory cytokine production and inflammatory cells into tissue (Cohen 2002). Also it acts as activator of the local and systemic inflammatory response, and its local accumulation can cause cardiovascular clinical symptoms, such as fever, swelling, pain and congestion. In vivo overall increase of the amount of TNF-α can decrease cardiac output, and then mediates tissue injury, such as formation of micro-thrombus and systemic capillary leak syndrome. As to regulating molecular pathway of the ischemia-reperfusion, TNF-α maybe one of the most significant factors (Niemann, Garner and Lewis 2004; Toledo-Pereyra et al. 2004). At the same time, TNF-α can also through activating other related cells secret cytokine such as IL-1, HMGB1, eicosanoids, NO, and reactive oxygen to enhance the role of inflammatory reaction and make this effect longer

(Tracey 2002). Thus it makes the tissue damage severely. It was found that serum TNF-α significantly elevated during the septic systemic inflammatory response syndrome (SIRS) (Kurt et al. 2007). In all , TNF-α can promote occurrence and development of the sepsis, the level of TNF-α increasing can lead to cell death, apoptosis, and several organs dysfunction (Niemann, Garner and Lewis 2004). Thus it is very important to inhibit the release of TNF. In the recently years, study found out UTI may significantly reduce the level of TNF-α to protect the tissue and organ during sepsis (Cao et al. 2010). Also a Japanese research team verify above conclusion in the other side, they found out in the pathogenesis of post-resuscitation syndrome, insufficient production of UTI for norepinephrine (NE) releasing, accompanied by persistent high TNF-α levels (Hayakawa et al. 2008). More over in vitro UTI can inhibit the release of TNF-a (Aosasa et al. 1998). In addition, in the liver injured animal model experiment also found the UTI can protect against severe liver injury through the suppression of TNF-a production(Takano et al. 2009). What is more, other scholar hold the opinion that UTI may reduce the NF-κB activation and then subsequently inhibit the expression of TNF-α , CXC-chemokines and vascular endothelial cell adhesion molecules to protect hepatic cells against ischemia and reperfusion(I/R) injury (Wu et al. 2009).

5.5 To decrease the level of Nitric oxide(NO) expression

The iNOS has been shown to be induced by mediators associated with sepsis, that is IL-1b, IL-2, IL-6, TNF and so on the pro-inflammatory cytokines and interferon endotoxin, and then increases production of NO in vitro (Nathan and Xie 1994). Early in 1991, there was evidence showed that NO synthase may play an important in the pathogenesis of septic shock and supressing activity of NO synthase may be a novel therapeutic option (Nathan 1992; Petros, Bennett and Vallance 1991). Recently, in several experimental models, research indicated that overproduction of NO under the regulation of cytokine-inducible NOS connects with the pathophysiology of microcirculatory failure and organs dysfunction during sepsis, and inhibiting NOS activity can prevent vessel from dilating and then hypotension happens were proved (Hollenberg et al. 2000; L¨®pez et al. 2004; Avontuur et al. 1998; Evans et al. 1994; Nava, Palmer and Moncada 1992; Tracey, Tse and Carter 1995; Kilbourn et al. 1994; Laszlo, Whittle and Moncada 1994). However, several experimental studies also showed that inhibition of NO synthesis has no effect or might be harmful (Robertson et al. 1994). Recently years, scholars found that in a model of sepsis mice induced by CLP which is deficient in nNOS didn't get the desired effect, in that they found out such mice had impaired bacterial clearance. Certainly, the survival rate also decreased. Ever more in wild-type mice under the regulation of a selective nNOS inhibitor had similar outcome. This suggested that nNOS wasn't always harmful to sepsis, it may have a protective role (Cui et al. 2007). Moreover, mice with a cardiomyocyte-specific over expression of endothelial nitric oxide synthase (eNOS) was also partially protected against both endotoxemia and polymicrobial sepsis (Ichinose et al. 2006).Why does these phenomenon happen? Maybe a research team can give a reasonable answer. They found out NO at different concentration levels has different actions. Such diversity of cellular responses to NO as followings (Table 1) (Thomas et al. 2008).

Generally, lower concentrations are good for cell survival and proliferation, and higher concentrations promote cell cycle arrest, senescence and apoptosis. As to the effects of UTI on NOS are still unknown. Chinese scholars discovered that UTI may through inhibiting the activity of NF-κB to down-regulate the expression of iNOS mRNA within intestinal muscle to protect the intestinal motility in rats with bacterial peritonitis. However whether UTI can through regulating the activity of the NOS to act as protection on cardiac myocytes and if different dose of UTI has different effects are still poorly understood.

NO concentrations	Effects
1 to 30 nM	Physiological actions under the regulation of cyclic GMP-mediated processes predominate
30 to 100 nM	Tissue repair processes including AKT phosphorylation
100 to 300 nM	Hypoxia inducible factor-1 stabilization
>1 μM	Nitrosative stress

Table 1.

5.6 Inhibit releasing of sCD14 to act as protective action

Soluble CD14 (sCD14), an endotoxin-related signal molecule, can enhance sensitivity of macrophage activation to endotoxin stimulation. In 1990, CD14 was recognized as a receptor for bacterial endotoxin (LPS) and it was the firstly described pattern-recognition receptor (Wright 1995). When combining with the conceptor TLR4, endotoxin-sCD14 complexes can lead to recruitment of IL-1–associated kinase under the endotoxin-induced, mediate nuclear factor (NF-κB) activation, then trigger the cytokine cascade. Evidence showed that sCD14 expression can be induced by interleukin (IL)-6 in liver cells and it was recognized as acute phase protein (Anas, van der Poll and De Vos 2010). It was also the key mechanism of the pathogenesis for multiple organ failure (Rahman et al. 2004). In the rat model of acute necrotic pancreatitis, the level of sCD14 in rats treatment with hyperbaric oxygen and UTI group significantly decreases compare to the ones hyperbaric oxygen only used. Combined with the level of IL-6, TNF-α and endotoxin, the author hold the opinion that UTI may decrease the level of sCD14 endotoxin related, and to function as inhibiting the pro-inflammatory reaction and then enhance the immunity function (Hou et al. 2010). Last year, Japanese scholars K Shirakawa. et al (K Shirakawa et al. 2010) hold the opinion that sCD14-ST may be the new sepsis marker. Thus UTI may through decreasing the level of sCD14 in blood to protect the heart tissue. But the reason is still unknown .

5.7 Enhance express of HSP-70 to act as protective action

Heat shock protein 70, which is conservative in organic evolution, was expressed constitutively in human erythroblasts that undergoes differentiation. It may suppress protein denaturation and keep the native conformation. It also acts as protective role in many physiological pathological and stress responses. Expression of HSP70 can be induced by fever, hypoxia, oxidative stress, endotoxin, cytokine, heavy metal ion and so on (Ribeil et al. 2007). Kustanova G. A. et al (Kustanova et al. 2006) found out the mortality of septic rat was decreased by supplying HSP-70 compare to control group. Another study suggested HSP-70 is closely associated with pulmonary biology, and has protective effects on lung injury, involves inhibition of NF-κB and pro-inflammatory gene expression. The mechanisms may concern that its anti-oxidation, anti-apoptosis, anti-inflammatory, stable transfection or molecular chaperone roles (Wheeler and Wong 2007). At the same time in animal experiments, Zhang, Z. et al (Zhang et al. 2010) found out the expression level of

HSP-70 in paraquat poisoning group under the treatment of UTI was higher than the control group. Then through associating the pathological section, they hold the opinion that UTI may regulate the expression of HSP-70 to increase the toxic tolerance and reduce lung injury. It was not reported that through up-regulation of HSP-70 in septic patients treated with UTI to increase the survival rate. However, based on the evidence above mentioned, we come out that UTI may has protective function of septic organs by up-regulating expression of HSP70.

5.8 Inhibit expression of HMGB1 to act as protective action

HMGB1, an intracellular protein, can translocate to the nucleus. It was originally described as a transcription factor which can bind DNA and regulate gene expression. HMGB1 is expressed by various cell types which contains a nucleus, such as macrophage, monocyte, neutrophil. The activating expression of HMGB1 may through non-transcriptional mechanisms and it is regulated by nuclear factor-κB (NF-κB) activation (Lotze and Tracey 2005). During inflammation, causative agent and pro-inflammatory mediator (such as TNF-a, IL-1b and IFNγ) can induce expression of HMGB1. Of course interaction of complement C5a and its acceptor C5L2 can also promote the expression of HMGB1 (Lotze and Tracey 2005). Signal transduction induced by HMGB1 is endowed with pleiotropic effects, that is pro-inflammatory and the role of potential damage to the endothelial barrier. Compare to other sepsis related cytokine, the peak level of HMGB1 occurs in the later stage and is regulated by autonomic nerves system(ANS) (Wang et al. 1999). Also a research found out that activation of the cholinergic anti-inflammatory pathway in sepsis can improve survival. The reason may be that this pathway can suppress macrophages secreting HMGB1 (Rittirsch, Flierl and Ward 2008). In the septic rat model experiment, HMGB1 is significantly decreasing in UTI treatment group than control group. In this experiment also found out UTI can suppress the generation of O_2^-, TNF-a, IL-6, and lactic acidosis. Thus the author hold the opinion that UTI may be helpful for sepsis through regulating expression of HMGB1 (Koga et al. 2010). As to whether UTI can decrease the level of HMGB1 express by activation of the cholinergic anti-inflammatory pathway is worth doing deep research.

5.9 Decrease the production of O_2^- then reduce the activity of NF-κB to act on anti-inflammatory effect

Activiting of macrophage, neutrophils and reactive oxygen species (ROS) play an important in the pathophysiology of endotoxemia. Following activating of macrophage, neutrophils, NADPH oxidase is activated, leading to excessive ROS generation to kill the bacteria. Of course, one coin two sides, at the sometime ROS also injuries lung tissue and lead to breathlessness, and then enhance SIRS (Tanaka et al. 2010). Japanese scholars found out the level of O_2^-, HMGB1, TNF-a, IL-6 were lower in UTI treatment group than the control group (Tanaka et al. 2010). Previous studies had shown that the expression of HMGB1, TNF-a, IL-6 was regulated by NF-κB. However NF-κB as a oxidation-reduction related sensitive transcription factor was under the regulation of ROS (Pahl 1999; Bar-Shai et al. 2008). It was known that UTI may through suppress the expression of HMGB1 (Koga et al. 2010), TNF-a (Cao et al. 2010), IL-6 (Park et al. 2010a) to act as anti-inflammatory. Thus they come up with the idea that anti-inflammatory of UTI may through reducing the production of O_2^- to decrease the activity of NF-κB for the first time.

5.10 As to regulating gene express of septic heart tissue under the UTI to act as protection

Certainly in gene level, UTI treatment in sepsis was also make progress. An research team led by Inoue, K (Inoue et al. 2005) found out that the lung and kedney damage are serious in UTI-/- sepsis rats than control group. At the sometime IL-1β, macrophage inflammatory protein (MIP)-1α, MIP-2 and monocyte chemotactic protein (MCP)-1 are also higher than the control group. At last they come out that UTI may has important protection on sepsis. In additional, our previous study suggest UTI may regulate exaggerated or expression of inflammatory response related gene (Agt, Ace, IL6r) to protect the function of septic heart and kidney (Y Lin et al. 2011; Y Liu et al. 2010).

5.11 At last UTI may also take important part in the energy metabolism during sepsis

Evidence showed that the activation of Na^+-K^+-ATP enzyme is associated with balance of electrolyte. It was known that well activation of Na^+-K^+-ATP enzyme is very important in absorption of edema fluid. Thus protection of Na^+-K^+-ATP enzyme can reduce cellular edema and play an important part in keeping the stability of cell membrane. Researching found out activation of Na^+-K^+-ATP enzyme remarkable decrease in ischemia tissue, while it was not obvious in UTI pretreatment ones (Xiaoqiao et al. 2004). Moreover, Na^+-K^+-ATP - ATPase and Ca_2^+-ATP-ATPase activity in UTI donor-pretreatment on liver graft group were higher than control group (Mao et al. 2011). In addition, a Japanese research team hold the opinion that UTI may maintain utilization of energy in the electron transfer system throughout the hypovolemic period and reperfusion and contributed to a rapid recovery of cardiac function after reperfusion, through reducing the damage of oxidative phosphorylation in mitochondria, even during hemorrhagic shock (Masuda et al. 2003). Those all suggest UTI may act as a protective role in organs of septic patients through maintaining the organs energy metabolism.

6. Concluding remarks

UTI may regulate the secretion of mediators of inflammation, related stress protein and related gene expression to release the inflammatory reaction, at the same time it can keep the activity of Na^+-K^+-ATP -ATPase to maintain the organs energy metabolism. It is derived from urine, as a endogenous anti-inflammatory drug, so it has little side effect and widely used in treatment of sepsis, acute pancreatitis, toxic shock, chronic pancreatitis. However, it's as a immunomodulators may along with other drug to treat sepsis in order to improve the survival rate.

7. References

Anas, A., T. van der Poll, and A.F. De Vos. 2010. "Role of CD14 in lung inflammation and infection." *Critical Care* 14(2):1-8.

Aosasa, S., S. Ono, S. Seki, E. Takayama, T. Tadakuma, H. Hiraide, and H. Mochizuki. 1998. "Inhibitory effect of protease inhibitor on endothelial cell activation." *Journal of Surgical Research* 80(2):182-87.

Astrup, T., and U. Nissen. 1964. "Urinary trypsin inhibitor (mingin): transformation into a new trypsin inhibitor by acid hydrolysis or by sialidase." *Nature* 203:255-57.

Atkins, E. 1960. "Pathogenesis of fever." *Physiological Reviews* 40(3):580.

Avontuur, J.A.M., R.P. Nolthenius, J.W. van Bodegom, and H.A. Bruining. 1998. "Prolonged inhibition of nitric oxide synthesis in severe septic shock: a clinical study." *Critical care medicine* 26(4):660.

Bae, H. B., C. W. Jeong, M. Li, H. S. Kim, and S. H. Kwak. 2011. "Effects of Urinary Trypsin Inhibitor on Lipopolysaccharide-Induced Acute Lung Injury in Rabbits." *Inflammation*.

Bar-Shai, M., E. Carmeli, P. Ljubuncic, and A. Z. Reznick. 2008. "Exercise and immobilization in aging animals: the involvement of oxidative stress and NF-kappaB activation." *Free Radic Biol Med* 44(2):202-14.

Bernard, G. R., A. P. Wheeler, J. A. Russell, R. Schein, W. R. Summer, K. P. Steinberg, W. J. Fulkerson, P. E. Wright, B. W. Christman, W. D. Dupont, S. B. Higgins, and B. B. Swindell. 1997. "The effects of ibuprofen on the physiology and survival of patients with sepsis. The Ibuprofen in Sepsis Study Group." *N Engl J Med* 336(13):912-8.

Borutaite, V., A. Matthias, H. Harris, S. Moncada, and G. C. Brown. 2001. "Reversible inhibition of cellular respiration by nitric oxide in vascular inflammation." *Am J Physiol Heart Circ Physiol* 281(6):H2256-60.

Brown, M. A., and W. K. Jones. 2004. "NF-kappaB action in sepsis: the innate immune system and the heart." *Front Biosci* 9:1201-17.

Cao, Y. Z., Y. Y. Tu, X. Chen, B. L. Wang, Y. X. Zhong, and M. H. Liu. 2010. "Protective effect of Ulinastatin against murine models of sepsis: Inhibition of TNF-alpha and IL-6 and augmentation of IL-10 and IL-13." *Exp Toxicol Pathol*.

Cao, Z., W.J. Henzel, and X. Gao. 1996. "IRAK: a kinase associated with the interleukin-1 receptor." *Science* 271(5252):1128.

Cao, Z., J. Xiong, M. Takeuchi, T. Kurama, and D.V. Goeddel. 1996. "TRAF6 is a signal transducer for interleukin-1."

Cohen, J. 2002. "The immunopathogenesis of sepsis." *Nature* 420(6917):885-91.

Cui, X., V. Besch, A. Khaibullina, A. Hergen, M. Quezado, P. Eichacker, and Z.M.N. Quezado. 2007. "Neuronal nitric oxide synthase deficiency decreases survival in bacterial peritonitis and sepsis." *Intensive care medicine* 33(11):1993-2003.

Cunnion, R. E., G. L. Schaer, M. M. Parker, C. Natanson, and J. E. Parrillo. 1986. "The coronary circulation in human septic shock." *Circulation* 73(4):637-44.

Danser, A. H., M. A. Schalekamp, W. A. Bax, A. M. van den Brink, P. R. Saxena, G. A. Riegger, and H. Schunkert. 1995. "Angiotensin-converting enzyme in the human heart. Effect of the deletion/insertion polymorphism." *Circulation* 92(6):1387-8.

Dinarello, C. A. 1997. "Proinflammatory and anti-inflammatory cytokines as mediators in the pathogenesis of septic shock." *Chest* 112(6 Suppl):321S-29S.

dos Santos, C.C., D.J. Gattas, J.N. Tsoporis, L. Smeding, G. Kabir, H. Masoom, A. Akram, F. Plotz, A.S. Slutsky, and M. Husain. 2010. "Sepsis-induced myocardial depression is associated with transcriptional changes in energy metabolism and contractile related genes: A physiological and gene expression-based approach*." *Critical care medicine* 38(3):894.

Evans, T., A. Carpenter, A. Silva, and J. Cohen. 1994. "Inhibition of nitric oxide synthase in experimental gram-negative sepsis." *Journal of Infectious Diseases* 169(2):343.

Exline, M. C., and E. D. Crouser. 2008. "Mitochondrial mechanisms of sepsis-induced organ failure." *Front Biosci* 13:5030-41.

Finfer, S., R. Bellomo, J. Lipman, C. French, G. Dobb, and J. Myburgh. 2004. "Adult-population incidence of severe sepsis in Australian and New Zealand intensive care units." *Intensive care medicine* 30(4):589-96.

Grocott-Mason, R. M., and A. M. Shah. 1998. "Cardiac dysfunction in sepsis: new theories and clinical implications." *Intensive Care Med* 24(4):286-95.

Grunfeld, C., C. Zhao, J. Fuller, A. Pollack, A. Moser, J. Friedman, and K.R. Feingold. 1996. "Endotoxin and cytokines induce expression of leptin, the ob gene product, in hamsters." *Journal of Clinical Investigation* 97(9):2152.

Hayakawa, M., A. Sawamura, Y. Yanagida, M. Sugano, N. Kubota, H. Hoshino, and S. Gando. 2008. "Insufficient production of urinary trypsin inhibitor for neutrophil elastase release after cardiac arrest." *Shock* 29(5):549.

Hernandez-Presa, M., C. Bustos, M. Ortego, J. Tunon, G. Renedo, M. Ruiz-Ortega, and J. Egido. 1997. "Angiotensin-converting enzyme inhibition prevents arterial nuclear factor-kappa B activation, monocyte chemoattractant protein-1 expression, and macrophage infiltration in a rabbit model of early accelerated atherosclerosis." *Circulation* 95(6):1532-41.

Hirata, Junichi. 2009. "Abstract P99: Oxidative Mitochondrial DNA Damage is Associated With Injury of Respiratory Chains, Especially Complex2 During Sepsis-induced Myocardial Dysfunction." *Circulation* 120(18_MeetingAbstracts):S1462-d-.

Hogaboam, C. M., N. W. Lukacs, S. W. Chensue, R. M. Strieter, and S. L. Kunkel. 1998. "Monocyte chemoattractant protein-1 synthesis by murine lung fibroblasts modulates CD4+ T cell activation." *J Immunol* 160(9):4606-14.

Hollenberg, S.M., M. Broussard, J. Osman, and J.E. Parrillo. 2000. "Increased microvascular reactivity and improved mortality in septic mice lacking inducible nitric oxide synthase." *Circulation research* 86(7):774.

Hou, J., M. W. Zhu, X. W. He, J. M. Wei, Y. G. Li, and D. N. Tang. 2010. "Effect of hyperbaric oxygen and ulinastatin on plasma endotoxin, soluble CD14, endotoxin-neutralizing capacity and cytokines in acute necrotizing pancreatitis." *Can J Surg* 53(4):241-5.

Ichinose, F., E.S. Buys, T.G. Neilan, E.M. Furutani, J.G. Morgan, D.S. Jassal, A.R. Graveline, R.J. Searles, C.C. Lim, and M. Kaneki. 2006. "Cardiomyocyte-specific overexpression of nitric oxide synthase 3 prevents myocardial dysfunction in murine models of septic shock." *Circulation research*:01. RES. 0000253888.09574. 7av1.

Inoue, K., H. Takano, H. Sato, R. Yanagisawa, and T. Yoshikawa. 2009. "Protective role of urinary trypsin inhibitor in lung expression of proinflammatory cytokines accompanied by lethal liver injury in mice." *Immunopharmacol Immunotoxicol* 31(3):446-50.

Inoue, K., H. Takano, A. Shimada, R. Yanagisawa, M. Sakurai, S. Yoshino, H. Sato, and T. Yoshikawa. 2005. "Urinary trypsin inhibitor protects against systemic inflammation induced by lipopolysaccharide." *Mol Pharmacol* 67(3):673-80.

K Shirakawa, K Naitou, J Hirose, M Nakamura, T Takeuchi, Y Hosaka, and S Furusako. 2010. "The new sepsis marker, sCD14-ST, induction mechanism in the rabbit sepsis models." *Critical Care* 14(Suppl 2):19.

Karin, M., and M. Delhase. 1998. "JNK or IKK, AP-1 or NF-¦ÊB, which are the targets for MEK kinase 1 action?" *Proceedings of the National Academy of Sciences* 95(16):9067.

Karlsson, S., M. Varpula, E. Ruokonen, V. Pettil , I. Parviainen, T.I. Ala-Kokko, E. Kolho, and E.M. Rintala. 2007. "Incidence, treatment, and outcome of severe sepsis in ICU-treated adults in Finland: the Finnsepsis study." *Intensive care medicine* 33(3):435-43.

Kilbourn, R.G., D.M. Cromeens, F.D. Chelly, and O.W. Griffith. 1994. "NG-methyl-L-arginine, an inhibitor of nitric oxide formation, acts synergistically with dobutamine to improve cardiovascular performance in endotoxemic dogs." *Critical care medicine* 22(11):1835.

Koga, Y., M. Fujita, R. Tsuruta, Y. Koda, T. Nakahara, T. Yagi, T. Aoki, C. Kobayashi, T. Izumi, S. Kasaoka, M. Yuasa, and T. Maekawa. 2010. "Urinary trypsin inhibitor suppresses excessive superoxide anion radical generation in blood, oxidative stress, early inflammation, and endothelial injury in forebrain ischemia/reperfusion rats." *Neurol Res* 32(9):925-32.

Kumar, A., C. Haery, and J. E. Parrillo. 2000. "Myocardial dysfunction in septic shock." *Crit Care Clin* 16(2):251-87.

Kurt, A. N., A. D. Aygun, A. Godekmerdan, A. Kurt, Y. Dogan, and E. Yilmaz. 2007. "Serum IL-1beta, IL-6, IL-8, and TNF-alpha levels in early diagnosis and management of neonatal sepsis." *Mediators Inflamm* 2007:31397.

Kustanova, G. A., A. N. Murashev, V. L. Karpov, B. A. Margulis, I. V. Guzhova, I. R. Prokhorenko, S. V. Grachev, and M. B. Evgen'ev. 2006. "Exogenous heat shock protein 70 mediates sepsis manifestations and decreases the mortality rate in rats." *Cell Stress Chaperones* 11(3):276-86.

L¨®pez, A., J.A. Lorente, J. Steingrub, J. Bakker, A. McLuckie, S. Willatts, M. Brockway, A. Anzueto, L. Holzapfel, and D. Breen. 2004. "Multiple-center, randomized, placebo-controlled, double-blind study of the nitric oxide synthase inhibitor 546C88: Effect on survival in patients with septic shock*." *Critical care medicine* 32(1):21.

Laszlo, F., BJ Whittle, and S. Moncada. 1994. "Time-dependent enhancement or inhibition of endotoxin-induced vascular injury in rat intestine by nitric oxide synthase inhibitors." *British journal of pharmacology* 111(4):1309.

Lin, J. D., Y. Cai, X. J. Xiao, and M. R. Lin. 2011. "Effect of ulinastatin preconditioning on gene expression profile of heart tissue in a rat sepsis model." *Zhongguo Wei Zhong Bing Ji Jiu Yi Xue* 23(1):688-92.

Liu, Y., J. D. Lin, X. J. Xiao, B. G. Ye, B. D. Wu, and Y. S. Lin. 2010. "Effect of ulinastatin preconditioning on gene expression profile of kidney tissue in a rat sepsis model." *Zhongguo Wei Zhong Bing Ji Jiu Yi Xue* 22(9):547-52.

Liu, Y., J. D. Lin, X. J. Xiao, B. L. Zhang, and H. Lin. 2009. "An investigation of changes in gene expression profile of heart tissue in a rat sepsis model." *Zhongguo Wei Zhong Bing Ji Jiu Yi Xue* 21(3):155-9.

Lotze, M. T., and K. J. Tracey. 2005. "High-mobility group box 1 protein (HMGB1): nuclear weapon in the immune arsenal." *Nat Rev Immunol* 5(4):331-42.

Mao, J., L. Zhang, A. L. Song, X. Chen, and Y. C. Zhang. 2011. "Effect of ulinastatin donor-pretreatment on liver graft during cold preservation in rats." *Chin Med J (Engl)* 124(4):574-80.

Martin, G.S., D.M. Mannino, S. Eaton, and M. Moss. 2003. "The epidemiology of sepsis in the United States from 1979 through 2000." *New England Journal of Medicine* 348(16):1546-54.

Masuda, T., K. Sato, C. Noda, K.M. Ikeda, A. Matsunaga, M.N. Ogura, K. Shimizu, H. Nagasawa, N. Matsuyama, and T. Izumi. 2003. "Protective effect of urinary trypsin inhibitor on myocardial mitochondria during hemorrhagic shock and reperfusion." *Critical care medicine* 31(7):1987.

Melamed, A., and F. Sorvillo. 2009. "The burden of sepsis-associated mortality in the United States from 1999 to 2005: an analysis of multiple-cause-of-death data." *Critical Care* 13(1):R28.

Messer, J. V., and W. A. Neill. 1962. "The oxygen supply of the human heart." *Am J Cardiol* 9:384-94.

Miller, L.S., R.M. O'Connell, M.A. Gutierrez, E.M. Pietras, A. Shahangian, C.E. Gross, A. Thirumala, A.L. Cheung, G. Cheng, and R.L. Modlin. 2006. "MyD88 mediates neutrophil recruitment initiated by IL-1R but not TLR2 activation in immunity against Staphylococcus aureus." *Immunity* 24(1):79-91.

Moriguchi, T., N. Kuroyanagi, K. Yamaguchi, Y. Gotoh, K. Irie, T. Kano, K. Shirakabe, Y. Muro, H. Shibuya, and K. Matsumoto. 1996. "A novel kinase cascade mediated by mitogen-activated protein kinase kinase 6 and MKK3." *Journal of Biological Chemistry* 271(23):13675.

Nakamura, H., S. Abe, Y. Shibata, M. Sata, S. Kato, H. Saito, T. Hino, H. Takahashi, and H. Tomoike. 1997. "Inhibition of Neutrophil Elastase-Induced Interleukin-8 Gene Expression by Urinary Trypsin Inhibitor in Human Bronchial Epithelial Cells." *International archives of allergy and immunology* 112(2):157-62.

Nathan, C. 1992. "Nitric oxide as a secretory product of mammalian cells." *The FASEB journal: official publication of the Federation of American Societies for Experimental Biology* 6(12):3051.

Nathan, C., and Q. Xie. 1994. "Nitric oxide synthases: roles, tolls, and controls." *Cell* 78(6):915.

Nava, E., RMJ Palmer, and S. Moncada. 1992. "The role of nitric oxide in endotoxic shock: effects of NG-monomethyl-L-arginine." *Journal of cardiovascular pharmacology* 20(6):132.

Niederbichler, A. D., L. M. Hoesel, M. V. Westfall, H. Gao, K. R. Ipaktchi, L. Sun, F. S. Zetoune, G. L. Su, S. Arbabi, J. V. Sarma, S. C. Wang, M. R. Hemmila, and P. A. Ward. 2006. "An essential role for complement C5a in the pathogenesis of septic cardiac dysfunction." *J Exp Med* 203(1):53-61.

Niemann, J.T., D. Garner, and R.J. Lewis. 2004. "Tumor necrosis factor-[alpha] is associated with early postresuscitation myocardial dysfunction." *Critical care medicine* 32(8):1753.

Ninomiya-Tsuji, J., K. Kishimoto, A. Hiyama, J. Inoue, Z. Cao, and K. Matsumoto. 1999. "The kinase TAK1 can activate the NIK-I¦ÊB as well as the MAP kinase cascade in the IL-1 signalling pathway." *Nature* 398(6724):252-56.

Pahl, H. L. 1999. "Activators and target genes of Rel/NF-kappaB transcription factors." *Oncogene* 18(49):6853-66.

Pallua, N., J. F. Low, and D. von Heimburg. 2003. "Pathogenic role of interleukin-6 in the development of sepsis. Part II: Significance of anti-interleukin-6 and anti-soluble interleukin-6 receptor-alpha antibodies in a standardized murine contact burn model." *Crit Care Med* 31(5):1495-501.

Park, J. H., S. H. Kwak, C. W. Jeong, H. B. Bae, and S. J. Kim. 2010a. "Effect of ulinastatin on cytokine reaction during gastrectomy." *Korean J Anesthesiol* 58(4):334-7.

Park, K.H., K.H. Lee, H. Kim, and S.O. Hwang. 2010b. "The Anti-Inflammatory Effects of Ulinastatin in Trauma Patients with Hemorrhagic Shock." *Journal of Korean medical science* 25(1):128.

Pathan, N., C. A. Hemingway, A. A. Alizadeh, A. C. Stephens, J. C. Boldrick, E. E. Oragui, C. McCabe, S. B. Welch, A. Whitney, P. O'Gara, S. Nadel, D. A. Relman, S. E. Harding, and M. Levin. 2004. "Role of interleukin 6 in myocardial dysfunction of meningococcal septic shock." *Lancet* 363(9404):203-9.

Pelleymounter, M.A., M.J. Cullen, M.B. Baker, R. Hecht, D. Winters, T. Boone, and F. Collins. 1995. "Effects of the obese gene product on body weight regulation in ob/ob mice." *Science* 269(5223):540.

Petros, A., D. Bennett, and P. Vallance. 1991. "Effect of nitric oxide synthase inhibitors on hypotension in patients with septic shock." *The Lancet* 338(8782-8783):1557-58.

Price, S., P. B. Anning, J. A. Mitchell, and T. W. Evans. 1999. "Myocardial dysfunction in sepsis: mechanisms and therapeutic implications." *Eur Heart J* 20(10):715-24.

Rahman, S. H., G. Salter, J. H. Holmfield, M. Larvin, and M. J. McMahon. 2004. "Soluble CD14 receptor expression and monocyte heterogeneity but not the C-260T CD14 genotype are associated with severe acute pancreatitis." *Crit Care Med* 32(12):2457-63.

Ribeil, J. A., Y. Zermati, J. Vandekerckhove, S. Cathelin, J. Kersual, M. Dussiot, S. Coulon, I. C. Moura, A. Zeuner, T. Kirkegaard-Sorensen, B. Varet, E. Solary, C. Garrido, and O. Hermine. 2007. "Hsp70 regulates erythropoiesis by preventing caspase-3-mediated cleavage of GATA-1." *Nature* 445(7123):102-5.

Ritter, C., M. Andrades, M. L. Frota Junior, F. Bonatto, R. A. Pinho, M. Polydoro, F. Klamt, C. T. Pinheiro, S. S. Menna-Barreto, J. C. Moreira, and F. Dal-Pizzol. 2003. "Oxidative parameters and mortality in sepsis induced by cecal ligation and perforation." *Intensive Care Med* 29(10):1782-9.

Rittirsch, D., M. A. Flierl, and P. A. Ward. 2008. "Harmful molecular mechanisms in sepsis." *Nat Rev Immunol* 8(10):776-87.

Robertson, F.M., P.J. Offner, D.P. Ciceri, W.K. Becker, and B.A. Pruitt Jr. 1994. "Detrimental hemodynamic effects of nitric oxide synthase inhibition in septic shock." *Archives of Surgery* 129(2):149.

Rolfe, D. F., and G. C. Brown. 1997. "Cellular energy utilization and molecular origin of standard metabolic rate in mammals." *Physiol Rev* 77(3):731-58.

Rook, G., R. Baker, B. Walker, J. Honour, D. Jessop, R. Hernandez-Pando, K. Arriaga, R. Shaw, A. Zumla, and S. Lightman. 2000. "Local regulation of glucocorticoid activity in sites of inflammation. Insights from the study of tuberculosis." *Ann N Y Acad Sci* 917:913-22.

Schulze-Osthoff, K., A. C. Bakker, B. Vanhaesebroeck, R. Beyaert, W. A. Jacob, and W. Fiers. 1992. "Cytotoxic activity of tumor necrosis factor is mediated by early damage of mitochondrial functions. Evidence for the involvement of mitochondrial radical generation." *J Biol Chem* 267(8):5317-23.

Sekido, N., N. Mukaida, A. Harada, I. Nakanishi, Y. Watanabe, and K. Matsushima. 1993. "Prevention of lung reperfusion injury in rabbits by a monoclonal antibody against interleukin-8." *Nature* 365(6447):654-7.

Sharma, A. C. 2007. "Sepsis-induced myocardial dysfunction." *Shock* 28(3):265-9.

Shirakabe, K., K. Yamaguchi, H. Shibuya, K. Irie, S. Matsuda, T. Moriguchi, Y. Gotoh, K. Matsumoto, and E. Nishida. 1997. "TAK1 mediates the ceramide signaling to stress-activated protein kinase/c-Jun N-terminal kinase." *Journal of Biological Chemistry* 272(13):8141.

Song, H.Y., C.H. R¨¦gnier, C.J. Kirschning, D.V. Goeddel, and M. Rothe. 1997. "Tumor necrosis factor (TNF)-mediated kinase cascades: bifurcation of nuclear factor-¦ÊB and c-jun N-terminal kinase (JNK/SAPK) pathways at TNF receptor-associated factor 2." *Proceedings of the National Academy of Sciences* 94(18):9792.

Suliman, H.B., K.E. Welty-Wolf, M.S. Carraway, L. Tatro, and C.A. Piantadosi. 2004. "Lipopolysaccharide induces oxidative cardiac mitochondrial damage and biogenesis." Cardiovascular research 64(2):279.

Takano, H., K. Inoue, A. Shimada, H. Sato, R. Yanagisawa, and T. Yoshikawa. 2009. "Urinary trypsin inhibitor protects against liver injury and coagulation pathway dysregulation induced by lipopolysaccharide/D-galactosamine in mice." *Lab Invest* 89(7):833-9.

Tanaka, R., M. Fujita, R. Tsuruta, K. Fujimoto, H. S. Aki, K. Kumagai, T. Aoki, A. Kobayashi, T. Izumi, S. Kasaoka, M. Yuasa, and T. Maekawa. 2010. "Urinary trypsin inhibitor suppresses excessive generation of superoxide anion radical, systemic inflammation, oxidative stress, and endothelial injury in endotoxemic rats." *Inflamm Res* 59(8):597-606.

Tatsumi, T., K. Akashi, N. Keira, S. Matoba, A. Mano, J. Shiraishi, S. Yamanaka, M. Kobara, N. Hibino, and S. Hosokawa. 2004. "Cytokine-induced nitric oxide inhibits mitochondrial energy production and induces myocardial dysfunction in endotoxin-treated rat hearts." Journal of molecular and cellular cardiology 37(3):775-84.

Thomas, D.D., L.A. Ridnour, J.S. Isenberg, W. Flores-Santana, C.H. Switzer, S. Donzelli, P. Hussain, C. Vecoli, N. Paolocci, and S. Ambs. 2008. "The chemical biology of nitric oxide: implications in cellular signaling." *Free Radical Biology and Medicine* 45(1):18-31.

Toledo-Pereyra, LH, AH Toledo, J. Walsh, and F. Lopez-Neblina. 2004. "Molecular signaling pathways in ischemia/reperfusion." *Experimental and clinical transplantation: official journal of the Middle East Society for Organ Transplantation* 2(1):174.

Tracey, K. J. 2002. "The inflammatory reflex." *Nature* 420(6917):853-9.

Tracey, W.R., J. Tse, and G. Carter. 1995. "Lipopolysaccharide-induced changes in plasma nitrite and nitrate concentrations in rats and mice: pharmacological evaluation of nitric oxide synthase inhibitors." *Journal of Pharmacology and Experimental Therapeutics* 272(3):1011.

van der Poll, T., and S. J. van Deventer. 1999. "Cytokines and anticytokines in the pathogenesis of sepsis." *Infect Dis Clin North Am* 13(2):413-26, ix.

Wang, H., O. Bloom, M. Zhang, J. M. Vishnubhakat, M. Ombrellino, J. Che, A. Frazier, H. Yang, S. Ivanova, L. Borovikova, K. R. Manogue, E. Faist, E. Abraham, J. Andersson, U. Andersson, P. E. Molina, N. N. Abumrad, A. Sama, and K. J. Tracey. 1999. "HMG-1 as a late mediator of endotoxin lethality in mice." *Science* 285(5425):248-51.

Watts, JA, JA Kline, LR Thornton, RM Grattan, and SS Brar. 2004. "Metabolic dysfunction and depletion of mitochondria in hearts of septic rats." Journal of molecular and cellular cardiology 36(1):141-50.

Wesche, H., W.J. Henzel, W. Shillinglaw, S. Li, and Z. Cao. 1997. "MyD88: an adapter that recruits IRAK to the IL-1 receptor complex." *Immunity* 7(6):837-47.

Westendorp, R. G., J. A. Langermans, T. W. Huizinga, A. H. Elouali, C. L. Verweij, D. I. Boomsma, and J. P. Vandenbroucke. 1997. "Genetic influence on cytokine production and fatal meningococcal disease." *Lancet* 349(9046):170-3.

Wheeler, A.P., and G.R. Bernard. 1999. "Treating patients with severe sepsis." *New England Journal of Medicine* 340(3):207-14.

Wheeler, D. S., and H. R. Wong. 2007. "Heat shock response and acute lung injury." *Free Radic Biol Med* 42(1):1-14.

Wright, S.D. 1995. "CD 14 and innate recognition of bacteria." *Journal of Immunology* 155(1):6-8.

Wu, Y.J., Q. Ling, X.H. Zhou, Y. Wang, H.Y. Xie, J.R. Yu, and S.S. Zheng. 2009. "Urinary trypsin inhibitor attenuates hepatic ischemia-reperfusion injury by reducing nuclear factor-kappa B activation." *Hepatobiliary Pancreat Dis Int* 8(1):53-58.

Xiaoqiao, Z., M. Rong, Y. Zhigang, D. Yong, F. Xihong, and S. Jingzhong. 2004. "Protective effect of ulinastatin against ischemia-reperfusion injury in rat small bowel transplantation." *Transplant Proc* 36(5):1564-6.

Yamaguchi, K., K. Shirakabe, H. Shibuya, K. Irie, I. Oishi, N. Ueno, T. Taniguchi, E. Nishida, and K. Matsumoto. 1995. "Identification of a member of the MAPKKK family as a potential mediator of TGF-β signal transduction." *Science* 270(5244):2008.

Zhang, Z., C. Zhou, Y. Luo, and H. Xiong. 2010. "Expression of heat shock protein 70 in lung tissues of acute paraquat poisoned rats and intervention of ulinastatin.pdf." *World J Emerg Med* 1(3):229-33.

17

Morphological Study of HDPE/Clay Hybrids Synthesized by an Alternative Compatibilization Path

Fernanda Elena Monasterio
Instituto de Investigaciones para la Industria Química (INIQUI)- CONICET, Avda, Bolivia, Salta, Argentina

1. Introduction

The control of the factors that influence the internal structure of polymeric materials (crystallinity, miscibility between phases, processing, etc.), is of paramount importance because it is closely related to the outcome of their final properties (Durmus et al., 2007; Sinha Ray & Okamoto, 2003; Franck, n.d.). Therefore, the morphological analysis is essential in order to know the structure-property relationship (Williams & Carter, 2009).

As far as particulate-filled polymer composites is concerned, there are many filling choices to make a material and the use of nanometric fillers such as clays is the most promising one (Sinha Ray & Okamoto, 2003; LeBaron et al., 1999). Due to the fact that these nanomaterials own a higher surface area for the polymer-filler interaction than conventional composites, the effect at the interface will promote changes in the morphology at a nanometric level which can be studied using conventional bulk characterization techniques (Sinha Ray & Bousmina, 2005).

The analysis by Transmission Electron Microscopy (TEM) has been of great help in the characterization of polymer layered nanocomposites (Sinha Ray & Bousmina, 2005; Michler, 2008; Durmus et al., 2007), given that the degree of affinity at the polymer-clay interface is exhibited by the nanosheets dispersion achieved in a TEM image. But this technique, by itself, does not allow a full interpretation of the composites final behavior and it is usually used in combination with other techniques. Among them it can be mentioned the differential scanning calorimetry (DSC) and the rheological analysis that widely reflect distinguishing material features such as miscibility and crystallinity degrees (Faker, 2008; Kim & Lee, 2008).

The DSC allows inferring changes in the polymers internal structure as it registers the transitions taking place in the materials when temperature varies, which is linked to the molecular mobility (Campbell & White, 1989; Rothon, 2003). Regarding clay filled polymers, these are usually superficially modified to reach miscibility leading to a highest intercalation/exfoliation of the multilamellar clay in the polymer matrix. Some authors

(Sinha Ray & Okamoto, 2003; Alexandre & Dubois, 2000; Mai & Yu, 2006) report that polymer structure variations are manifested by changes in the crystallization temperature (T_c) or in the glass transition temperature (T_g) which can be detected by DSC measurements.

On the other hand, rheological tests extensively are used in polymer characterization (Zhang Q. et al., 2004) and they bring important information about processing optimization (C. Wan et al., 2004; Shah & Paul, 2004). Particularly, the structural characterization of filled polymer through low-amplitude oscillatory experiments allows determining the linear viscoelastic properties that describe the material behavior in a condition where the structure remains practically unchanged. These properties are sensitive to the interactions that take place among phases, and to the degree of miscibility reached (Okamoto, 2005; Bretas & D'Avila, 2000; Cassagnau, 2008), due to the fact that the applied strain deformations are small enough to affect the material properties.

In this chapter, several hybrids based on high density polyethylene (HDPE) and montmorillonite are studied. Due to the differences between the hydrophilic natures of these materials, a clay surface modification using difunctional silanes was chosen in order to generate oligomers with hydrophobic lateral groups around the lamellas.

Furthermore, to determine if the final materials behaviors were affected by alterations in the chemical nature of the interphase, alkyl groups were alternated with aryl groups.

On the other hand, transport properties are analyzed aiming to characterize the outcome of the materials at the macroscopic level. In this way, it would be able to know how the use of the structural analysis via TEM can help to understand these behaviors and how, if combined with the two techniques mentioned above, allow a better analysis of polymer composites.

Finally, the results made possible a better understanding of the barrier properties, and some quantitative relations with the structure of the synthesized materials were found.

2. Materials

In polymer/clay systems, the structure at nanometric level (intercalation/exfoliation) depends on the matrix polarity, being more feasible to get better miscibility in polar matrixes. Consequently, compatibilization is required in order to obtain a polyolefin nanocomposite. This implies using copolymers or clay modification by trifunctional silanes or organic ions rendering the nanolamellas more hydrophobic (Durmus et al., 2007; Okamoto, 2005; Malucelli, 2007; L. Wang & J. Sheng, 2005).

Polymer additives were synthesized by treating an organophilic clay (oC) with difunctional silanes, aiming to obtain surface oligomers with side groups compatible to the high density polyethylene matrix, while their backbones share the same clay chemistry. This procedure is proposed as an alternative path for the obtaining of polyolefin/clay nanocomposites and it is based on the methods reported by (X. Zheng & Wilkie, 2003; J. Zhang et al., 2005; C. Zhao et al., 2004).

The molecular models of the employed silanes are described on Figure 1:

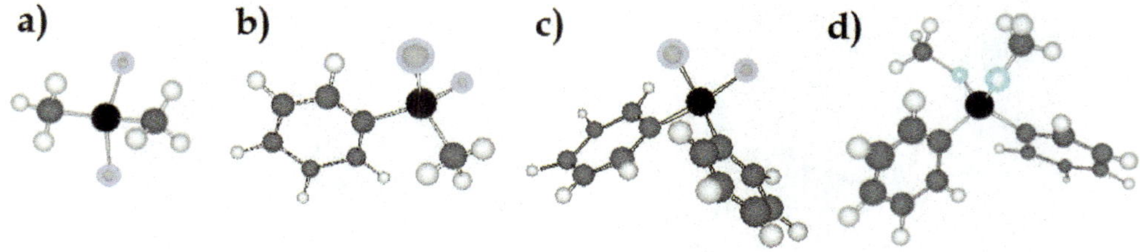

Fig. 1. Molecular simulation of: a) dichlorodimethylsilane (DM)
b) dichloromethylphenylsilane (MP); c) dichlorodiphenylsilane (DP_{Cl}) and
d) dimethoxydiphenylsilane (DP_{OCH_3}).

From these monomers, by means of an in situ hydrolysis, four clays with their respective oligosiloxanes were obtained, named as: oC/DM, oC/MP, oC/DP_{Cl} y oC/DP_{OCH_3}. Subsequently, they were mixed with HDPE by melting process thus each composite contained ~3%wt clay. The preparation conditions and the results from X-ray diffraction, Fourier transform infrared spectroscopy and non-oxidative thermal degradation of the clays and the polymeric materials can be found in Monasterio et al. (2010, 2011) and Monasterio & Destéfanis (2011).

To systematize the analysis of the HDPE/clay materials, the collected data was split in two groups according with the criteria described on Table 1:

Group	Criteria	Hybrid
I	Oligomers obtained from the same reactive group (-Cl) with different side groups	PE-oC/DM PE-oC/MP PE-oC/DP_{Cl}
II	Oligomers with equal side groups ($-C_6H_5$) obtained from different reactive groups	PE-oC/DP_{Cl} PE-oC/DP_{OCH_3}

Table 1. Polymeric systems studied in this chapter.

In the following sections it must be taken into account that the prepared PE composites contain equal clay percentage, and since it is the same polymer matrix and the same inorganic multilayered material, the changes detected by the above mentioned techniques will be associated to the effects obtained by the modifying species present in the different interphases generated on each clay.

To perform a more detailed analysis of the oligomers effects in the interphase, the hybrids with sC and oC are included.

3. Structural characterization via TEM

Before incorporating the additives (clays) into the HDPE, they were analyzed by TEM. Each of the samples was sonicated in ethanol. The resulting suspension of the product in the solvent was settled for approximately 20 minutes and then, a drop of the dilute suspension was placed onto a coated copper grid. The solvent evaporated from the grid, leaving a thin-layer of each of the products on the grid. Figure 2 shows the photographs of three of the used clays.

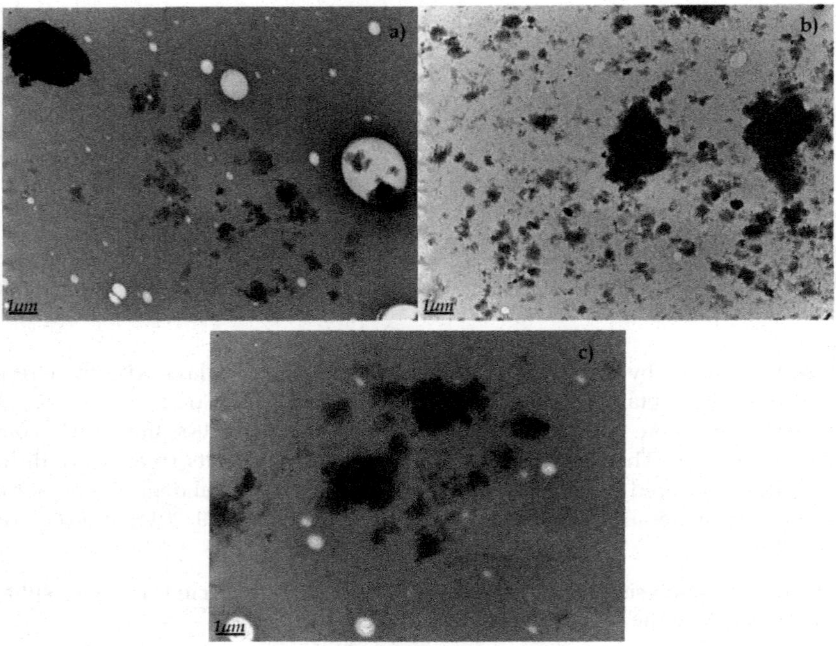

Fig. 2. TEM images of a) oC, b) oC/DM y c) oC/DP_{OCH_3} performed on a JEOL100 CX II (10000×). The whiter ovals are part of the defects on the support films.

In general, particulate materials of different sizes were visualized. In addition, tonality changes were observed which are related to the mass-thickness contrast, pointing out differences in the particle thickness or the composition. The thinner particles (light grey) were associated with clay sheets.

The clays containing oligosiloxanes tend to form clusters. In all cases, agglomerates superior to 1 µm were observed but they presented different appearances. In Figure 2a), it has a black color while similar sized agglomerates in the other samples are surrounded by lighter grey particles. This can be the result of higher clay dispersion caused by the in situ generation of oligomers into the organophilic clay. These results are in agreement with previous scanning electron microscopy studies of these additives (Monasterio et al., 2010).

Thin sections (~60nm) cut by means of a cryoultramicrotome were obtained for the HDPE/clay hybrids morphological analysis. Cryoultramicrotomy is a standard method often used with low T_g polymers (e.g. HDPE T_g=-125°C) (Kaufman & Falcetta, 1977), because it provides thin sections that are free from artefacts. Hence it is possible to carry out a correct interpretation of the actual composite morphology (Michler, 2008). The visualization was performed at high acceleration voltage (100kV), which improves the resolution and reduces the scattering cross-section to protect the specimens which are sensitive to electronic radiation (Williams & Carter, 2009). It must be mentioned that SEM photographs of these composites (Monasterio et al., 2011; Monasterio & Destéfanis, 2011) revealed that clays (with or without surface treatment) achieved a good distribution inside

HDPE. In order to guarantee representativeness, TEM images with different magnification were captured (10000×, 40000× and 50000×).

Figure 3 is showing that modification of the chemical nature of siloxane oligomers used as compatibilizer induced differences in the hybrids internal structure.

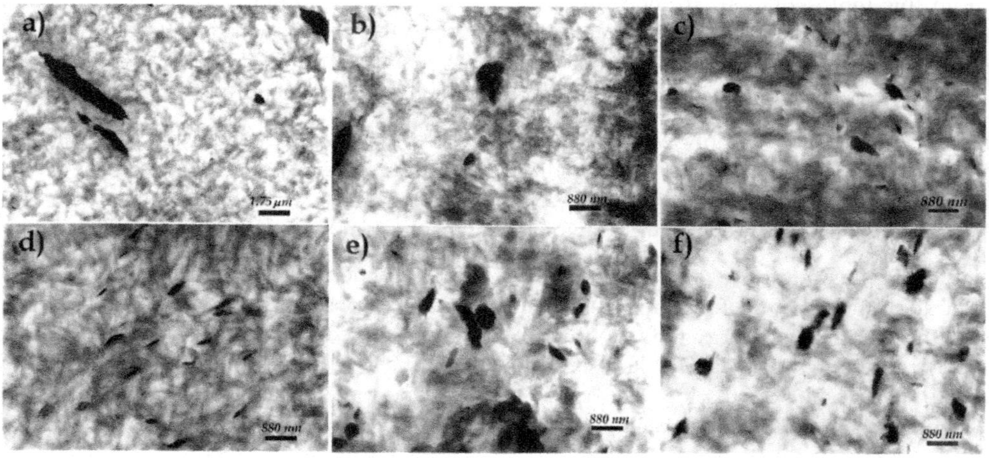

Fig. 3. TEM micrographs of a)PE-sC, b) PE-oC, c) PE-oC/DM, d) PE-oC/MP, e) PE-oC/DP_{Cl} and f) PE-oC/DP_{OCH_3} (Reprinted from Monasterio et al., Copyright (2010) The Mineralogical Society, U.K.).

Dispersed particles of micrometric size were observed in PE-sC and PE-oC. While oligosiloxanes modified clays exhibited a major stacks disaggregation. Taking into account group I (Table 1), PE-oC/MP presents the best dispersion/distribution (particles thickness < 200nm) while the rest of the composites show agglomerations. For group II, where the treated clay composition is basically the same, PE-oC/DP_{OCH_3} exhibited better dispersion of the filler (particles thickness < 500 nm).

On the other side, X-ray diffractograms reported before (Monasterio et al., 2011; Monasterio & Destéfanis, 2011) showed intercalation/exfoliation for PE-oC/MP and PE-oC/DP_{OCH_3}, and they were coextensive with that observed in TEM photographs. Although PE-oC/DM and PE-oC/DP_{Cl}, had better delamination than PE-sC y PE-oC, they exhibited 'stacking recovery'.

In conclusion, the decreasing dispersion order was PE-oC/MP, PE-oC/DP_{OCH_3}, PE-oC/DM, PE-oC/DP_{Cl}, PE-oC, PE-sC. Specially, the morphology obtained in PE-oC/MP is similar to that reported by TEM images by Zhang et al. for PE and PP nanocomposites (J. Zhang et al., 2005). Consequently, the proposed oligosiloxane treatment could be considered as an alternative for compatibilization.

4. Structural characterization via rheological analysis

Rheological behaviors in oscillatory regime were determined in a rheometer TA AR2000 of parallel plates (D=25 mm, h= 1 mm) at 200°C in N_2 atmosphere. Before measuring, each sample underwent a relaxation at an angular frequency of 5 rad/s. Results were plotted using the 'frequency domain phasor representation' (Figure 4). The locus of a phasor is known as the 'digital fingerprint' of a material. This plot summarizes the values of storage (G') and loss (G'') moduli and the phase angle (δ).

From this representation, an increase of clay-polyolefin compatibility was inferred. It can also be noticed that PE was above the other values, whereas PE with sodium clay values were below the other hybrids. PE-oC behavior is near of that described by PE-sC while materials containing oligosiloxanes presented a behavior closing to pure PE.

On the other hand, not only were noticed changes for the materials with different reactive leaving groups (group II) but also for those with different side groups (group I). At low frequencies, composites containing oligomers presented lower G' and G'' values than PE. Phase angle were between 30 and 60° which denotes the composites viscoelastic nature.

Figure 5 illustrates the curve of viscosity as a function of the frequency where a pseudoplastic behavior for pure HDPE and hybrids was observed. This is due to the fact that the input shear energy tends to line up random oriented molecules or particles and to disaggregate any large clay stacks, thereby reducing the overall hydrodynamic drag, which in turn reduces the dissipation of energy in the fluid and the viscosity (Punnarak et al., 2006; D. Chen et al., 2005; TA Instruments, (RN-9B)).

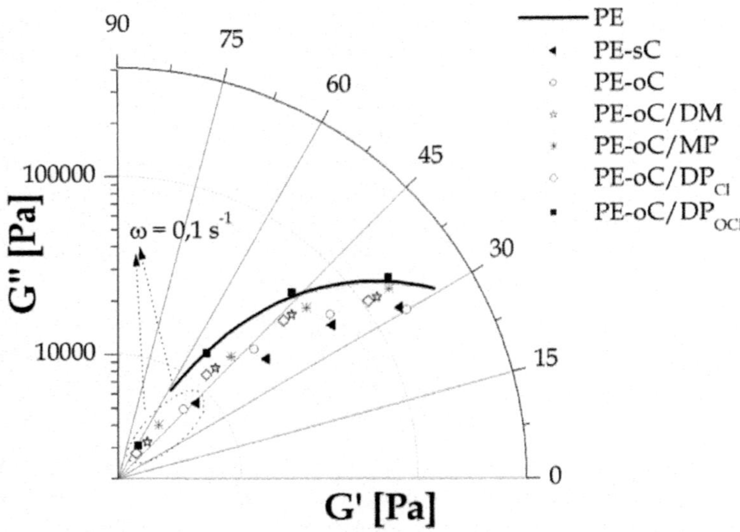

Fig. 4. Phasor representation of pure PE and PE/clay composites at ω=0.1,1,10,100 rad/s.

Unlike other works describing an increase in viscosity due to the addition of filler into the polymeric matrix (Cassagnau, 2008), the materials modified with clays showed lower values than that corresponding to HDPE. In the case of PE-sC and PE-oC, one of the factors that could give rise to this behavior would be the diminution among PE macromolecules interactions due to the presence of clay tactoids.

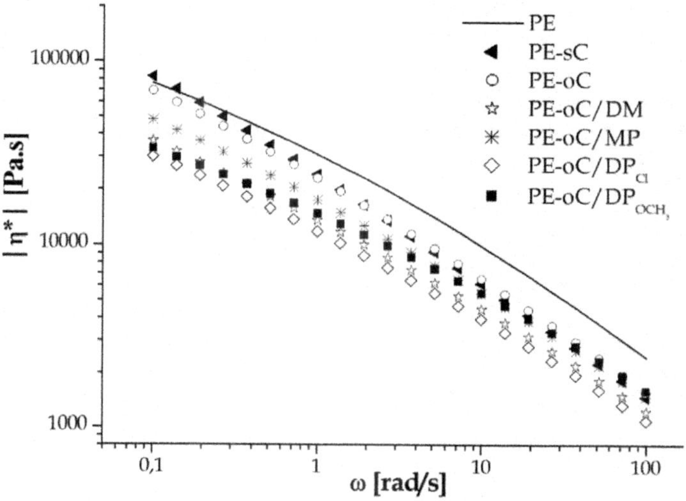

Fig. 5. Complex viscosity vs frequency for neat and clay-filled HDPE.

For hybrids filled with silane treated clays, oligomers would diminish the average molecular weight affecting the viscosity (Billmeyer, 1988). Among siloxane treated hybrids, it is observed that at low frequencies the highest viscosity is for PE-oC/MP, followed by PE-oC/DM and finally by PE-oC/DP$_{OCH_3}$ and PE-oC/DP$_{Cl}$. Probably this is a consequence of the effects caused by oligomeric species obtained during the in situ hydrolysis of each monomer introduced into oC.

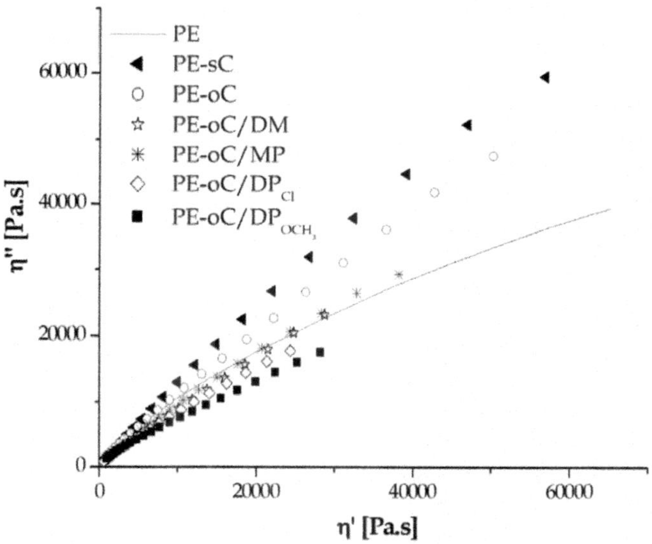

Fig. 6. Cole-Cole plot showing clays effect in PE hybrids miscibility.

The Cole-Cole viscosity representation (Figure 6) was also analyzed to know the miscibility among the hybrids components. The curves proximity to a semicircular shape indicates higher miscibility and the deviations respond to a component interaction reduction (Ahmed et al., 2010). In this way, it was detected that both sC and oC did not reach a good compatibility with the polymer matrix, while the other siloxane treated clays composites showed miscibility with PE.

A deep structural analysis can be performed by using the van Gurp-Palmen plot which provides information about the length and amount of branches in a sample, and the molecular weight distribution (Lohse et al., 2002; Schlatter et al., 2005).

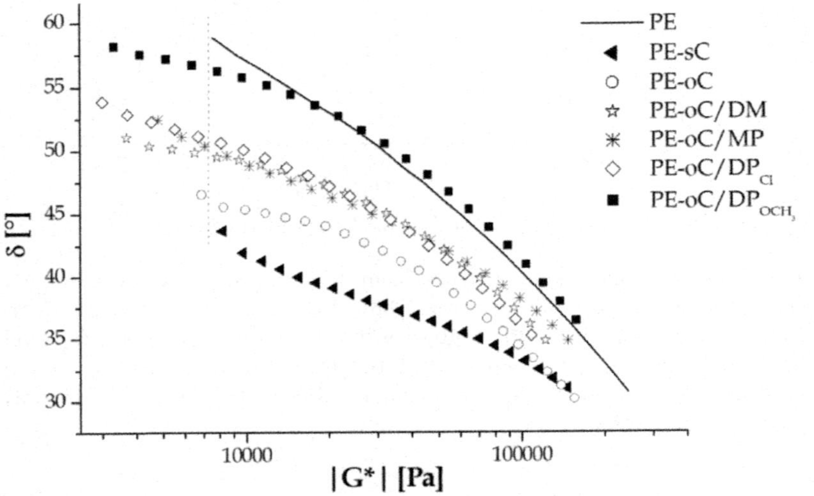

Fig. 7. Phase angle (δ) vs. the absolute value of the complex modulus |G*| of HDPE/clay hybrids.

The dashed line in Figure 7 was pointed out for polydispersion comparison taking pure PE as reference. As was expected for PE-sC and PE-oC there are not variations. Nevertheless, for group I, it was observed the following polydispersion decreasing order: PE-oC/MP, PE-oC/DM and PE-oC/DP_{Cl}. Regarding group II, PE-oC/DP_{OCH_3} and PE-oC/DP_{Cl} exhibited similar behaviors.

These results are in agreement with the GPC results (Monasterio et al., 2010, 2011) (Table 2) and they explain the above mentioned viscosity decrease (Figure 5). Interestingly in group II, dimers were synthesized and the chemical nature of the interaction was similar. However, different behaviors were detected. In one of the cited papers, it was made a comparative analysis between PE-oC/DP_{Cl} and PE-oC/DP_{OCH_3} behaviors. From it, it was possible to conclude that the differences between the mechanisms during the synthesis of the oligosiloxane surface modified clays affected their further interactions when being added to PE. Basically, the difference between both additives lies in the degree of coating achieved on clay platelets.

Further analyses were carried out and their values are shown in Table 2. Even though the above graphics give an idea of the information of each material behavior, the Table data offer specific quantitative results of each system which later will be associated with transport properties.

	G_{cop} [kPa]	G_N^0 [kPa]	G′ slope	PDI
PE	38.1	841.1	0.73	
PE-sC	5.5*	792.6	0.52	
PE-oC	8.6	612.1	0.56	
PE-oC/DM	23.3	523.6	0.60	1.10
PE-oC/MP	20.3	681.4	0.63	1.32 and 2.37
PE-oC/DP_{Cl}	22.1	477.7	0.66	1.26
PE-oC/DP_{OCH_3}	45.9	638.8	0.69	1.08

Table 2. Rheological characteristics of HDPE and HDPE/clay hybrids. (*Extrapolated value. G_N^0: Plateau modulus calculated from extrapolation on van Gurp-Palmen plot (García-Franco et al., 2006). PDI: polydispersity index of the oligomers extracted from the oligosiloxane clays before their incorporation to PE. Values obtained by GPC measures).

The crossover point modulus (G_{cop}) indicates the transition from a solid-like behavior to a liquid-like behavior. Its value can provide the qualitative information about the average molecular weight (M_w), the degree of branching and the molecular mass distribution (Yang, Q. et al., 2009; Ahmed et al., 2010). Reminding that the matrix is the same in the analyzed materials (HDPE), it can be observed that G_{cop} has had a significant change according to the surface treatment applied on each clay. When polydipersion of the modifying oligomers increase, the G_{cop} values decrease. The low values presented by PE-sC and PE-oC are associated to poor interfacial adhesion.

Group I hybrids presented similar values among them and a more elastic (liquid-like) behavior than PE, but in group II it is evident that PE-oC/DP_{OCH_3}, in spite of its similarity with PE-oC/DP_{Cl}, it shows a value even higher than neat PE, which it make expect differences in the internal structures of these materials.

It was also analyzed the effect of the clays on the plateau modulus (G_N^0). This value is defined by the Equation 1 (Rose et al., 2007):

$$G_N^0 = \frac{b.R.T.\rho}{M_e} \tag{1}$$

where:

b: coefficient equal to 1, according to Ferry, and equal to 4/5, according to Doi and Edwards,
R: is the ideal gas constant,
T: is the absolute temperature,
ρ: is the polymer density, and
M_e: is the entanglement molecular weight.

So G_N^0 bring information about the entanglement spacing of a polymer melt and the entanglement molar mass. In the case of filled materials, this value could be associated with temporary crosslink between the surface of the particles and the polymer (Gacitua E. et al., 2005). In PE-sC the clay has not any organic species as compatibilizer, so the lower value it might be because no interaction takes place between its surface and the long HDPE chains. It is evident that the presence of intergallery organic ions in oC is not enough to improve the interaction with the matrix. Not only does the G_N^0 not increase, but even more it is inferior to PE-sC value. This behavior could be due to oC tendency to easily delaminate (compared to sC), which would increase the amount of low compatibility particle dispersed in the matrix.

The treatments with silanes had different effects, being PE-oC/DP_{OCH_3} and PE-oC/MP the composites with higher values taking PE-oC as reference, whereas PE-oC/DM and PE-oC/DP_{Cl} had the lowest values. Evidently, these behaviors are not directly affected by the chemical nature of the side groups. It would be more appropriate to take into account the coating generated by the dissimilar species over the clays to explain these results. Therefore one can think in an oligomer network generated around the sheets (or stack of sheets).

In oC/MP there are trimers and longer chains that can lead to an intricate coating, just like oC/DP_{OCH_3} which has a higher proportion of dimers outside of clay galleries (compared to oC/DP_{Cl}). So, inside HDPE matrix is more probable that these networks give rise to temporary anchoring that would improve its interaction. In the case of oC/DP_{Cl}, the low dimers concentrarion on the surface would prevent the network formation, and although oC/DM has longer oligomers (mostly pentamers), the rigidity and spatial complexity of the -CH_3 group is smaller than the phenyl group, which would explain that this clay does not have such capability to 'entangled' the PE macromolecules. This would also explain the PE-oC/DM and PE-oC/DP_{Cl} 'stacking recovery' tendency mentioned above.

Nevertheless, the G_N^0 values (Table 2) were below the pure PE value, evidencing that despite of the improvements got at the interphase there would be more free volume.

G' values at the terminal zone give information about matrix-filler interaction (Wang, K.H. et al., 2002). Due to PE-sC hybrid did not have any compatibilization it is expected to show the lowest value, as in the system coexist hydrophilic and hydrophobic components.

The Na^+ exchange with hexadecyltrimethylammonium ion provided a slight improvement in the interface interaction. Only after the second treatment of the clays in order to generate oligosiloxanes, it was possible that these additives increase their slopes, which is associated to an increase in compatibility. Moreover, the G' slope for PE-oC/DP_{OCH_3} presented a value near to the pure polymer, which is in agreement with the behaviors observed in Figure 6 and which would indicate an improvement in polymer-clay interactions.

5. Structural characterization by DSC

Semicrystalline polymers like HDPE give information about the behaviors of the amorphous and crystalline phases through T_g (glass transition temperature) and T_m (melt temperature) or ΔH_m (melting enthalpy), respectively (Kaufman & Falcetta, 1977). In the analyzed materials, measurements were carried out in a temperature range above the T_g (-50 to 150°C).

The determinations were performed on a DSC Q200 equipment from TA Instruments (coupled with a refrigerated cooling system), equipped with modulated DSC software, which allows the application of a sinusoidal heating programme superimposed to the normal linear temperature ramp (Verdonck et al., 1999; TA Instruments (TS-25)).

$$\frac{dH}{dt} = C_p \frac{dT}{dt} + f(T,t) \tag{2}$$

where:

$\frac{dH}{dt}$: is the Total Heat Flow due to the linear heating rate.
C_p: is the Heat Capacity of the Reversing Heat Flow Component.
$\frac{dT}{dt}$: is the measured heating rate, which has both a linear and sinusoidal (modulated) component.
$f(T,t)$: is the Kinetic Component of the Total heat flow and is calculated from the difference between the Total signal and Heat Capacity Component. (Non-reversing component)
$C_p \frac{dT}{dt}$: is the Reversing Heat Flow Component of the Total Heat Flow.

This analysis (Equation 2) allows the deconvolution of the total heat flow signal into 'reversing' and 'non-reversing' contributions. Total Heat Flow due to the underlying heating rate it is equivalent to standard DSC signal at the same average heating rate (De Meuter et al., 1999), and the results obtained from these curves are described in Table 3. Compared with pure polymer, hybrids did not presented significant changes for T_m, T_{onset} and degrees of crystallinity, which are in agreement with that reported by other authors in HDPE/clay systems (Gupta & Bhattacharya, 2008).

	T_m [°C]	T_{onset} [°C]	ΔH_m [J/g]	Crystallinity degree [%]
PE	129.86	123.85	213.9	73.8
PE-sC	130.09	121.82	193.0	68.6
PE-oC	130.40	121.81	192.0	68.2
PE-oC/DM	130.31	122.17	197.4	70.2
PE-oC/MP	129.75	122.36	202.4	71.9
PE-oC/DP_{Cl}	130.47	122.09	188.8	67.1
PE-oC/DP_{OCH_3}	130.10	122.18	194.1	69.0

Table 3. Results from DSC non-isothermal analysis.

However, regarding the degree of crystallinity it is noticed that clay addition (with or without oligomers), had a slight plasticizing effect as it affected the chain regularity, inhibiting crystallization.

To have a better insight of composites structure, each component of the Total Heat Flow was analyzed. The advantage of modulated signals is the possibility of separate overlapping phenomena (overlapping transitions such as melting/recrystallization in semi-crystalline materials). This complementary information enables a more detailed study of complicated material systems (Verdonck et al., 1999; Coleman & Craig, 1996). Special care must be taken in the selection of modulation technique as it will affect data interpretation (Schawe, 1995), here it was adopted the model proposed by TA Instruments (Thomas, 2005).

It must be pointed out that the melting process is partially retrieved in the reversing heat flow as well as in the non-reversing heat flow signal, and moreover the fraction of melting in both of these signals depends on the experimental conditions (Coleman & Craig, 1996; Verdonck et al., 1999). For that reason, experimental conditions (heating rate and period) for the hybrids studied were chosen guaranteeing enough modulation cycles during the transition of interest.

In order to avoid degradative processes overlapping during result interpretation, TG-DTG analyses over the assessed temperature range were analyzed (Monasterio et al., 2011; Monasterio & Destéfanis, 2011) and no degradation was observed.

Signal deconvolution shows that the main contribution corresponds to the non-reversing component, which exhibited a similar behavior to that of Total Heat Flow signal, even the maxima from both curves coincided. Regarding reversing signals (Figure 8), the curve corresponding to pure polymer showed a small exothermic peak above T_m and an endothermic peak below the T_m. They can be associated to simultaneous melting-crystallization occurring during heating, until complete melting.

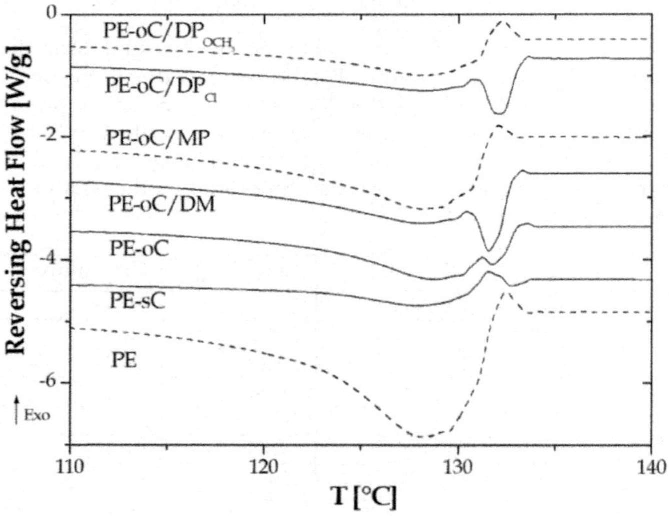

Fig. 8. PE and PE/clay hybrids 'reversing' components. Heating rate was 1°C/min and was modulated by a sinusoid with amplitude ±0,16°C and period 60 s. Nitrogen was used as a purge gas. Samples thermal history was erased by a first heating-cooling cycle.

On the other hand, despite their low clay percentage, composites showed changes compared with the curve above mentioned. Evidently, there are alterations into molecular ordering linked to the different features provided by the interphases that lead to different heat absorption-release mechanisms in the zone near to 130°C. In general it was observed a decrease in the endotherm below to 130°C and in most cases it was seen an endothermic peak above this temperature. PE-oC/MP (group I) and PE-oC/DP$_{OCH_3}$ (group II) behaviors must be underlined because they are similar to PE behavior, and for that reason they were represented by dashed lines.

Based on this, it would be expected that values in a Table 3 would be closer for these hybrids. Nonetheless, values did not match with the tendencies described by reversing curves, as crystallinity degree estimation involves information provided by conventional DSC curves. In this way, it is observed that crystallinity does not entirely reflect the material information related to the variation of the interphase characteristics.

Comparing Table 3 with the behaviors registered in Figure 8, a better interpretation of results can be carried out. PE-oC and PE-sC presented similar crystallinity values, thus alkylammonium chains are not enough to produce significant changes. Meanwhile, it is observed that composites containing oligomers with methyl side groups (similar to HDPE) presented a closer value to pure polymer. Moreover, the highest crystallinity of PE-oC/MP can be explained by the phenyl side group which increases free volume and according to the spatial configuration acquired, it could give to HDPE less spatial restrictions.

Finally, PE-oC/DP_{Cl} and PE-oC/DP_{OCH_3} had low flexibility oligomeric species, which would explain the crystallinity decrease because it would be a hindrance to achieve oligomer-HDPE interfacial ordering. The highest value of PE-oC/DP_{OCH_3} can be the result of the already mentioned difference in the clay coating, allowing a better ordering with HDPE chains at the interphase.

A better understanding of oligomeric species mobility can be performed by reversing heat capacity (Rev Cp) curves (Figure 9) as there are associated to structural changes (Coleman & Craig, 1996).

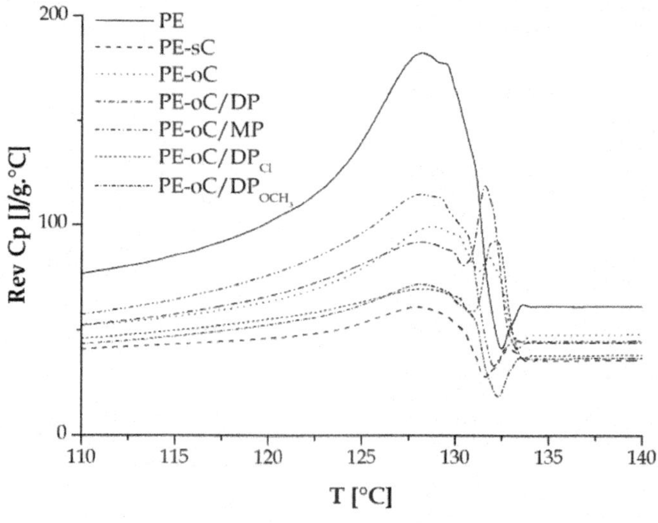

Fig. 9. Reversing Cp comparison of the studied composites in the range of 110 to 140°C.

As it was pointed out, DRX analysis of these polymers (Monasterio et al., 2011; Monasterio & Destéfanis, 2011) corroborated the dispersion exhibited in Figure 3 (TEM) showing exfoliation for PE-oC/MP and intercalation for PE-oC/DP_{OCH_3}. Both composites show similar behaviors to that of HDPE, perhaps the cause is that oligomeric species provide tridimensional arrangements that permit macromolecules to achieve highly free movements near the dispersed particles interphase. For the remaining hybrids, dispersions and interphase features were different.

Therefore, the peaks above 132°C for PE-oC, PE-oC/DM and PE-oC/DP_{Cl} would be the result of clay confined species that only can manifest themselves after complete HDPE melting.

6. Transport properties

Until now, due to alterations at PE-clay interphase, changes were detected in the structure of the studied hybrids and therefore, it will be analyzed if such changes have led to any effect on barrier properties. There are several papers where these macroscopic properties could be modified by handling polymer morphological features such as orientation and degree of crystallinity, as both are linked to chain packing efficiency (Koros, 1990).

Bearing in mind that HDPE is a semicrystalline material, it must be outstood that transport through these polymeric systems are usually analyzed on the basis of a simple two phase model, consisting of a dispersed impermeable crystalline phase in a permeable amorphous matrix (Koros, 1990). In addition, a more complex scenario is given by filled composites as they have clay, which effect will depend on the aspect ratio provided according to the degree of tactoids delamination.

Other factors that influence these properties are the size and chemical nature of penetrating species (Koros, 1990). For the following experiences, apolar molecules of different sizes were employed.

6.1 Gas permeation

Gas permeation measurements were performed on standard permeation equipment. Permeability values of CH_4 and CO_2 were determined at a pressure of 4 bar and a temperature of 30°C.

Table 4 shows higher CO_2 permeability compared to CH_4 permeability, due to higher condensability of CO_2. For unfilled PE, the CO_2/CH_4 permeability rate is near to four, similar to that reported by Stancell (Tobolsky & Mark, 1971).

Sample	CH_4		CO_2	
	P [barrer]	P/P_{PE}	P [barrer]	P/P_{PE}
PE	0.772	*1*	3.433	*1*
PE-sC	0.623	*0,807*	1.980	*0,577*
PE-oC	1.512	*1,959*	6.566	*1,913*
PE-oC/DM	0.835	*1,082*	2.551	*0,743*
PE-oC/MP	0.786	*1,018*	3.667	*1,068*
PE-oC/DP_{Cl}	0.666	*0,863*	2.324	*0,677*
PE-oC/DP_{OCH_3}	0.620	*0,803*	2.357	*0,687*

Table 4. Transport properties for PE and PE/clay hybrids. (P_{PE}: Permeability of pure HDPE)

From a simplified model proposed by Nielsen (1967) (Equation 3), it is observed that the permeability ratios described in Table 4 are inversely proportional to the tortuosity factor.

$$\frac{P_F}{P_U} = \frac{\Phi_P}{\tau} \tag{3}$$

where:

P_F: filled polymer permeability,
P_U: unfilled polymer permeability,
Φ_P: volume fraction of the polymer, and
τ: tortuosity factor

Considering that the volumetric fraction of the polymer on each system is the same, the possible causes that give rise to these behaviors were analyzed.

At first sight, it is noticed that transport properties improvements were not according to the order of dispersion described on TEM analyses.

PE-oC hybrid showed similar permeability rate for both gases, indicating that the phenomenon was indifferent to the penetrant gas characteristics. Evidently, this is a consequence of free volumes generated by poor interfacial adhesion. PE-oC/MP also showed similar ratios for both gases, but permeabilities values were near to pure PE values, which points out that interphase species developed a more tortuous path than the former clay.

The CH_4 permeability ratios of PE-sC, PE-oC/DP_{Cl} and PE-oC/DP_{OCH_3} systems were similar, despite of the differences at the PE/clay interphase. Among them, they can be mentioned the clay aspect ratio and the oligomeric species coating clays. In Figure 3, PE-sC presents an aspect ratio higher than fillers from group II, which after different treatments (cationic exchange and in situ oligosiloxane synthesis), exhibited a lower aspect ratio. Consequently, their efficiency as barrier material has been reduced. However, at the group II hybrids interphase, there are dimers that evidently equate the effects of tortuosity until reaching PE-sC permeability value.

CO_2 permeability ratios for PE-sC, PE-oC/DP_{Cl} and PE-oC/DP_{OCH_3} were alike, too. But permeability reduction of CO_2 was more significant than that observed to CH_4 transport. These results are probably associated with differences in gas molecule sizes or with the possible interactions that can take place between CO_2 and the organic groups at the filler surface.

About PE-oC/DM, the CH_4 permeation had a similar value to that on PE and PE-oC/MP. Nevertheless, this hybrid at CO_2 permeation exhibited an improvement in the barrier properties. Therefore, in spite of having the same interphase characteristics (free volume and oligomers), gas molecules had presented different behaviors, probably due to reasons already cited on the above paragraph.

6.2 Pervaporation

A conventional pervaporation system was used for membrane performance testing. The permeation cell consisted of two sections separated by the membrane. A porous disk was placed to support the membrane on the low pressure side of the membrane cell. The solvent vapor permeated through the membrane because of the very low pressure (vacuum) on the permeate side. The permeate was condensed in a trap cooled in liquid nitrogen. Measurements were carried out at 50, 60 and 70°C.

First, an analysis of PE-filler interaction was performed, and for that task values at 50°C were employed (Figure 10), as it is considered that the interactions effect at the interface will predominate over the temperature effect. PE-oC/DP_{Cl} showed similar values to that of unmodified polymer. While PE-sC and PE-oC/DM exhibited better barrier properties. The remaining composites showed permeabilities above PE. In particular, PE-oC/MP and PE-oC/DP_{OCH_3}, which in TEM micrographs presented higher dispersion degree, showed the highest permeabilities. These results reveal that the inorganic filler dispersion is not the only factor that affects solvent transport through a polymeric membrane, but also interphase species that interact with cyclohexane molecules (Nielsen, 1967).

This analysis was extended by assessing the composite behaviors at different temperatures. Results are displayed in Figure 10.

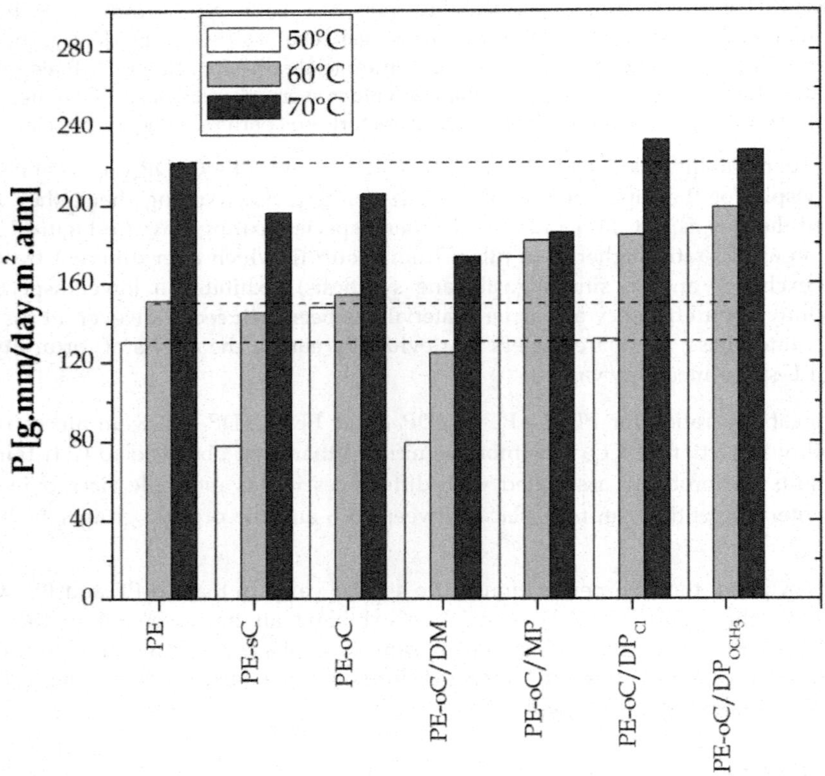

Fig. 10. Permeability coefficients of the prepared composites, compared to PE polymer (Monasterio & Destéfanis, 2011).

In order to compare the temperature effects, dashed lines taking PE values as references were pointed out. By doing so, it is possible to remark that, for each temperature, permeabilities do not increase at the same proportion for each system. This evidences different temperature sensitivities and these behaviors are closely linked to Activation Energy of Permeation (E_P). To complement this information, solubility values were determined at different temperatures which allowed calculating the Heat of Solution (ΔH_S). Finally, the Activation Energy of Difusion (E_D) is obtained by Equation 4 (Koros, 1990):

$$E_P = E_D + \Delta H_S \tag{4}$$

With the values summarized in Table 5 it is possible to split the information hidden in E_P.

	E_P [kJ/mol]	ΔH_S [kJ/mol]	E_D [kJ/mol]
PE	24.45	-11.51	35.96
PE-sC	42.68	-19.20	61.88
PE-oC	14.99	-16.42	31.41
PE-oC/DM	36.00	-12.46	48.46
PE-oC/MP	3.14	-12.64	15.78
PE-oC/DP_{Cl}	26.32	-19.16	45.48
PE-oC/DP_{OCH_3}	8.87	-14.50	23.37

Table 5. Values of activation parameter, E_D, E_P and ΔH_S.

All hybrids showed an exothermic behavior for cyclohexane solubilization. By means of ΔH_S it was observed that PE/filler systems containing oligosiloxane treated clays with side group $-CH_3$ had responses similar to PE. However, pristine clay (sC) and oC/DP_{Cl} clay, which contain low concentration of oligomeric species at the surface, presented a similar behavior when interacting with the solvent. The remaining hybrids exhibited intermediate values.

The highest E_D value corresponds to the polymer with sodium clay, probably due to the tortuosity effect already mentioned in gas permeation. Hybrids with high degree of clay dispersion (PE-oC/MP y PE-oC/DP_{OCH_3}) presented the lowest Ed values. This tendency is associated with that proposed on reversing Cp analysis (Figure 9), where the behaviors observed where related to the mobility at the interphase, which according to the curves would be higher for PE-oC/MP and PE-oC/DP_{OCH_3}.

7. Discussion

In view of these results it can be stated that transport properties were affected by the structural changes on each material. According to observations in polymer-clay composites, barrier properties diminish when clay dispersion increases (Sinha Ray & Okamoto, 2003). However, although some of the hybrids herein studied presented good dispersion, they do not always present transport properties improvements.

On the other side, through DSC and rheological analysis it was noticed a plasticizing effect that was detected by the low crystallinity percentage and the decrease in viscosity, respectively. In the papers mentioned above (X. Zheng & Wilkie, 2003; J. Zhang et al., 2005; C. Zhao et al., 2004), the treatments carried out on clay also give rise to a plasticizing effect on the polymeric matrices. Consequently, it would be expected that hybrids barrier properties were below than HDPE values. Nonetheless, behaviors were observed which did not follow this tendency and that required the consideration of the incidence of other factors for the composites analysis.

Hybrids presented similar behaviors for gas permeation, but these tendencies were not the same for cyclohexane. Specifically, PE-oC/MP and PE-oC/DP_{OCH_3} seem to have reached better polymer-filler compatibility according to Cole-Cole plot (Figure 6), polar coordenates representation (Figure 4), the van Gurp-Palmen plot (Figure 7) and DSC curves (Figure 8), where those hybrids exhibited similar behaviors to that of pure PE, and both composites showed better dispersion/distribution on TEM images. Nevertheless, regarding gas transport, PE-oC/DP_{OCH_3} exhibited improvements in barrier properties, while PE-oC/MP had values near to PE. But for C_6H_{12} transport both hybrids showed poor barrier properties, compared to unmodified polymer.

These responses can be explained by taking into account ´tortuosity' (τ) and 'chain immobilization' (β) factors, which in spite of being defined for pure polymers, will be applied to the composites herein studied. τ value considers flow hindrance provoked by the increased effective length path as well as variations in the cross sectional area of the transport regions.

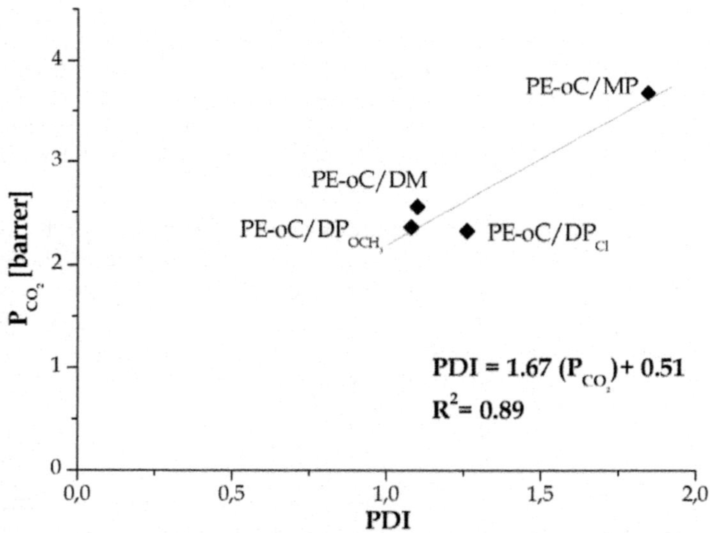

Fig. 11. Oligomers polydispersion effects on CO_2 permeation (For PE-oC/MP, PDI is an average value).

While β indirectly reflects the amount of increase in activation energy of diffusion caused by 'anchoring' effect from crystalline phase dispersed in amorphous phase (Koros, 1990). The diffusing species size will affect these factors in different proportions. As the sizes of the molecules employed to determine transport properties are different (Volumes [Å^3] CH_4=27.0 < CO_2=30.7 < C_6H_{12}=101.0) (Laso & Perpète, 2006), it would be expected that for small molecules (e.g. CH_4) predominates the tortuosity effect, and while the molecule size increases the chain immobilization factor will gain importance.

Due to the hybrids contain several components, τ and β will also experiment changes related to the size and shape of oligomer side groups, and to the oligomer-oligomer and oligomer-PE interactions. According to chemical nature and spatial arrangement (or oligomer flexibility), they will affect the hybrids responses in different ways. In HDPE, crystalline domains had a 'crosslinking' effect (Koros, 1990) which prevents chain mobility on the amorphous phase.

When PE is modified with clays, although the inorganic filler inhibit chain mobility, the resulting holes from the low adhesion or the structural differences of oligomeric species would promote the amorphous phase. All factors here mentioned would explain the reason why in spite of being the same PE/clay hybrid, the transport properties behaviors in front of different molecules are not the same.

With the data obtained from each material internal structure, an analysis of the possible relations among them it will be performed. Several models have been proposed to explain and predict polymers transport properties (Koros, 1990). Nevertheless, permeability through filled polymers is very complex and the possibility of developing a model that contemplates all factors is remote (Nielsen, 1967). Thereby, they are proposed correlation that could be used in further researches.

In the hybrids containing silane modified clays, dependence between CO_2 permeability and oligosiloxane polydispersions it was found (Figure 11). Unlike CH_4 and C_6H_{12} permeabilities which would indicate that these molecules transport through these hybrids does not depend on PDI, or that this factor is uncovered by others.

For cyclohexane transport, a correlation between the crystallinity percentage and the Enthalpy of Solubility was observed. Figure 12 exhibits the mentioned correlation. Some authors (Koros, 1990) have established a link between crystallinity and solubility. Changes in the degree of crystallinity for a semicryslalline material modifies solubility because both values depend on the cohesion achieved by the species on the system (Kaufman & Falcetta, 1977), and by means of these results it was observed that the fillers used affected the molecular interactions and they were manifested on the solvent solubilization process.

Finally, the possible relations among transport properties and rheological parameters were evaluated. There were not relations describing cyclohexane behavior. However, gas permeabilities offered interesting responses for hybrids containing oC clay, as it is shown in Figure 13. Values of G′ slope at the terminal zones were more related to CH_4 permeability than CO_2 permeability.

Figure 14 shows the relations detected between relative permeabilities and G_N^0/G_{cop}. In the paper of García-Franco et al. (2006), it was observed that G_N^0/G_{cop} value allows splitting the PE rheological response from copolymer composition and tends to a constant value that can be implicated with the matrix behavior in front of species of the same chemical nature. From the proposed correlations, it could be said that slope and y intercepts provide information about gas transport through the matrix and dispersed inorganic filler. G_N^0/G_{cop} would bring information about structural changes caused by compatibilizing species (changes at the interphase).

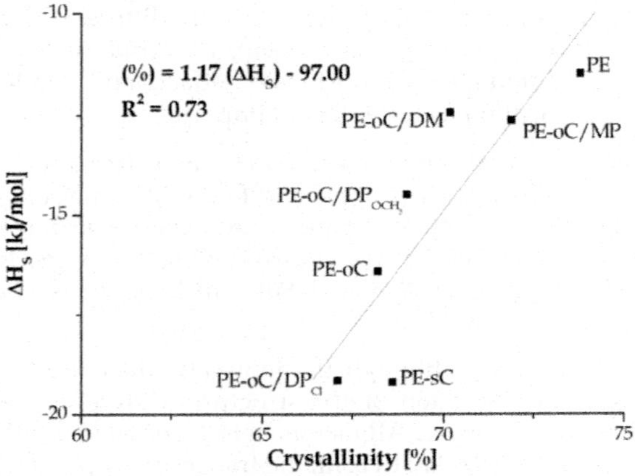

Fig. 12. Degree of crystallinty vs. Enthalpy of Solubility.

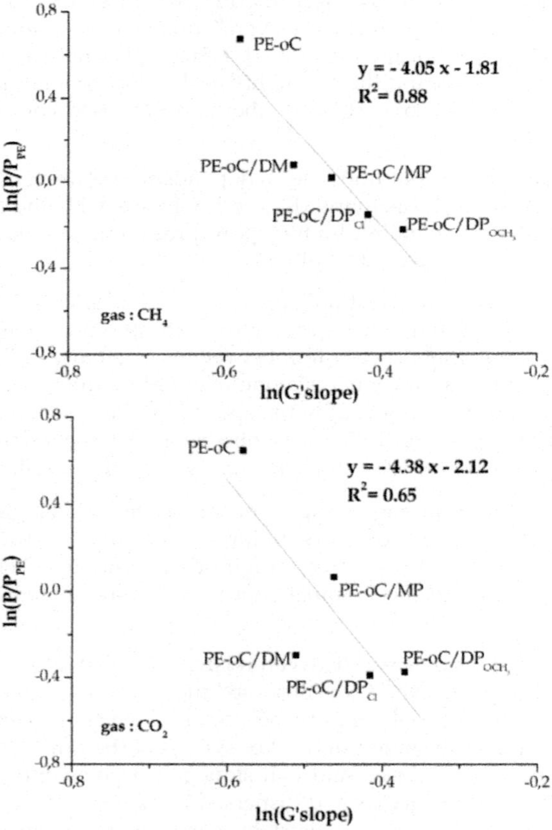

Fig. 13. Comparison of CH_4 and CO_2 permeabilities with G′ slopes.

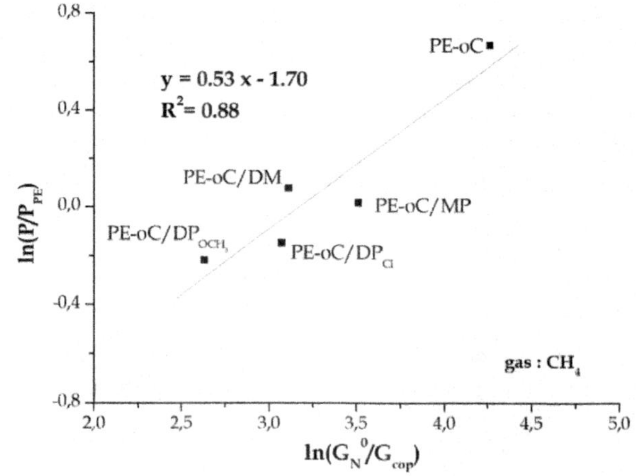

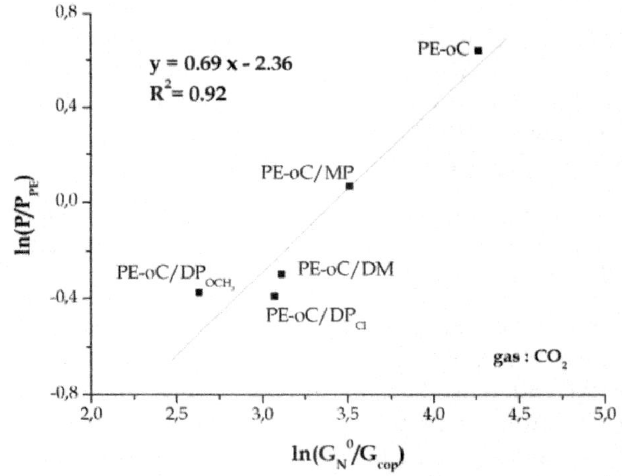

Fig. 14. Correlations founded between transport properties and rheological parameters for both gases.

As relative permeabilities are related with the tortuosity factor, a power law relationship (Equation 5) (Koros, 1990) was taken as reference aiming to explain the results.

$$\tau = \Phi^{-n} \tag{5}$$

where:

τ: tortuosity factor
Φ: amorphous volume fraction
n: exponent factor

In this expression, the crystallinity degree (related to Φ) would be describing the structural changes in the system. But, as it was noticed by modulated DSC, this value does not entirely describe the hybrids global behavior. Therefore, it is proposed the adimensional value G_N^0/G_{cop}, as each value involves several features of the systems. Taking into account that chain immobilization factor (β) can improve the interpretation of permeation in semicrystalline polymers, the following expression is suggested (Equation 6):

$$\tau.\beta = \left(\frac{G_N^0}{G_{cop}}\right)^{-n} \tag{6}$$

Certainly this equation is very simply considering the amount of factors above mentioned and it must be assessed in other system to prove their true validity.

8. Conclusions

By means of the characterizations obtained from TEM, rheology and DSC, changes in the original internal structure of HDPE were registered and this is due to the modifications caused on PE- clay intephases.

Although Transmission Electron Microscopy is a powerful technique in the detection of internal structures of filled polymers, when it was used with rheological and DSC analyses, it allow to improve the interpretation of the factors influencing the final behavior at a macroscopic level. Especially, when studying the barrier properties, as their components get into contact with the new species that diffuse through the system.

Particularly, for the hybrids from this work (groups I and II), the modulated DSC analyses and the dynamic oscillatory shear measurements lead to detect a plasticizing effect that was not perceived by TEM photographs.

Therefore, barrier properties of a system with a high degree of clay dispersion (corresponding to high filler-polymer compatibility) depend on a factor that it is not directly seen by TEM and it is the nanostructural behavior at the interphase (chain arrangement and chemical affinity between matrix and the species used as compatibilizers). This was evidenced by the group II composites (PE-oC/DP_{Cl} and PE-oC/DP_{OCH_3}), since both interphases contain oligomers of the same length and chemistry. However, differences during silane treatment affected further matrix interactions (macromolecules arrangement) and this led to differences in the responses obtained from the applied techniques.

In addition, the use of MDSC technique allowed to distinguish changes that were not detected by the conventional DSC signal.

Finally, taking into account that clay dispersion into apolar matrices (polyolefins) has been considered a critical problem, the proposed compatibilization path offers an alternative to traditional procedures. With the advantage that, by changing oligosiloxane side groups, it is possible to modify interphase behavior using low filler percentage to achieve tailor-made materials with different bulk properties.

9. Acknowledgements

The CCT (Bahía Blanca) is gratefully acknowledged for TEM analyses. The author thanks Dr. M.G. García (UNSL) for the use of permeation equipment.

10. References

Ahmed, J., Varshney, S.K., Auras, R. & Hwang, S.W. (2010). Thermal and Rheological Properties of L-Polylactide/Polyethylene Glycol/Silicate Nanocomposites Films. *Journal of Food Science*, Vol. 75, No. 8, pp. 97-108.

Alexandre, M. & Dubois, P. (2000). Polymer-layered silicate nanocomposites: preparation, properties and uses of a new class of materials. *Materials Science and Engineering*, Vol. 28, pp. 1-63.

Billmeyer, F. (Ed.). (1988). *Ciencia de los polímeros*, Reverté S.A., España.

Bretas, R.E.S. & D'Avila, M.A. (2000). *Reologia de Polímeros Fundidos*, Editorial UFSCar, Brasil.

Campbell D. & White, J. R. (1989). *Polymer Characterization, Physical Techniques*. Chapman and Hall, ISBN: 0412271605, UK.

Cassagnau, P. (2008). Melt rheology of organoclay and fumed silica nanocomposites. *Polymer*, Vol. 49, pp. 2183-2196.

Chen, D., Yang, H., He, P. & Zhang, W. (2005). Rheological and extrusion behavior of intercalated high-impact polystyrene/organomontmorillonite nanocomposites. *Composites Science and Technology*, Vol. 65, pp. 1593-1600.

Coleman, N.J. & Craig, D.Q.M. (1996). Modulated temperature differential scanning calorimetry: a novel approach to pharmaceutical thermal analysis. *International Journal of Pharmaceutics*, Vol. 135, pp. 13-29.

De Meuter, P., Rahier, H. & Van Mele, B. (1999). The use of modulated temperature differential scanning calorimetry for the characterisation of food systems. *International Journal of Pharmaceutics*, Vol. 192, pp. 77–84.

Durmus, A., Kasgoza, A. & Macosko, C.W. (2007). Linear low density polyethylene (LLDPE)/clay nanocomposites. Part I: Structural characterization and quantifying clay dispersion by melt rheology. *Polymer*, Vol. 48, No.15, pp. 4492-4502.

Faker, M., Razavi Aghjeh, M.K., Ghaffari, M. & Seyyedi, S.A. (2008). Rheology, morphology and mechanical properties of polyethylene/ethylene vinyl acetate copolymer (PE/EVA) blends. *European Polymer Journal*, Vol. 44, pp. 1834-1842.

Franck, A. (n.d.). Mixing Rules for Complex Polymer Systems, AN008e, *TA Instruments*.

Gacitua E., W., Ballerini A., A. & Zhang J. (2005). Polymer nanocomposites: synthetic and natural fillers a review. *Ciencia y tecnología*, Vol. 7, No.3, pp. 159-178.

García-Franco, C.A., Harrington, B.A. & Lohse, D.J. (2006). Effect of Short-Chain Branching on the Rheology of Polyolefins. *Macromolecules*, Vol. 39, pp. 2710-2717.

Gupta, R.K. & Bhattacharya, S.N. (2008). Polymer-clay Nanocomposites: Current Status and Challenges. *Indian Chemical Engineer*, Vol. 50 No. 3, pp. 242-267

Kaufman, H.S. & Falcetta, J.J. (1977). *Introduction to Polymer Science and Technology* (1st edition), John Wiley & Sons Inc, ISBN: 978-0471014935, USA.

Kim, B.C. & Lee, S.J. (2008). Silicate dispersion and rheological properties of high impact polystyrene/organoclay nanocomposites via in situ polymerization. *Korea-Australia Rheology Journal*, Vol. 20, No. 4, pp. 227-233.

Koros, W.J. (Ed.).(1990). *Barrier Polymers and Structures*, American Chemical Society, USA.

Laso, M & Perpète, E.A. (2006). *Multiscale modeling of polymer properties*, Elsevier, UK.

LeBaron, P.C., Wang, Z., & Pinnavaia, T.J. (1999). Polymer-layered silicate nanocomposites: an overview. *Applied Clay Science*, 15, pp. 11–29

Lee, I.B., Son, H.H., Um, C.M. (2003). Rheological properties of flowable, conventional hybrid, and condensable composite resins. *Dental Materials*, Vol. 19, pp. 298–307.

Lohse, D. J., Milner, S. T., Fetters, L. J., Xenidou, M., Hadjichristidis, N., Mendelson, R. A., García-Franco, C. A. & Lyon, M. K. (2002). Well-Defined, Model Long Chain Branched Polyethylene. 2. Melt Rheological Behavior. *Macromolecules*, Vol. 35, pp. 3066-3075.

Mai, Y.-W. & Yu, Z.-Z. (Eds.). (2006). *Polymer nanocomposites*, CRC Press, USA.

Malucelli, G., Ronchetti, S., Lak, N., Priola, A., Dintcheva, N.T. & La Mantia, F.P. (2007). Intercalation effects in LDPE/o-montmorillonites nanocomposites. *European Polymer Journal*, Vol. 43, pp. 328-335.

Michler, G.H. (2008). *Electron Microscopy of Polymers*, Springer, Germany.

Monasterio, F., Rodriguez Pita, V., Lopes Dias, M., Erdmann, E. & Destéfanis, H. (2011). Thermal and rheological properties of polyethylene composites based on poly(diphenylsiloxanes)/organoclay hybrids obtained from two different silanes. *Macromolecular Symposia*, Vol. 299/300, pp. 81–91.

Monasterio, F., Lopes Dias, M., Rodriguez Pita, V., Erdmann, E. & Destéfanis, H. (2010). Effect of the organic groups of difunctional silanes on the preparation of coated clays for olefin polymer modification. *Clay Minerals*, Vol. 45, pp. 489-502.

Monasterio, F.E. & Destéfanis, H.A. (2011). Synthesis of additives from montmorillonite to modify high density polyethylene final properties. *Macromolecular Symposia*, Vol. 301, No.1, pp. 104–113.

Nielsen, L.E. (1967). Models for the permeability of filled polymer systems. *Journal of Macromolecular Science*, Vol. A1, No.5, pp. 929-942.

Okamoto, M. (2005). Biodegradable Polymer/Layered Silicate Nanocomposites: A Review. In: *Handbook of Biodegradable Polymeric Materials and Their Applications*, Mallapragada, S. & Narasimhan, B., pp. 1-45, American Scientific Publishers, ISBN: 1-58883-053-5, USA.

Punnarak, P., Tantayanon S. & Tangpasuthadol, V. (2006). Dynamic vulcanization of reclaimed tire rubber and high density polyethylene blends. *Polymer Degradation and Stability*, Vol. 91, pp. 3456-3462.

Rose, J.M., Cherian, A.E., Lee, J.H., Archer, L. A., Coates G.W. & Fetters, L.J. (2007). Rheological Behavior of Chain-Straightened Poly(α-olefin)s. *Macromolecules*, Vol. 40, No.19, pp 6807–6813.

Rothon, R.N. (2003). *Particulated-filled polymer composites (2nd edition)*. Smithers Rapra Publishing, ISBN: 978-1859573822, USA.

Schawe, J.E.K. (1995). A comparison of different evaluation methods in modulated temperature DSC. *Thermochimica Acta*, Vol. 260, pp. 1-16.

Schlatter, G., Fleury, G. & Muller, R. (2005). Fourier Transform Rheology of Branched Polyethylene: Experiments and Models for Assessing the Macromolecular Architecture. *Macromolecules*, Vol. 38, pp. 6492-6503.

Shah, R.K. & Paul, D.R. (2004). Nylon 6 nanocomposites prepared by a melt mixing masterbatch process. *Polymer*, Vol. 45, pp. 2991–3000.

Sinha Ray, S., Bousmina, M. (2005). Biodegradable polymers and their layered silicate nanocomposites: In greening the 21st century materials world. *Progress in Materials Science*, Vol. 50, pp. 962–1079

Sinha Ray, S. & Okamoto, M. (2003). Polymer/layered silicate nanocomposites: a review from preparation to processing. *Progress in Polymer Science*, Vol. 28, pp. 1539–1641.

TA Instruments (n.d.). Measurement of Aging Effects on Amorphous Pet, TS-25.

TA Instruments (n.d.). Rheology Software Models (Flow), RN-9B. Rheology applications note.

Thomas, L.C. (2005). Why Modulated DSC? An Overview and Summary of Advantages and Disadvantages Relative to Traditional DSC. *TA Instruments*

Tobolsky, A.V. & Mark, H. (Eds.). (1971). *Polymer Science and Materials*, John Wiley & Sons, Inc., USA.

Verdonck, E., Schaap, K. & Thomas, L.C. (1999). A discussion of the principles and applications of Modulated Temperature DSC (MTDSC). *International Journal of Pharmaceutics*, Vol. 192, pp. 3–20.

Wan, C., Zhang, Yong & Zhang, Yinxi (2004). Effect of alkyl quaternary ammonium on processing discoloration of melt-intercalated PVC-montmorillonite composites. *Polymer Testing*, Vol. 23, pp. 299-306.

Wang, K.H., Choi, M.H., Koo, C.M., Xu, M., Chung, I.J., Jang, M.C., Choi, S.W. & Song, H.H. (2002). Morphology and Physical Properties of Polyethylene/Silicate Nanocomposite Prepared by Melt Intercalation. *Journal of Polymer Science: Part B: Polymer Physics*, Vol. 40, pp. 1454–1463.

Wang, L. & Sheng, J. (2005). Preparation and properties of polypropylene/org-attapulgite nanocomposites. *Polymer*, Vol. 46, pp. 6243-6249.

Williams,D. & Carter, C.B. (2009) *Transmission Electron microscopy* (2nd), Springer, ISBN, New York.

Yang, Q., Chung, T.-S., Weber, M. & Wollny, K. (2009). Rheological investigations of linear and hyperbranched polyethersulfone towards their as-spun phase inversion membranes' differences. *Polymer*, Vol. 50, pp. 524-533.

Zhang, J., Jiang, D. & Wilkie, C. (2006). Thermal and flame properties of polyethylene and polypropylene nanocomposites based on an oligomerically modified clay. *Polymer Degradation and Stability*, Vol. 91, No.2, pp. 298-304.

Zhang, Q., Yang, H. & Fu, Q. (2004). Kinetics-controlled compatibilization of immiscible polypropylene/polystyrene blends using nano-SiO2 particles. *Polymer*, Vol. 45, pp. 1913–1922.

Zhao, C., Feng, M., Gong, F., Qin, H. & Yang, M. (2004). Preparation and characterization of polyethylene-clay nanocomposites by using chlorosilane-modified clay, *Journal of Applied Polymer Science*, Vol. 93, pp. 676-680.

Zheng, X. & Wilkie, C.A. (2003). Nanocomposites based on poly (ε-caprolactone) (PCL)/clay hybrid: polystyrene, high impact polystyrene, ABS, polypropylene and polyethylene. *Polymer Degradation and Stability*, Vol. 82, pp. 441–450.

18

Investigation on Structure and Behaviours of Proton Exchange Membrane Materials by TEM

Zhe Wang[1], Chengji Zhao[2], Hongzhe Ni[1], Mingyao Zhang[1] and Huixuan Zhang[1]
[1]*School of Chemical Engineering, Changchun University of Technology,*
[2]*Alan G. MacDiarmid Institute, College of Chemistry, Jilin University, Changchun, P. R. China*

1. Introduction

H_2/O_2 fuel cells and direct methanol fuel cells (DMFC) are one of the most promising efficient and environmental friendly energy generation systems because of their low emissions and high working efficiency, which are extensively applied in stationary power, automobiles and portable electrical source[1-2]. The proton exchange membrane (PEM), the key component of fuel cell, acts as a separator to prevent the mixing of the fuel and the oxidant and severs as an electrolyte to transfer protons from the anode to the cathode [3]. The PEM must be satisfied with following peculiarities such as high proton conductivity, low electronic conductivity, long-term chemical stability, good mechanical strength under fuel cell operation conditions and low cost. In the early 1960s, sulfonated polystyrene copolymers membranes were used as the PEM in proton exchange membrane fuel cells (PEMFCs) [4]. However, short lifetime and relatively high price limited their further applications. In the later 1960s, Nafion® perfluorosulfonated membranes produced by DuPont company have been extensively used because of their high conductivity and chemical stability [5-7]. However, there are some drawbacks, such as high methanol permeability, relatively low proton conductivity at high temperature and high price, which baffle their further developing. Recently many researchers have been devoted to developing new hydrocarbon PEM materials that are of good mechanical performance and thermo-oxidative stability, low price and high proton conductivity. Nowdays, the developed novel polymer electrolyte membranes include sulfonated poly(arylene ether sulfone)s[8-10], sulfonated poly(arylene ether ketone)s[11-15], sulfonated polyimides[16-19], sulfonated polybenzimidazoles[20-22], block sulfonated copolymers and their composite membranes[23-28], etc. Sulfonated polymers may be prepared either by postsulfonation of the polymers or by polycondensation of sulfonated monomer with other non-sulfonated monomers [29]. The postsulfonation polymers were prepared by electrophilic aromatic sulfonation using concentrated sulfuric acid, fuming sulfuric acid, chlorosulfonic acid or sulfur trioxide as sulfonating agents. The postsulfonation reaction usually has some drawbacks such as not ease to precisely control on the degree of sulfonation (DS) and the location of sulfonic groups, the possibility of side reaction, and the degradation of polymer. The direct polycondensation sulfonated copolymers with different sulfonation degrees were synthesized by varying the ratios of the sulfonated monomers to the non-sulfonated monomers [30-35]. The latter method has been proved to be more advantageous than postsulfonation. Some of the advantages are: (1) the position and the content of sulfonated groups can be easily controlled and (2) cross-linking and other side reactions can be avoided. In order to investigate more availably the relationship between structures and properties, sulfonated aromatic copolymers were obtained by aromatic nucleophilic substitution polycondensation reaction in most literatures. As PEM materials, these sulfonated aromatic copolymers showed relatively good properties including high proton conductivity, good water transport properties, thermo-oxidative stability, good mechanical performance and methanol permeability. To further develop these sulfonated polymers, which can be potentially suitable for applications in proton exchange membrane fuel cell or direct methanol fuel cell systems, some researchers have modified sulfonated polymers. The sulfonated polymers/heteropolyacid and sulfonated polymers/TiO_2 composite membranes exhibited high proton conductivity at high temperature, which are satisfied with the requirement of fuel cell operation at high temperature[36-38]. The sulfonated polymers/polyanine and sulfonated polymers/polypyrrole composite membranes decreased methanol permeability with a little lower proton conductivity change, thus showing very good prospective in DMFC application [39, 40]. The cross-linked sulfonated aromatic polymer membranes displayed higher mechanical strength, thermo-oxidative stability and lower methanol diffusion coefficient than those of non-crosslinked membranes, which were particularly attractive as the PEM for DMFC. The important parameters of proton exchange membranes included ion-exchange capacity, water uptake, water diffuse coefficient, thermo-oxidative stability, proton conductivity, methanol permeability and mechanical performance, which directly decide proton exchange membrane behaviors. These parameters are determined by their chemical structures and morphology. In this paper, series of sulfonated aromatic polymers with different chemical structures were synthesized. The measured and studied methods about PEM were detailed discussed as well as the relationship between structure and properties were investigated by transmission electron microscope (TEM).

2. Experiment

2.1 The synthesis of sulfonated copolymers

It has been proved that direct copolymerization have some advantage, for example, it can enhance the stability and the acidity of sulfonic groups due to introducing two sulfonic groups in each repeating unit. A series of sulfonated poly(aryl ether sulfone)s, sulfonated poly(aryl ether ketone)s and sulfonated poly(aryl ether ketone sulfone) copolymers with different DS were synthesized by aromatic nucleophilic substitution polycondensation. The detailed procedure is described as following: sulfonated monomer, non-sulfonated monomer, double hydroxybenzene monomer and potassium carbonate were added into a 500 mL three neck round bottom flask with an atmosphere of nitrogen inlet, a mechanical stirrer, and a Dean-Stark trap with a reflux condenser. Tetramethylene sulfone (TMS) was used as a solvent and toluene was used as an azeotroping agent. The reaction mixture was refluxed at 145 °C for 4 h until water was fully removed from the reaction system by azeotropic distillation, and then the excess toluene was distilled out. The reaction temperature was then slowly raised to 200 °C and kept at this temperature for another 10 h. After the reaction mixture became highly viscous, TMS was added to dilute the solution, and then the solution was cooled to room temperature. The cooled copolymers with low DS solutions were directly poured into deionized water. While the copolymers with high DS solutions were poured into acetone. The precipitated copolymers were washed by boiling deionized water several times and dried at 80 °C for 48 h. This process is shown in **Scheme 1** and **Scheme 2**. The sulfonated aromatic polymers with different main chain structure were named as SPAEKS-M and the sulfonated aromatic polymers with side chain structure were defined as SPAEKS-S.

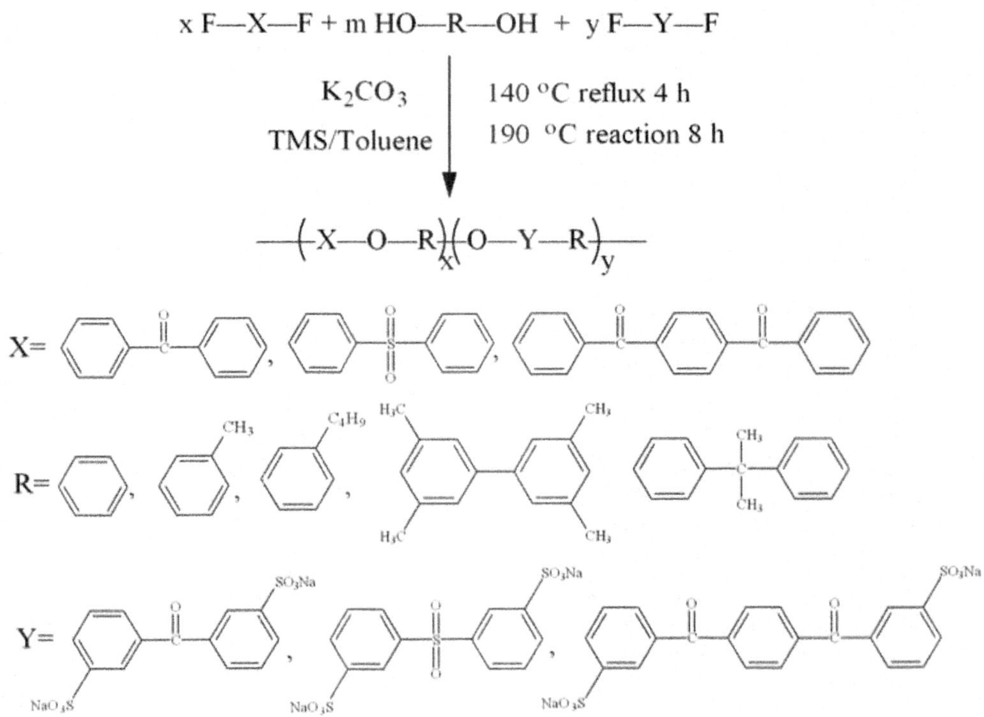

Scheme 1. A series of sulfonated aromatic polymers with different main chain structures.

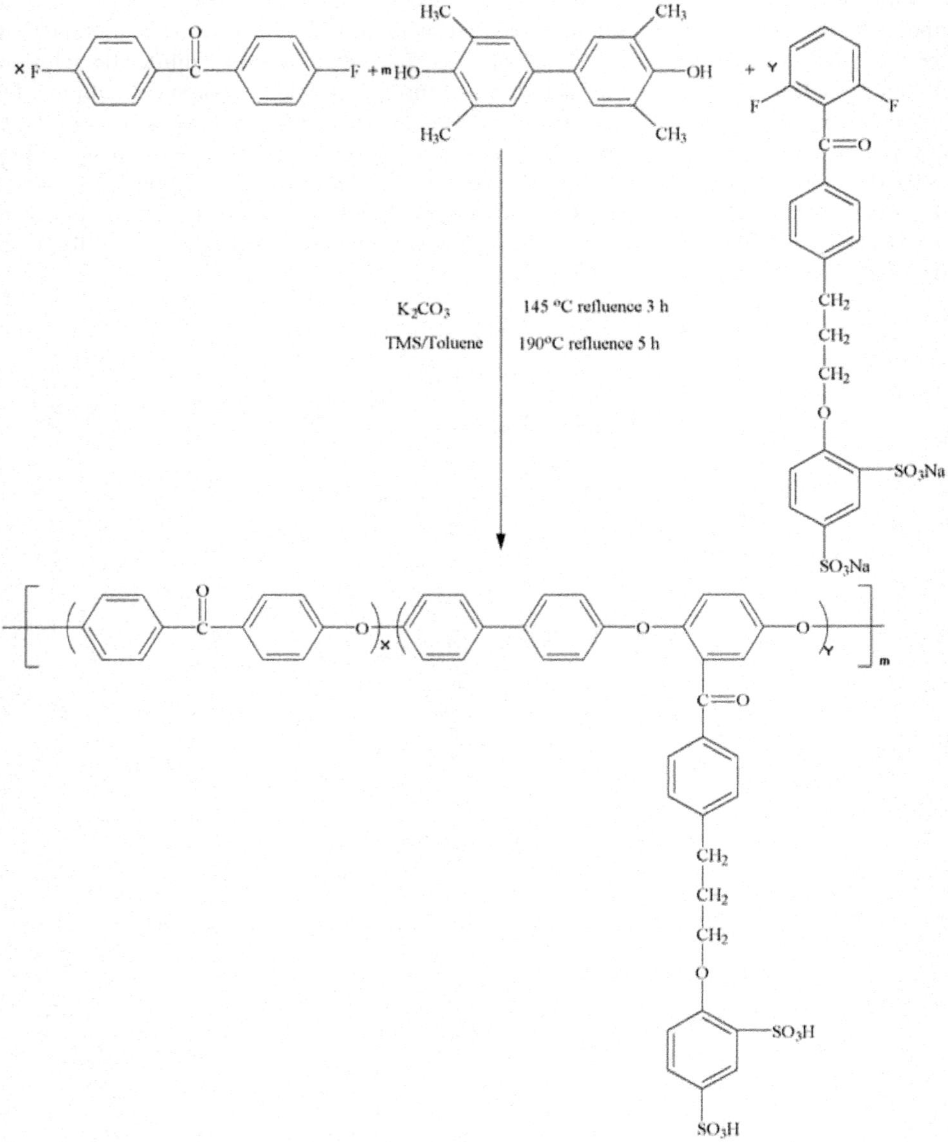

Scheme 2. The sulfonated aromatic polymer with a side chain structure.

3. Results and discussion

3.1 The FT-IR characterization of SPAEKS copolymers

The chemical structure of the SPAEKS copolymers was verified by fourier transform infrared (FTIR) spectroscopy using a Perkin Elmer recording in 4 cm^{-1} of spectral resolution in the range 400-4000cm^{-1}. The FTIR spectra of SPAEKS copolymers are presented in **Fig.1**, where the peaks at 1078 cm^{-1}, 1020 cm^{-1} and 1028 cm^{-1} corresponded to the stretching of the sodium sulfonate groups and at 1660cm^{-1} was assigned to double bond of phene stretching. The S=O stretching band of sodium sulfonate groups was observed at 1243 cm^{-1}. These results confirmed that the sodium sulfonate groups were successfully incorporated into the copolymers as expected.

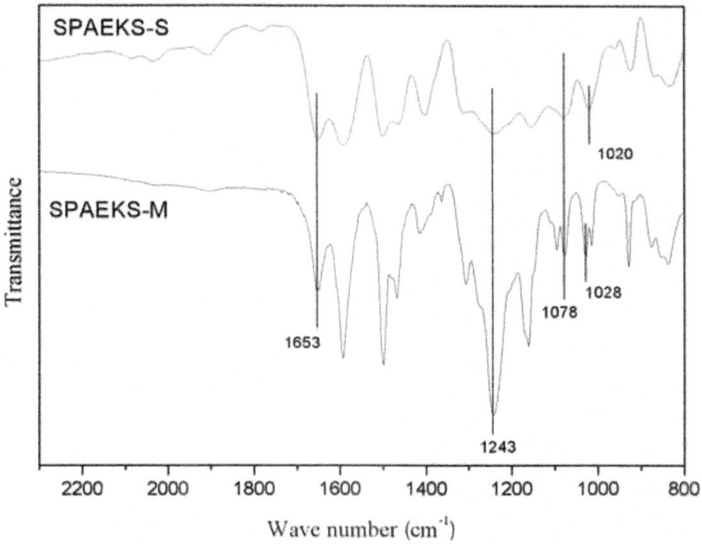

Fig. 1. The FTIR spectra of SPAEKS copolymers.

3.2 Ion exchange capacity

Ion exchange capacity (IEC) is usually defined as the mole of fixed SO_3 sites per gram copolymer, which indicates the ability of proton exchange membrane transferring proton. The higher IEC values are, the better membranes transfer proton. The IEC values of membranes were determined by the titration and calculation. The detailed procedures are descibled as following: Membranes in acidic form were immersed into a 50 ml 1 M NaCl solution for 24 h in order to achieve complete ion exchange of H^+ with Na^+. The released H^+ was titrated with 0.1 M NaOH solution, in which phenolphthalein was used as an indicator. The IEC values were calculated from the titration data via the following equation.

$$\text{IEC}\left(\text{mequiv.}/\text{g}\right) = \frac{\text{consumed NaOH} \times \text{molarity NaOH}}{\text{weight dried membrane}}$$

The IEC values measured by titration increased with DS increasing and accorded with the calculated IEC values. It indicated that IEC values could be readily controlled by varying the feed ratio of sulfonated monomers to non-sulfonated monomers. This has the advantage over the postsulfonated reaction of introducing sulfonic groups into the copolymers without a sulfonation degradation reaction on the polymer chain by direct copolymerization.

3.3 Water transport properties

It is well known that water plays a critical role for proton exchange membrane applications. On the one hand, adequate water uptake is desired to maintain good proton conductivity; on the other hand, water sorption should be minimized to ensure the membrane mechanical performance and the dimensional stability. Nafion with low water retention at high temperatures resulting in conductivity falls is one of the drawbacks, which limited its further commercial application. Water uptake and swelling ratio of membrane were calculated using the following formula.

$$\text{Water uptake} = \frac{W_w - W_d}{W_d} \times 100\%$$

The swelling ratio was defined as:

$$\text{Swelling ratio} = \frac{L_w - L_d}{L_d} \times 100\%$$

Where W_d and L_d are the weight and length of dried membrane, respectively. The dried membranes were immersed into deionized water until constant weight and length were obtained, which was recorded as W_w and L_w.

Water retention of proton exchange membrane has significant effects on its proton conductivity, especially at high temperature. In our work, water retention of membrane was expressed by water diffusion coefficient, which TGA recorded the weight change with time. The relationships between the water desorption and time might follow the Ficker diffusion law. The water diffusion coefficient of membranes was calculated according to the following formula.

$$\frac{M_t}{M_\infty} = 4\left(\frac{D_t}{\pi l^2}\right)^{1/2}$$

Where D is the water diffuse coefficient; l is the membrane thickness; M_t/M_∞ is the water desorption, which can be obtained by the slope of water desorption curves because the plots of M_t/M_∞ versus $t_{1/2}$ are linear for Ficker diffusion [39].

The state of water in the sulfonated copolymers has a critical influence on the electrochemical and physical properties of the membranes, which may be very important for PEM fuel cells. The water within sulfonated aromatic polymers membranes exists into two states: the loosely bound water and the tightly bound water. The loosely bound water content can be calculated by DSC instrument [3]. Thermo-gravimetric analysis as an excellent method was used to study the tightly bound water. It is well known that the relative weight loss before sulfonic groups decompose due to water evaporation corresponded to water retention capacity of the membrane. The temperature of water desorption best fasten recording as T_{max} in TGA differential curves served as an indication of the water retention capacity of a given membrane. The higher the T_{max} was, the better the membrane hold water. In this paper, we select two samples with main chain structure and side chain structure at the same IEC values as comparing. The results showed that the water uptake, swell ratio and water diffusion coefficient of SPAEKS-S were higher than those of SPAEKS-M at the same DS.

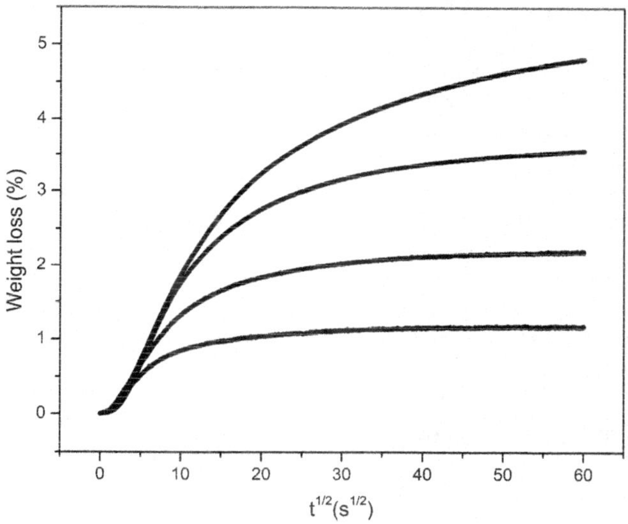

Fig. 2. The water desorption curves of SPAEKS membrane at 80 °C.

Samples	IEC (mmol/g)	Water uptake (%)	Swelling ratio (%)	Water diffusion coefficient (cm^2/s)
SPAEKS-M	1.88	35	9	1.21×10^{-9}
SPAEKS-S	1.90	41	11	8.96×10^{-8}

Table 1. The data of SPAEKS-M and SPAEKS-S membrane.

3.4 Thermal properties, thermo-oxidative stability, mechanical strength of membrane

Proton exchange membrane must hold good thermal properties due to fuel cell operation at a relatively high temperature. In our previous works, thermal properties were performed on thermo-gravimetric analysis (TGA) and different scanning calorimetry (DSC) analyzer. It can be seen from TGA curves that the membrane in acid form has two degradation steps. The first weight loss about range from 160 °C to 200 °C was assigned to the elimination of sulfonic groups. The second weight loss around 400 °C was attributed to the decomposition of polymer backbone. The onset weight loss temperature (T_{oneset}) of membrane in acid form gradually deceased with increment of DS. Thermo-oxidative stability may be very decisive with respect to proton exchange membrane life span. Sulfonated aromatic polymers are amorphous and only one glass transition temperatures (T_g) due to rigid polymer backbone, which is proved by differential scanning calorimetry (DSC). All sulfoanted polymers exhibited relatively high T_g that increased with DS increasing. It was well explained that the introduction of sulfonate groups increased intermolecular interaction and molecular bulkiness. Both effects hinder the internal rotation, which leads to increase T_g of sulfonated copolymers.

Thermo-oxidative stability was studied by immersing the membrane into Fenton's reagent (3% H_2O_2 containing 2 ppm $FeSO_4$), which is a routine testing method for oxidative stability. The consumed time that the membrane started to break into pieces was used to evaluate oxidative stability of membrane. Good mechanical strength is necessary for a proton exchange membrane material. Aromatic polymers including poly(aryl ether sulfone)s, poly(aryl ether ketone)s and polyamides as a series of typical thermal plastics, are well known for their excellent thermal and mechanical properties as well as their resistance to oxidation and their stability under acidic conditions. Sulfonated aromatic polymers also exhibited good mechanical strength, which the tensile strength and tensile modulus of membrane reached up to 40 MPa and 1 GPa, respectively. It indicates that they are tough enough to be performed as a functional membrane material. The mechanical strength of membrane can be obtained by electro-tensile machine at a tensile speed of 2 mm/min under ambient conditions, the size of samples is 15 mm×4 mm. Three samples of each membrane were analyzed and average value was obtained.

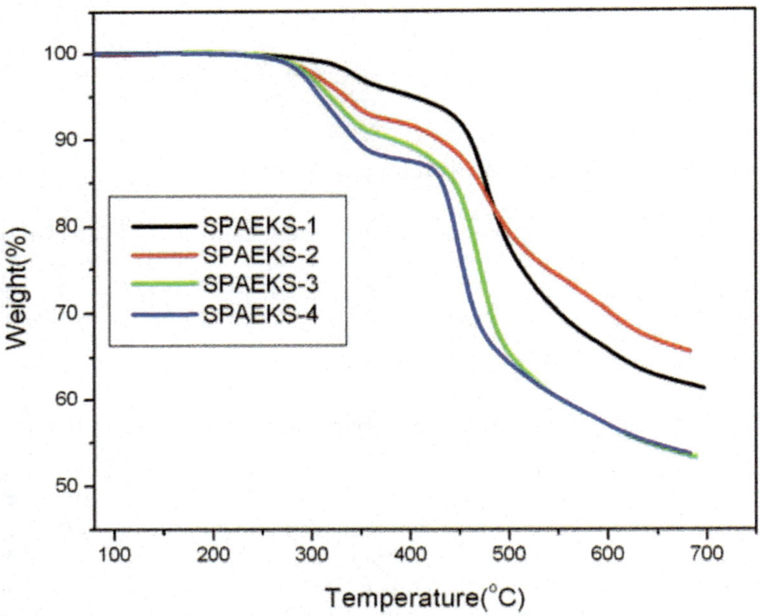

Fig. 3. The TGA curves of SPAEKS copolymers.

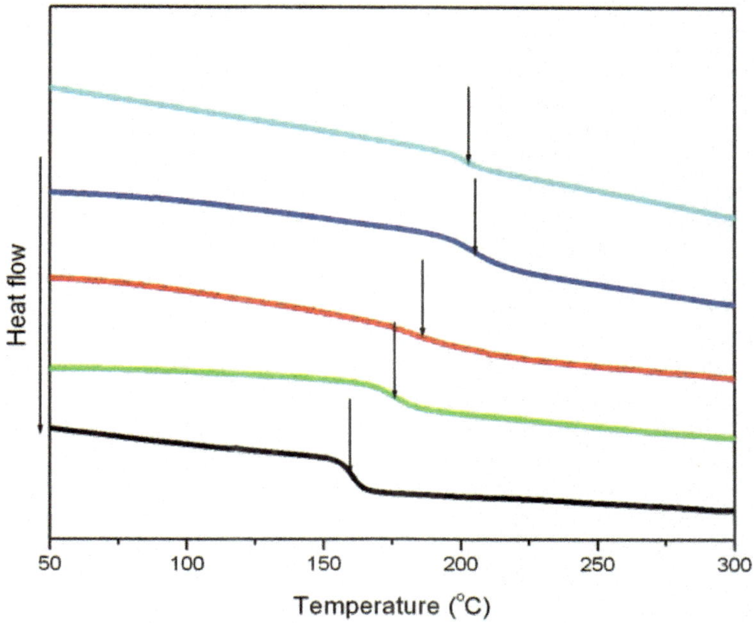

Fig. 4. The DSC curves of SPAEKS copolymers.

Sample	T_g (°C)	T_d (°C)	Oxidative time (h)	Tensile strength (MPa)	Tensile modulus (GPa)	Maximum elongation (%)
SPAEKS-M	160	252	225	47	1.15	8.2
SPAEKS-S	153	227	198	45	1.02	10.1

Table 2. The data of SPAEKS-M and SPAEKS-S membrane.

3.5 Morphology of the proton exchange membrane

The physical and electrochemical properties of proton exchange membrane are related to their microstructure and morphologies, especially the spatial distribution of ionic sites. The sulfonic groups in SPAEKS copolymers aggregated into hydrophilic ionic clusters are related to water uptake and proton transport. The transporting of proton in PEM occurs through three mechanisms, Grotthus, vehicular and surface transport. The Grotthus mechanism is that protons transfer via structural diffusion (proton charge carriers dissociate from water molecular to next water molecular). However, proton transport through the movement of larger cations (such as $H_5O_2^+$, $H_9O_4^+$) was defined as vehicular mechanism. It is likely that proton transport predominate through Grotthus and vehicular mechanisms in SPAEKS membrane with relatively high water contents. Surface mechanism refers that proton transmission occurs between sufonic groups located on the walls of hydrophilic channels. Surface mechanism becomes increasingly important at low humidity. The distribution of hydrophilic and hydrophobic domains may determine the proton conductivity of SPAEKS membrane. The water serving as proton carrier plays an important role in determining the proton conductivity of PEM. Some researchers have used Atomic Force Microscopy (AFM), Small-angle X-ray scattering (SAXS) and TEM to study the distribution of hydrophilic and hydrophobic domains in PEM. In order to find the distribution of sulfonic groups in copolymers matrix, in this paper, TEM experiments were performed on a JEM-2000EX microscope with an acceleration voltage of 200 kV. The detailed operation procedures are described as following: The SPAEKS copolymers in acidic form were obtained by immersing the SPAEKS copolymers in sodium form into a 2M HCl solution for 24 h. Sulfonated copolymers in acid form were immersed in $AgNO_3$ solution for 24 h in order to make Na^+ forms convert into Ag^+ form because Ag^+ particles can provide a good contrast. Copolymers in Ag^+ form were filtrated and vacuum dried at 80 ^{0}C overnight, and then were dissolved in DMF forming 5% DMF solutions and dropped on carbon coated copper grids for TEM investigation. The sulfonic groups in sulfonated polymers aggregate into hydrophilic ionic transport channels in copolymers matrix contributing to transport proton and the nonsulfonated copolymers backbone form hydrophobic domains providing their mechanical strength. The TEM pictures of Ag^+ chelated SPAEKS membranes were shown in Figure 1. The dark domains (Ag^+ particles) in pictures represent hydrophilic sulfonic groups, while the light domains refer to hydrophobic copolymer backbone. It can also be seen that the sulfonic groups randomly disperse in sulfonated copolymer matrix. With an increment in DS from SPAEKS-10 (DS=0.2) to SPAEKS-60 (DS=1.2) membrane, the density and the size of sulfonic groups become larger. The diameter of sulfonic groups increases from 5 nm to 10 nm. Thus, it is concluded that the ionic groups could aggregate to relatively large transport pathways at high contents of sulfonic acid groups. This may be likely to enhance proton or ion transport properties of SPAEKS membranes. The conclusion about the density of Ag^+ particles can not be definitely made due to the different thickness of the TEM samples. The size of Ag^+ particles will denote the size of proton conductive channels in SPAEKS membranes. It is also important to note that the membranes observed under TEM do not hydrate, so in a practical application, the ionic clusters would probably be more evenly distributed. When this hydrophilic domain is hydrated, protonic charge carriers dissociate from the SO_3^- functional groups (Grotthus mechanism) and proton transfer by H_3O^+ (vehicular mechanism). The micrograph of sulfonated copolymers with side chain sulfonic acid groups was shown in Figure 1(k). It can be seen that the sulfonic acid groups spreading as a cloud-like belt were also observed for side chain sulfonated copolymers. These wide and well-connected ionic domains can provide large proton transport channels, which are responsible for fast proton transport and high proton conductivity. From here we see that sulfonic groups disperse throughout the polymer matrix and phase separation does not happen for random SPAEKS copolymers. While for block and graft sulfonated copolymers, sulfonic groups aggregated into hydrophilic ionic clusters and form phase separation. The morphology of PEM explained why the water uptake and proton conductivity of membrane increased with increasing DS. The relationship between micro-structure and properties is established by TEM.

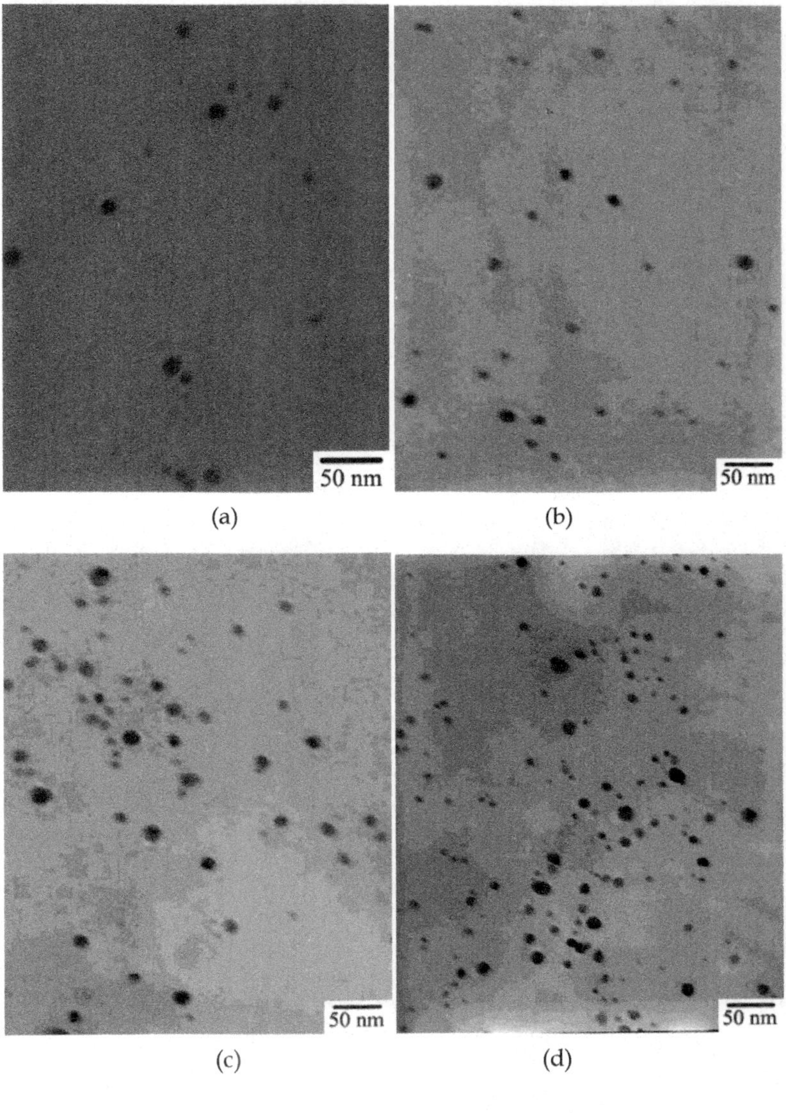

(a) (b) (c) (d)

Fig. 5. The micrographs of SPAEKS membranes: (a) SPAEKS-M1-10; (b) SPAEKS-M1-20; (c) SPAEKS-M1-30; (d) SPAEKS-M1-40; (e) SPAEKS-M1-50; (f) SPAEKS-M1-60; (g) SPAEKS-M2-20; (h) SPAEKS-M2-30; (i) SPAEKS-M2-40; (j) SPAEKS-M2-50; (k) SPAEKS-S-50.

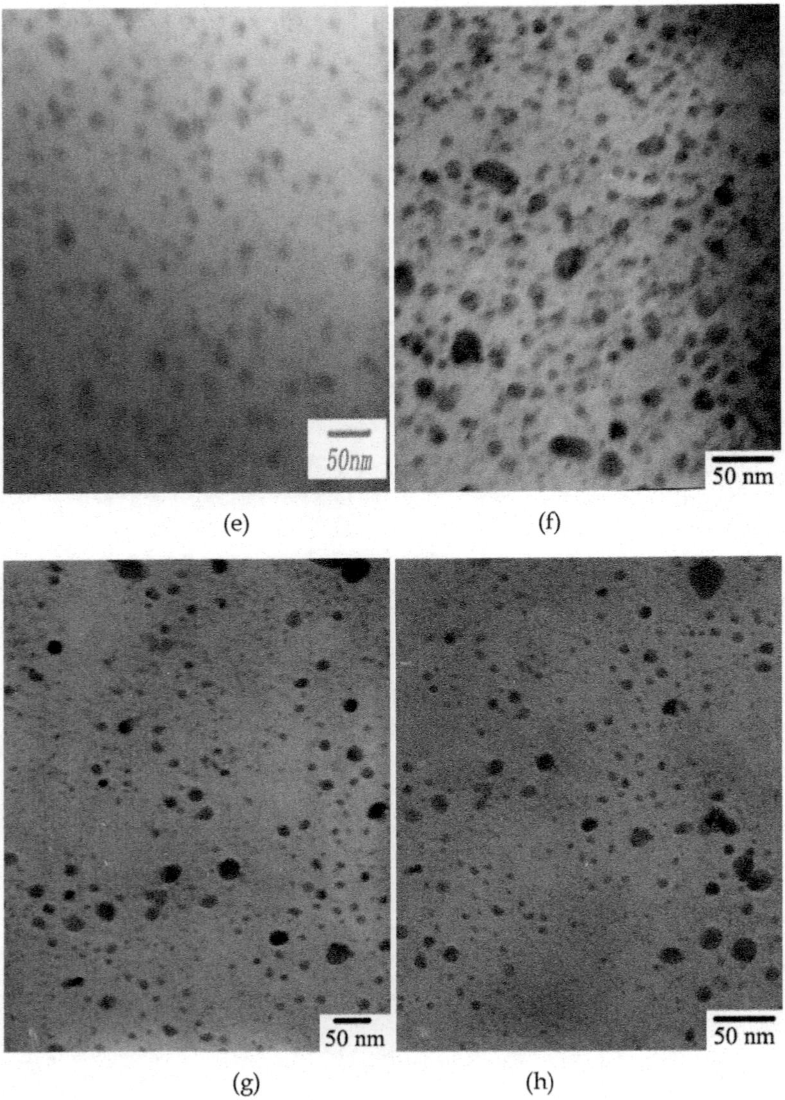

Fig. 5. The micrographs of SPAEKS membranes: (a) SPAEKS-M1-10; (b) SPAEKS-M1-20; (c) SPAEKS-M1-30; (d) SPAEKS-M1-40; (e) SPAEKS-M1-50; (f) SPAEKS-M1-60; (g) SPAEKS-M2-20; (h) SPAEKS-M2-30; (i) SPAEKS-M2-40; (j) SPAEKS-M2-50; (k) SPAEKS-S-50. (Continuation)

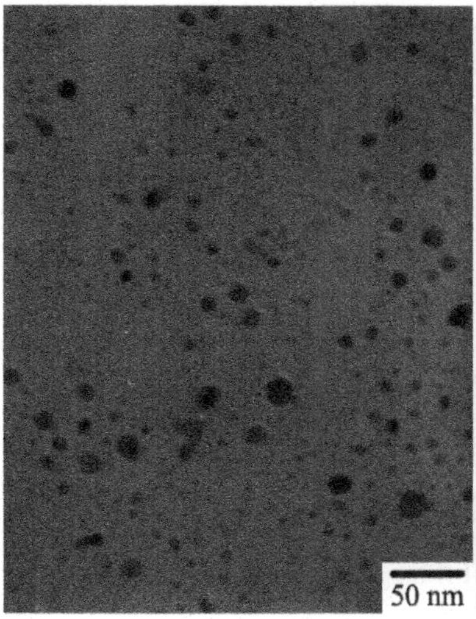

(i)

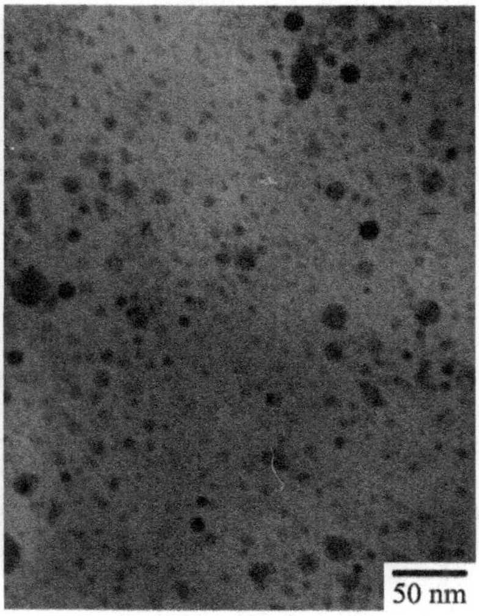

(j)

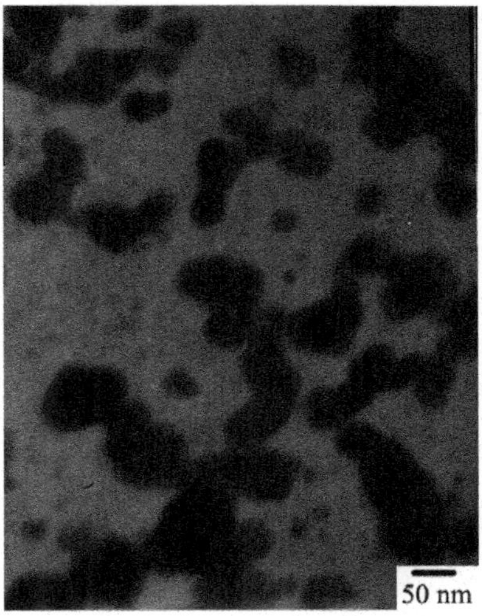

(k)

Fig. 5. The micrographs of SPAEKS membranes: (a) SPAEKS-M1-10; (b) SPAEKS-M1-20; (c) SPAEKS-M1-30; (d) SPAEKS-M1-40; (e) SPAEKS-M1-50; (f) SPAEKS-M1-60; (g) SPAEKS-M2-20; (h) SPAEKS-M2-30; (i) SPAEKS-M2-40; (j) SPAEKS-M2-50; (k) SPAEKS-S-50. (Continuation)

3.6 Proton conductivity

Proton conductivity is a particularly important parameter for proton exchange membranes. Without sufficient proton conductivity, a membrane will never be applicable to H_2/O_2 fuel cells or direct methanol fuel cells. Proton conductivity may have difference results with different measured approaches and instruments. In our work, Nafion 117 was selected as a reference to compare the proton conductivity of membrane. The proton conductivity (σ) of membrane was determined by AC impedance spectroscopy over a frequency range of 10 Hz-1 MHz. The membrane in acid form was immersed in deionized water for at least 24 h prior to the test. Conductivity measurements of fully hydrated membranes were carried out with the cell immersed in liquid water (water resistance is 18.2 MO; 100% RH). The resistance value related to the proton conductivity of membranes was determined from the low intersect of the high frequency semicircle on a complex impedance plane with the real axis. High temperature conductivity measurements were performed in water vapour. This experimental set up allowed the membranes to equilibrate with saturated water vapour at the desired temperature. The temperature was controlled by a wrap-around resistance heater and feed back temperature controller. The proton conductivity of membranes was calculated from the resistance based on the following formula:

$$\sigma = l/RS$$

Where l is the distance between the two electrodes, R is the membrane resistance, and S is the cross-sectional area of membrane. For a series of sulfoanted copolymers membrane, it can be found from measurement results that the proton conductivity increased with DS and temperature. This might be attributed to the increasing of size and density of ion domains

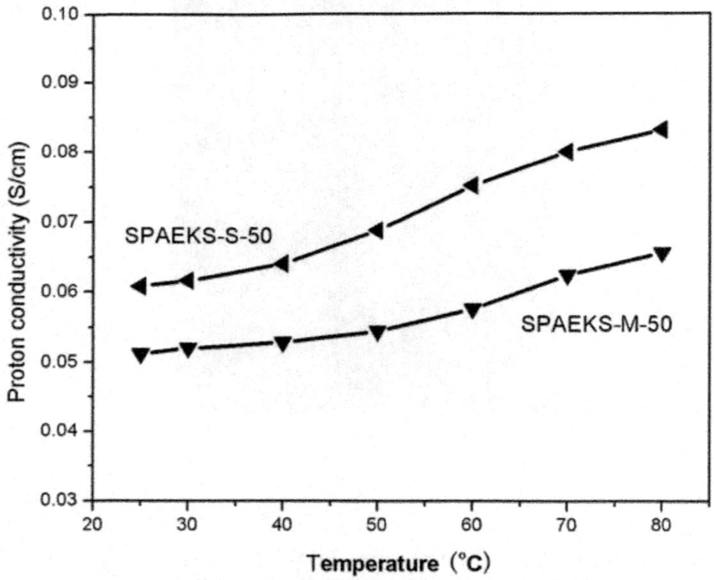

Fig. 6. The proton conductivities of SPAEKS-M-50 and SPAEKS-S-50 membranes from 25 °C to 80 °C.

when DS increases, which can be concluded from the structure and morphology of membrane (TEM, SAXS and AFM). Also, the speed of H^+ transmitting in membrane was fastened with increasing temperature, which greatly improved the proton conductivity. The proton conductivity of membrane was closely related to their microstructure. In order to improve proton conductivity, McGrath and co-works have synthesized a series of block sulfonated copolymers, which shows higher proton conductivity than that of random sulfonated copolymers at the same of degree of sulfoantion and IEC values [52-54]. This is attributed to the different distribution of ion domains in membranes. In this paper, the proton conductivity of SPAEKS-M membrane was lower than that of SPAEKS-S at the same Ds. This is because that the existence of the two distinct hydrophobic and hydrophilic regions may lead to the micro-phase separation structure. Also, the concentrated sulfonic groups may provide much larger proton transport channels, which are beneficial to enhance the proton conductivity of the membranes. However, for the random sulfonated copolymers, the sulfonic groups were dispersed throughout the copolymers matrix and not easily phase separated, which may lead to the relativity lower proton conductivity than the side chain ones.

3.7 Methanol permeability

Methanol permeability of membranes is one of the most significant roles for DMFC, which directly affects the work efficiency of a DMFC. The high methanol diffusion coefficient of Nafion membrane in the DMFC application is one of the most drawbacks, which limits its further development in DMFC applications. The methanol permeability of membrane was expressed by methanol diffusion coefficient, which was measured using a cell that consists of two half cells divided by the membrane. Pure methanol was placed on one side of the diffusion cell (A cell) and water was placed on the other side (B cell). In order to ensure solution uniformity, magnetic stirrer was used in each compartment. The concentration of the methanol in the cell B was determined by a gas chromatograph. The methanol diffusion coefficient was calculated by following formula.

$$C_{B(t)} = \frac{A}{V_B}\frac{DK}{L}C_A\left(t - t_0\right)$$

Where DK is the methanol diffusion coefficient (cm^2/s), $C_{B(t)}$ is the concentration of methanol in B cell (mol L^{-1}), CA is the concentration of methanol in A cell (mol L^{-1}), A is the membrane area (cm^2) and L is thickness of membrane (cm).

The methanol diffusion coefficients of sulfoanted aromatic polymers were lower than that of Nafion membrane and increased with degree of sulfonation increasing. This may be becaused that the sulfonic groups in Nafion distribute on the side chains of polymers, which may easily form a phase-separation structure. Furthermore, these sulfonic groups can provide much larger methanol transport channel than aromatic polymers.

	Proton conductivity (S/cm)		Methanol permeability
	25°C	80°C	(cm^2/s)
SPAEKS-M-50	0.052	0.062	4.65×10^{-7}
SPAEKS-S-50	0.061	0.083	7.36×10^{-7}

Table 3. The data of SPAEKS-M and SPAEKS-S membrane.

It can be seen from Table 3 that the methanol permeability of SPAEKS-S-50 was higher than that of SPAEKS-M-50 ($4.65\times10^{-7}cm^2/s$). This may be because the sulfonated groups of SPAEKS-M copolymers are dispersed throughout the copolymer matrix and not easily lead to phase separation. While sulfonate groups of SPAEKS-S copolymers are located on the side chains of polymers, which may easily form phase separation, thus providing much larger transport channel than SPAEKS-M copolymers. The separation of copolymer chains generated unoccupied volume in the form of pores lined with sulfonic acid groups and formed transport pathways, which should absorb more methanol.

4. Conclusions

This paper has attempted to introduce the basic research method for evaluating the performance of proton exchange membrane and reveal the relationship between structure and properties by TEM. In our previous works, a series of random sulfoanted poly(arylene ether sulfone)s, sulfonated poly(arylene ether ketone)s and sulfonated poly(aryle ether ketone sulfone)s were synthesized by the direct copolymerization. These sulfonated copolymers membrane exhibited relatively high proton conductivity, good thermal-oxidative stability and mechanical performance. The newly synthesized side chain sulfonated poly(ether ether ketone) copolymer membranes showed higher proton conductivity than that of random sulfonated poly(ether ether ketone) membranes at the same of IEC values, which was attribute to a greater degree of phase separation. All these results relating to their micro-structure and morphology have been characterized by TEM. It is assured that new proton exchange membranes materials with low cost, high proton conductivity, low water uptake and methanol diffusion coefficient, good mechanical performance and thermal-oxidative stability, long life-span and good fuel cell performance will appear at short time.

5. References

[1] Hickner, M. A.; Ghassemi, H.; Kim, Y. S.; Einsla, B. R.; McGrath, J. E. Chem. Rev. 2004, 104, 4587.
[2] Harrision, W. L.; Hicker, M. A.; Kim, Y. S.; McGrath, J. E. Fuel Cell 2005, 5, 201.
[3] Hickner, M. A.; Pivovar, B. S. Fuel Cells 2005, 5, 213.
[4] Borroni-Bird, C. E. J. Power Sources 1996, 1, 33.
[5] Holze, R.; Ahn. J. J. Membr.Sci. 1992, 73, 87.
[6] Xu, G. Polymer 1993, 25, 397.
[7] Lee, E. M.; Thomas, R. K.; Burgess, A. N.; Barnes, D. Y.; Soper, A. K.; Rennil, A.R. Macromolecules 1992, 25, 3106.
[8] Kim, D. S.; Shin, K. H.; Park, H. B.; Chung, Y. S.; Nam, S. Y.; Lee, Y. M. J. Membr. Sci. 2006, 278, 428.
[9] Wang, Z.; Ni, H. Z.; Zhao, C. J.; Li, X. F.; Zhang, G.; Shao, K.; Na, H. J. Membr. Sci. 2006, 285, 239.
[10] Li, Y. X.; Wang, F.; Yang, J.; Liu, D.; Roy, A.; Case, S.; Lesko, J.; McGrath, J. E. Polymer 2006, 47 ,4210.
[11] Li, X .F.; Zhang, G.; Xu, D.; Zhao, C. J.; Na, H. J. Power Sour. 2007, 165, 701.
[12] Kim, D. S.; Guiver, M. D. J. Polym. Sci. Part A Polym. Chem. 2008, 46, 989.

[13] Liu, B. J.; Robertson, G. P.; Kim, D. S.; Guiver, M. D.; Hu, W.; Jiang, Z. H. Macromolecules 2007, 40, 1934.
[14] Sun, Y. M.; Wu, T. C.; Lee, H. C.; Jung, G. B.; Guiver, M.; Gao, D. Y.; Liu, Y. L.; Lai, J. Y. J. Membr. Sci. 2006, 265, 108.
[15] Shang, X. Y.; Li, X. H.; Xiao, M.; Meng, Y. Z. Polymer 2006, 47, 3807.
[16] Li, Y. H.; Jin, R. Z.; Cui, Z. M.; Wang, Z.; Xing, W.; Qiu, X. P.; Ji, X. L. Gao, L. X. Polymer 2008, 49, 715.
[17] Zhang, G. M.; Guo, X. X.; Fang, J. H.; Chen, K. C.; Okamoto, K. I. J. Membr. Sci. 2009, 326, 708.
[18] Li, N. W.; Cui, Z. M.; Zhang, S. B.; Li, S. H. Polym. Sci. Part A Polym. Chem. 2008, 46, 2820.
[19] Zhang, F.; Lee, N. W.; Cui, Z. M.; Zhang, S. B.; Li, S. H. J. Membr. Sci. 2008, 314, 24–32.
[20] Qing, S. B.; Huang, W.; Yan, D. Y. Reactive and Functional, Polymers 2006, 66, 219.
[21] Qing, S. B; Huang, W.; Yan, D. Y. Euro. Polym. J. 2005, 141, 1589.
[22] Zhang, H. Q.; Li, X. F.; Zhao, C. J.; Fu, T. Z.; Shi, Y. H.; Na, H. J. Membr. Sci. 2008, 308, 66.
[23] Yu, X.; Roy, A.; Dunn, S.; Badami, A. S.; Yang, J.; Good, A. S.; McGrath, J.E. J. Polym. Sci. Part A Polym. Chem. 2009, 47, 1038.
[24] Badamil, A. S.; Roy, A.; Lee, H. S.; Li, Y. X.; McGrath, J. E. J. Membr. Sci. 2009, 328, 156.
[25] Roy, A.; Yu, X.; Dunn, S.; McGrath, J. E. J. Membr. Sci. 2009, 327, 118.
[26] Li, X. F.; Liu, C. P.; Xu, D. ; Zhao, C. J.; Wang, Z.; Zhang, G.; Na, H.; Xing, W. J. Power Sour. 2006, 162, 1.
[27] Zhao, C. J.; Li, X. F.; Wang, Z. ; Dou, Y ; Zhong, S. L ; Na, H. J. Membr. Sci. 2006, 280, 643.
[28] Li, X. F.; Chen, D. J.; Zhao, C. J.; Wang, Z.; Lu, H.; Xu, D.; Na, H. J. Membr. Sci. 2006, 275, 134.
[29] Harrision, W. L.; Wang, F.; Mecham, J. B.; Bhanu, V. A.; Hill, M.; Kim, Y. S.; McGrath, J. E. J. Polym. Sci. Part A: Polym. Chem. 2003, 41, 2264.
[30] Li, X. F.; Zhao, C. J.; Lu, H.; Wang, Z.; Na, H. Polymer 2005, 46, 5820.
[31] Li, X. F.; Liu, C. P.; Lu, H.; Zhao, C. J.; Wang, Z.; Xing, W.; Na, H. J. Membr. Sci. 2005, 255, 149.
[32] Wang, Z.; Li, X. F.; Zhao, C. J.; Ni, H. Z.; Na, H. J. Power Sour. 2006, 160 , 969.
[33] Wang, Z.; Li, X. F.; Zhao, C. J.; Ni, H. Z.; Na, H. J. App. Polym. Sci. 2007, 104, 1443.
[34] Wang, Z.; Ni, H. Z.; Zhao, C. J.; Zhang, M. Y.; Na, H. J. App. Polym. Sci. 2009, 112(2), 858.
[35] Gil, M.; Ji, X. L.; Li, X. F.; Na, H.; Hampsey, J. E.; Lu, Y. F. J. Membr. Sci. 2004, 234, 75.
[36] Wang, Z.; Ni, H. Z.; Zhao, C. J.; Li, X. F.; Fu, T. Z.; Na, H. J. Polym. Sci. Part B: Polym. Phy. 2006, 44, 1967.
[37] Wu, X. M.; He, G. H.; Gu, S.; Hu, Z. W.; Yao, P. J. J. Membr. Sci. 2007, 295, 80.
[38] Chuang, S. W.; Chung, S. L.; Liu, Y. H. J. Membr. Sci. 2007, 305, 353.
[39] Xiong, Y.; Lin, Q. L.; Zhu, A. M.; Huang, S. M.; Zeng, Q. H. J. Power Sour. 2009, 186, 328.
[40] Yang, Y. S.; Shi, Z. Q.; Holdcroft, S. Macromolecules 2004, 37, 1678.
[41] Eisenberg, A. Macromolecules 1970, 3, 47.
[42] Fujimura, M.; Hashimoto, T. J.; Kawai, H. Macromolecules 1981, 14, 1309.
[43] Fujimura, M.; Hashimoto, T. J.; Kawai, H. Macromolecules 1982, 15 , 136.
[44] Mo, Z. S.; Zhang, H. F. Science Publishing Company, China, October 2003, pp. 307–311.

[45] Gebel, G.; Moore, R. B. Macromolecules 2002, 33, 4850.
[46] Gebel, G.; Lambard, J. Macromolecules 1997, 30, 7914.
[47] Yang, Y. S.; Shi, Z. Q.; Holdcroft, S. Macromolecules 2004, 37, 1678.
[48] Wang, F.; Hickner, M.; Kim, Y. S.; Zawodzinski, T. A.; McGrath, J. E. J. Membr. Sci. 2002, 197, 231.
[49] Ding, J. F; Chuy, C.; Holdcroft, S. Adv. Funct. Mater. 2002, 12, 389.
[50] Kim, Y. S.; Wang, F.; Hickner, M.; Mccartnry, S.; Hong, Y. T.; Harrision, W.; Zawodzinski, T. A.; McGrath, J. E. J. Polym. Sci. Part B: Polym Phys. 2003, 41, 2816.
[51] Wang, F.; Hickner, M.; Kim, Y. S.; Zawodzinski, T. A.; McGrath, J. E. J. Membr. Sci. 2002, 197, 231.
[52] Lee, S.; Roy, A.; Lane, O.; Dunn, S.; McGrath, J. E. Polymer 2008, 49, 715.
[53] Li, Y. X.; Roy, A.; Badami, A. S.; Hill, M.; Yang, J. J. Power Sour. 2007, 172, 30-38.
[54] Lee, H. S.; Badami, A. S.; Roy, A.; McGrath, J. E. J. Polym. Sci. Part A Polym. Chem. 2007, 45, 4879.
[55] Kim, H. K.; Chang, H. J. Membr. Sci. 2007, 288, 188.
[56] Hill, M. L.; Kim, Y. S.; Einsla, B. R.; McGrath, J. E. J. Membr. Sci. 2006, 283, 102.
[57] Fu, Y. Z.; Manthiram, A.; Guiver, M. D. Electrochemistry Communications 2006, 8, 1386.
[58] Li, T.; Yang, Y. J. Power Sour. 2009, 187, 332.
[59] Xiong, Y.; Lin, Q. L.; Zhu, A. M.; Huang, S. M.; Zeng, Q. H. J. Power Sour. 2009, 186, 328.
[60] Zhang, Y.; Zhang, H. M.; Bi, C.; Xiao, X. B. Electrochimica Acta. 2008, 53, 4096.
[61] Yamada, M. I. J. Phys. Chem. B 2004, 108, 5522.
[62] Zhao, C. J.; Wang, Z.; Lin, H. D.; Shao, K.; Fu, T. Z.; Zhong, S. L.; Na ,H. Polymer 2007, 48, 3090.
[63] Swier, S.; Shaw, M. T.; Weiss, R. A. J. Membr. Sci. 2006, 270, 22.

19

Ceramic-Metal Joining Using Active Filler Alloy-An In-Depth Electron Microscopic Study

Abhijit Kar[1*] and Ajoy Kumar Ray[2]

[1]J.B.Centre of Excellence, Jagadis Bose National Science Talent Search,
[2]Material Science & Technology Division,
National Metallurgical Laboratory (CSIR),
India

1. Introduction

Joining of materials provides a means of fabricating structures, where difficulty is encountered to make one piece directly. Very often joining can be considered to be less expensive than making single piece structure for many intricate shaped components. Brazing is one of the most important techniques for joining various materials especially ceramics. To fabricate near net shape joined component or to make prototypes of intricate shapes, joining of ceramics to ceramics/metals, brazing has been considered as the most frequently used technique. Due to their excellent high temperature strength, resistance to corrosion and wear, application of ceramics in structural components, has received extensive attention in recent decades. However difficulties on joining ceramics with metals restrict their use in many occasions. The ability to produce a reliable ceramic-ceramic/metal and composite joint is a key enabling technology for many productions, prototype and advanced developmental items and assemblies. Thus it becomes an interesting challenge to the researchers for ceramic-metal joining.[1,2] Amongst several ceramic joining processes, *active metal brazing* is one of the most extensively used joining techniques for metal-ceramic joining. In this process, bonding is promoted by the use of an active filler alloy. The active filler alloy containing a small amount of an active element which is capable of reacting with the ceramic substrate facilitates joining.[3-12]. Characteristics of filler alloy play a significant role in obtaining unique joining properties. Filler alloy should have the liquidus temperature, below the melting point of the substrate to be joined and also must be capable of producing joint at a temperature where the properties of base materials are not degraded

Joining of two materials, whether homogeneous or heterogeneous almost always causes changes in the microstructure and mechanical properties in the vicinity of joint. A better understanding of the microstructure and mechanical property relationships of the braze joints will give valuable feed back to the materials developmental activities both in conventional and new material areas.[13] In case of metal-ceramic joining, interfaces exhibit

* Corresponding Author

abrupt discontinuity of properties, e.g. crystallographic, electronic, mechanical, thermodynamic and thermo-chemical.[3,14-16] Successful joining depends mainly on the interface characteristics; namely its mechanical behaviour, which in turn is highly dependent on the microstructural morphology and on how the discontinuity of the properties is accommodated by the interface. Therefore, the knowledge of the reaction product nature, their distribution and its chemical and mechanical characteristics are of fundamental interest.

2. Mechanism of ceramic brazing

Brazing of ceramic substrate to a metallic substrate is less studied. In case of joining ceramics, many attempts have been made to improve the wettability between ceramic to filler alloy by reducing contact angle (θ) between *solid-liquid*. Metalizing process improves the wettability of ceramic surfaces with conventional filler alloys. To ensure the production of reliable ceramic to metal seals, most of the cases ceramic surfaces are coated with Ni, Cu and other metals for metallization. The coating is usually done by electrolytic plating, gas-phase precipitation, thermal spraying, plasma spraying, ionic plating, electron and laser-beam techniques.[17,18] The widely used method for joining ceramic to metal is the multi-step moly-manganese process, where a moly-manganese coating is applied on the ceramic surface to induce wetting properties and the assembly is joined by brazing,[19-23].

One of the basic problems in brazing of ceramics without metallization lies in their poor wetting by the conventional brazing alloys. The wetting behaviour of the ceramic surfaces by the liquid metals has long been the subject of study. It is well known that in many systems the wetting process depends on the chemical reactions occurring at the solid-liquid interface. Normally, the active filler alloys used for direct joining contain reactive elements such as Ti, Zr, Ta etc. These elements react with the ceramic and thus provide a good joint.[24] The mechanism may be describe that at elevated temperatures, the active element reacts with the non-metallic component of the ceramic to form a complex interfacial layer that is wettable by other constituents of the brazing alloy.

Many researchers used laminated filler alloys [25-27] to join ceramics. In this process the active element layer (e.g. Ti foil) in general is kept at the sandwich condition between two layers of eutectic alloy (e.g. Ag-Cu). In general this is a suitable process where the gap between the two substrates are substantial, Since there are at least three layers, in all practical purposes it is to use very difficult to precisely maintain the dimension of the foils.

Considering all the above mentioned limitations now a day the most widely used method is active metal brazing technique. In this chapter readers should consider active filler metal and active filler alloy is synonymous

3. Selection of the filler alloy

Some of the Ag-based active filler metals are ductile and adaptable to braze materials such as Al_2O_3, Si_3N_4, SiC, partially stabilized zirconia (PSZ) as well as many other refractory ceramics.[28,29] One of the most commonly used active constituents is Ti, this can facilitate wetting of the majority of engineering ceramics. The reactivity and wetting behaviour of filler alloys are considerably increased by small additions of active metal such as Ti or Zr. The driving force for the reactivity is the high oxidation potential of Ti, which results in a redox reaction with the ceramic. Merely increasing the concentration of titanium in the alloy prone to increase the liquidus temperature and the likelihood of alloy embrittlement by the formation of excessive amount of intermetallics / oxides of titanium. Therefore, the amount of Ti in the Ag-Cu base alloy should be optimized before application.

The choice of the reactive element depends on several other factors [30]. Fox and Slaughter had developed a few active metal alloys, like Ti-28Ag-4Be and Ti-49Cu-2Be for joining graphite to metal and these are mainly used for nuclear reactor technology.[31] Mizuhara et al.[32] and Xu et al.[33] had prepared Ag-Cu eutectic braze alloy containing up to 5% Ti. These alloys were prepared either as a tri-foil, consisting of a titanium sheet that is roll-clad with the silver-copper alloy, or as a silver-copper alloy wire with a titanium core. Similarly in certain processes, Ti vapour coating on the ceramic surface prior to brazing improved their wetting characteristics and permitted braze filler metals to adhere strongly to the ceramics. [34, 35]

Currently the active filler alloys have been identified as potential materials for ceramic brazing, however the details about the identification of the phases in the reaction layers and the extent to which microstructure in the reaction layers influences wetting, adhesion properties and bond strength, are still largely unresolved.[36]

3.1 Experimental procedure

In the current chapter we have considered TiCuSil *[97(72Ag28Cu) 3Ti]* as the active filler alloy in order to braze Al_2O_3 with 304SS. Active filler alloy was prepared in melting and casting route. Conventional chemical analysis shows that the concentration of Ag, Cu, and Ti in the brazing alloy is 71.8 wt%, 25.1 wt% and 3.1 wt% respectively. Substrates and the alloys were polished and cleaned ultrasonically using ethanol and acetone. The roughness of the cleaned substrates was measured by a profilometer (Taylor Hobson precision, Taylsurf Series 2) with a resolution of 16nm. The dimensions and surface roughness of the materials used for joining were as follows:

Al_2O_3 = 10(l) x 8(b) x 5(t) mm;
304 SS = 10(l) x 8(b) x 5(t) mm;
filler alloy = 9(l) x 7(b) x 0.4(t) mm;
roughness (Ra) of the Al_2O_3 was 0.46 μm and that of the SS was 0.16 μm.

The filler alloy was sandwiched between the alumina and the stainless steel substrates. The assembly was placed in a graphite resistance furnace (ASTRO, Thermal Inc., USA). A fixed load of ~8MPa was applied over the sample to keep the assembly aligned properly during brazing. This ensured a proper contact between the surface of the substrates and the filler alloy. This load also facilitates some interfacial reactions. The furnace was heated at the rate of $6^{o}C$/min. till it reached the maximum brazing temperature of $1000^{o}C$. Before the cooling starts it was kept at $1000^{o}C$ for 15 minutes. The cooling rate was maintained at $3^{o}C$/min. upto $200^{o}C$ and thereafter it was furnace cooled. Before heating, the furnace was purged with argon gas (IOLAR I), and throughout the brazing cycle ~50-millitorr pressure was maintained through a rotary pump.

It is also important to note that, microstructures presented and described in this chapter are mostly from the samples brazed at 1000°C. It was found that for the above considered specifications of the substrates and fillers, brazing carried out at 1000°C produced the best result.

4. Interfacial reaction products

Joining of ceramics requires reaction between the substrate and the filler alloy to form an interface. The formation of reaction products provides the bond strength for the brazed sample. The reaction depends on the wetting behaviour of the ceramic and the filler alloy.

The nature of the reaction product depends on several factors. Most importantly concentration of the filler alloy used and the reactivity of the filler alloy with the substrates (ceramic /metal) apart from other different physical factors such as time, temperature profile, atmosphere ect. The reaction products that are formed mostly the active element bearing phases. Often the quantity of these phases is also affect the quality of the brazed joint. Thus it is the most vital factor in terms of qualitative and quantitative analysis of these phases, in order to determine the quality of the overall brazed interface.

Ti is most widely used as an active element of different filler alloys. The reactive Ti wets the ceramic/alumina and metal surfaces by reducing them, hence the reaction products are formed which facilitate the further bonding between the two substrates. [37] Wetting depends on the interfacial reactions and the thermodynamic activity of the Ti (a_{Ti}). The a_{Ti} depends on the mole fraction of the Ti (χ_{Ti}) in the melt. [38]

For alumina brazing, once Ti diffuse towards the alumina substrate, it reduces the alumina, hence forms the different reaction products. Sequentially the reactions that are believed to be taken place between the alumina and the active filler alloy are as follows,

$$4Al_2O_3 + 6Ti = 8Al + 6TiO + 6O \tag{1}$$

$$2Al + Ti + 5O = Al_2TiO_5 \tag{2}$$

$$3Ti + 3Cu + O = Cu_3Ti_3O \tag{3}$$

It is interesting to note that all the reaction products found at the alumina interface is obviously Ti bearing, thus the diffusion zone is considered with respect to the variation of the concentration gradient of the Ti across the interface which has been observed in EPMA concentration penetration profile (CPP) [39-46].

The formation and growth of the reaction layer are controlled mainly by diffusion of titanium through the continuous reaction layer. From the EPMA-CPP of Al_2O_3 to the 304SS, it is observed that the Ti diffusion mostly follow a Gaussian path [39]. By transmission electron microscopy (TEM) analysis two different phases like Cu_3Ti_3O (Figure 1. TEM) and Al_2TiO_5 (Figure 2. TEM has been confirmed apart from TiO (Figure 3. TEM) on the interface adjacent to alumina In fact the presence of these phases is more feasible as Ti has better chemical affinity towards Al, oxygen and Cu rather than Ag. It is interesting to note that so far no other report for the existence of Al_2TiO_5 phase within the interface adjacent to Al_2O_3.

Presence of these phases has been proved to be beneficial for the interface design Table 1. (Thickness and strength). Exhibits how reaction products and their thickness changes with temperature and its effect on mechanical property.

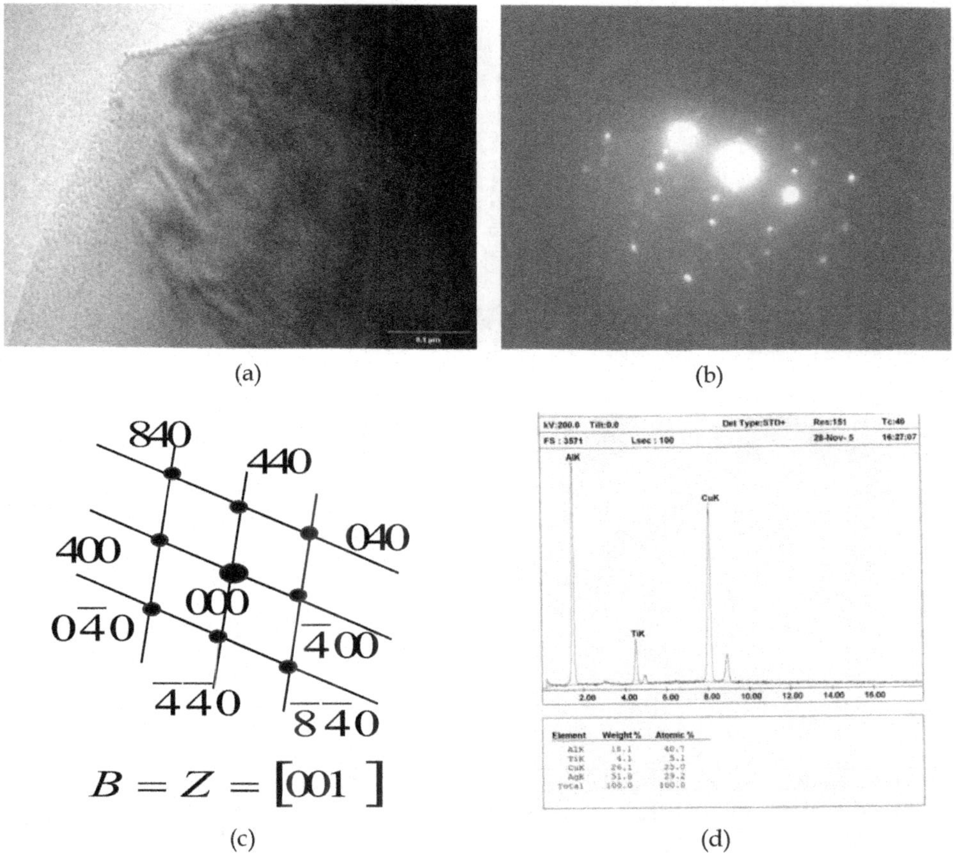

	Observed/ Experimental lattice spacing values (d)	Standard values at JCPDS file	*(hkl)*	α_1 experimental from the SADP	α_1 Theoritical	α_2 experimental from the SADP	α_2 Theoritical
R_1	$d_1 = 2.79$	2.814	400	34°	36°	29°	31°
R_2	$d_2 = 1.26$	1.258	840				
R_3	$d_3 = 2.0$	1.99	440				

Fig. 1. (a) TEM micrograph (b) SAD pattern (c) Schematic diagram of the diffraction array (d) corresponding quantitative EDS analysis and the alumina interface; confirms the presence of cubic $\mathbf{Cu_3Ti_3O}$ phase in the interface.

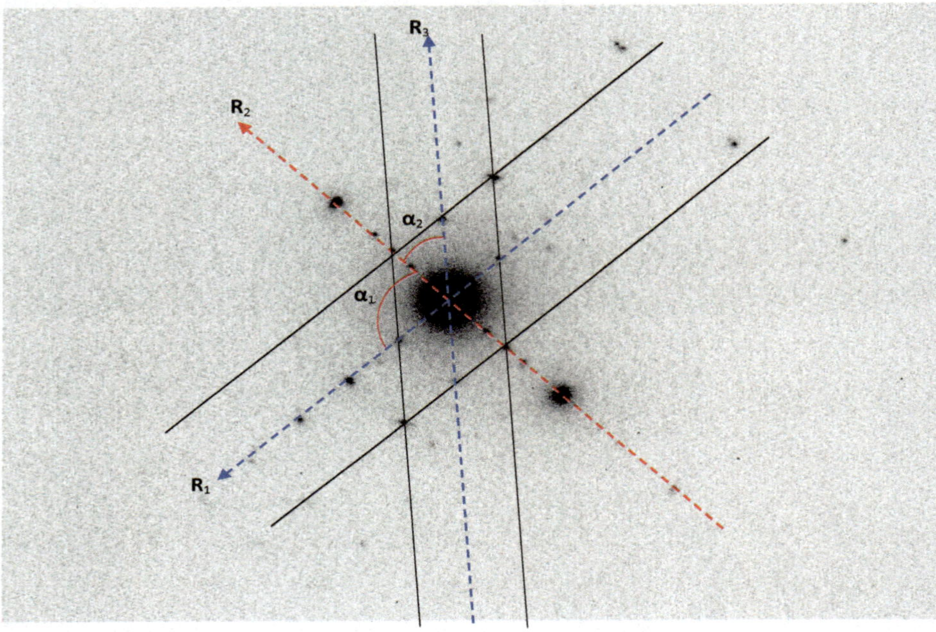

	Observed/ Experimental lattice spacing values (d)	Standard values at JCPDS file	(hkl)	α_1 experimental from the SADP	α_1 Theoritical	α_2 experimental from the SADP	α_2 Theoritical
R_1	$d_1 = 1.76$	1.784	$\bar{0}25$	81°	79.8°	44.5°	46°
R_2	$d_2 = 1.544$	1.585	222				
R_3	$d_3 = 1.308$	1.305	$24\bar{3}$				

Fig. 2. Selected Area Diffraction (SAD) pattern and Schematic diagram of the diffraction array; table below exhibits corresponding lattice parameters, theoretical and experimental values; confirms the presence of orthorhombic $\mathbf{Al_2TiO_5}$ phase in the interface. Here, the diffraction array has been marked ($\mathbf{R_1}$, $\mathbf{R_2}$, $\mathbf{R_3}$) and angles have been denoted by α_1, α_2

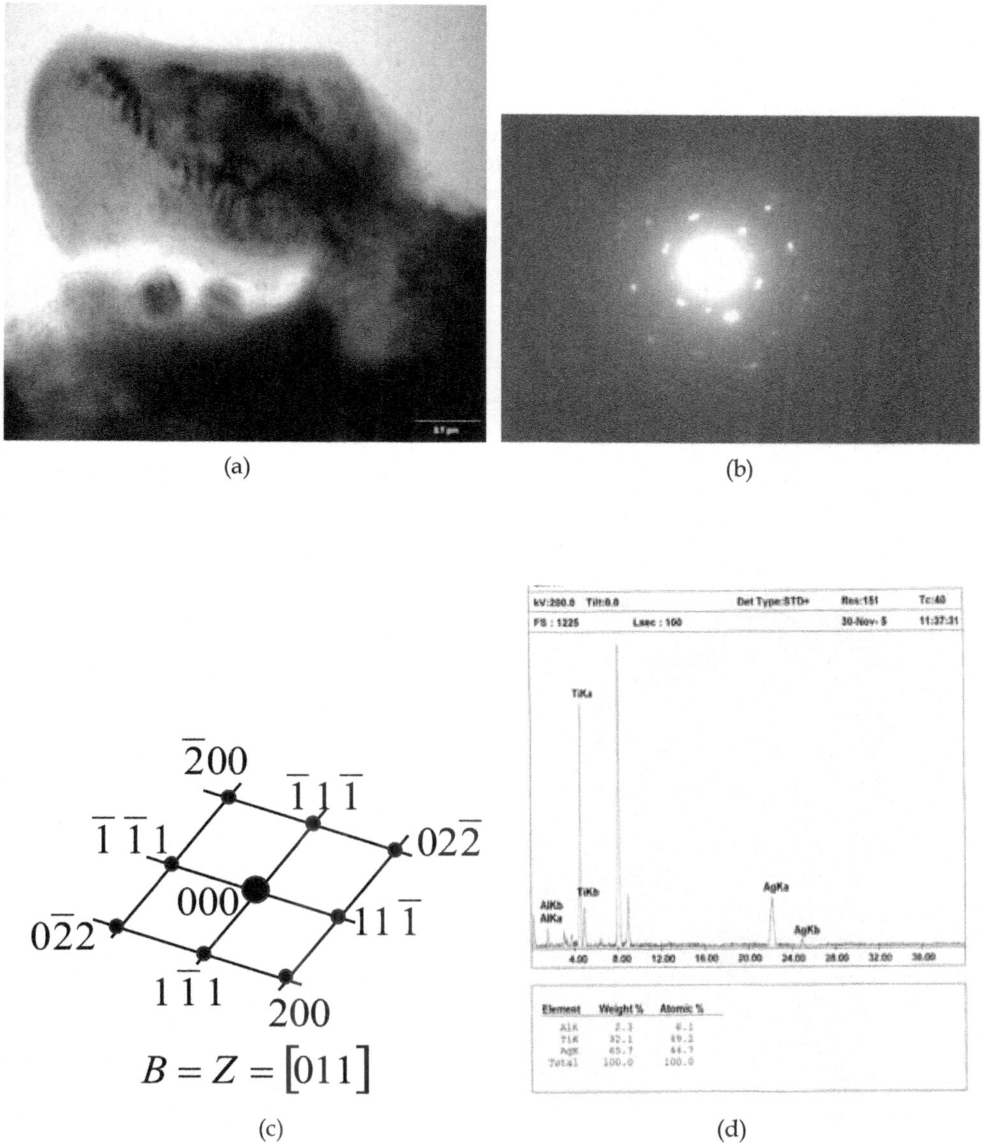

Fig. 3. (a) TEM micrograph (b) SAD pattern (c) Schematic diagram of the diffraction array (d) corresponding quantitative EDS analysis and the alumina interface; confirms the presence of cubic **TiO** phase in the interface.

Location	Phase/layer of reaction products	**Average Atomic Number**	Approximate Thickness **(µm)**			
			800 °C	**900 °C**	**1000 °C**	**1100 °C**
Ceramic Substrate	Al_2O_3	10	--	--	--	--
Alumina Interface	TiO (L_1)	15	0.8	3.0	2.3	Not Distinguishable
	Al_2TiO_5 (L_2)	11	--	Scattered Islands	6.3	
	Cu_3Ti_3O (L_3)	23	3.0	4.5	5.7	
	Total Thickness (µm)	--	3.8	7.5	14.7	4.2
Residual Filler alloy	Ag-Cu-Ti	--	484	350	280	85
SS Interface	$Fe_{35}Cr_{13}Ni_3Ti_7$ (L_4)	25.17	3.0	4.0	5.5	3.9
	FeTi (L_5)	24	4.8	7.5	9.5	7.7
	Total Thickness (µm)	--	7.8	13.7	15	11.6
Metal Substrate	Major constituent, Fe	26	--	--	--	--
Total Joint thickness	--	--	495	370	310	215
Shear strength of the joint (MPa)	--	--	**57**	**64**	**94**	**46**

Table 1. Thickness of different reaction product layers, residual filler, total joint, and shear strength of the brazed samples prepared at different temperatures.

Kar et al. [40] showed that in case of Al_2O_3-304SS brazed couple, it consists of two different interfaces. One is alumina interface and the other is SS interface. This complies with the fact that the reaction products formed in this interface i.e. TiO, Cu_3Ti_3O and Al_2TiO_5 [13,40] exhibits gradually decrease in hardness. The SS interface on the other hand found to be consists of two different reaction product layers, e.g. FeTi and $Fe_{35}Cr_{13}Ni_3Ti_7$ [47,40] all these phases have been identified and confirmed by X-ray diffraction analysis (Figure 4 (a, b). XRD) and transmission electron microscopy (Figure 5. TEM FeTi) and Figure 6. TEM $Fe_{35}Cr_{13}Ni_3Ti_7$).. In general, the oxides are more brittle than intermetallics; hence the hardness of the interface at the ceramic side is more than that of, at the metal side. As a result, ceramic interface becomes more susceptible to thermal or mechanical shock.

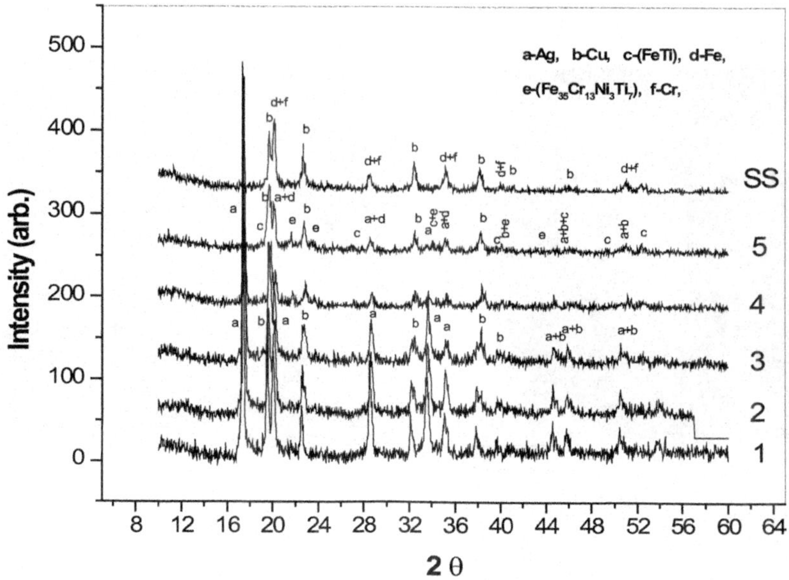

Fig. 4. (a) X-Ray diffraction patterns (XRD) of Al_2O_3 -304 SS brazed joint. Some of the most relevant diffraction patterns taken from the SS interface are depicted. MoK_α radiation

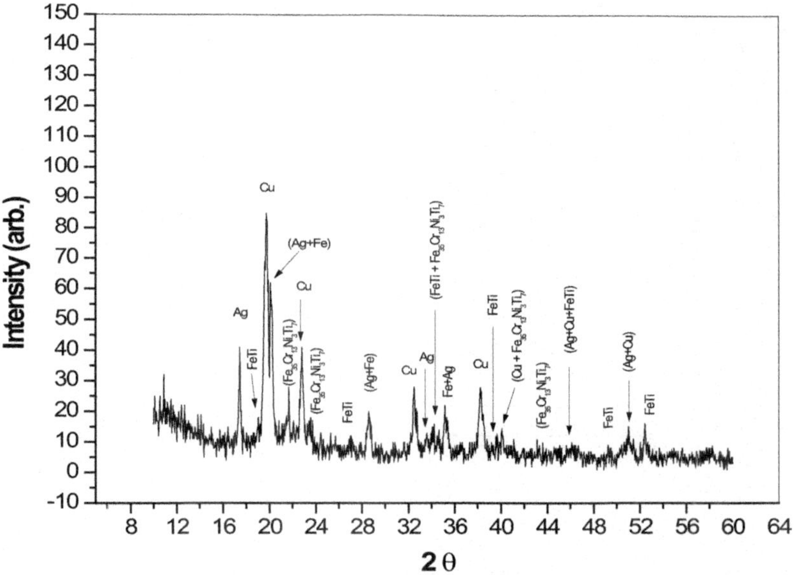

Fig. 4. (b) X-ray diffraction patterns (XRD) of Al_2O_3– SS brazed joint: pattern '5' of the above figure adjacent to SS interface. MoK_α radiation, Figure identifies the presence of FeTi and $Fe_{35}Cr_{13}Ni_3Ti_7$

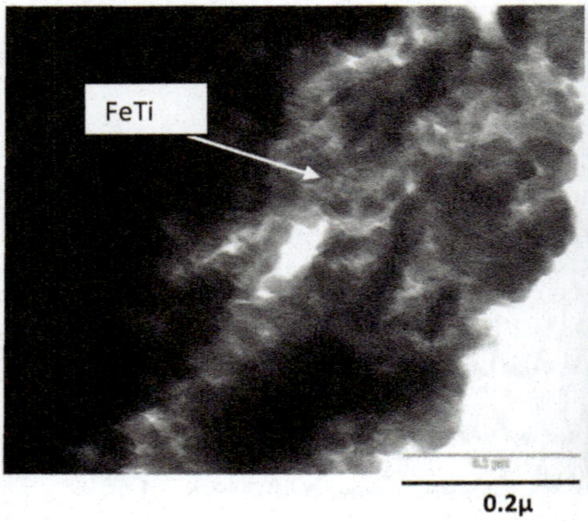

(a)

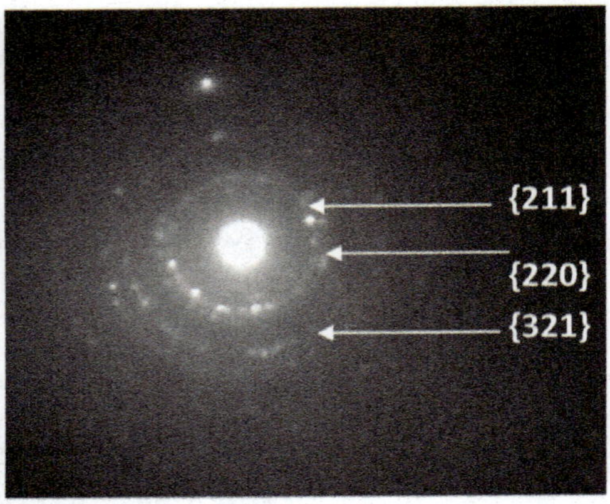

(b)

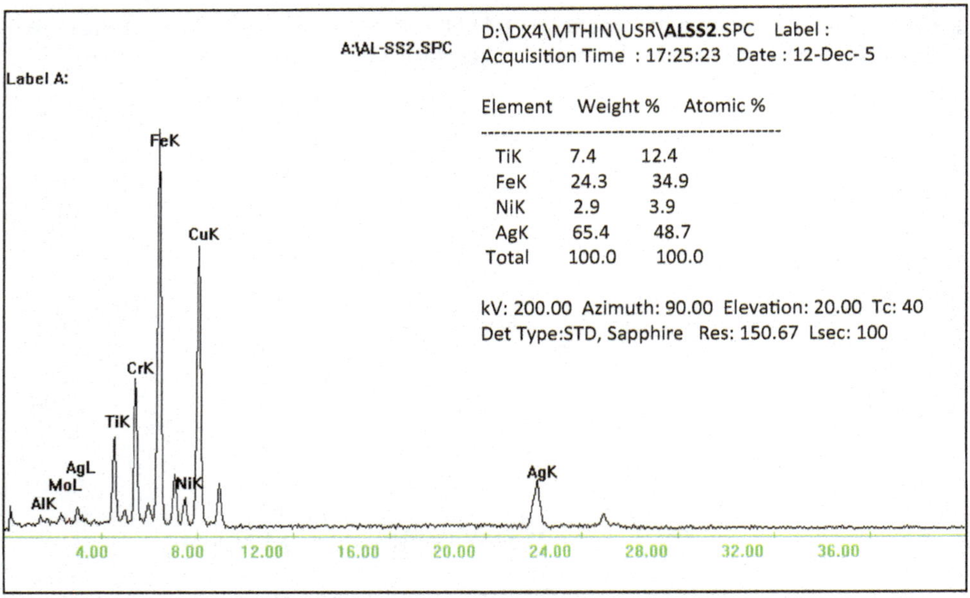

(c)

Fig. 5. TEM micrograph and SAD patterns obtained from the SS interface (a) TEM micrograph and (b) diffraction pattern of **FeTi** phase. (c) Exhibits the EDS analysis obtained from the area from which the diffraction pattern has been recorded. Spots marked by circle in the SAD patterns are from Ag with zone axis B=z= 122

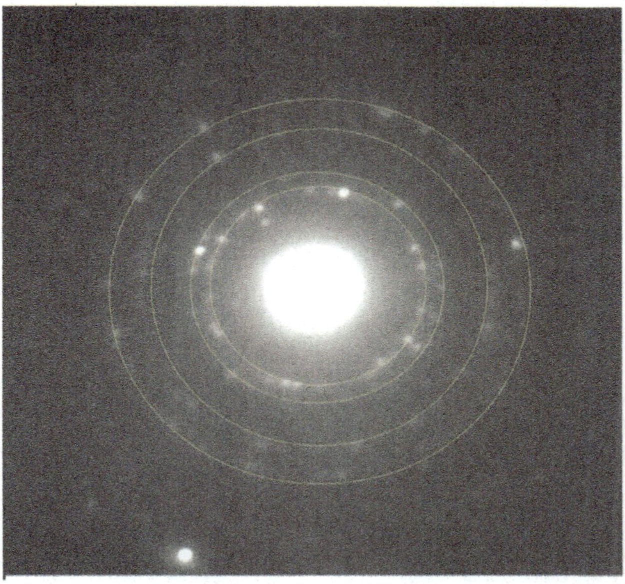

(a)

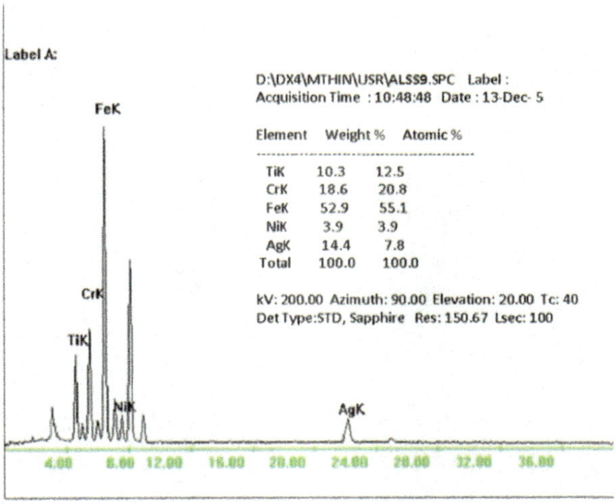

(b)

Fig. 6. TEM- EDS analysis of the cubic $\mathbf{Fe_{35}Cr_{13}Ni_{3}Ti_{7}}$ phase. Diffraction array has been marked by dotted circular line

As the thicknesses of the interfaces are quite narrow and often there is no sharp boundary, it is always an extremely difficult task to know particularly about the volume and relative positions. Recently Kar et al. [48] showed that the high resolution back scattered SEM images

(Figure 7. SEM Alu & Figure 8. SEM SS interface) of the interface can be used to determine the respective location of different phases present in the interfacial reaction products zone. From the principle of electron microscopy, it is known that the elements with higher atomic number experience more interaction with the bombarded electrons, thus the number of scattered electrons is also high; as a result bright images are formed. On the other hand elements with lower atomic number, will have lesser electron-mass interaction, as a result the amount of electron scattered will also be less, hence produces darker images. Therefore, this principle of analyzing the different contrasts in the images and correlating the corresponding EDS results, the relative position of the phases present in the interface can be identify. Based on this it is concluded that TiO is formed adjacent to Al_2O_3, Cu_3Ti_3O is adjacent to the filler alloy and Al_2TiO_5 is between TiO and Cu_3Ti_3O phases. Using same technique they suggested that the formation of Al_2TiO_5 was initiated around 900°C and grown fully at 1000°C.[48] (Table 1)

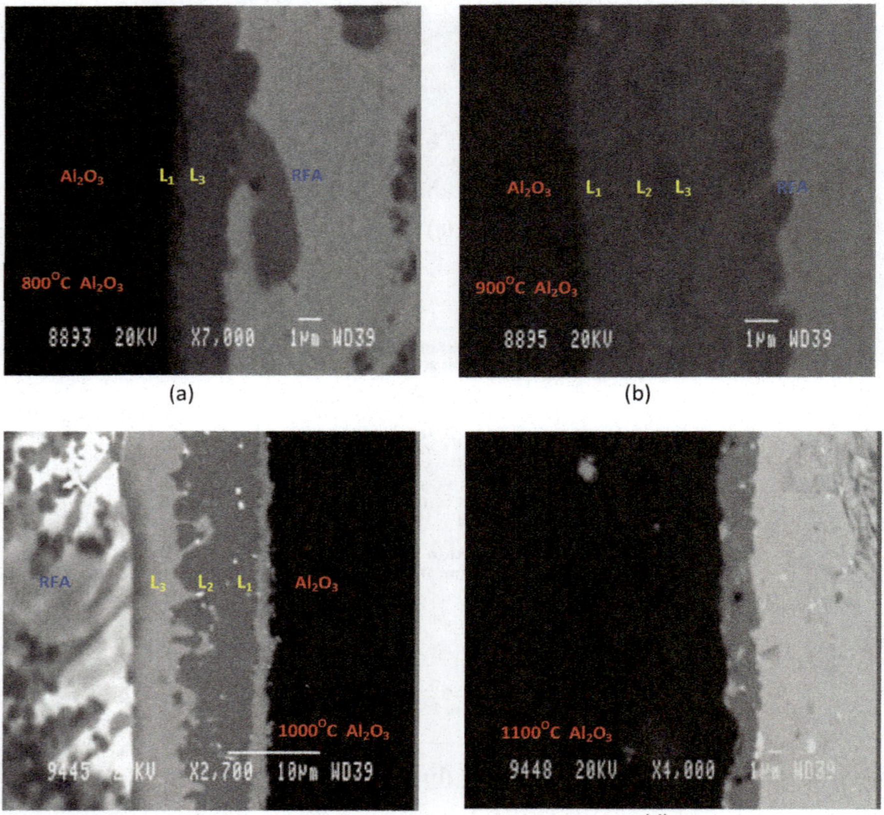

(c) (d)

Fig. 7. SEM -BS images of the Al_2O_3 interface. Different reaction product layers can be clearly distinguished from the contrasts. Clock wise from the top left corner it is, sample brazed at (a) 800°C, (b) 900°C, (c) 1000°C, (d) 1100°C. *For abbreviation* L_1, L_2, L_3 please see table 1.

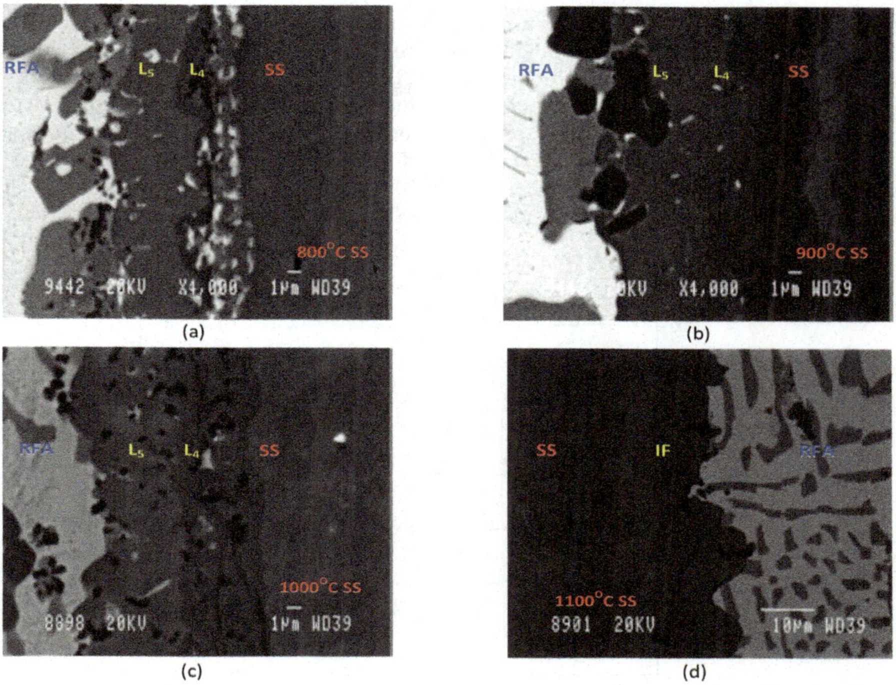

Fig. 8. SEM –BS images of the SS interface. Different reaction product layers can be clearly distinguished from the contrasts. Clock wise from the top left corner it is, sample brazed at 800°C; 900°C; 1000°C; 1100°C. *For abbreviation* L_4, L_5 *please see table 1.*

5. Phase identification

For any system, the presence of phases may be confirmed either by XRD, TEM or by Neutron diffraction studies; other methods can only help identify the existence of different phases within the system. To know the chemistry of the interface formed during brazing, it is very much necessary to determine the reaction products/phases present. As the thickness of the interfaces varies in the range of 7 to 15μm eventually the volume fraction of the reaction products are very small, hence the major peaks obtained by XRD are mostly either from ceramic, residual filler alloy or metal substrates. One of the easiest ways to get reasonably identifiable peaks either of the substrates was sliced orthogonally to the joint axis, approximately 0.2mm away from the interfacial reaction product zone. This enables to reach the interfacial reaction product layer by further mechanical grinding. This process gives better exposure to the interface, for the phase identification by XRD analysis. X-ray diffraction patterns can be recorded after removing each layer step in successive steps. The tentative thickness of each removed layer may be in the order of ≤1μm. The process is tedious and time consuming, but it could identify the reaction products present in the interface, The XRD patterns of the Al_2O_3 interface of the Al_2O_3-SS joint has been exhibited in the Figure 9. (a-c) represent mostly Al_2O_3 substrate, Figure 9(d,e) represents reaction products along with Al_2O_3, Ag, Cu. Whereas, Figure 19(f) represents residual filler alloy (mostly Ag-Cu after the diffusion of Ti). The interface adjacent to alumina was very thin and the concentration of reaction products were also less in quantity, as a result, intensity of the XRD peaks observed are small. The small peaks that are observed in Figure 9(d) are identified as TiO, Al_2TiO_5 and Cu_3Ti_3O at the alumina interface. The crystal structure of Cu_3Ti_3O and Cu_2Ti_4O are quite similar and both have *Fd3m* space group. The lattice spacing (d) values of both the phases are also more or less identical although they differ in their relative intensities. Before confirming the presence of Cu_3Ti_3O instead of Cu_2Ti_4O, the "d" values and their relative intensities should be very carefully analyzed. In this regard EDS or EPMA results also provided some conclusive information as the Cu:Ti ratio are 1:1 and 1:2 for Cu_3Ti_3O and Cu_2Ti_4O respectively.[13]

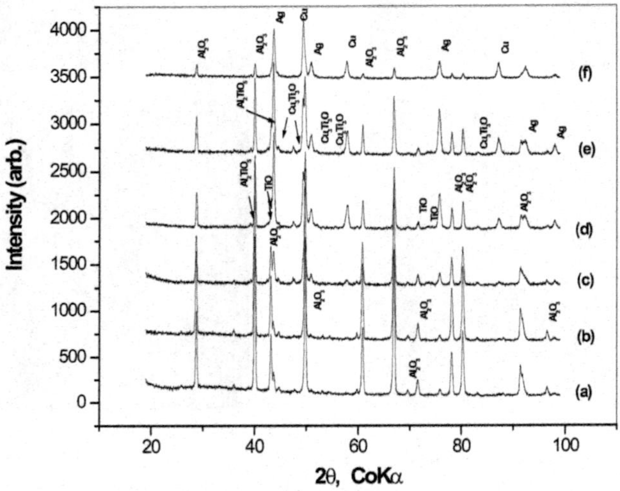

Fig. 9. XRD analyses around the Al_2O_3 interface of Al_2O_3-304 SS braze joint; (a-c) represent mostly Al_2O_3 substrate, (d & e) represent mostly interface along with Al_2O_3, Ag and Cu, (f) represents mostly the residual filler alloy.

It was suggested that Ti_3Cu_3O phase be developed through precipitation process. Karlsson et al.[49,50] suggested that Ti_3Cu_3O was isomorphous with cubic Fe_3W_3C and the phase is likely to be metallic in nature. The Ti_3Cu_3O structure was the same as that of many other transition-metal borides, carbides, nitrides and oxides, collectively known as η phases.[102] The Ti_3Cu_3O-phase layer may also provide a more gradual transition in physical properties and help to minimize the effect of local strains developed due to thermal expansion coefficient mismatches on adhesion.

The XRD peak intensities for the Al_2TiO_5 phase were also observed to be very low. This is because some of the peaks are merged with Al_2O_3 peaks. In the XRD patterns, many a times it is observed that some of the peak positions of two different phases superimposed to each other. This happens because of the similarities of the lattice parameters of those phases. In that case it is difficult to conclude the presence of a particular phase. In such cases, in order to conclude the reaction products present in the interface, it is recommended to verify its presence using different experimental, methods and instrumental techniques.

However, because of the higher volume fraction of the reaction products present in case of SS interface of Al_2O_3-SS joint; their identification is rather easier by XRD technique. Kar et al. suggested the presence of two phases namely $Fe_{35}Cr_{13}Ni_3Ti_7$ and FeTi in the SS interface using

Ag-Cu-Ti filler alloy, Figures 4(a,b). The phases are well distinguishable from the EDS and contrast in the BS-SEM images, Figure 8. The respective lattice spacing for different reaction products ('d' value) are tabulated in Tables 2 and 3 for Al_2O_3 and SS interfaces respectively.

Phase	Lattice spacing (d) Experimental	Lattice spacing (d) Standard in JCPDS	Diffraction plane (*hkl*)
TiO	3.30	3.32	
NaCl -type	2.41	2.41	
	2.10	2.098	
	2.08	2.069	*NaCl-type*
	1.48	1.485	structure
	1.47	1.467	
	1.35	1.365	
Al_2TiO_5	2.6	2.6	0 2 3
S.G.:Cmcm	2.35	2.35	0 4 0
Orthorhombic	2.13	2.14	0 2 4
	2.15	2.17	0 4 2
Cu_3Ti_3O	2.59	2.58	3 3 1
S.G.:Fd3m *Cubic*	2.30	2.298	4 2 2
	2.18	2.17	5 1 1
	1.83	1.87	4 4 2
	1.34	1.33	6 6 0
	1.26	1.259	8 4 0

Table 2. Comparative lattice spacing values for different phases identified in the Al_2O_3 interface.

Phase	Lattice spacing (d) Experimental	Lattice spacing (d) Standard in JCPDS	Diffraction plane (*hkl*)
$Fe_{35}Cr_{13}Ni_3Ti_7$	1.886	1.88	3 3 2
Cubic	1.80	1.808	4 2 2
SG: 143m	1.741	1.736	5 1 0
	1.207	1.205	7 2 1
	1.09	1.09	7 4 1
FeTi	2.06	2.09	1 1 0
Cubic	1.43	1.48	2 0 0
SG: Pm3m	1.207	1.21	2 1 1
19-0636	1.09	1.052	2 2 0
	0.802	0.85	2 2 2

Table 3. Comparative lattice spacing values for different phases identified in the SS interface.

As it is mentioned that, although XRD analysis is the most commonly used technique to identify phases, for a smaller volume fraction of a phase, to render precise information about the reaction products formed in the interface, combined TEM studies (microstructure, diffraction and EDS) are necessary.

So far only one or two reports are available suggesting presence of such (Ti-Al-O) type of phase, but not specifically Al_2TiO_5. The presence of $Ti_3(Cu_{0.76}Al_{0.18}Sn_{0.06})_3O$ was reported by Santella et al.,[36] whereas, (Ti, Al)$_4$Cu$_2$O was reported by Paulasto et al..[52] However, Ohuchi et al. confirmed the formation of Ti_3Al phase while Ti was deposited on alumina at 1000^oC.[53] Ohuchi also reported that Ti reduces alumina and formed a stable Ti-Al compound with a fixed composition.

In case of the SS interface, presence of at least two distinct phases i.e., FeTi, and $Fe_{35}Cr_{13}Ni_3Ti_7$ has been confirmed by TEM analyses.[48] (Figure 5 and Figure 6).Three distinct rings from {211}, {220} and {321} are depicted in Figure 25 as observed in the SADP. Analysis of this SAD pattern confirms the presence of polycrystalline FeTi phase in the SS interface, local chemical composition this phase was also checked by EDS analysis. The zone axis of this cubic FeTi phase was found to be $[\bar{1}11]$. The crystal structure and also the lattice parameter of FeTi and NiTi are similar. However, higher concentration of Fe (~67 wt.%) than Ni (~10 wt.%) in the 304 SS substrate increase the possibility of FeTi formation rather than the metastable NiTi. TEM-EDS analysis as well as the Kikuchi pattern obtain from SEM-EBSD also confirm the presence of FeTi phase Figure 10.[54]

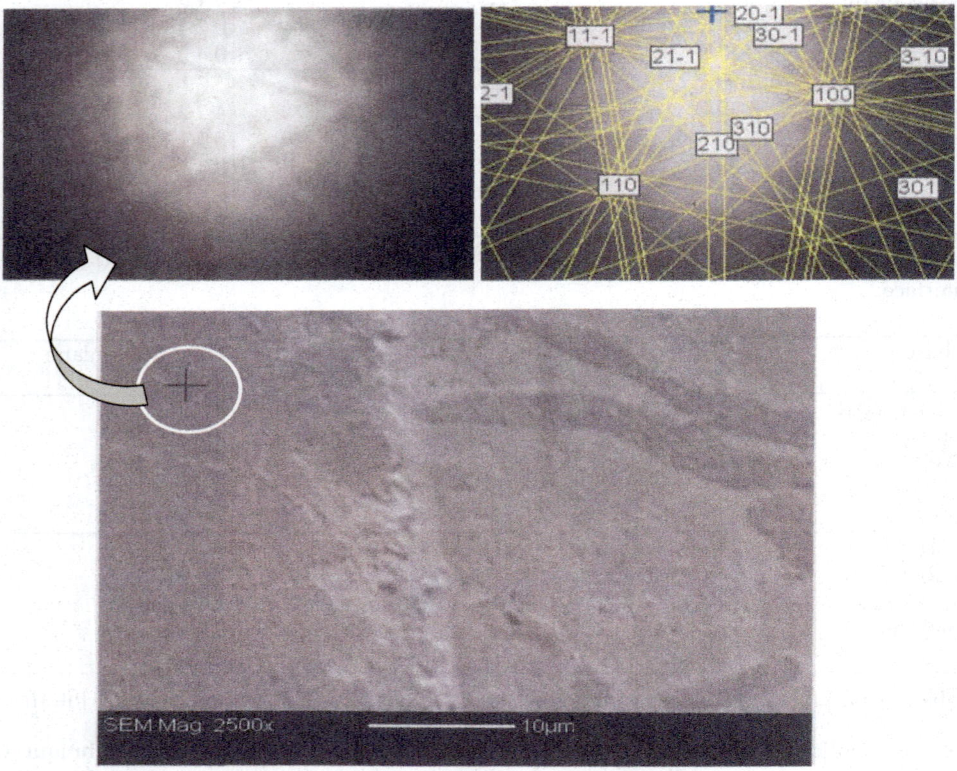

Fig. 10. SEM-EBSD Kikuchi patter confirms the presence of **FeTi** in the SS interface. Micrograph showing the position from which Kikuchi pattern has been recorded.

The presence of cubic $Fe_{35}Cr_{13}Ni_3Ti_7$ phase was also confirmed by the TEM analyses Figure 6.[48] Ti from the Ag-Cu-Ti filler alloy has been diffused towards SS side and reacted with the constituents present in the SS to form phases like FeTi and $Fe_{35}Cr_{13}Ni_3Ti_7$. Formation of these phases causes bonding between SS and the residual filler alloy. Since these phases are considered to be intermetallics, they are harder than that of the substrate. The phases present in SS interface was polycrystalline in nature, whereas the SAD patterns recorded from the Al_2O_3 interface were observed to be single crystal spot pattern. Such variations in the particle sizes in the interface may be explained by their formation mechanisms. It has already been mentioned that at the juncture of Al_2O_3 and filler alloy, there should be unequal heat dissipation during cooling due to of the large difference of their CTEs. Such inhomogeneous cooling allowed the TiO, Al_2TiO_5 or Cu_3Ti_3O phases to grow into a bigger crystallite, after nucleation. Whereas the SS interface experience comparatively fast and uniform cooling, as in case of metal they have comparable CTEs. Such cooling does not provide time to the $Fe_{35}Cr_{13}Ni_3Ti_7$ or FeTi particles to grow and form bigger crystals. However, in most of cases secondary spots are observed in the diffraction SAD patterns and identified as either Ag or Cu.

6. Microstructure of the interface

Many investigators have suggested that the reaction products and the microstructure affect the quality of a braze joint. In the literature, there are different suggestions and opinions about the reaction layers within the interface of Al_2O_3 - (Ag-Cu-Ti)- SS braze joints. Santella et al.[36] observed that adjacent to Al_2O_3, there was Ti-rich layer having face-centered-cubic (fcc) crystal structure with a lattice parameter of 0.423 nm and the phase was identified as oxygen-deficient γ-TiO. The width of that layer was 0.1 to 0.2μm. The next layer was ~3μm thick and separated the TiO layer from the metallic phases of the filler metal. The structure of this layer was identified as diamond cubic unit cell of space group Fd3m with a lattice parameter 1.137nm and the composition of the phase was Ti_3Cu_3O. Similarly, Janickovic et al.[55,56] confirmed that the layer adjacent to Al_2O_3 is fcc TiO with lattice parameter of 0.423 nm. Moreover, they had suggested that the next layer is a mixture of hexagonal Ti_2O_3 and Cu_2O. Based on thermodynamic study,[57] the high temperature γ-TiO can be stabilized with Ti_3Cu_3O at local thermodynamic equilibrium in the Ti-Cu-O ternary system, when the chemical activity of titanium is low. Santella et al.[36] and Yang et al.[58] reported that aluminum concentration in the layer containing Ti_3Cu_3O phase, was relatively high and confirmed that the reduction of alumina surface layer occurred during the brazing process because the braze alloy originally contained no aluminum. As a result of this redox reaction, γ-TiO formed directly on the alumina surface and the reduced aluminum formed a solid solution with Ti_3Cu_3O in the next reaction layer. It was found that the Ti activity and the activity coefficient in the eutectic Ag-Cu melts increases with the increase of Ti concentration and showed a negative deviation from ideal solution at 1000°C. However, Pak. et al.[59] found that the equilibrium oxide phase formed by the reaction of Ti in the Ag-Cu alloy melts with an Al_2O_3 crucible was Ti_2O, and the activity coefficient of Ti showed a positive deviation from ideal solution behaviour at 1000°C. The activity coefficient of Ti at infinite dilution is about 0.076 relative to pure solid Ti and it decreases with the increment of Cu concentration and increases with the increment of Ag concentration.[60]

The interfacial reaction between Al_2O_3 and Ag-Cu-Ti alloy was investigated by Hongqi et al..[61] They reported that the joining conditions have a larger influence on the interfacial reaction. As the temperature and holding time are increased, the reaction layer thickness was increased but the rate of increment decreased when the temperature and time increased further. This may be due to the decrease in viscosity of the filler alloy in liquid state. With the increase of processing temperature, this less viscous filler alloy squeezed out from the intermediate place between the two substrates and results thinner interface.[54] The layer thickness was mainly controlled by the diffusion of titanium through the reaction layers. The reaction products were Cu_2Ti_4O and AlTi at or below 950°C K. However, in the interface there were two distinct layers at or above 900°C. One layer in the vicinity of the Al_2O_3 consisting mainly of Ti_2O and TiO.

According to Hongqi et al. analyses of interfacial microstructures and morphologies, a lower or a higher brazing temperature and shorter or a longer holding time were disadvantageous for stable and reliable joined interface. However, there was no evidential or supporting experimental information made by Hongqi et al.. In general it may be mentioned that the lower processing temperature does not provide sufficient driving energy to produce a strong bonding interface. As it is mentioned earlier that at higher temperature, the molten filler alloy could not sustain to produce sufficient reaction products for a reliable joint. There should be an optimum brazing temperature and holding time for sound joint.

Kritsalis et al. found that there were two distinct layers between the Al_2O_3 and the alloy in the Cu-Ti/Al_2O_3 system.[62] One layer near the Al_2O_3 was titanium monoxide, $TiO_{1\pm x}$, with $x \leq 0.05$ and another layer near the filler alloy corresponds to the compound Cu_2Ti_2O. Cho et al. found that δ-TiO and γ-TiO phases were present at the interface of Al_2O_3 / Ag-Cu-0.5Ti joints.[63]

Loehman et al. analyzed the cross section of an interface results due to the reaction between Ag-33.5Cu-1.5Ti alloy with a high purity Al_2O_3 for 30 min at 900°C.[64] They found that, a complex sequence of reactions and elemental diffusion occurred between the braze alloy and the Al_2O_3. Adjacent to the Al_2O_3 an uneven 1-2μm thick layer of titanium oxide was found which contains some Al. The slight relative shift in the Ti and O profiles suggests a variable stoichiometry of titanium oxide with a decreasing O / Ti ratio away from alumina. The successive region, moving away from alumina, is a thick layer of ~5μm. It is primarily Cu-Ti intermetallics, followed by another 5μm thick layer that is mixed Cu-Ti alloy and Cu-Ag alloy. At the extreme end, Cu-Ag eutectic structure was observed. They had also reported that all of the Ti in the braze alloy was segregated to the alloy/Al_2O_3 interface. A thin layer of that Ti reacted with Al_2O_3 and the remainder formed a Cu-Ti alloy. With longer reaction times (30 min.), TiO_x reaction zone would be greater and that of Cu-Ti would be smaller. Although Loehman et al. did not suggest any results correspond to the joint strength. So it is difficult to conclude or suggest on the effect of longer holding time or formation of thicker TiO layer.

7. Conclusion

The interfaces, of any braze joint are the key areas to control the quality of a joint. It consists of different reaction products; hence, in order to get a good braze joint it is essential to characterize the interface and optimize the amount of reaction product formed in the interface. It is observed that presence of different reaction products at the interface of similar system has been reported by different researchers. This indicates, either the formation of the interface is highly sensitive to the processing parameters or difficult task to characterize the interface properly. Non-availability of the systematic analysis of metal-ceramic brazed interface data is also a constrain.

Here we have shown the results of a very systematically analyzed interface of alumina-304SS brazed sample. Electron microscopy and simultaneous X-ray diffraction analysis is found to be the most authentic tool to be used for characterization of such interface, where within a few micron interface, consists of multiple layered reaction products.

Hence we conclude that in depth microstrutural characterization using transmission electron microscopy and simultaneous XRD and electron microscopy can provide good and reliable result which help designing interface for even better mechanical and functional property for any metal ceramic brazing system

8. References

[1] A. Murari, H. Albrecht, A. Barzon, S. Curiotto, L. Lotto, *Vacuum*, 2003, *68*, 321.

[2] Jian Chen. Pan Wei, Qiang Mei, Yong Huang, *J. Euro. Ceram. Soc.*, 2000, *20*, 2685.

[3] A. Guedes, A. M. P. Pinto, M. Vieira, F. Viana, *Mater. Sci. and Engg. A*, 2001, 301 118.

[4] M. R. Rijnders and S. D. Peteves, *Scripta Mater.*, 1999, *41* 1137.

[5] M. A. Groeber, B. K. Haley, M. D. Uchic, D. M. Dimiduk, S. Ghosh, *Mater. Charact.*, 2006, *57*, 259.

[6] Caroline S. Lee, Sung-Hoon Ahn, Lutgard C. DeJoghe, Gareth Thomas, *Mater. Sci. and Engg. A*, 2006, *434*, 160.

[7] Michikazu Kinsho, Yoshio Saito, Zenzaburo Kabeya, Keisuke Tajiri, Tomaru Nakamura, Kazuhiko Abe, Taketoshi Nagayama, Daiji Nishizawa, Norio Ogiwara, *Vacuum*, 2004, *73*,187.

[8] Braze Alloys- Silver Solders; 'Brazing Tungsten Carbide and Ceramics for Saws, Tools, and Wear Analysis-Carbide Processors, Inc., Chapter-12 (Northwest Research Institute, Inc. / Carbide Processors, Inc. 3847 S. Union Ave. Tacoma, WA. 98409 800 346-8274).

[9] A. H. Elsaway, M. F. Fahmy, *J. Mater. Process. Techno.*, 1998, *77*, 266.

[10] Chia-Hasiang Chiu, Chien-Cheng Lin, *J. Am. Ceram. Soc.*, (2006), *89*, 1409.

[11] Kun-Lin Lin, Chien-Cheng Lin, *J. Am. Ceram. Soc.*, (2006), *89*, 1400.

[12] A. Meier, P. R. Chidambaram, G. R. Edwards, *Acta Mater.*, 1998. *46*, 4453.

[13] S. Mandal, *"Studies on the physical and metallurgical properties of silver copper base system for brazing applications"*; Ph.D. thesis, Jadavpur University, Kolkata, India, 2004.

[14] K. Scott Weil, Christopher A. Coyle, Jens T. Darsell, Gordon G. Xia, John S. Hardy, *J. Power Sources*, 2005, *152*, 97.

[15] W. Cao, J. Zhu, Y. Yang, F. Zhang, S. Chen, W.A. Oates, Y.A. Chang, *Acta Mater.*, 2005, *53*, 4189.

[16] S. Schmidt, S. Beyer, H. Knabe, H. Immich, R. Meistring, A. Gessler, *Acta Austronautica*, 2004, *55*, 409.

[17] M. M. Schwartz, in *Brazing for the engineering technologist*, Publ. Chapman & Hall, London, UK, 1995.

[18] M. M. Schwartz, in *Ceramic Joining*, ASM International, Materials Park, Ohio, USA, 1990.

[19] A.G. Pincus, *Ceram. Age*, 1954, *63*, 16.

[20] K. Suganama, T. Okamoto and M. Shimada, *Commun. Am. Ceram. Soc.*, July 1983, C-117-118.

[21] J. F. Burgess and C. A. Neugebauer, *"Direct bonding of metal to ceramics for electronic applications"*; Advances in. Joining Technology, Burke et al. Eds., 1974.

[22] H. J. Nolte, *Metallized ceramic, U.S. Patent 2,667,432*, Jan26, 1954.

[23] K. Ettre, *Ceram. Age*, 1965, *81*, 57.

[24] M. Vila, C. Prieto, P. Miranzo, M. I. Osendi, A. E. Terry, G. B. M. Vaughan, *Nucl. Instruments and Methods in Phys. Research B*, 2005, *238*, 119.

[25] Yutai Katoh, M. Kotani, A. Kohyama, M. Montorsi, M. Salvo, M. Ferraris, *J. Nucl. Mater.*, 2000, *283-287*, 1262.

[26] J. W. Park, P.F. Mendz, T.W. Eager, *Acta Mater.*, 2002, *50*, 883.

[27] A.M. Glaeser, *Composite Part B*, 1997, *28B*, 71.

[28] R. Morrell, in *Joining to other components, Part I, an introduction for engineer and designer, Handbook of properties of technical and engineering ceramics*, London, 1985.
[29] C. R. Weymueller, *Weldg. Des. & Fab.*, 1987, 45.
[30] A. Xian and Z. Si, *J. Mater. Sci. Lett.*, 1991, *10*, 1315.
[31] C. W. Fox and G. M. Slaughter, *Weld. J.*, 1964, *43*, 591.
[32] H. Mizuhara and K. Mally, *Wed. J.*, 1985, *64*, 27.
[33] S. Xu and J. E. Indacochea, *J. Mater. Sci.*, 1994, *29*, 6287.
[34] E. F. Brush, Jr. and C. M. Adams, Jr., *Weld. J.*, 1968, *47*, 106.
[35] M. L. Santella, *Adv. Ceram. Matls.*, 1988, *3*, 457.
[36] M. L. Santella, J. A. Horton and J. J. Pak, *J. Am. Ceram. Soc.*, 1990, *73*, 1785.
[37] Coralie Valette, Marie-Francoise Devismes, Rayisa Voytovych, Nicholas Eustathopoulos, *Scripta Mater.*, 2005, *52*, 1-6.
[38] M. G. Nicholas, *Br. Ceram. Trans. J.*, 1986, *85*, 144.
[39] Abhijit Kar, S. Mandal, K. Venkateswarlu and Ajoy Kumar Ray, *Mater. Charact.*, 2007, *58*, 555.
[40] Abhijit Kar, Sanjay Chaudhuri, Pratik. K. Sen, Ajoy Kumar Ray, *Scripta Mater.*, 2007, *57*, 881.
[41] J. Zhang, Y.L. Guo, M. Naka and Y. Zhou, *Ceram. Intern.*, 2008, *34*, 1159.
[42] Abhijit. Kar, S. Mandal, R.N. Ghosh, T. K. Ghosh and A. K. Ray; *J. Mater. Sci.*, 2007, *42(14)*, 5556.
[43] Ulrich E. Klotz, Chunlei Liu, Peter J. Uggowitzer, Jörg F. Löffler, *Intermetallics*, 2007, *15*, 1666.
[44] P. He, J. C. Feng, H. Zhou, Mater. Charact., 2005, 54, 338.
[45] Yutaka Hiraoka, Susumu Nishikawa, *Internat. J. Refrac. Metals and Hard Mater.*, *14*, 1996, 311.
[46] Feng Gao, Hui Zhao, Dusan P. Sekulic, Yiyu Qian, Larry Walker, *Mater. Sci. and Engg A*, 2002,*337*,228
[47] Abhijit Kar and Ajoy Kumar Ray, *Mater. Lett.*, 2007, *61*, 2982.
[48] Abhijit Kar, Mainak Ghosh, Ashok Kumar Ray and Ajoy Kumar Ray, *Mater. Sci. Engg. A*,2008,*498*,283.
[49] N. Karlsson, *Nature*, 1951, *168*, 558.
[50] N. Karlsson, *J. Inst. Met.*, 1951, *79*, 391.
[51] J. W. Edington, in *Practical Electron Microscopy in Materials Science*, Vol. 2 (N.V. Philips) Eindhoven, 1975.
[52] M. Paulasto, J. Kivilahti, *J. Mater. Res.*, 1998, *13*, 343.
[53] F. S. Ohuchi and M. Kohyama, *J. Am. Ceram. Soc.*, 1991, *74*, 1163.
[54] Abhijit Kar, *"Studies on interfacial characterization of ceramic-ceramic/metal brazed joints"*; Ph.D. thesis,Jadavpur University, Kolkata, India, 2004
[55] D. Janickovic. P. Sebo, P. Duhaj and P. Svec, *Mater. Sci. and Engg. A*, 2001, *304-306*,569.
[56] P. Duhaj, P. Sebo, P. Svec and D. Janickovic, *Mater. Sci. and Engg. A*, 1999, *271*, 181.
[57] G. P. Kelkar, K. E. Spear and A. H. Carim, *J. Mater. Res.*, 1994, *9*, 2244.
[58] P. Yang, B. N. Turman, S. J. Glass, J. A. Halbleib, T. E. Voth, F. P Gerstle, B. Mckenzie, J. R. Clifford, *Mater. Chem. and Phys.*, 2000, *64*, 137.
[59] J.J. Pak, M.L. Santella and R.J. Fruehan, *Metall. Trans. B*, 1990, *12*, 349.
[60] L. Rongti, P. Wei, C. Jian and L. Jie, *Mater. Sci. and Engg. A*, 2002, *335*, 21.
[61] H. Hongqi, W. Yonglan, J. Zhihao and W. Xiaotian, *J. Mater. Sci.*, 1995, *30*, 1233.
[62] P. Kritsalis, L. Coudurier and N. Eustathopoulos, *J. Mater. Sci.*, 1991, *26*, 3400
[63] H.C. Cho and J. Yu, *Scripta Mater.*, 1992, 26, 797.
[64] R. E. Loehman and A. P. Tomsia, *Am. Ceram. Soc. Bull.*, 1988, *67*, 375.

20

In-Situ Mechanical Testing of Nano-Component in TEM

Takashi Sumigawa and Takayuki Kitamura
Department of Mechanical Engineering and Science, Kyoto University, Japan

1. Introduction

MEMS (Micro electro mechanical system) and NEMS (Nano electro mechanical system), which are composed of large number of submicron- or nano-scale dissimilar components, often experience extrinsic loads due to operational and environmental conditions and these bring about local fracture. In order to make an accurate life prediction, the mechanical properties of components (e.g., deformation, fracture, and fatigue behaviour) have to be understood well. Especially, there are numerous uncertainties reported in the mechanical properties of micro-nano scale components, which are different from those of macroscopic one.

1. Smaller metals than a micrometer possess different plastic deformation properties from those of the bulk (Vinci et al., 1995; Xiang et al., 2006) because small volume components require a large driving force for dislocation generation and glide due to the dislocation source starvation (Greer & Nix, 2006; Hou et al., 2008; Liu et al., 2009) and the mirror force from free surfaces (J. Weertman & J.R Weertman, 1964). Furthermore, since the components in device are usually surrounded by dissimilar materials, the resulting deformation constraint affects the plastic behaviour as well.
2. Dissimilar interface is one of the most favorable potential crack initiation sites in a micro- and nano-scale component because it often has weak bonding (low resistance to fracture) and stress concentration (high driving force to fracture) due to the deformation mismatch between dissimilar materials (Bogy, 1968). In the component, the region governed by the stress concentration is confined in the nano-scale. In such case, the applicability of continuum mechanics to the crack initiation is questionable.
3. It is well known that irreversible cyclic deformation results in characteristic dislocation structures (vein, ladder-structure, and cell) in fatigue of bulk metals (Mughrabi, 1978; Thompson et al., 1956; Winter, 1978). In particular, the ladder and cell structures cause crack initiation owing to the strain localization because they are much softer than the matrix (Suresh, 1998). These fatigue substructures have a size in the scale of a few micrometers (Mughrabi, 1978). However, since the strain-concentrated region in nano-components is a few nanometers or at most a few tens of nanometers, typical fatigue substructure cannot form there.

Although we have to specify the mechanical properties of nano-component, the strength evaluation experiment of nano-component involves some major challenges in I.

Observation, II. Experimental set-up, III. Specimen fabrication, and IV. Specimen gripping and alignment. Consequently, we have to use special devices and methods described below for the investigation.

I. Observation

For observation of nano-components, electron microscopes (e.g., transmission electron microscope (TEM) and scanning electron microscope (SEM)) are required because nano-components are invisible at optical wavelength. TEM is better in terms of image resolution while SEM possesses the depth of focus in the surface observation.

II. Experimental set-up

Piezoelectric device (Keller et al., 1996; Liu et al., 2009; Tsuchiya et al., 2000) and electromagnetic actuator (Kim & Huh, 2011; Komai et al., 1998), which can control nanometer order displacement, are used as an actuator. Load is measured by means of capacitive load cell (Zhu et al., 2005). Laser displacement meter (Burdess et al., 1997; Yi et al., 2000) and image analysis (Cheng et al., 2005; Hua et al., 2007) can detect specimen displacement with nanometer-scale resolution.

III. Specimen fabrication

Specimen is usually fabricated by means of wet-etching (Li et al., 2005; Namazu et al., 2002), dry-etching (Ryu et al., 2006 ; Tripathy et al., 2007), and focused ion beam (FIB) processing (Dimiduk et al., 2005; Motz et al., 2005). Wet-etching is often applied to single crystal Si, which is an important material in MEMS devices. Dry-etching and FIB processing are used for processing of multi-layered and composite materials.

IV. Specimen gripping and alignment

Although tensile testing is popular for evaluation of mechanical properties, it is difficult to apply it to nano-component because it requires precise alignment and reliable gripping (Haque & Saif, 2002; Huang & Spaepen, 2000; Tsuchiya et al., 1998). Especially, bending (Motz et al., 2005; Namazu et al., 2000) and compression (Dimiduk et al., 2005 ; Moser & Wasmer, 2007) methods are often adopted. Bending one has an advantage in terms of examination on the fracture behaviour.

In this chapter, we introduce bending experiment of nano-component in a TEM for *in-situ* observation. The material is composed of silicon (Si), copper (Cu), and silicon nitride (SiN), and the focus is on the deformation of nano-metal (Cu) and the fracture along the Si/Cu interface.

2. Mechanical testing with *in-situ* TEM observation (Sumigawa et al., 2010a, 2010b, 2010c, 2011a, 2011b)

Figure 1 schematically illustrates specimen configuration and loading scheme. The tested parts of Si, Cu, and SiN are in 100 nm order, and Si/Cu and Cu/SiN interfaces are vertically arranged in the cantilever. Load is applied to the SiN layer away from the Cu portion by means of a diamond tip. The residual stress in each layer is examined beforehand (See Table

1), and the influence is included in the finite element method (FEM) stress analysis in the following sections.

	Residual stress [MPa]
20 nm-thick Cu film (Sumigawa et al., 2010a)	760 MPa
200 nm-thick Cu film	150 MPa
500 nm-thick SiN layer (Sumigawa et al., 2010a)	-290 MPa

Table 1. Residual stresses of 20 nm-thick, 200 nm-thick Cu films and SiN layer.

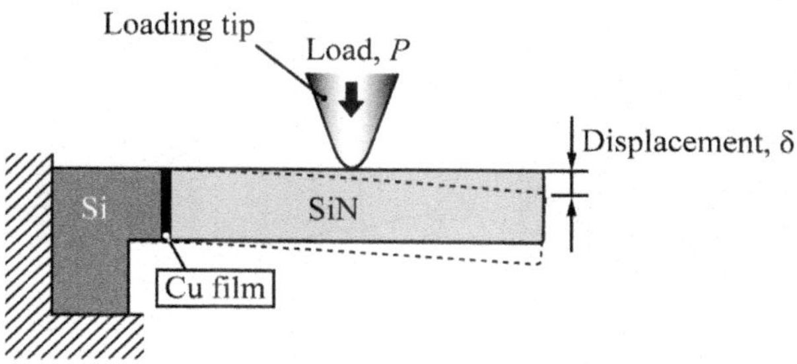

Fig. 1. Schematic illustration of bending experiment.

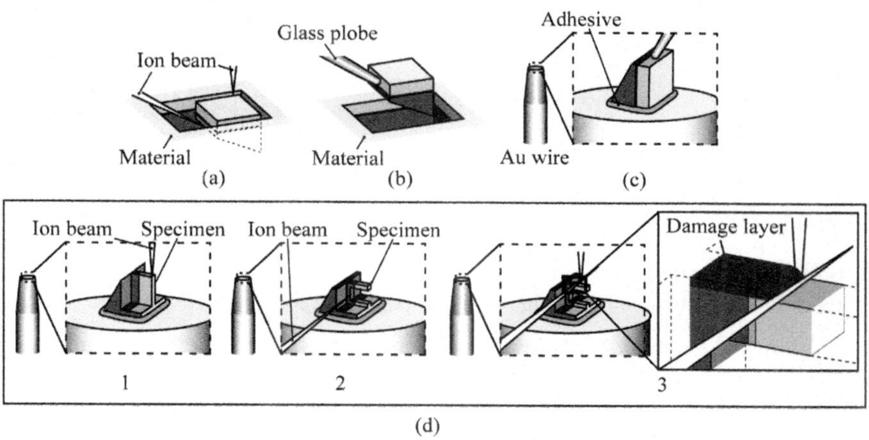

Fig. 2. Specimen fabrication procedure.

The specimen is carved out by means of focused ion beam (FIB) processing from a multilayer material, where Cu and SiN are deposited on a Si substrate by magnetron sputtering. The fabrication procedure of specimen is as follows.

1. A block with a size of about 10×10×10 μm^3 is carved out from the multilayer material. (Fig. 2(a))
2. The block is picked up by a micro glass probe using electrostatic force. (Fig. 2(b))
3. A gold (Au) wire with a flat top is prepared, and the block is mounted using an adhesive. (Fig. 2(c))

4. The block is formed into a cantilever-shaped specimen. (Fig. 2(d) from 1 to 3)
    1. The side planes are processed by a vertical FIB. (Fig. 2(d)-1)
    2. The top and the bottom planes of the cantilever are processed by a lateral FIB. (Fig. 2(d)-2)
    3. The surfaces are finished to remove the layer damaged by the crude processing, and the shape of the interface edge is flattened (Fig. 2(d)-3)

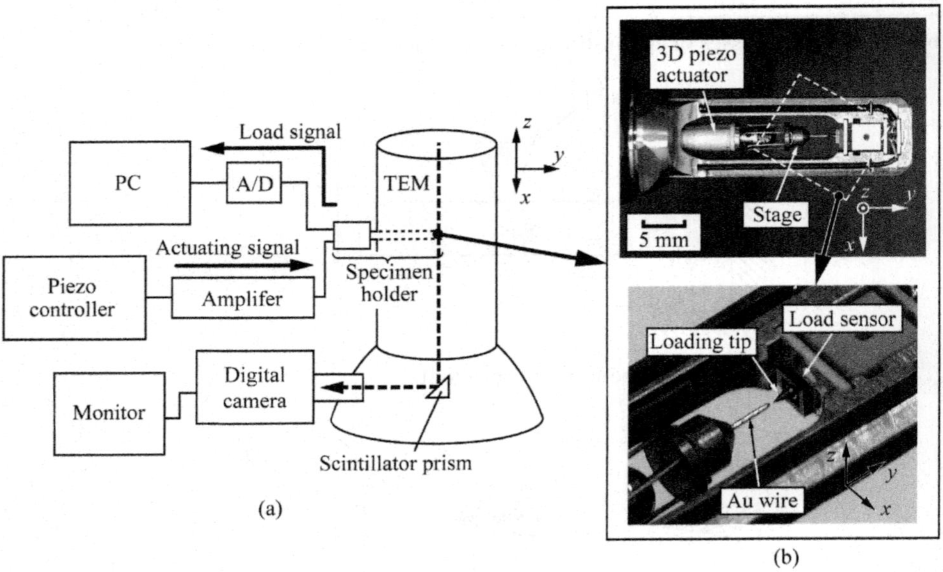

Fig. 3. (a) System of loading experiment with *in-situ* TEM observation, and (b) detail of loading apparatus.

The output voltage of the FIB processing is kept at 30 kV, and the currents in the crude processing and in the surface finishing are 5-10 nA and 10 pA, respectively.

Figure 3(a) shows a schematic illustration of experimental set-up. It consists of a TEM (JEOL, JEM-2100), a specimen holder with a mechanical loading apparatus (Nanofactory Instruments AB, SA2000N), a piezo controller, and a digital camcorder. Figure 3(b) shows the loading apparatus, which consists of a stage which can move in three orthogonal directions and a conical diamond tip with a capacitance load sensor. The measurement range and accuracy of the load are 0-100 μN and ±0.1 μN, respectively.

An Au wire, which holds the specimen, is connected to the stage, and a load is applied by pressing the specimen onto the diamond tip. The applied load, P, is directly detected by means of the load sensor attached to the back of the tip. The displacement at the specimen end, δ, (See Fig. 1) is measured using a TEM with the loading device installed. TEM observation has the advantages of precise measurement of displacement and precise identification of exact locations of the interfaces and contact area of the loading tip. The observations are carried out with an accelerating voltage of 200 kV under a vacuum of 1.5×10^{-5} Pa.

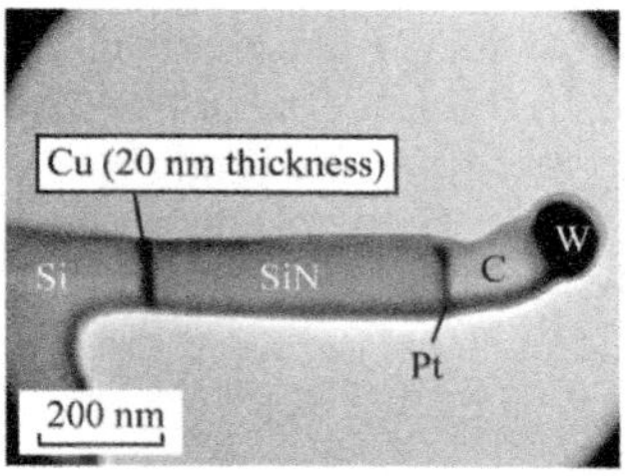

Fig. 4. Bright-field TEM image of cantilever specimen.

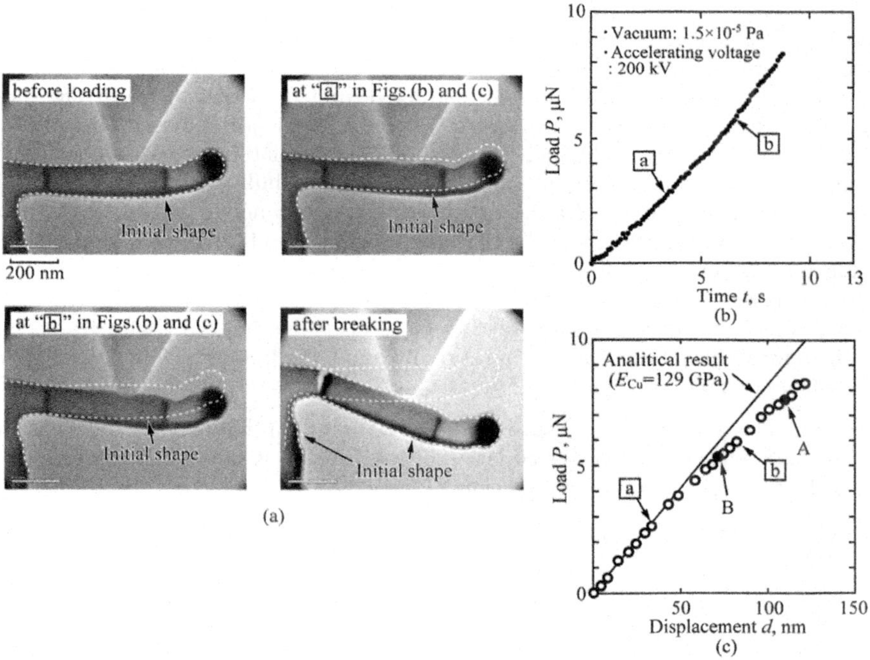

Fig. 5. Deformation behaviour of cantilever specimen: (a) bright-field TEM images, (b) load-time curve, and (c) load-displacement curve.

3. Mechanical properties of nano-component

3.1 Plasticity in 20 nm-thickness Cu film (Sumigawa et al., 2010a)

3.1.1 Research target and experimental detail

Although some researchers have clarified the plastic property of micron-thickness thin film by means of the indentation method (Dao et al., 2001; Huang et al., 2000), it is extremely difficult to apply the method to nano-components because measurement error becomes large owing to the identification of exact tip shape and sample roughness. Thus, the target in this section is to investigate the plastic behaviour of nano-metal by using the bending experiment under the *in-situ* TEM observation.

Figure 4 shows the bright-field TEM image of specimen. The cantilever arm is extended in order to magnify the deformation in Cu. The difference in transmission rates of the dissimilar materials enables us to recognize the location of dissimilar interfaces in the TEM micrograph. The target material to specify the plasticity is Cu with the dimension of 280 nm × 125 nm × 20 nm, which is sandwiched by Si and SiN.

3.1.2 Inverse analysis

Figure 5 shows (a) bright-field TEM images before and during loading, (b) the relationship between the applied load, P, and the time, t, (P-t relationship), and (c) the relationship between P and the displacement at the cantilever tip (arm end of specimen), δ, (P-δ relationship). The loading point and the Cu film are clearly identified by the TEM, and the displacement at the arm end can be quantitatively measured from the images. The P-t relationship (Fig. 5(b)) indicates smooth loading during the experiment. In Fig. 5(c), the displacement increases linearly with increasing load at low loading levels. The solid line in Fig. 5(c) shows the relation obtained by elastic FEM simulation, where all materials are assumed to be linear-elastic and the Young's modulus of the bulk material (E_{Cu} = 129 GPa) is used for the Cu film. The good agreement with the experimental results at low loading levels indicates that the structure of the film has little effect on the Young's modulus. The P-δ relationship shows clear nonlinear behaviour beyond P = 3 μN, where the deviation greatly exceeds the measurement smallest limit (about 5 nm). Since the TEM image after the breaking points out that the Si and the SiN parts keep the original shape and size, they elastically deform during the loading. In short, the nonlinearity in the P-δ relationship is due to the plasticity of Cu film.

The Cu portion is too thin to measure the deformation quantitatively by the TEM image. Thus, the elasto-plastic constitutive equation of the Cu film is derived by inverse analysis using the P-δ relation and a three-dimensional FEM.

Using the Von Mises equivalent stress and strain, the deformation behaviour of the Cu film is described by

$$\sigma = \begin{cases} E\varepsilon & , for\ \sigma \le \sigma_y \\ R\varepsilon^n & , for\ \sigma \ge \sigma_y \end{cases}, \tag{1}$$

where E, R, n, ε, and σ_y are the Young's modulus, strength coefficient, strain hardening exponent, total effective strain, and yield stress, respectively. At the yield point, as we have

$$\frac{\sigma_y}{R} = \left(\frac{\sigma_y}{E}\right)^n \tag{2}$$

R is given by the quantities E, σ_y, and n. Since we can use the bulk elastic constants as shown in Fig. 5(c), the unknown parameters are only σ_y and n.

The derivation procedure of σ_y and n is as follows:

1. Arbitrary points (points A and B) are picked up in the non-linear portion of the load-displacement curve (Fig. 6(a)).
2. The combination of σ_y and n, which fulfills the displacement at the point A as measured by the experiment, is found using elasto-plastic FEM by trial-and-error (Fig. 6(b)). Since two parameters cannot be determined at once from one experimental point, all sets of (σ_y, n), which satisfy the non-linear deformation at the point A, have a relationship schematically shown in Fig. 6(b)
3. By the same way, the combination of σ_y and n at point B is derived through FEM (Fig. 6(c)). The broken line in Fig. 6(c) shows the relationship determined from point B.
4. The intersection between the two lines gives the adequate combination of σ_y and n (Fig. 6(c)) for the constitutive equation of nano-Cu.

The points A (109.5 nm, 7.6 μN) and B (71.0 nm, 5.3 μN) in the *P-d* relationship are picked up for the inverse analysis.

For the accurate analysis, it is important to identify the exact shape of the specimen. The FEM model is prepared using commercial three-dimensional CAD (computer aided design) software. The model is divided into meshes as shown in Fig. 7, where the region near the interface has finer mesh. A perfect constraint condition is imposed on the left and bottom edges of the model, and the elastic constants in the analysis are summarized in Fig. 7.

Figure 8(a) shows the σ_y-n relationships at points A and B (Fig. 5(c)) obtained by the inverse analysis. The intersection gives the following elasto-plastic constitutive equation for the 20 nm-thick Cu film:

$$\sigma = \begin{cases} 129000\varepsilon & , for\ \sigma \leq 765 \quad [MPa] \\ 3316\varepsilon^{0.3} & , for\ \sigma \geq 765 \quad [MPa] \end{cases} . \tag{3}$$

Figure 8(b) shows the *P*-δ relationship (solid line) obtained by the FEM on the basis of Eq. (3). This demonstrates that Eq. (3) consistently reproduces the nonlinearity of the Cu film.

Figure 9 shows the tensile behaviour obtained and those of a 200 nm-thick Cu film (Hirakata et al., 2007) and a Cu bulk material (average diameter of grains is 58 μm) (ASM Handbook, 2000) for comparison. The yield stress of the bulk material is the lowest (σ_y = 60 MPa), and increases with decreasing thickness of the film (200 nm-thick: σ_y = 272 MPa, 20 nm-thick: σ_y = 765 MPa).

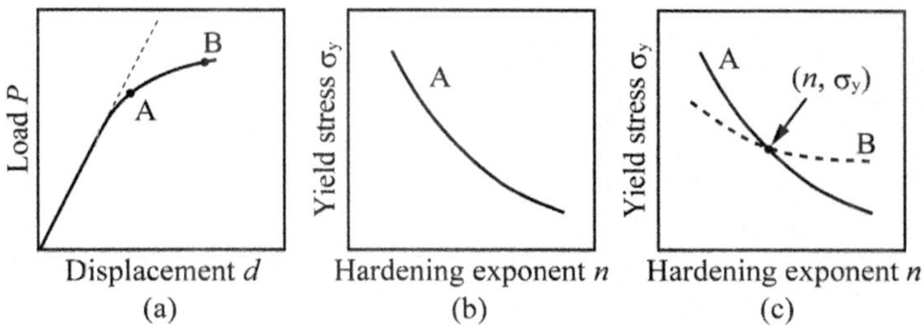

Fig. 6. Procedure for inverse analysis.

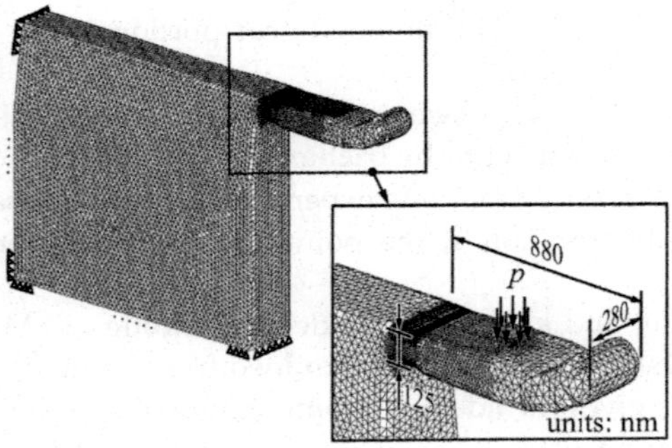

Material	E (GPa)	ν
SiN	197	0.27
Pt	168	0.377
C	10	0.3
W	407	0.17

Material	C_{11}(GPa)	C_{12}(GPa)	C_{44}(GPa)
Si	167.4	65.23	79.57

Fig. 7. Analytical model and material constants in FEM analysis.

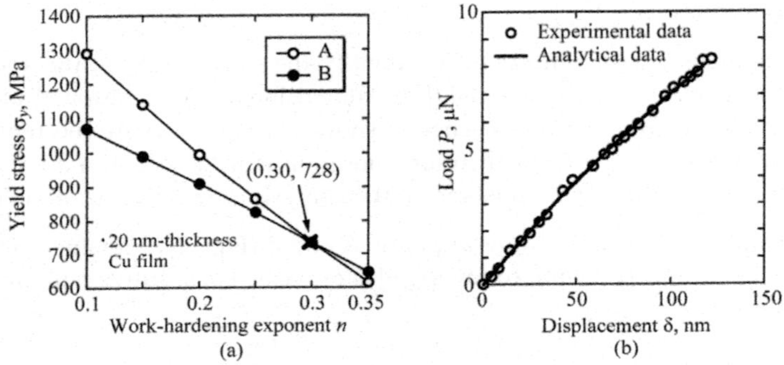

Fig. 8. (a) σ_y-n relationships of points A and B, and (b) loading curves obtained from the inverse analysis.

If the mean grain diameter is assumed to be film thickness, the Hall-Petch relation of spattered Cu films (20 nm thickness) (Ruud et al., 1993) gives the yield stress of about 1 GPa. Similarly, it gives about 250 MPa for the grain diameter of 200 nm. Although the yield stress obtained for the 20 nm-thickness film and the 200 nm-one has a slight deviation, they approximately follow the Hall-Petch relation.

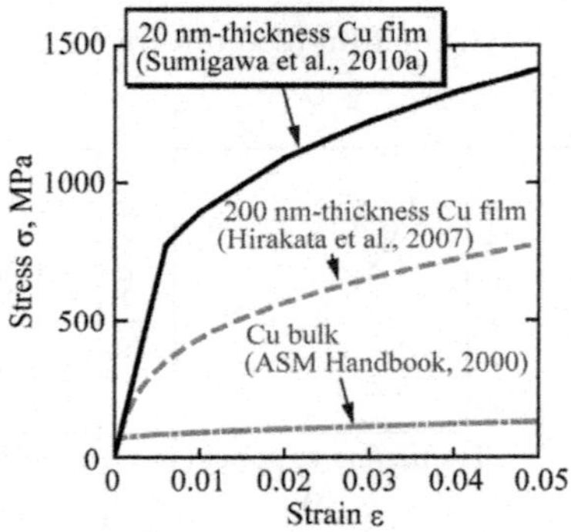

Fig. 9. Dependence of stress-strain relationship on film thickness. The solid line shows Eq. (3).

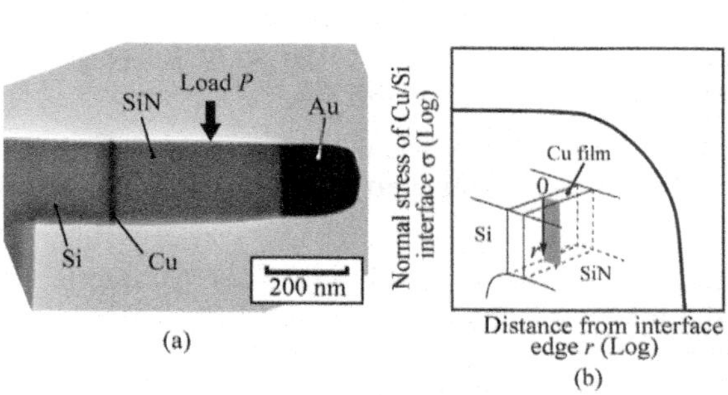

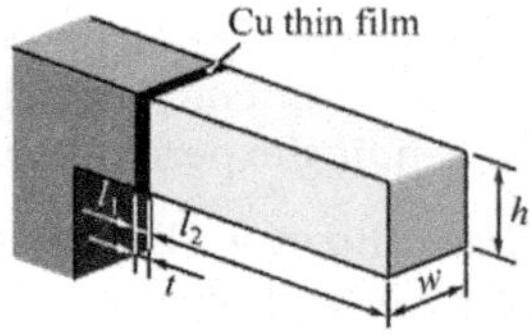

	l_1	l_2	h	w	t
Specimen N1	220	715	250	434	20
Specimen N2	120	700	233	308	20
Specimen N3	84	767	642	415	20

(c)

Fig. 10. (a) TEM image of a specimen (Specimen N1), (b) schematic illustration of stress distribution along the Si/Cu interface, and (c) dimensions of specimens.

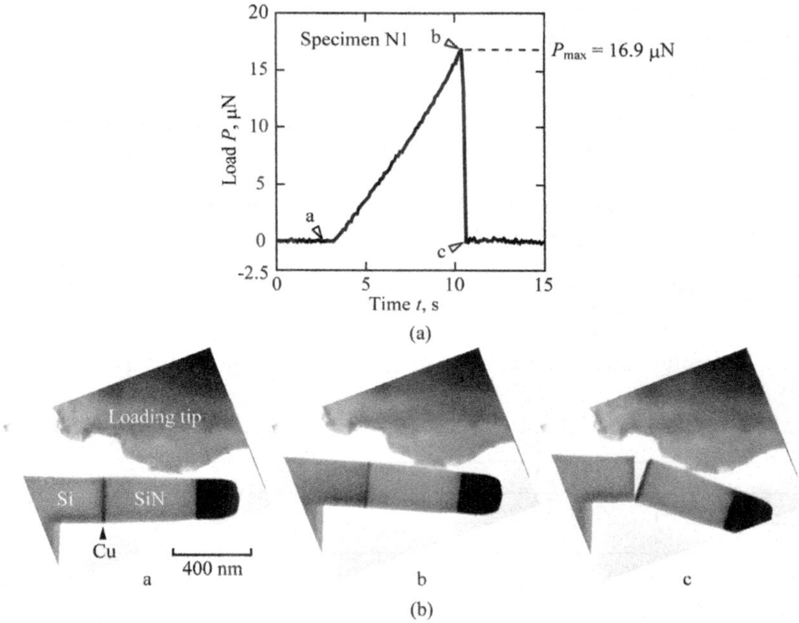

Fig. 11. (a) Load-time relationship, and (b) *in-situ* observation of the interfacial fracture of Specimen N1.

3.2 Crack initiation at the interface edge between Si substrate and 20 nm-thickness Cu film (Sumigawa et al., 2010b)

3.2.1 Research target and experimental detail

At a dissimilar interface edge, stress is concentrated owing to the free-edge effect (Bogy, 1968). Although an interfacial crack may appear at the top interface edge in the cantilever specimen, the cracking criterion on the basis of continuum mechanics such as "critical stress" is questionable in a nano-component because only a few tens or a few hundreds atoms exist in the stress concentrated region. The target in this section is to examine the dominant factor of crack initiation at the dissimilar interface edge in nano-components by using the bending experiment under the *in-situ* TEM observation.

Figure 10(a) shows a TEM image of specimen for interfacial cracking experiment. Because the strength of interface between Si and Cu is weaker than that between Cu and SiN, the Si/Cu interface is broken preferentially.

Figure 10(b) schematically illustrates the stress distribution along the Si/Cu interface in the specimen. A crack is initiated at the interface edge on the top surface. Three specimens (Specimen N1, N2, and N3) with different size are prepared in order to change the stress distribution along the Si/Cu interface. The size of specimens is summarized in Fig. 10(c).

3.2.2 Criterion of crack initiation

Figure 11 (a) shows the loading curve (*P-t* relationship) of Specimen N1. The load, P, monotonically increases up to a peak magnitude of 16.9 μN (point b) and then abruptly drops to 0 μN (point c). Figure 11(b) shows bright-field TEM images corresponding to the points a, b and c on the *P-t* curve. The crack is initiated at the point b at the top interface edge and instantly propagates along the interface. Detailed observation after the experiment demonstrates that no dissimilar material remained on either delaminated surfaces, confirming pure interface-cracking. Similar behaviour is observed in Specimens N2 and N3. Thus, the peak loads for interfacial crack initiation, P_c, are critical, and Table 2 lists the magnitudes of P_c. There are significant differences in P_c among the specimens.

The stress in the specimen is analyzed by the FEM, in which an individual model is prepared for each experimental specimen. Figure 12(a) shows the analytical model of Specimen N1, where the configuration is reconstructed on the basis of the SEM and TEM micrographs. The constitutive equation obtained in previous section (Eq. (3)) is used for Cu film.

	P_c (μN)
Specimen N1	16.9
Specimen N2	10.3
Specimen N3	85.3
Specimen S1	57.9

Table 2. Critical load for crack initiation at Si/Cu interface.

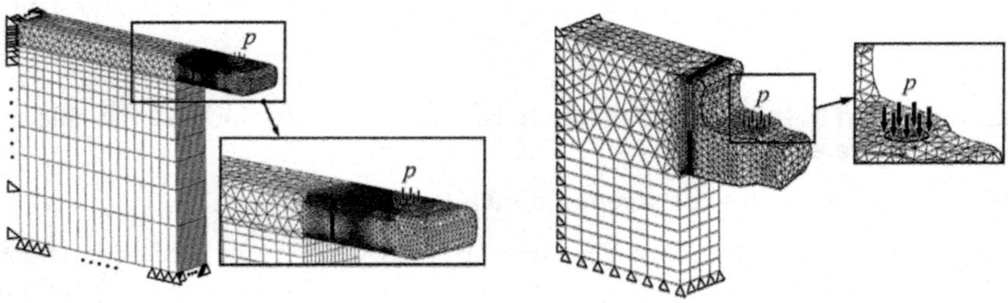

(a) (b)

Fig. 12. Mesh division in FEM analysis: (a) Specimen N1 and (b) Specimen S1.

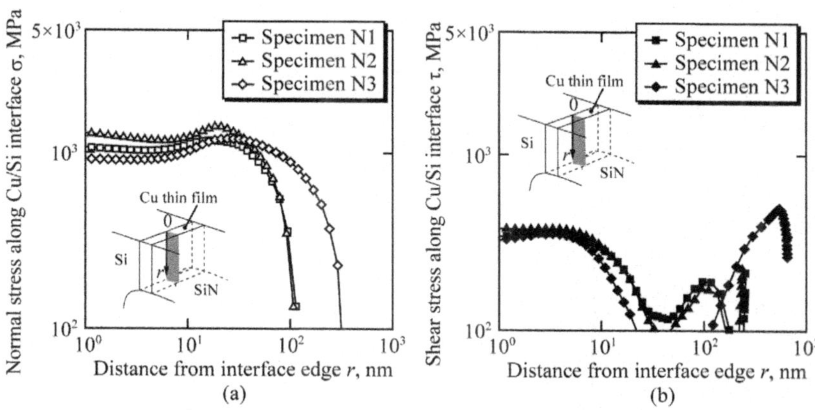

Fig. 13. Elasto-plastic stress distributions along Si/Cu interface near the edge: (a) normal stress σ and (b) shear stress τ.

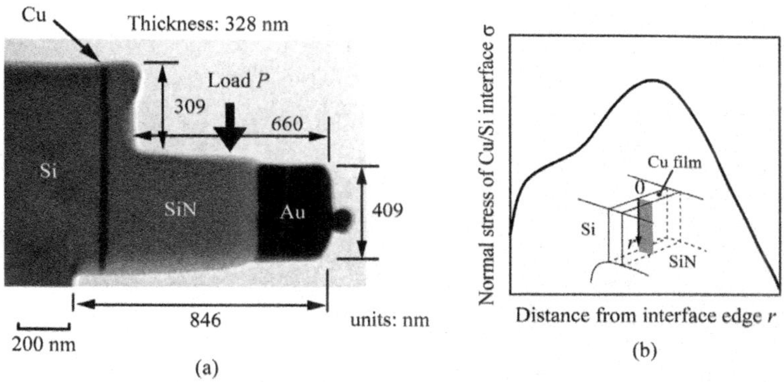

Fig. 14. (a) TEM image and dimensions of Specimen S1, and (b) stress distribution along the Si/Cu interface.

Figure 13 shows the distribution of normal and shear stresses, σ and τ, along the Si/Cu interface near the edge at the critical load P_c. Although the stresses away from the edge in Specimens N1 to N4 differ from each other, they show fairly good agreement in the range of 20-30 nm near the edge. This indicates that crack initiation is governed by the concentrated stress. The high normal stress of about 1 GPa in a region of a few tens of nanometers near the interface edge, which is approximately three-times larger than that of the shear stress ($\tau \approx 300$ MPa), governs the crack initiation.

3.3 Crack initiation in the interior of the interface between Si substrate and 20 nm-thickness Cu film (Sumigawa et al., 2010b)

3.3.1 Research target and experimental detail

Stress concentration does not always take place at the interface edge because of the complex shapes of nano-components. Because the edge has a peculiar atomic arrangement, the fracture resistance is expected to be different from the one of interface inside of the component. This suggests that interface crack initiation is dependent upon the location in the component. In this section, an experiment where a crack is initiated in the interior of the dissimilar interface is conducted, and the difference in the crack initiation criteria along the interface at the edge and in the interior is discussed.

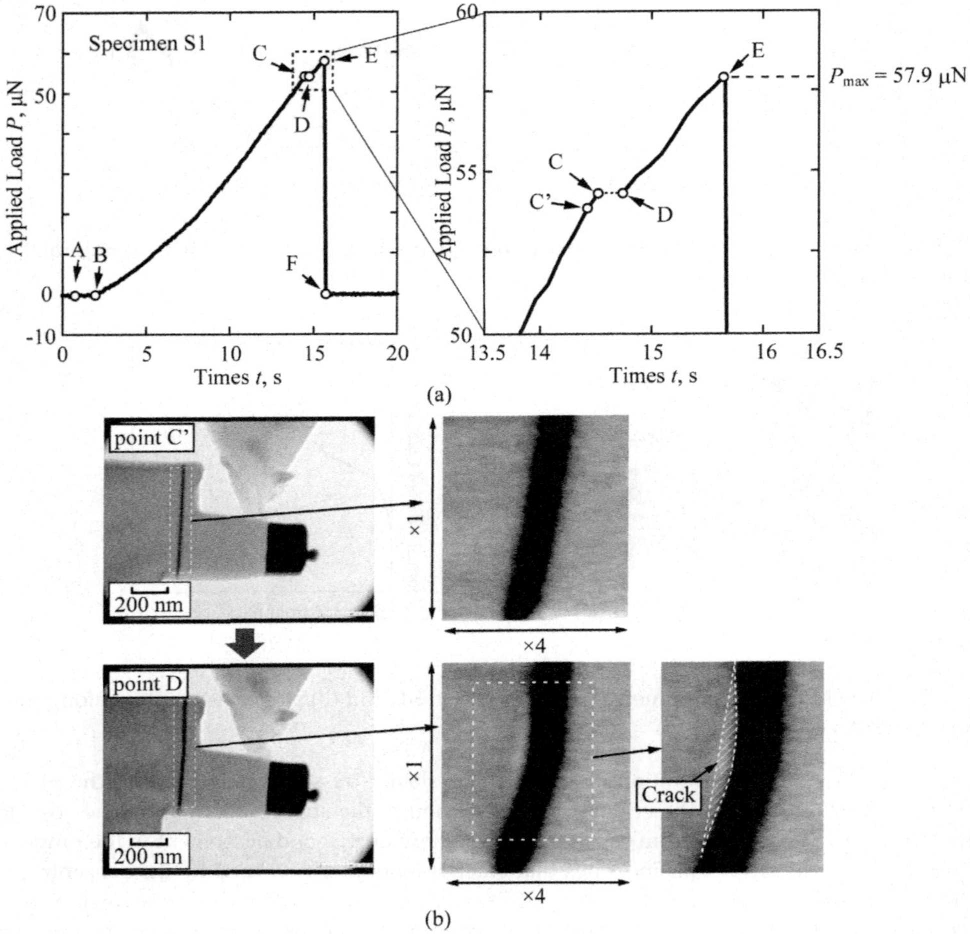

Fig. 15. (a) Load-time relationship of Specimen S1 and (b) detailed TEM images before and after the crack initiation.

Figure 14(a) shows a TEM image of specimen with a step in the SiN portion of cantilever. Figure 14(b) schematically illustrates the stress distribution along Si/Cu interface in the specimen. While there is no stress concentration at the interface edge, the stress has the maximum in the central region of the Si/Cu interface. This suggests that an interfacial crack is initiated in the interior of the Si/Cu interface.

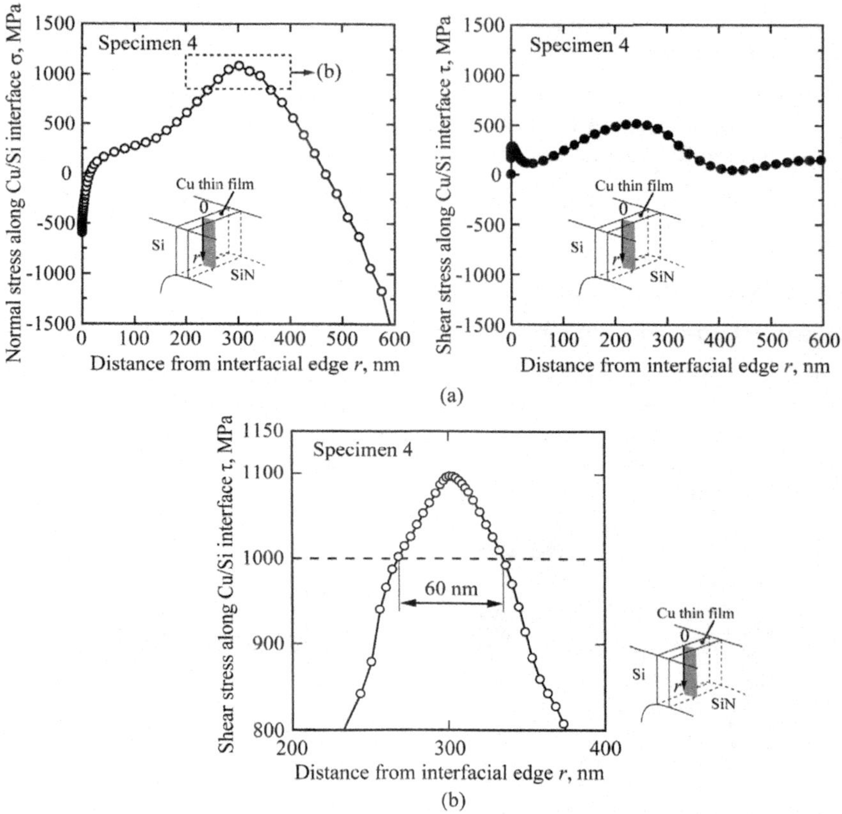

Fig. 16. Stress distributions along the Si/Cu interface (Specimen S1).

3.3.2 Criterion of crack initiation

Figure 15(a) shows the *P-t* relationship for Specimen S1. *P* monotonically increases up to P = 54.3 μN (point C), and it jumps from C to D. Then, the load begins to increases again before the sharp drop at the point E. Figure 15(b) shows sequential TEM micrographs corresponding to the points on the *P-t* curve. The interface is suddenly bent at the jump C-D, and a crack shown in the lower photo in Fig. 15 (b) appears at the location of r = 300 nm from the interface edge on the upper surface. Finally, the crack instantaneously propagates along the interface at point E (P = 57.9 μN).

Figure 16(a) shows the distributions of normal stress, σ, and shear stress, τ, at P = 54.3 μN along the Si/Cu interface, which are obtained by using an FEM model shown in Fig. 12(b). σ has a peak concentration at $r \approx 300$ nm, which corresponds to the experimentally observed portion of crack initiation. Because σ at $r \approx 300$ nm is much larger than that of τ, the former is the dominant factor in the crack initiation. Figure 16(b) shows a magnified view of the σ-distribution near r = 300 nm. The maximum stress is about 1.1 GPa, and the region of $\sigma > 1$ GPa extends about 60 nm.

Comparing the results shown in Figs. 13(a) and 16(b), crack initiation in the interior has the following characteristics:

1. The critical stresses in both cases are of almost the same magnitude, about 1 GPa.
2. The high-stress regions in both cases are several tens of nanometers in width, but the interior crack initiation requires a slightly wider region.

In short, there is little difference between the crack initiation criteria along the interface at the edge and in the interior.

3.4 Fatigue cracking (Sumigawa et al., 2010c)

3.4.1 Research target and experimental detail

Since typical fatigue substructures in a bulk metal, of which size is in the scale of a few micrometers (Mughrabi, 1978), cannot form in a nano-component. However, the fatigue of nano-component has not been clarified yet because of the experimental difficulty. The target in this section is to examine the fatigue behaviour of metal in a nano-component using the bending cantilever under the *in-situ* TEM observation.

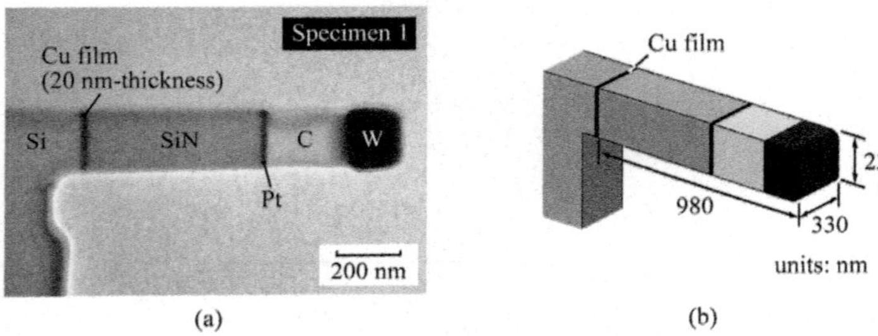

Fig. 17. (a) Bright-field TEM image of specimen, and (b) dimensions of specimens.

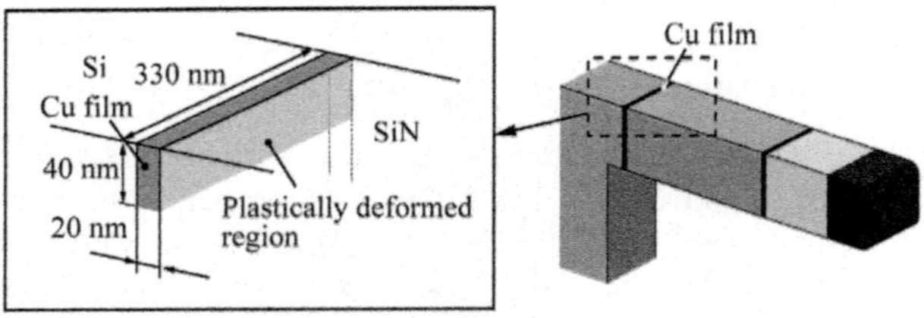

Fig. 18. Fatigue region (plastically deformed region) in the Cu film.

Figure 17 shows (a) a bright-field TEM image of specimen and (b) its size. It should be noted that the stress concentration near the Si/Cu interface obtained by the FEM analysis points out the fatigue region is about 40 nm×20 nm×330 nm in the Cu film as schematically illustrated in Fig. 18.

The fatigue experiment is conducted under a constant load range, $\Delta P = P_{max}$-P_{min}, with a load ratio of $P_{min}/P_{max} = 0$ (P_{min}: minimum load, P_{max}: maximum load). The initial load range is 8 μN, which is increased to 9 μN after 100 cycles. This process of fatigue (100 cycle) and load increase (1 or 1.5 μN) is repeated until the specimen breaks under $\Delta P = 18$ μN (See Fig. 19).

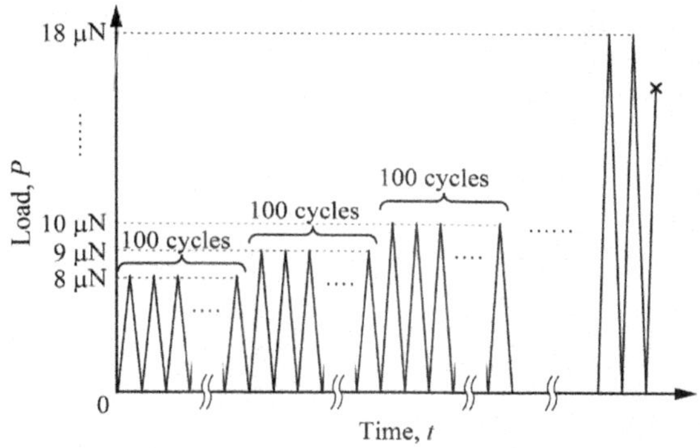

Fig. 19. Schematic illustration of the loading condition of fatigue experiment with a constant load range.

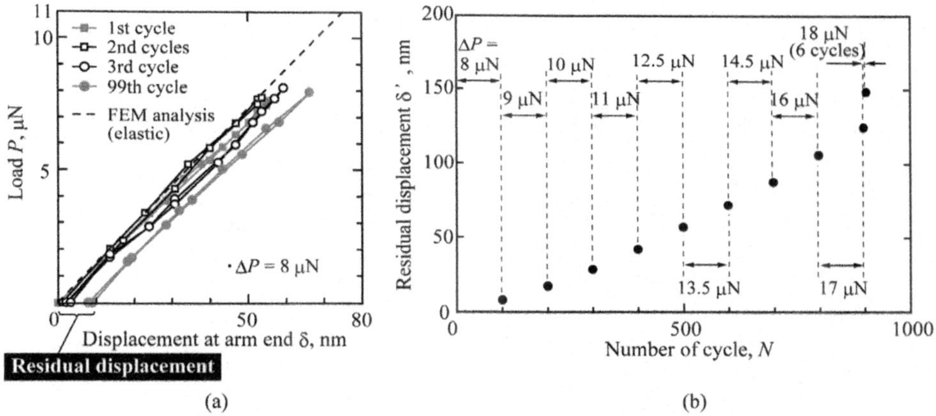

Fig. 20. (a) Load-displacement curves at 1st, 2nd, 3rd, and 99th cycles under the initial load range (ΔP = 8 μN), and (b) residual displacement, δ', with respect to the number of cycles, N.

3.4.2 Fatigue of 20 nm-thick Cu and cracking along Si/Cu interface

Figure 20(a) shows the load-displacement (P-δ) curves in the 1st, 2nd, 3rd, and 99th cycles under the initial load range (ΔP = 8 μN). In the figure, the broken line shows the relation obtained from elastic FEM analysis. The linear and reversible behaviour in each cycle indicates that the specimen, including the Cu layer, elastically deforms under the cyclic loading. The fact that the P-δ curves experimentally obtained are in a good agreement with the analytical one indicates the sufficient accuracy in the control/measurement of load and displacement. The gradient of the P-δ curve does not change until the last cycle. The elastic deformation ensures the reliability of our system in fatigue, where long-term stability of the experimental system is essential.

Careful examination of the P-δ relation reveals that irreversible residual displacement, δ', of about 10 nm is accumulated until the last cycle. Considering the measurement precision, this signifies the accumulation of plastic strain in the specimen though the magnitude in each cycle cannot be individually identified. Figure 20(b) shows δ' in the last cycle under each load range revealing that irreversible (plastic) strain accumulates with the progress of cyclic deformation. In particular, δ' rapidly increases under ΔP = 18 μN.

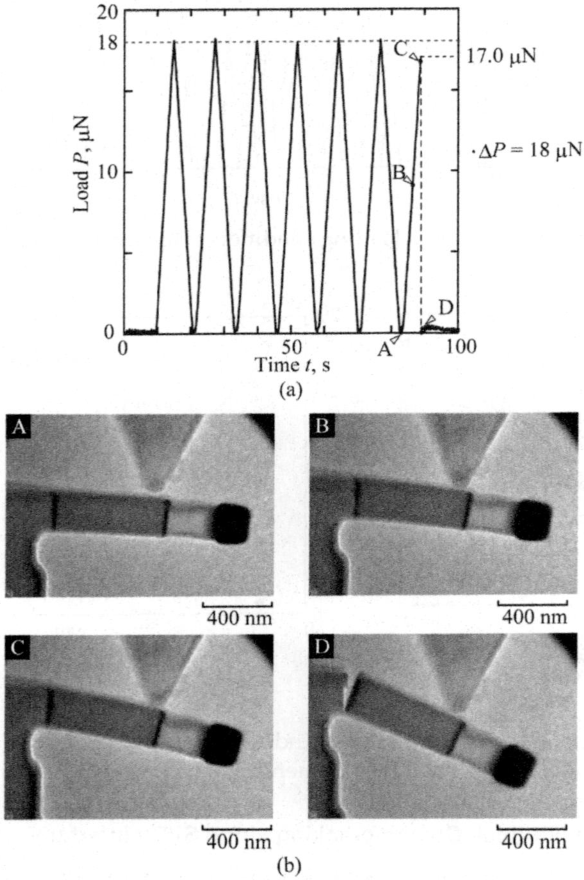

Fig. 21. (a) Load-time relationship under the load range of 18 μN, and (b) TEM micrographs in the 7th cycle.

The specimen breaks in the 7th cycle under ΔP = 18 μN. Figure 21 shows (a) the load-time (P-t) relationship under ΔP = 18 μN and (b) bright-field TEM images corresponding to the alphabets in the P-t relationship. The specimen is cyclically loaded until the 6th cycle without any damage observed on the interface. In the 7th cycle, the load suddenly drops before the maximum load (17.0 μN; point C in Fig. 21(a)), and the specimen breaks along the Si/Cu interface. Since the load at C is about 6 % smaller than P_{max} = 18 μN, it is clearly different from the fracture in a monotonic loading. The TEM images in Fig. 21(b) show that there is no remarkable damage in the Si/Cu interface before the break.

Figure 22 shows the *P*-δ curves in each cycle. The nonlinear behaviour and distinct hysteresis loop indicate the cyclic plasticity of the Cu film because the Si substrate and the SiN layer elastically deforms at this load level. After the 2nd cycle, reverse yielding appears during the unloading process. These suggest the development of a cyclic substructure in the Cu film.

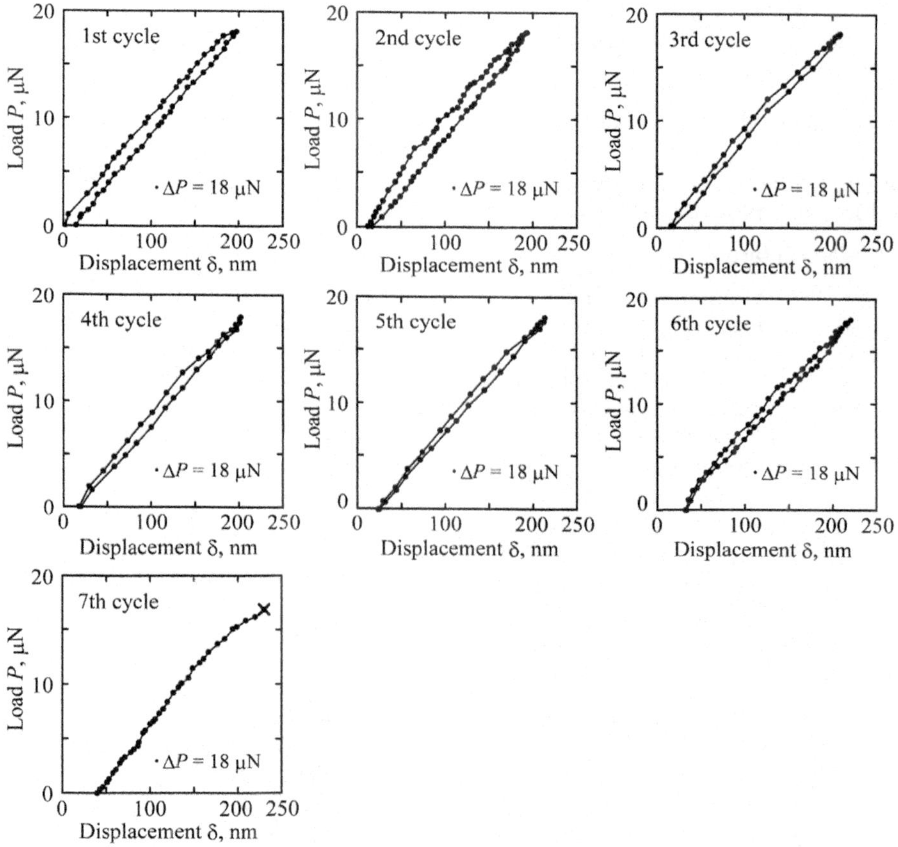

Fig. 22. Load-displacement curves under the load range of 18 μN.

As shown in Fig. 20(b), the residual displacement increases rapidly after the load range is increased to ΔP = 18 μN. The average in one cycle is 4.0 nm, which is more than one-order larger than that under ΔP = 17 μN. This drastic softening of the Cu film may also be due to the transition and development of the cyclic substructure. The width of the cyclic *P*-δ curve becomes narrower with increasing number of fatigue cycles. Figure 23 shows the half bandwidth of the cyclic loop, $w_{1/2}$, with respect to the number of cycles. $w_{1/2}$ gradually decreases with the progress of cyclic deformation, and the magnitude in the 6th cycle is about half of that in the 1st cycle. This indicates cyclic hardening in the Cu film due to the transition of the substructure.

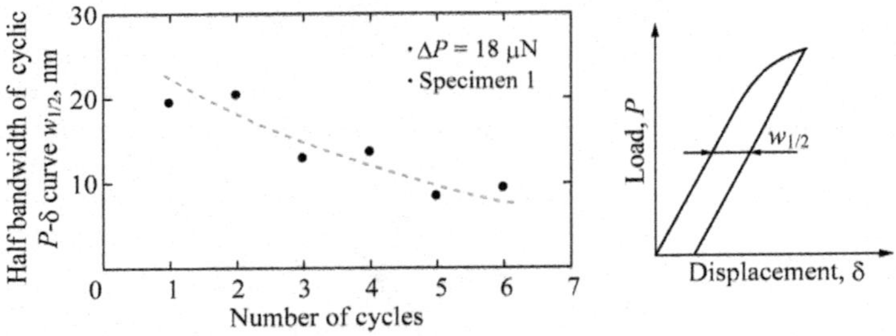

Fig. 23. Change of half bandwidth of the P-δ loop under the load range of 18 μN.

3.5 *In-situ* transmission observation of plasticity in 200 nm-thickness Cu film near Si/Cu interface edge (Sumigawa et al., 2011a)

3.5.1 Research target

High-voltage TEM observation enables us to investigate the understructure at the nano- or atomic scale in materials. However, it is not convenient to handle and requires heavy economic load to the researches. If we can conduct the *in-situ* observation by a general-purpose TEM, it is great help for the researcher to seek the understanding of precise mechanism in a nano-scale component. For the sake of *in-situ* observation by a general-purpose TEM, the testing section must be thinned to about 100 nm. In order to observe the understructure of the specimen, a trial is discussed in this section.

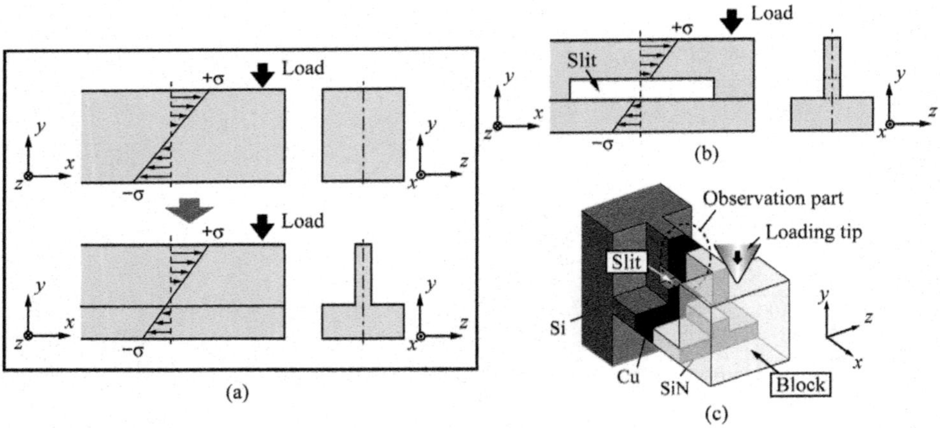

Fig. 24. (a) Cantilever specimen with the cross section of inverted-T-shape, (b) introduction of a horizontal through slit, and (c) final concept of specimen configuration and loading experiment.

3.5.2 Specimen and fabrication procedure

In bending test, the specimen buckles at the thin part under the compressive stress. In order to prevent the buckling at the lower part, the specimen with an inverted-T-shaped cross-section, where the neutral plane of bending deformation locates near the bottom, is adopted in order to confine the compressive stress to the thicker portion (Fig. 24(a)). Moreover, by introducing a horizontal through-slit in the center of the thinned part, we can surely apply tensile stress in the experimental area (Fig. 24(b); thin part). A block with high rigidity is added at the end of the cantilever for loading by a diamond tip. The final concept of the specimen configuration is schematically illustrated in Fig. 24(c).

The specimen is fabricated by means of a FIB processing. The fabrication procedure is described in detail below.

1. Gold (Au) with a thickness of 500 nm is deposited on the multi-layered plate (Si substrate/Cu (200 nm thickness)/SiN (500 nm thickness)) to protect the SiN layer from the ion beam.
2. A 10 μm×10 μm×10 μm block is carved out from the multi-layered plate, and is mounted on the flat top of an Au wire (wire diameter: 250 μm, and diameter of flat top: 100 μm) with an adhesive (Figs. 25(a)-(c)) so that the dissimilar interfaces (Si/Cu and Cu/SiN interfaces) are vertically aligned on the wire top (Fig. 25(c)).
3. A part of the block is thinned to less than 1500 nm in the thickness by a z-direction beam (beam current: 50-1000 pA) (Fig. 25(d)).
4. The thinned part is processed to about 500 nm in the thickness by a weak beam (beam current: 30 pA) (Fig. 25(e)).
5. After the horizontal through-slit is introduced in the specimen by a y-direction beam (beam current: 4 pA) (Fig. 25(f)-(i)), the thinned part is processed to less than 100 nm in the thickness by a z-direction beam (beam current: 4 pA).
6. The upper surface is finished from the x-direction (beam current: 4 pA) (Figs. 25(f)-(ii)), and the bottom is processed from the y-direction to complete the cantilever specimen (Fig. 25(f)-(iii)).

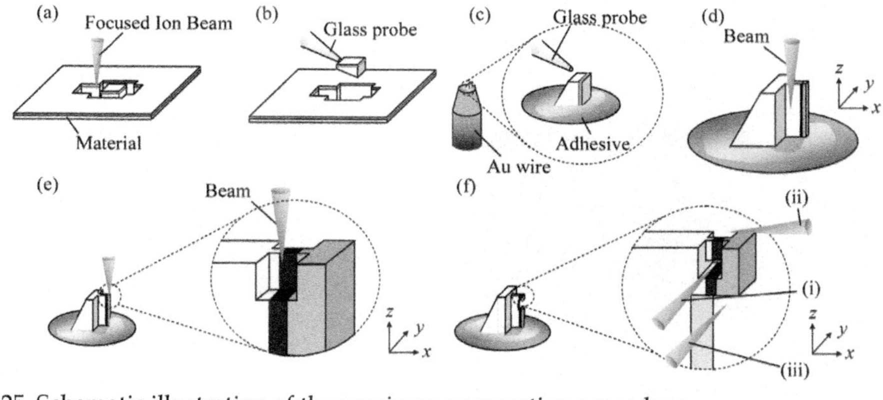

Fig. 25. Schematic illustration of the specimen preparation procedure.

3.5.3 Observation of plastic region

Figure 26(a) shows SEM images of fabricated specimen, in which the thickness of the thinned portion is about 75 nm. The block size at the cantilever end is 550 nm×310 nm×700 nm. Au at the specimen end is a protection layer for FIB processing. Figure 26(b) shows the configuration and size of the specimen without the Au passivation layer.

Figure 27 shows a bright-field TEM image of the specimen. The magnified view in Fig. 27 shows no grain boundary near the Si/Cu interface edge though the Cu film is polycrystalline material. As the average diameter of grins is about 300 nm, the region near the edge is occupied by a single crystal. In general, elastic deformation and lattice distortion due to a dislocation brings about inhomogeneous contrast in a TEM image. Since there is no contrast difference observed near the Si/Cu interface edge in Fig. 27 before loading, the initial dislocation density is almost constant in this region.

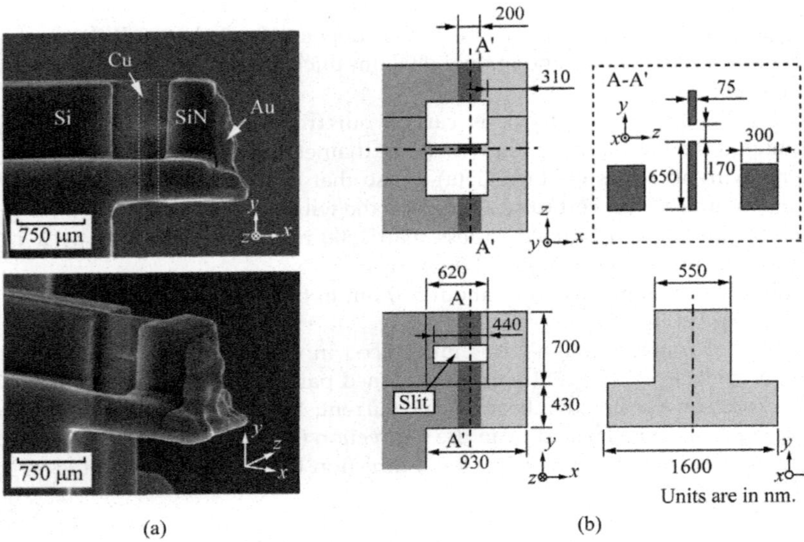

Fig. 26. (a) SEM image, and (b) size of specimen.

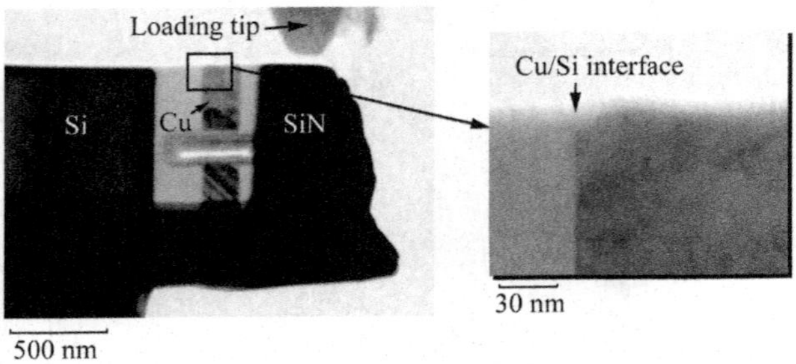

Fig. 27. TEM image of specimen.

Figure 28 shows the specimen model for the FEM analysis, which is precisely reconstructed following the actual specimen shape. Although microscopic structures such as grains would affect the deformation of the Cu and SiN, isotropic homogeneity is assumed because the focus is on an approximate evaluation of the stress distribution here.

Figure 29 shows a distribution of the normal stress σ_x along the Si/Cu interface near the upper edge. The stress is normalized by a reference stress σ' which is σ_x at the center of the upper surface in the Cu film (indicated by an × in Fig. 28). There is no compressive stress in the test area, and tensile stress prevails as designed. The stress concentrates within about several tens of nm from the interface edge because of the deformation mismatch between the Cu film and the Si substrate.

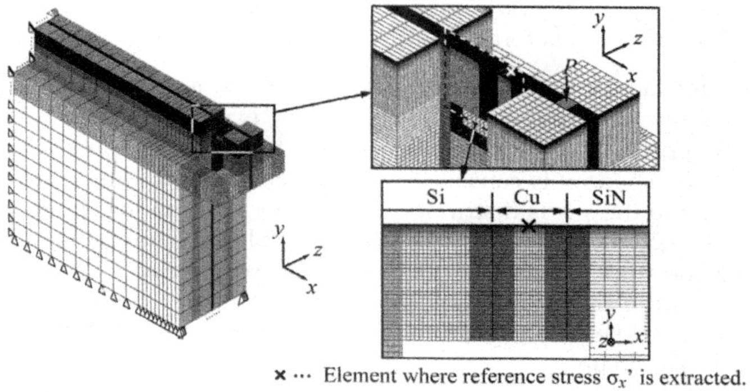

Fig. 28. Model for finite element analysis.

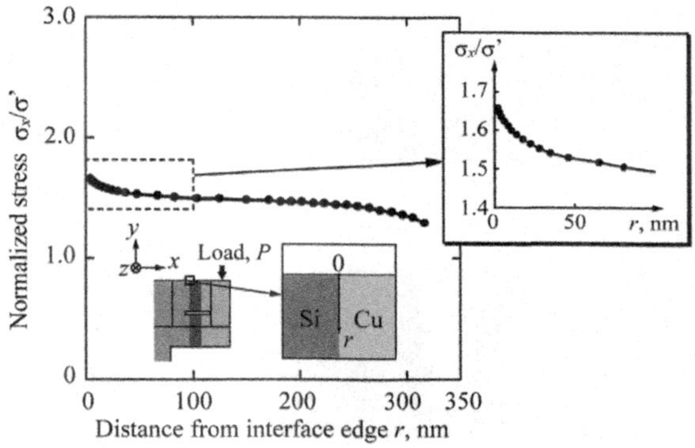

Fig. 29. Distribution of normalized reference stress, σ_x/σ_x', along the Si/Cu interface.

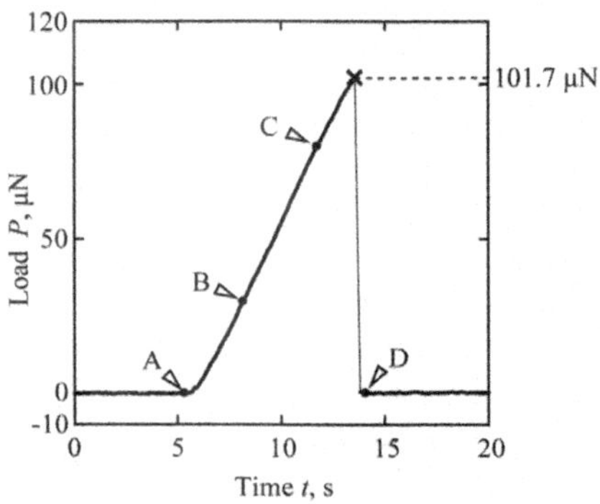

Fig. 30. Load-time relationship.

Figure 30 shows the load-time relationship obtained from the bending experiment. After the load P linearly increases, it suddenly drops at $P = 101.7$ μN.

Figure 31 shows a series of TEM images during the experiment. These indicate that the loading tip firmly touches the block at the cantilever end without slip at the contact. The specimen breaks along the Si/Cu interface when the load reaches $P = 101.7$ μN. No buckling takes place in the specimen before breaking.

The dark area indicated by a dashed circle in Fig. 31 (B) appears near the interface edge in the Cu film at about $P = 20$ μN. The area gradually expands as the load increases (See Fig. (C)). The location corresponds to the stress-concentration region predicted by FEM analysis (See Fig. 29). If the shadow is caused by a local elastic deformation, the boundary of dark area is vague due to the stress distribution. However, the shadow in the images has a clear boundary. These indicate that the area is caused by plastic deformation. The shadowed area expands along the upper surface of the specimen with increasing the applied load. The plastic region near the Si/Cu interface edge is determined by image processing, and Fig. 33 shows the change in the area. A definite plastic region appears at $P = 20$ μN and reaches about 920 nm² at $P = 80$ μN.

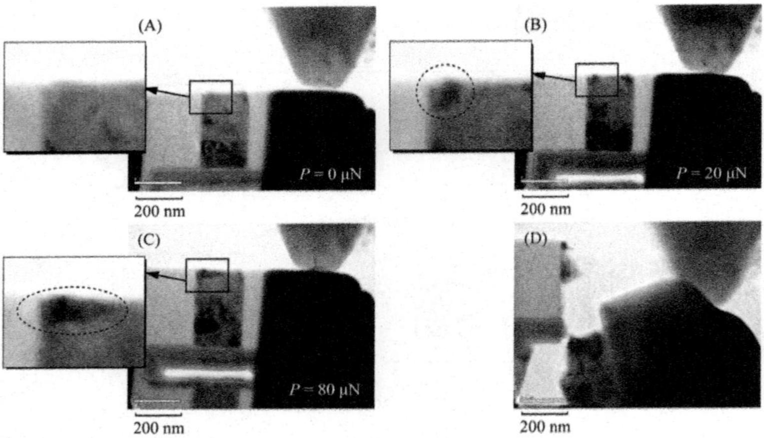

Fig. 31. *In-situ* TEM images during bending deformation.

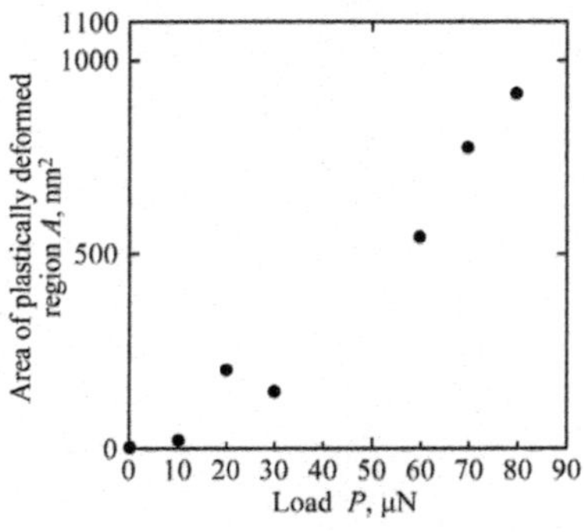

Fig. 32. Change of area where plastic deformation occurs.

4. Conclusion

This chapter highlights bending experiments on the strength of nano-scale interface with *in-situ* TEM observation. Cantilever specimens are carved out of a multi-layered material where a 20 nm-thick Cu film is sandwiched between a SiN layer and a Si substrate by means of the FIB processing. The results are summarized as follows.

- TEM image enables us to quantitatively measure the displacement at the arm end of specimen during the bending experiment.
- The elastic-plastic constitutive equation of the Cu portion can be estimated by an inverse analysis using the load-displacement relationship.
- Crack initiation at Si/Cu interface edge under monotonic loading is governed by the normal stress of about 1 GPa in a region of 20 - 30 nm.
- A Specimen with a step in the SiN portion induces a crack initiation at the interior Si/Cu interface owing to the stress concentration. There is little difference between the crack initiation criteria along the interface at the edge and in the interior.
- Under cyclic loading, the Cu portion evidently shows fatigue behaviour, and then, the specimen breaks along the Si/Cu interface by the fatigue.
- Cantilever specimen with an inverted-T-shaped cross-section, a through-slit in the thinned part, and a rigid block at the arm end is designed for the *in-situ* observation inside of nano-component. It enables us to conduct a bending experiment with *in-situ* transmission observation, which can observe the plastic zone in the Cu portion.

The mechanical properties in nano-components are described in detail in a book "Fracture nanomechanics" (Kitamura et al., 2011).

There are further challenging topics on the mechanical behaviour of nano-component remained. For example, although the Section 3.2 demonstrates that continuum mechanics is still valid in the range of 20-30 nm and the stress concentration can describe fracture characteristics, the applicability in smaller regions is unclear. Moreover, the mechanism of plasticity and the cyclic substructure in nano-component are not clarified. Since the ratio of the surface to the volume in nano-components is remarkably large, the mechanical properties are strongly affected by corrosion and surface diffusion. Consequently, the deformation and fracture behaviour under an environment and a high-temperature is of great interest. The technique described in this chapter may greatly contribute to future research on the mechanical property of nano-components.

5. Acknowledgment

We acknowledge financial support by a Grant-in-Aid for Scientific Research (S)(No.21226005), from the Japan Society for the Promotion of Science, and by a Grant-in-Aid for Young Scientists (A)(No. 21686013), from the Ministry of Education, Culture, Sports, Science and Technology, Japan.

6. References

ASM Handbook (2000). *Mechanical Testing and Evaluation Vol.8*, ASM International, Metals Park, Ohio, p. 103

Bogy, DB. (1968). Edge-bonded Dissimilar Orthogonal Elastic Wedges under Normal and Shear Loading. *Journal of Applied Mechanics*, Vol.35, pp. 460-466

Burdess, J.S.; Harris, A.J.; Wood, D.; Pitcher, R.J. & Glennie, D. (1997). A System for the Dynamic Characterization of Microstructures. *Journal of Microelectromechanical Systems*, Vol.6, pp. 322–328

Cheng, Y.-W.; Read, D.T.; McColskey, J.D,; Wright, J.E. (2005). A Tensile-testing Technique for Micrometer-sized Free-standing Thin Films. *Thin Solid Films*, Vol.484, pp. 426-432

Dao,M.; Chollacoop, N.; Vliet, K.J.V.; Venkatesh, T.A. & Suresh, S. (2001). Computational Modeling of the Forward and Reverse Problems in Instrumented Sharp Indentation. *Acta materialia*, Vol.49, pp. 3899–3918

Dimiduk, D.M.; Uchic, M.D. & Parthasarathy, T.A. (2005). Size-affected Single-slip Behavior of Pure Nickel Microcrystals. *Acta materialia*, Vol.53, 4065–4077

Greer, J.R. & Nix, W.D. (2006). Nanoscale Gold Pillars Strengthened through Dislocation Starvation. *Physical Review B*, Vol.73, article number 245410

Haque, M.A. & Saif, M.T.A. (2002). In-situ Tensile Testing of Nano-scale Specimens in SEM and TEM. *Experimental Mechanics*, Vol.42, No.1, pp. 123-128

Hou, C.; Li, Z.; Huang, M. & Ouyang, C. (2008). Discrete Dislocation Plasticity Analysis of Single Crystalline Thin Beam under Combined Cyclic Tension and Bending. *Acta Materialia*, Vol.56, pp. 1435–1446

Hua, T.; Xie, H.; Pan, B.; Qing, X.; Dai, F. & Feng, X. (2007). A New Micro-tensile System for Measuring the Mechanical Properties of Low-dimensional Materials — Fibers and Films. *Polymer Testing*, Vol.26, pp. 513–518

Huang, H. & Spaepen, F. (2000). Tensile Testing of Free-standing Cu, Ag and Al Thin Films and Ag/Cu Multilayers. *Acta Materialia*, Vol.48, pp. 3261-3269

Huang, Y.; Xue, Z.; Gao, H.; Nix, W.D. & Xia, Z.C. (2000). A Study of Microindentation Hardness Tests by Mechanism-based Strain Gradient Plasticity. *Journal of Material Research*, Vol.15, No.8, pp. 1786-1796

Hirakata, H.; Takahasi, Y. ; Truong, D.V. & Kitamura, T. (2007). Role of Plasticity on Interface Crack Initiation from a Free Edge and Propagation in a Nano-component. *International Journal of Fracture*, Vol.145, No.4, pp. 261-271

Keller, R.R.; Phelps, J.M. & Read, D.T. (1996). Tensile and Fracture Behavior of Free-standing Copper Films. *Materials Science and Engineering A*, Vol.214, pp. 42-52

Kim, J.S. & Huh, H. (2011). Evaluation of the Material Properties of an OFHC Copper Film at High Strain Rates Using a Micro-Testing Machine. *Experimental Mechanics*, Vol.51, pp. 845–855

Kitamura, T.; Hirakata, H.; Sumigawa, T. & Shimada, T. (2011). *Fracture Nanomechanics*, Pan Stanford Publishing Pte. Ltd., Singapore

Komai, K.; Minoshima, K. & Inoue, S. (1998). Fracture and Fatigue Behavior of Single Crystal Silicon Microelements and Nanoscopic AFM Damage Evaluation. *Microsystem Technologies 5*, pp. 30-37

Li, X.; Kasai, T.; Nakao, S.; Tanaka, H.; Ando, T.; Shikida, M. & Sato, K. (2005). Measurement for Fracture Toughness of Single Crystal Silicon Film with Tensile Test. *Sensors and Actuators A*, Vol.119, pp. 229–235

Liu, Z.L.; Liu, X.M.; Zhuang, Z. & You, X.C. (2009). A Multi-scale Computational Model of Crystal Plasticity at Submicron-to-nanometer Scales. *International Journal of Plasticity*, Vol. 25, pp. 1436–1455

Namazu, T.; Isono, Y. & Tanaka, T. (2000). Evaluation of Size Effect on Mechanical Properties of Single Crystal Silicon by Nanoscale Bending Test using AFM. *Journal of Microelectromechanical Systems*, Vol.9, No.4, pp. 450-459

Namazu, T.; Isono, T. & Tanaka, T. (2002). Plastic Deformation of Nanometric Single Crystal Silicon Wire in AFM Bending Test at Intermediate Temperatures. *Journal of Microelectromechanical Systems*, Vol.11, No.2, pp. 125-135

Moser, B. & Wasmer, K. (2007). Strength and Fracture of Si Micropillars: A New Scanning Electron Microscopy-based Micro-compression Test. *Journal of Material Research*, Vol.22, No.4, pp. 1004-1011

Motz, C.; Schoberl, T. & Pippan, R. (2005). Mechanical Properties of Micro-sized Copper Bending Beams Machined by the Focused Ion Beam Technique. *Acta Materialia*, Vol.53, pp. 4269–4279

Mughrabi, H. (1978). The Cyclic Hardening and Saturation Behaviour of Copper Single Crystals. *Materials Science and Engineering*, Vol.33, pp.207-223

Ruud J.A.; Josell, D. & Spaepen, F. (1993). A New Method for Tensile Testing of Thin Films, *Journal of Materials Research*, Vol.8, pp. 112-117

Ryu, J.; Kim, J.-H.; Chu, S.; Lee, S. & Moon, S. (2006). Fabrication and Mechanical Characterization of Micro Electro Mechanical System based Vertical Probe Tips for Micro Pad Measurements. *Japanese Journal of Applied Physics*, Vol.45, No.12, pp. 9238–9243

Sumigawa, T.; Shishido, T.; Murakami, T,; Iwasaki, T. & Kitamura, T. (2010a). Evaluation on Plastic Deformation Property of Copper Nano-film by Nano-scale Cantilever Specimen. *Thin Solid Films*, Vol.518, pp. 6040–6047

Sumigawa, T.; Shishido, T.; Murakami, T. & Kitamura, T. (2010b). Interface Crack Initiation due to Nano-scale Stress Concentration. *Materials Science and Engineering A*, Vol.527, pp. 4796–4803

Sumigawa, T.; Murakami, T.; Shishido, T. & Kitamura, T. (2010c). Cu/Si Interface Fracture due to Fatigue of Copper Film in Nanometer Scale. *Materials Science and Engineering A*, Vol.527, pp. 6518–6523

Sumigawa, T.; Kitagawa, Y. & Kitamura, T. (2011a). Development of In-situ TEM Observation Method on Plasticity in Nanoscale Component. *Journal of Solid Mechanics and Materials Engineering*, Vol.5, No.3, pp. 128-137

Sumigawa, T.; Kitamura, T. & Murakami, T. (2011b). Fatigue Strength of the Cu/Si Interface in Nano-components. *Materials Science and Engineering A*, Vol.528, pp. 5158–5163

Suresh, S. (1998). *Fatigue of Materials - Second Edition*, Cambridge University Press, Cambridge, England

Thompson, N.; Wassworth, N. & Louat, N. (1956). The Origin of Fatigue Fracture in Copper. *Phlosophical Magazine*, Vol.1, No.2, pp. 113-126

Tripathy, S.; Lin, V.K.X; Vicknesh, S. & Chua, S.J. (2007). Micro-Raman Probing of Residual Stress in Freestanding GaN-based Micromechanical Structures Fabricated by a Dry Release Technique. *Journal of Applied Physics*, Vol.101, article numver 063525

Tsuchiya, T.; Tabata, O.; Sakata, J. & Taga, T. (1998). Specimen Size Effect on Tensile Strength of Surface-Micromachined Polycrystalline Silicon Thin Films. *Journal of Microelectromechanical Systems*, Vol.7, No.1, pp. 106-113

Tsuchiya, T.; Inoue, A. & Sakata, J. (2000). Tensile Testing of Insulating Tthin Films; Humidity Effect on Tensile Strength of SiO Films. *Sensors and Actuators*, Vol.82, pp. 286–290

Vinci, R.P.; Zienlinski, E.M. & Bravaman, J.C. (1995). Thermal Strain and Stress in Copper Thin Films. *Thin Solid Films*, Vol.262, pp. 142-153

Weertman, J. & Weertman, J.R. (1964). *Elementary Dislocation Theory*. The Macmillan Company, New York, Collier-Macmillan Limited, London

Winter, A.T. (1978). Nucleation of Persistent Slip Bands in Cyclically Deformed Copper Crystals. *Philosophical Magazine A*, Vol.37, No.4, pp. 457-463

Xiang, Y., Tsui, T.Y. & Vlassak, J.J. (2006). The Mechanical Properties of Freestanding Electroplated Cu Thin Films. *Journal of Material Research*, Vol.21, No.6, pp.1607-1618

Yi, T.; Li, L. & Kim, C.-J. (2000). Microscale Material Testing of Single Crystalline Silicon: Process Efects on Surface Morphology and Tensile Strength. *Sensors and Actuators*, Vol.83, pp. 172–178

Zhu, Y.; Moldovan, N. & Espinosa, H.D. (2005). A microelectromechanical load sensor for *in situ* electron and x-ray microscopy tensile testing of nanostructures. *Applied Physics Letters*, Vol.86, article number 013506

Permissions

The contributors of this book come from diverse backgrounds, making this book a truly international effort. This book will bring forth new frontiers with its revolutionizing research information and detailed analysis of the nascent developments around the world.

We would like to thank Dr Maaz Khan, for lending his expertise to make the book truly unique. He has played a crucial role in the development of this book. Without his invaluable contribution this book wouldn't have been possible. He has made vital efforts to compile up to date information on the varied aspects of this subject to make this book a valuable addition to the collection of many professionals and students.

This book was conceptualized with the vision of imparting up-to-date information and advanced data in this field. To ensure the same, a matchless editorial board was set up. Every individual on the board went through rigorous rounds of assessment to prove their worth. After which they invested a large part of their time researching and compiling the most relevant data for our readers. Conferences and sessions were held from time to time between the editorial board and the contributing authors to present the data in the most comprehensible form. The editorial team has worked tirelessly to provide valuable and valid information to help people across the globe.

Every chapter published in this book has been scrutinized by our experts. Their significance has been extensively debated. The topics covered herein carry significant findings which will fuel the growth of the discipline. They may even be implemented as practical applications or may be referred to as a beginning point for another development. Chapters in this book were first published by InTech; hereby published with permission under the Creative Commons Attribution License or equivalent.

The editorial board has been involved in producing this book since its inception. They have spent rigorous hours researching and exploring the diverse topics which have resulted in the successful publishing of this book. They have passed on their knowledge of decades through this book. To expedite this challenging task, the publisher supported the team at every step. A small team of assistant editors was also appointed to further simplify the editing procedure and attain best results for the readers.

Our editorial team has been hand-picked from every corner of the world. Their multi-ethnicity adds dynamic inputs to the discussions which result in innovative outcomes. These outcomes are then further discussed with the researchers and contributors who give their valuable feedback and opinion regarding the same. The feedback is then collaborated with the researches and they are edited in a comprehensive manner to aid the understanding of the subject.

Apart from the editorial board, the designing team has also invested a significant amount of their time in understanding the subject and creating the most relevant covers. They scrutinized every image to scout for the most suitable representation of the subject and create an appropriate cover for the book.

The publishing team has been involved in this book since its early stages. They were actively engaged in every process, be it collecting the data, connecting with the contributors or procuring relevant information. The team has been an ardent support to the editorial, designing and production team. Their endless efforts to recruit the best for this project, has resulted in the accomplishment of this book. They are a veteran in the field of academics and their pool of knowledge is as vast as their experience in printing. Their expertise and guidance has proved useful at every step. Their uncompromising quality standards have made this book an exceptional effort. Their encouragement from time to time has been an inspiration for everyone.

The publisher and the editorial board hope that this book will prove to be a valuable piece of knowledge for researchers, students, practitioners and scholars across the globe.

List of Contributors

Rajiv Kumar Mandal
Department of Metallurgical Engineering, Centre of Advanced Study, Institute of Technology, Banaras Hindu University, Varanasi, India

Katsuhiro Sasaki, Kotaro Kuroda and Hiroyasu Saka
Department of Quantum Engineering, School of Engineering, Nagoya University, Japan

Hidekazu Murata
Department of Electrical and Electronic Engineering, Meijyo University, Japan

Jian Li
CANMET-Materials Technology Laboratory, Hamilton, Ontario, Canada

M. Bieda, K. Sztwiertnia, A. Korneva and J. Kawalko
Institute of Metallurgy and Materials Science PAS, Krakow, Poland

David B. Carlson and James E. Evans
Department of Molecular and Cellular Biology, University of California at Davis, USA

Saeed Shokri
Anatomy Department, School of Medicine, Zanjan University of Medical Sciences, Zanjan, Iran

Masoud Hemadi
Fertility and Infertility Research Center, Imam Khomeini Hospital, Faculty of Medicine, Ahvaz Jundishapur University of Medical Sciences, Ahvaz, Iran

Robert John Aitken
ARC Centre of Excellence in Biotechnology and Development, Discipline of Biological Sciences, University of Newcastle, NSW, Australia

S. Lawrence Selvaraj and Takashi Egawa
Research Center for Nano-Device and System, Nagoya Institute of Technology, Japan

Hiroo Omi
NTT Basic Research Laboratories, NTT Corporation, Japan

Kelvii Wei Guo
MBE, City University of Hong Kong, Hong Kong

Yekaterina D. Bedoshvili and Yelena V. Likhoshway
Limnological Institute of the Siberian Branch of the Russian Academy of Sciences, Russia

Łukasz Major and Jerzy Morgiel
Institute of Metallurgy and Materials Science Polish Academy of Sciences, Poland

Jürgen M. Lackner
Joanneum Research Forschungsges. m. b. H., Institute for Surface Technologies and Photonics, Austria

Nora Elizondo, Ricardo Obregón, Sergio Belmares-Perales and Manuel García-Méndez
Facultad de Ciencias Físico-Matemáticas, México

Arquímedes Cruz-López, Odilón Vázquez-Cuchillo and Antonio A. Zaldívar
Facultad de Ingeniería Civil, Universidad Autónoma de Nuevo León, San Nicolás de los Garza, N. L., México

Donald H. Galván
Centro de Nanociencias y Nanotecnología, Universidad Nacional Autónoma de México, Ensenada, México

Lorena Álvarez-Contreras
Departamento de Química de Materiales, Centro de Investigación en Materiales Avanzados, S. C., Chihuahua, México

Ran Tel-Vered
Institute of Chemistry, the Hebrew University of Jerusalem, Jerusalem, Israel

Lanzhuang Chen
Faculty of Environmental and Horticultural Science, Minamikyushu University, Tateno, Miyakonojo city, Miyazaki, Japan

Liming Guan
Faculty of Education and Culture, University of Miyazaki, Gakuen Kibanadai Nishi, Miyazaki, Japan

Hajivand Shokrollah
Department of Plant and Seed Improvement, Agriculture and Natural Resources Research Center of Qazvin, Qazvin, Iran

Thohirah Lee Abdullah
Departments of Crop Science, Faculty of Agriculture, University Putra Malaysia, Serdang, Malaysia

Kamaruzaman Sijam
Department of Plant Protection, Faculty of Agriculture, University Putra Malaysia, Serdang, Malaysia

Naoki Yamamoto
Physics Department, Tokyo Institute of Technology, Tokyo, Japan

Jian-Dong Lin and Ming-Rui Lin
Department of Intensive Care Unit, The First Affiliated Hospital of Fujian Medical University, Fuzhou, Fujian, China

Fernanda Elena Monasterio
Instituto de Investigaciones para la Industria Química (INIQUI)- CONICET, Avda, Bolivia, Salta, Argentina

Zhe Wang, Hongzhe Ni, Mingyao Zhang and Huixuan Zhang
School of Chemical Engineering, Changchun University of Technology, P. R. China

Chengji Zhao
Alan G. MacDiarmid Institute, College of Chemistry, Jilin University, Changchun, P. R. China

Abhijit Kar
J.B.Centre of Excellence, Jagadis Bose National Science Talent Search, India

Ajoy Kumar Ray
Material Science & Technology Division, National Metallurgical Laboratory (CSIR), India

Takashi Sumigawa and Takayuki Kitamura
Department of Mechanical Engineering and Science, Kyoto University, Japan

CPSIA information can be obtained
at www.ICGtesting.com
Printed in the USA
BVOW07*0748160817
492093BV00021B/4/P

9 781632 382832